Ecological Studies, Vol. 220

Analysis and Synthesis

For further volumes see list at the end of the book and:
http://www.springer.com/series/86

Rainer Matyssek • Hans Schnyder • Wolfgang Oßwald • Dieter Ernst • Jean Charles Munch • Hans Pretzsch
Editors

Growth and Defence in Plants

Resource Allocation at Multiple Scales

Springer

Editors
Professor Dr. Rainer Matyssek
Technische Universität München
Chair of Ecophysiology of Plants
Freising, Germany

Professor Dr. Wolfgang Oßwald
Technische Universität München
Phytopathology of Woody Plants
Freising, Germany

Professor Dr. Jean Charles Munch
Helmholtz Zentrum München
Institute of Soil Ecology
Neuherberg, Germany

Professor Dr. Hans Schnyder
Technische Universität München
Lehrstuhl für Grünlandlehre
Freising, Germany

Dr. Dieter Ernst
Helmholtz Zentrum München
Institute of Biochemical Plant Biology
Neuherberg, Germany

Professor Dr. Hans Pretzsch
Technische Universität München
Chair of Forest Growth and Yield Science
Freising, Germany

ISSN 0070-8356
ISBN 978-3-642-30644-0 ISBN 978-3-642-30645-7 (eBook)
DOI 10.1007/978-3-642-30645-7
Springer Heidelberg New York Dordrecht London

Library of Congress Control Number: 2012948418

Printed on acid-free paper

Springer is part of Springer Science+Business Media (www.springer.com)

To our late colleague and friend, Prof. Dr. Heinrich Sandermann jr. († August 18, 2009), who was one of the initiators and members of the interdisciplinary research program SFB 607 "Growth and Parasite Defence in Plants – Competition for Resources in Economic Plants from Agronomy and Forestry", giving rise to this book project, but to whom it was not granted to see its completion.

Prologue

A New View on Systems Biology: Information, Knowledge, Understanding

Cost/Benefit Scenarios and the Dilemma of Growth and Defence

Cost/benefit relations determine the functions of plants in natural ecosystems as well as anthropogenically formed agro- and forestry ecosystems. Under the influence of abiotic and biotic cues, they regulate the dynamics in the realm of the opposing forces of growth and defence (Chap. 1). In the competition for resources the dilemma has the costs of productivity on the one hand and of survival of stress on the other hand.

Talking of a dilemma we must ask the question if there are in fact options of ways out of it. Evidently, like all living organisms plants are also open systems subject to non-linear dynamics in space and time. This implies that spatio-temporal developments under a plethora of external cues and the opposing forces of growth and defence inevitably lead to branching points where alternatives or options are given, but where chance may eventually also lead the way into deterministic chaos. The latter is an inherent problem in evaluations of the degrees of possible predictability in all systems with non-linear dynamics (Schuster 1995). Options in spatio-temporal responses of plants under the dilemma of growth and defence and potential predictability are the central themes underlying the hypotheses and concepts evaluated in this book.

Scenarios are dominated by

- different life forms and types of habitats, viz. grassland and forest,
- competition (Chaps. 17 and 18), abiotic stress (ozone, elevated CO_2, nitrogen nutrition), pathogens (Chaps. 3 and 5), herbivores (Chap. 4), symbionts (Chaps. 5 and 10),
- ontogenetic development (Chap. 11),

- types of resources, e.g. originating from the soil (Chaps. 9 and 10), from aboveground components, from the atmosphere, of the solar radiation (Chap. 8), of space itself as a resource (Chap. 12, Grams and Lüttge 2010),
- allocation, partitioning and consumption of resources in allometry relations (Chaps. 6, 7, 9, 11–18).

These scenarios are elements of the formation of integrative networks within various scalar levels which are also connected between each other over both ascending and descending scales and thus can form supra-networks. This determines a new view on systems biology in an ecological context (Chap. 19), which must advance from information via knowledge towards understanding (Liessmann 2008), i.e. from purely descriptive compilation of data via correlative interpretation of processes towards reflection and causal assessment. Combining the approaches of experiment, theory and modelling develops predictive power on the basis of understanding observed past events.

An Architectural Metaphor

In a metaphoric way we may visualise the architecture of the Sonderforschungsbereich (SFB) "Wachstum und Parsitenabwehr – Wettbewerb um Ressourcen in Nutzpflanzen aus Land- und Forstwirtschaft", from whose work the present volume is emerging, like the knots and edges of networks. We may consider the experimental approaches on two basically different levels as two columns or towers (knots) which by long wings (edges) are connected to theory and modelling (knot).

I cannot imagine a better way of depicting this than choosing the Castle of Hirschberg above the city of Beilngries in the valley of the Altmühl in Central Bavaria. For more than a decade the SFB had regular internal meetings there for brainstorming and strategy planning. When we arrive at the castle we pass the two towers and we see the two wings of the building guiding our view in a remarkably suggestive way towards the entrance (Fig. 1). The architect managed this by having the two wings in an angle and much wider apart from each other at the start than at the end with the entrance. Let us consider the two towers as the two sections of the SFB. One of the towers (one knot) of the SFB or its section A (Chaps. 2–5) is linking molecular biology with biochemistry and physiology (Chap. 2) and, thus, is developing molecular ecology of

- gene induction and transcription,
- proteins and enzyme activities,
- primary and secondary metabolism,
- signalling by phytohormones and metabolites.

The other tower (knot) of the SFB, its section B (Chaps. 6–14), is developing physiological ecology at the organismic, stand and ecosystem levels, with

Fig. 1 Hirschberg Castle of the bishopric of Eichstätt above the city of Beilngries in the valley of the Altmühl (photographs: K.-H. Häberle)

- primary production, allocation, partitioning and allometry,
- occupation of space above and below ground,
- the use of reserves under stress of resource availability, ozone pollution, elevated atmospheric CO_2, pathogen attack.

With the two wings the two towers are connected to the entrance, section C (Chaps. 15–19) of the SFB. What is this an entrance to? Transduction of the information and knowledge sensu Liessmann (2008) from the two sections, towers or knots A and B via the wings or edges into section or knot C enters the wealth of observations (information) and interpretations (knowledge) into understanding sensu Liessmann (2008) as probed and advanced by

- statistical modelling (Chap. 16),
- mechanistic modelling (Chaps. 14, 15, 17 and 18),
- integration via abstraction.

Thus, metaphorically we can see that, as in the most compelling way in which the two wings of the castle guide our anticipation to arrive towards the entrance, sections A and B of the SFB funnel our attention towards the integration by theory and modelling of section C.

What Is a "System" in Biology?

The extreme view is Gaia considering the whole biosphere as one single system or one mega-organism (Lovelock 1979, 2009).

In the life sciences the term "systems biology" originated from describing vast accumulations of data in "genomics", and in a noteworthy inflationary fashion it now covers all kinds of "-omics", such as transcriptomics, proteomics, metabolomics and channelomics. The present book presents examples of such databases (Chaps. 2–5). Remarkably comprehensive sets of data on mineral macro- and micro-nutrients in plants have also been obtained (Chap. 9). In the SFB this has been called "mineralomics". The term "ionomics" is somewhat of a misnomer in the literature (Salt et al. 2008) as long as it refers only to inorganic ions and does not comprise on the one hand the wealth of organic ions important in the functioning of organisms and realise on the other hand that many minerals in organisms occur and function in non-ionic bound or chelated states. In the vein of a certain omics-inflation we may also allude to the necessity of considering structural or functional complements, i.e. "anatomics" or "functionomics".

However, are such data bases "systems"? They are really nothing more than primary information. They constitute a necessary basis but themselves they have nothing to do with knowledge which must be advanced from the information.

Another way of looking from a molecular point of view to systems is to consider specific key functions which are often expressed by many different isogenes coding different isoenzymes. Interestingly abiotic stress due to ozone and biotic stress due to pathogen attack may elicit very similar reactions at the molecular level, viz. transcription (Chap. 2). The translational and post-translational regulation and modulation of isoenzyme activities as well as their localisation and compartmentation can characterise the spatio-temporal functioning of the systems of whole plants or even higher integrated consortia, such as host/parasite or host/symbiont associations. In the present book we find examples of this with respect to genes and enzymes involved in the growth/defence dilemma (Chaps. 2–4). Here, at levels of increasingly larger scale, we recognise systems as e.g. organelles or compartments of cells, cells, whole organisms, interspecies associations, ecosystems and so on. It is remarkable in this context that the term "ecosystem" coined as early as 1935 by Sir Arthur C. Tansley already refers to "system". With respect to the term anatomics suggested above we realise that in most complex regulation networks stress may not simply inhibit growth but rather redirect growth. Thus, a large diversity of different stresses via a plethora of pathways of hormonal regulation may result in remarkably similar morphological responses (Potters et al. 2009).

This already moves us forward from purely descriptive compilations towards correlative interpretation. Evidently the proof of a correlation is not yet showing a causal relationship, but conversely if there is no correlation there is no relationship. From another angle we can advance from pure information to knowledge if we fathom the processes governing the spatio-temporal functioning of systems, where we need to integrate a wealth of different approaches required to assess the dynamics of systems.

The climax realising effective understanding of systems must be seen in the combination of theory and modelling integrating past and present observation for future prediction (Chaps. 15–18). This comprises reflection opening a path towards abstraction which leads to deeper understanding (Lüttge and Hütt 2009). It is essential to note, however, that similar to empirical observation, theory is never completely definite. A continuous and iterative ping-pong like mutual input is required between experiment/observation on the one side and theory/modelling on the other side.

Scalar Levels of Systems

We have seen above that when considering systems we inevitably arrive at changing levels of scaling. Ecology must cover a vast scope of systems in time and in space (Fig. 2). For an integrative view it is important to both descend and ascend the stairs or ladders of scaling levels or even jump up and down between levels. In the present volume throughout the various chapters this is documented by consideration of

- molecules and cells,
- plant tissues and organs,
- whole plants,
- intraspecific relations, e.g. relations between individual plants of a given species in stands,
- interspecific relations, e.g. relations between plants of different species in stands (Chaps. 12–14, 18), host/symbiont, host/parasite and host/herbivore relations (Chaps. 3–5, 10),
- ecosystems, e.g. grassland, forest (Chap. 12).

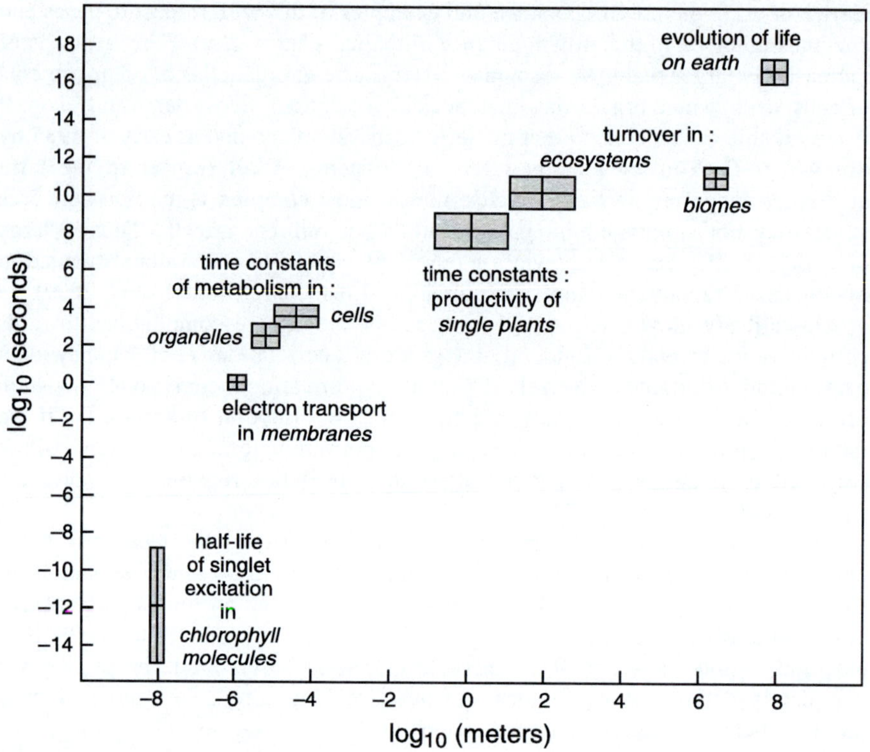

Fig. 2 Scalar levels of life in time (*y*-axis) and space (*x*-axis). Ecologically relevant levels range from molecules to the ecosphere (from Lüttge and Hütt 2009)

It is important to be aware of the fact that reductionism is not a reduction of scale size, e.g. from ecosystem to molecule. Complexity can be similarly overwhelming at any scale. Reductionism occurs at any given scale when the degrees of freedom are restricted, which is associated with a restriction of the number of different elements sampled and analysed or incorporated in mathematical models.

Processes and the Dynamics of Systems

Essential processes must be assessed comprehensively ("processomics"). The processes effective in creating and modulating the dynamics of systems in relation to the growth/defence dilemma treated in this volume are

- allocation and resource capture (Chaps. 12–14),
- partitioning and allometry (Chaps. 12 and 13),
- using the space as a resource (Chap. 12),

- competition and cooperation (Chaps. 2–5, 12),
- information biology:
  - Electrical signals (Lautner et al. 2005; Grams et al. 2009)
  - Chemical signals in the liquid and in the gas phase (Chaps. 4 and 5)
  - Light signals (Chap. 8)
  - Mechanical signals in space occupation

Perturbation of Systems by Investigation

Any researcher performing experiments or making measurements on empirical systems knows and the wave-particle dualism of quantum physics tells us most dramatically that any such investigation unavoidably always disturbs and perturbs the system studied. As also seen in the work presented in this book there are several degrees of severity of perturbation, which also cause methodological limitations:

- Perturbation is strong in experimental approaches primarily by direct manipulation of plant systems studied by the experimenters. However, this also includes implementing modes of access (cranes, towers, canopy walkways), e.g. affecting penetration of irradiance or introducing foreign substances (e.g. zinc from the metal of towers; Chap. 16), or using transplantations (controlled climate chambers, lysimeters).
- Perturbation is weak when non-invasive sensors are installed.
- Perturbation is weak when small size samples are taken en route or when sampling is performed destructively only at the termination of the study.
- Perturbation is close to inexistent in modelling and theory where modelling and its parameterisation on the one hand dwell on experiments and observations having perturbed the system under investigation and mathematical operations on the other hand play around and "perturb" the models, but modelling in itself never directly perturbs the empirical systems studied.

At all stages in a comprehensive approach as in the project at large covered in this volume perturbation needs to be evaluated carefully, which again documents the necessity of combining the three legs of the magic tripod experiment—mathematical model—theory.

Hypotheses and Predictability

In the foregoing we have seen that point after point we encounter a multiplicity of view points: scenarios, scalar levels and processes. The major hypothesis underlying the work presented in this book, derived from the growth–differentiation balance theory (GDB, regarded as a theory sensu Chap. 1; Herms and Mattson 1992), was:

Independent of the type of scenarios and their effective factors the plant regulates its allocation of resources in a way that increased growth and competitiveness lead to reduced defence of stress and pathogen attacks

(Chaps. 1, 19 and 20). Sub-hypotheses were developed. Among these hypotheses when looking at the growth defence dilemma we find three categories:

1. Hypotheses which are unanimously accepted from most or even all view points
2. Hypotheses which are clearly falsified
3. Hypotheses which are accepted at particular spatiotemporal levels but rejected at other levels

The reader will find detailed discussions within the volume. Among (1) we have statements such as

- that increased susceptibility to pathogen attack is reflected in different reactivity of primary and secondary metabolism and their genetic control,
- that independent of herbaceous or woody life form analogous metabolites and genetic and physiological mechanisms of regulation determine the degree of fitness,
- with slight reservations that efficiency in competitiveness and stress tolerance increase with carbon allocation to the rhizosphere and to mycorrhizae,
- with slight reservations that pressures of competition, scarcity of resources and pathogen attack increase the efficiency of the occupation of space by roots and shoots and the acquisition of resources,
- that competition, pathogen attack and symbiotic interaction in the mycorrhizosphere influence allocation by individual plants and biomass production as well as occupation of space.

Among (2) we find

- that an increase in primary metabolism is prerequisite for an increased intraspecific efficiency of competitiveness between neighbouring individual plants, which is rejected as space occupation and allocation are more decisive.

Among (3) we note

- that with increased resource availability species with low substrate turnover are supported in their competitiveness over species with high turnover but simultaneously more sensitive to parasite attack,
- that increased *C/N* ratios related to reduced nitrogen availability and to ontogenetic development increase pathogen defence,
- that pathogen sensitivity increases with increased stimulation of primary metabolism due to CO_2-concentration in the atmosphere as well as increased N-availability, while ozone treatment increases resistance due to increased secondary metabolism.

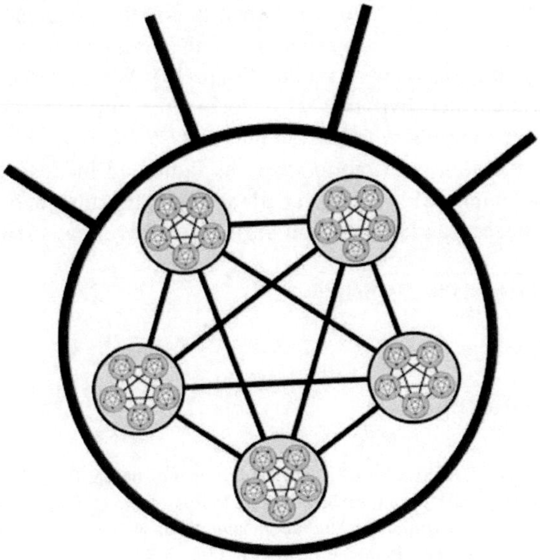

Fig. 3 Networks—macro networks—supra networks—mega networks, etc. where each *point* or *circle* is a system or knot at a particular scaling level and where at each level the output of its individual system is more than the sum of its parts. *Lines* are the connections or edges

The major hypothesis above belongs to category (3): Trade-off between growth and defence is dependent on specific mechanisms and scenarios, and such trade-off is not always expressed. It is an essential result of the work compiled in this book that GDB may basically continue to serve as a basis provoking observations, experiments and thought but that at the same time it needs critical re-evaluation, modification and extension. It is also shown where and how the critique of GDB is becoming effective and where future study is needed. This is achieved by Chap. 1 introducing GDB and Chaps. 19 and 20 putting it under scrutiny with the flesh given by Chaps. 2–18 in between.

That there are hypotheses, including the major hypothesis, which receive positive, negative or neutral answers from different view points or at different scalar levels must not be seen as a failure of the work. On the contrary, it is an intrinsic consequence of the fact that the performance of a whole integrated system is usually not just reflecting the simple sum of its parts. If we consider the impact of the output of a single system in a network of several systems placed at a certain scalar level, we may receive a different answer than when we are seeing the output of several systems of that particular scalar level integrated as one of several parts in a new supra-system at the next higher scalar level. Graphically this can be illustrated by considering the knots (individual systems) of a network at a certain scalar level and unite them into a new macro-knot of a macro-network on the next higher level and taking that macro-network as a knot for a supra-network at the next higher level advancing to a mega-network and so on (Fig. 3; Watts 1999).

Hence, we really must expect that we encounter many scenarios where we receive different answers at different scalar or hierarchical levels as it were for such systems as investigated here. In addition to non-linear behaviour and branching or options between alternative mechanisms, this reflects plasticity and diversity. This is relevant to the assessment of the problem to which extent predictions can be made (Chap. 19). With all success of attempts for predictions (confirmed hypotheses) it documents intrinsic in-determination of systems (open hypotheses), which often makes systems more robust with respect to perturbations by abiotic and biotic stress as fathomed in this volume. In-determination also is keeping the systems challenging, intriguing and interesting. Research is never finished and closed, and curiosity is never satisfied.

Darmstadt, Germany Ulrich Lüttge

References

Grams TEE, Lüttge U (2010) Space as a resource. Prog Bot 72:349–370

Grams TEE, Lautner S, Felle H, Matyssek R, Fromm J (2009) Heat-induced electrical signals affect cytoplasmic and apoplastic pH as well as photosynthesis during propagation through the maize leaf. Plant Cell Environ 32:319–326

Herms DA, Mattson WJ (1992) The dilemma of plants: to grow or defend. Q Rev Biol 67:283–335

Lautner S, Grams TEE, Matyssek R, Fromm J (2005) Characteristics of electrical signals in poplar and responses in photosynthesis. Plant Physiol 138:2200–2209

Liessmann PK (2008) Theorie der Unbildung. Die Irrtümer der Wissensgesellschaft. Piper, München, Zürich

Lovelock J (1979) Gaia. A new look at life on earth. Oxford University Press, Oxford

Lovelock J (2009) A final warning: the vanishing face of Gaia. Penguin, London

Lüttge U, Hütt M-Th (2009) Talking patterns: communication of organisms at different levels of organization – an alternative view on systems biology. Nova Acta Leopoldina NF 96, 357:161–174

Potters G, Pasternak TP, Guisez Y, Jansen MAK (2009) Different stresses, similar morphogenic responses: integrating a plethora of pathways. Plant Cell Environ 32:158–169

Salt DE, Baxter I, Lahner B (2008) Ionomics and the study of the plant ionome. Ann Rev Plant Biol 59:709–733

Schuster HG (1995) Deterministic chaos. An introduction. VCH, Weinheim

Watts DC (1999) Small worlds. The dynamics of networks between order and randomness. Princeton University Press, Princeton NJ

Preface

Resource manager—this might be the "profession" of a plant, if one should assign one. The resources to be managed are carbon, nutrient elements, water and energy. Management here means distribution of resources to vital needs and to "arrange with" (i.e. acclimate to) the environment. Such needs imply to stay operational and competitive, to survive abiotic and biotic stress, to augment biomass and to reproduce. Management notably also comprises preventing premature loss of resources to consumers (pathogens, herbivores) as a prerequisite for meeting all other needs. Clearly, such kind of distribution requires priorities in management, i.e. the plant must make "decisions". The decision policy depends both on the plant's current developmental and metabolic status and on the environmental scenario, i.e. the site conditions. As the internal and external settings can be variable, so can apparently be the plant's response. This phenotypical variability in plant performance is termed "plasticity", and the prioritisation in the distribution policy is reflected in resource allocation and partitioning. In this book, allocation is defined as the process, and partitioning as the result of the resource management. To survive the plant requires to "modulate", i.e. regulate, the different needs through allocation by evaluating the diverse sources and sinks in resource fluxes *versus* the constraints associated with them. This regulatory process is complex, given the multifactorial world and impacts which set the stage for the individual plant's ecology, economy and for the process of evolution. Although one is aware that plants do not ponder on their regulation, the introductory allegory accentuates their challenge for survival and, hence, for empirical and theoretical assessment.

In such terms, how do plants cope in their resource management with complexity? This question is vital for the plant's persistence and fitness and so is the clarification of the functional grounds of resource allocation and its priorities in regulation (Mooney et al. 1991). Apparently, plants have been successful in operating their resource management, as can be concluded from their evolutionary history. However, how intense is the challenge on them in operating? Are they "jacks-of-all-trades, masters of all", as they have been termed in doing their "job" (Koricheva et al. 2004), seemingly in a virtuous way in spite of the complexity of the task? Or, do they encounter "dilemmas" in their "policy" on resource allocation

(Herms and Mattson 1992)? Then, prioritising would appear to be cumbersome. In either case, however, a mechanistic comprehension of resource allocation and its regulation at the whole-plant level evidentially is the prerequisite for understanding the existence and fitness of plants (Stamp 2003a). Nevertheless, attaining such an understanding still poses a major challenge for plant science (Bazzaz and Grace 1997).

The conventional view on the issue outlined above conceives plants as facing dilemmas, *sensu* predicaments in choosing from two or more vexing options, when prioritising the needs to be covered in resource allocation. A prominent, seeming dilemma is the one between the needs to grow to staying competitive in resource acquisition and to defend against stress for retaining the resources once acquired. Mostly stress by consumers such as herbivores or pathogens, but also by abiotic factors, is typically considered (Herms and Mattson 1992). Conversely, the awareness has also grown that allocation can be understood only in response to the continuum of biotic and abiotic impacts (Matyssek et al. 2005). This latter insight was originally founded on the assumed physiological trade-off in allocation at the individual plant level between growth and herbivore defence, represented by secondary metabolism—becoming part of the "growth–differentiation balance hypothesis" (GDB).

In its core, GDB claims such kind of trade-off in plant-internal resource allocation to materialise between growth and defence (Herms and Mattson 1992). Differentiation here means resource investment into chemical and structural modifications of biomass as opposed to growth, which represents irreversible biomass increment. As detailed in Chap. 1, increasing resource availability is presumed, according to GDB, to promote gross primary productivity (GPP) towards a maximum level (cf. Matyssek et al. 2005). In parallel, defence is claimed to be favoured at low resource availability at the expense of growth and growth to be favoured at high availability when defence is low. At severe resource limitation, defence may be constrained by GPP.

More explicitly, nutrient (and water) availability was claimed to have a parabolic effect on secondary metabolites, resulting in a unimodal optimum function with maximum concentration at about medium supply. At limitation, a positive correlation is predicted between growth and secondary metabolism, whereas the correlation should turn negative towards saturation (i.e. high carbohydrate investment into growth rather than defence). Having existed by now for about 60 years (Loomis 1953), GDB has experienced several extensions towards reaching a broad ecologically and evolutionarily relevant scope. As a result, GDB appears to possess, in comparison with other related, partly competing hypotheses, high integrative strength in incorporating a plethora of findings and strong potential for theory development (Herms and Mattson 1992; Koricheva et al. 2004; Stamp 2003a). On such grounds, GDB will be viewed in the remainder of this book as a theory on resource allocation in plants, being still on the way towards maturation (*sensu* science theory, given the unabated demand for integrating the challenging plasticity in the plant's biology and ecology; Stamp 2003b).

Given the current stage of GDB, a comprehensive, mechanistically founded treatment and timely update are missing. Evidence has increased, in addition, that plants appear to regulate resource allocation also beyond the scope of GDB. A holistic view, never presented before, is required, therefore, to integrate spatio-temporal process scaling (i.e. across hierarchical dimensions in structure and time) between cells, organs, whole plants and stands (cf. Ehleringer and Field 1993; Schulze 1994) along with ontogenetic stages and transition between controlled and field conditions of growth (Sandermann and Matyssek 2004). Such a perspective needs to account for links between molecular and biochemical/physiological processes and is reflected in the title of this book as *Growth and Defence in Plants: Resource Allocation at Multiple Scales*. Allocation control is to be clarified as an intrinsic component of interrelated plant–plant, plant–pathogen and plant–mycorrhizosphere interactions in approaching an extended mechanistic understanding. Regarding the biotic interactions, space-related cost/benefit relationships in resource turnover (i.e. investments vs. returns) will be highlighted to arrive at common underlying principles of resource allocation and to examine their validity across plant and interaction types, ontogeny and growth scenarios.

The three stages in theory building will be covered:

1. Examination in view of new empirical evidence from a spectrum of ecological scenarios
2. Exploration of conflicts and validity ranges
3. Extension followed by re-examination of the theory

In such a way, based on methodological advancement and recent gains in evidence, data analysis and modelling, the ultimate aim of this book is a rigorous validation of GDB that may result in a mechanistically founded revision or extension of this theory within ecophysiologically relevant contexts.

An integrative and unique view across forest and orchard trees, herbaceous crop plants and grassland species will be developed on this research issue. Conceptual links will be demonstrated and emphasised between empirical and theoretical approaches as powerful means for hypothesis building and evaluating and theory development. Covered ecological scenarios include competitors, pathogens, herbivores, mycorrhizae, soil microorganisms, CO_2/O_3 regimes, N and light availabilities, as well as drought.

Given the mechanistic perspective and ecological scope of this book volume, the presented new evidence is relevant for the biology of both wild and economic plants. Basic knowledge is augmented as a starting point for applied research on food production and quality, plant breeding and disease control, production of renewable resources and plant system management, altogether within contexts of changing environmental conditions. On these grounds, the focused aims of the book are to

- gather a timely understanding of resource allocation and its regulation in herbaceous and woody plant systems, linking molecular with biochemical and physiological process levels,

- clarify allocation control as an intrinsic component of plant–plant, plant–pathogen and plant–mycorrhizosphere interactions,
- integrate ontogeny and contrasting growth scenarios into spatio-temporal scaling,
- clarify extents of common underlying mechanisms in resource allocation across plant types, ontogeny and growth scenarios,
- evaluate the potential for advanced mechanistic and ecophysiologically relevant theory development as one result of the integrative analyses and hypotheses testing in relation to GDB.

The review character of this book profits from the outcome of interdisciplinary case studies on the subject, e.g. of SFB 607 (an integrated research centre, supported by the German research funding agency, DFG, from 1998 through 2010 in the Munich area/Germany, on *Growth and Parasite Defence – Competition for Resources in Economic Plants from Agronomy and Forestry*) and from the contributions of invited external experts: C. Anderson (Corvallis, USA), J. Bohlmann (Vancouver, Canada), R. Hampp (Tübingen, Germany), J. Koricheva (London, UK) and C. Mathews (Palmer Stone North, New Zealand). Their valuable contributions to this book project are highly appreciated.

Book publications reviewing and comprehensively updating knowledge on resource allocation in plants have been missing for more than one decade. None of the preceding books had pursued a comparably holistic and focused rationale towards theory maturation on resource allocation in plants, inherently addressing joint mechanisms of resource flux and turnover across plant–plant, plant–pathogen and plant–mycorrhizosphere interactions. Part I of the present book elucidates the theoretical grounds of resource allocation between growth and defence. This sets the stage for Part II, presenting the new evidence. Part III then strives to arrive at an integration of the achieved state of knowledge, promoting theory development and introducing into the conclusions of Part IV. To readers who prefer to obtain an overview on the essentials elaborated by this book volume as a whole, a glance into the summarizing Chaps. 19 and 20 is recommended before visiting the detailed explorations of the other chapters. Prominent intention of the book is the reconsideration of research strategies towards a mechanistic and ecologically relevant understanding of the plants' "resource husbandry".

Freising and Neuherberg, Germany

Rainer Matyssek
Hans Schnyder
Wolfgang Oßwald
Dieter Ernst
Jean Charles Munch
Hans Pretzsch

References

Bazzaz FA, Grace J (1997) Plant resource allocation. Academic, San Diego, pp 303

Ehleringer JR, Field CB (1993) Scaling physiological processes – leaf to globe. Academic, San Diego, pp 387

Herms DA, Mattson WJ (1992) The dilemma of plants: to grow or defend. Q Rev Biol 67:283–335

Koricheva J, Nykänen H, Gianoli E (2004) Meta-analysis of trade-offs among plant antiherbivore defenses: are plants Jacks-of-all-trades, masters of all? Am Nat 163:E65–E75

Loomis WE (1953) Growth and differentiation – an introduction and summary. In: Loomis WE (ed) Growth and differentiation in plants. Iowa State College Press, Ames, pp 1–17

Matyssek R, Agerer R, Ernst D, Munch J-C, Osswald W, Pretzsch H, Priesack E, Schnyder H, Treutter D (2005) The plant's capacity in regulating resource demand. Plant Biol 7:560–580

Mooney HA, Winner WE (1991) Partitioning response of plants to stress. In: Mooney HA, Winner WE, Pell E (eds) Response of plants to multiple stresses. Academic, San Diego, pp 129–141

Sandermann H, Matyssek R (2004) Scaling up from molecular to ecological processes. In: Sandermann H (ed) Molecular ecotoxicology of plants. Ecological studies, vol 170. Springer, Heidelberg, pp 207–226

Schulze E-D (1994) Flux control in biological systems. Academic, San Diego, pp 494

Stamp N (2003a) Out of the quagmire of plant defenses hypotheses. Q Rev Biol 78:23–55

Stamp N (2003b) Theory of plant defensive level: example of process and pitfalls in development of ecological theory. Oikos 102:672–678

Contents

Part I
Introduction

Chapter 1
The Balance Between Resource Sequestration and Retention: A Challenge in Plant Science

R. Matyssek, J. Koricheva, H. Schnyder, D. Ernst, J.C. Munch, W. Oßwald, and H. Pretzsch

1.1 Setting the Stage

Plants like all other organisms require sustaining a state of structural and functional order, i.e. to prevent loss of control on internal entropy, and by this warrant the crucial pre-requisite for—what is called—life processes. Such grounds represent energetic pseudo steady-states which are established by a continuous flux of energy and matter through plants as open systems, mirroring dynamic equilibria between

R. Matyssek (✉)
Chair of Ecophysiology of Plants, Technische Universität München,
Hans-Carl-von-Carlowitz-Platz 2, 85354 Freising, Germany
e-mail: matyssek@wzw.tum.de

J. Koricheva
School of Biological Sciences, Royal Holloway, University of London, Egham,
Surrey TW20 0EX, UK

H. Schnyder
Lehrstuhl für Grünlandlehre, Technische Universität München, Alte Akademie 12,
85350 Freising, Germany

D. Ernst
Institute of Biochemical Plant Biology, Helmholtz Zentrum München,
Ingolstädter Landstr. 1, 85764 Neuherberg, Germany

J.C. Munch
Institute of Soil Ecology, Helmholtz Zentrum München, Ingolstädter Landstr. 1, 85764
Neuherberg, Germany

W. Oßwald
Phytopathology of Woody Plants, Technische Universität München,
Hans-Carl-von-Carlowitz-Platz 2, 85354 Freising, Germany

H. Pretzsch
Chair of Forest Growth and Yield Science, Technische Universität München,
Hans-Carl-von-Carlowitz-Platz 2, 85354 Freising, Germany

R. Matyssek et al. (eds.), *Growth and Defence in Plants*, Ecological Studies 220,
DOI 10.1007/978-3-642-30645-7_1, © Springer-Verlag Berlin Heidelberg 2012

intake of resources (energy, carbon, water, inorganic nutrients) *versus* release back into the environment after temporary use (Bazzaz and Grace 1997). Upon anabolizing the resources, usage is facilitated through complex metabolic processes and eventually leads into catabolism, unless biomass is shed and becomes subject to decomposition, as does the whole plant body at the end of its life span. Hence, a resource turnover exists, which is operated through input/output balances along time scales specific to the different plant functions. Resource gains from the environment are invested internally to ensure growth, survival and reproduction, and for warranting physiological acclimation and genetic adaptation to the environment. However, the input/output balances are constantly at risk from abiotic (e.g. wind, fire, frost, drought) as well as biotic stressors (competitors, pathogens, herbivores). Therefore, plants must also preserve some of their resources for stress defence, i.e. preventing impediment of uptake and loss. Additional resources are invested into symbionts (like mycorrhizal fungi or N-fixing bacteria) or beneficial soil micro-organisms (van Dam and Heil 2011; Vannette and Hunter 2011) or insects in tritrophic settings (cf. Chap. 4), which together with the plant form the "holobiont" as the co-evolutionarily effective unity (cf. Zilber-Rosenberg and Rosenberg 2008) that determines both resource gain and retention (e.g. Ericsson et al. 1996). The priorities of the different plant functions are dynamic (typically driven by fluctuations in the most growth-limiting factor) so that regulation of the internal resource flux is required. For preventing critical limits in regulation, i.e. ensuring resource supply to vital functions, a buffering component must be sustained, which is the plant's reserve storage. The latter can be intrinsic to defence and reproduction, but is particularly important to plants with prolonged life spans (Schulze 1982).

The individual plant's success in growth, survival and reproduction in relation to competitors is associated with cost/benefit relationships in resource turnover (Schwinning 1996). It is conceivable, therefore, that marginal fluctuations in such cost/benefit relationships, i.e. in the efficiencies in resource management, are crucial for plant competitiveness. Such an "economic" view on the plant's existence (e.g. Givnish 1986 and review articles therein) reflects the core of ecology, as expressed by Ernst Haeckel, a founder of the research discipline, *sensu* ecology as the economy of organisms (Haeckel 1870). In such terms, survival of the fittest during the evolutionary process (Darwin 1859) results from efficient resource use, which might be more decisive than maximum resource sequestration relative to competitors (e.g. Schulze et al. 1986; Matyssek and Schulze 1987; Küppers 1994; Schwinning 1996; Grams and Andersen 2007).

Given the ecological need for efficient resource use, the plant faces the challenge that any resource can only be spent once at a given instant, although "recycling" for same or different functions is possible. Examples are storage, which postpones the ultimate investment, and metabolites with rapid turnover or precursors of several usages (Stitt and Schulze 1994). Therefore, plants may encounter a dilemma in resource allocation to various concurrent needs, giving rise to potential trade-offs, i.e. favouring some functions at the expense of others in terms of inverse relationships. A crucial trade-off is associated with investment into growth (for ensuring competitiveness) *versus* that into defence against stress with risks of resource loss. In such terms, the plant has to balance resource uptake and incorporation *versus* resource retention (Matyssek et al. 2005). This balance is the expression of individual plant fitness by providing the ability and extent of reproduction through the capacities in competitiveness and defence as crucial prerequisites. The balance is evidently fed from the whole-plant resource pool with all its metabolites (regardless of being conventionally classified as "primary" or "secondary"; Schwachtje and Baldwin 2008). It is tempting to conceive underlying mechanisms through which the required balance is accomplished and the potential dilemma in resource allocation is resolved.

In the following, we will first highlight theoretical concepts that give guidance to understanding resource allocation in plants. "Theory maturation" will be examined in view of the available knowledge *prior to* the recent progress reported in this book—or, in other terms, of capacities for hypothesis formulation and falsification in promoting and consolidating knowledge. Empirical aspects will then be viewed both in terms of constraints on and potential for theory development. The stage will be set for demonstrating recent empirical and theoretical progress on the outlined subject in the subsequent book chapters.

1.2 Theories on Whole-Plant Resource Allocation

Amongst analytical concepts which view resource availability as a driver of whole-plant allocation, three prominent ones focus on the "*growth–differentiation-balance*" (GDB; Herms and Mattson 1992), the "*carbon–nutrient balance*" (CNB, Bryant et al. 1983) or on the role of protein synthesis ("*protein competition model*", PCM; Jones and Hartley 1999). Conceiving allocation by different regulatory principles, these will be featured in this section *prior to* also introducing the concept of "*optimal defence*" (OD, Rhoades and Cates 1976), which is based on the value of organs for plant fitness *versus* their risk of loss. Such concepts, each of them claiming to reflect specific evolutionary outcome, have been termed hypotheses, although it is debatable of whether they may also be viewed as theories. Classification of the above concepts as theories appears to be justified to the extent that guidance to experimental clarification and mechanistic explanation is provided (Stamp 2004), as will be elucidated in the following. On such grounds, the introduced concepts will be termed as "theories" in the remainder of this chapter, while being aware of their present "immature" state (Stamp 2003a; also see Sect. 1.3).

1.2.1 Growth–Differentiation Balance Theory

As introduced by Loomis (1953; also see Lorio 1988) and extended by Herms and Mattson (1992) and Matyssek et al. (2002, 2005), GDB states a trade-off in plant

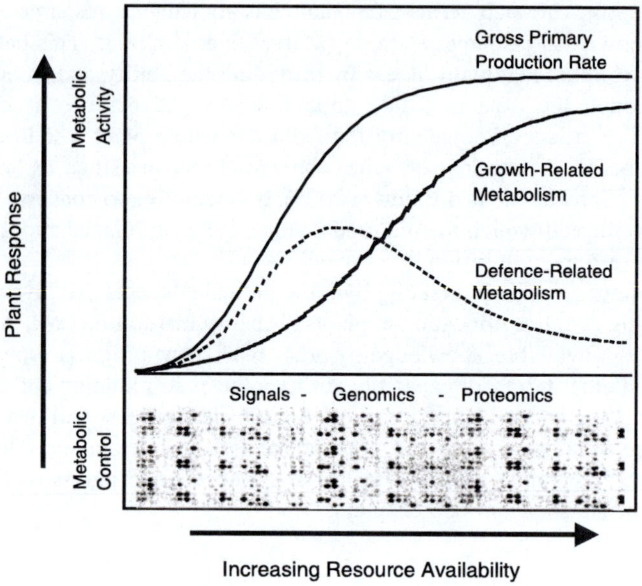

Fig. 1.1 Plant response in relation to increasing resource availability, expressed as variations in gross primary productivity along with such growth and defence-related metabolism. Note indicated trade-offs between growth (and competitiveness, see text) *versus* defence, according to the "*growth–differentiation balance theory*" (cf. Herms and Mattson 1992). Metabolic activity (biochemical and physiological process level) is linked with metabolic control at the molecular (gene) level through signalling, genomics and proteomics (from Matyssek et al. 2005, with staining pattern symbolizing genomics as macro-array based transcript analysis)

internal resource allocation between growth and defence. Differentiation means resource investment into the chemical and structural modification of biomass as opposed to growth, which represents irreversible biomass increment (Potters et al. 2009). Differentiation can serve both mechanical stability (being conducive to growth) and biochemical defence (Arnold and Targett 2003). Hence, differentiation implies that the functional transition between growth and defence is gradual. On such grounds, Fig. 1.1 schematically approximates the core of GDB, according to the extended view of Matyssek et al. (2005), in that increasing resource availability promotes gross primary productivity (GPP) towards a maximum level. In parallel, growth and defence-related metabolism respond in complementary ways to each other, in terms of a trade-off. This means, favoured defence at low resource availability at the expense of growth, but favoured growth at high availability when defence is low. Uncertainty may arise at severe resource limitation, which may constrain both the growth and defence-related metabolism (Glynn et al. 2007).

The processes associated with the three resource-driven functions of Fig. 1.1 are controlled by gene regulation (through signalling, transcription and protein synthesis). Haugen et al. (2008) underlined that the plant's capability of expressing trade-offs in growth/defence-related allocation is genotype-specific. The molecular basis indicated in Fig. 1.1 (signals–genomics–proteomics) reminds of deficits in the understanding of the mechanistic link between metabolic control and metabolic activity (Matyssek et al. 2005; Ballhorn et al. 2008). In comparison to previous visualizations of GDB (Herms and Mattson 1992), choosing gross primary production rate (GPP) instead of net assimilation rate (NAR) as one measure of plant productivity stresses the paramount importance of respiratory demands in driving the trade-off (cf. Bolton 2009). Replacing, in addition, the previously used terms "primary" and "secondary" with "growth-related" and "defence-related" metabolism, respectively, overcomes the conceptual restriction of the conventional classification in that metabolites (or their precursors) may serve growth and/or defence regardless of their chemical nature (cf. Arnold and Targett 2003). As these two functions have to be achieved by the plant simultaneously, clarification of the role of metabolites in resource uptake (through competitive growth) and/or retention (through defence; Riipi et al. 2002) is crucial. Nevertheless, distinguishing between different types of secondary metabolites is important, because they reflect diversity acquired during evolutionary history, providing plants with the capacity of responding to specific ecological challenges (i.e. particular herbivores, pathogens). This evolutionary component in combination with the assumption of allocation trade-offs makes GDB a framework that allows understanding of departure from the core of the theory. In a strict sense, GDB views the availabilities of the resources (e.g. water and nutrients) as affecting the ratio between the pool size of photosynthate (i.e. the carbon pool) and the demand of growth for photosynthate (i.e. the sink strength for biomass formation; Koricheva et al. 1998). Resource limitation, predominantly by nitrogen (N) *sensu* GDB, but also by phosphorus (P; Sampedro et al. 2011), curtailing the sink strength of growth may lead to an accumulation of carbon, which then is available to differentiation processes including defence. Notwithstanding such kind of regulation, allocation is conceived to serve growth at a higher priority than differentiation.

1.2.2 Carbon–Nutrient Balance Theory

In a sense that fluctuations in source/sink ratios as drivers of allocation trade-offs affect the relative availabilities between carbon and other resources, GDB indicates some conceptual proximity to CNB (Bryant et al. 1983). The latter, however, basically conceives growth/defence allocation trade-offs to be driven through changes in the carbon/nutrient ratio. Hence, rise of this ratio, as occurring under nitrogen limitation or at high light or elevated CO_2, is believed to favour differentiation. Mattson et al. (2005) suggested that climate change-associated atmospheric CO_2 increase will favour defence (e.g. as based on phenolics; cf. Koricheva et al. 1998), given the indications of a prevalently carbon-saturated metabolism, in particular, in trees (Körner 2003). Conversely, a decreasing ratio may curtail differentiation and defence. In addition, the sizes of C and nutrient pools relative to each other should determine, according to CNB, the chemical quality of defence metabolites in view of N demand during synthesis. Overall, similar working hypotheses may be derived from GDB and CNB, in that, for example, plants regulate their resource allocation in a way that increase in growth and competitiveness leads to constraints on stress defence, in particular, of pathogens and herbivores (Matyssek et al. 2002, 2005).

CNB has been criticized recently for its inability to cover a wide range of ecological scenarios (Hamilton et al. 2001; Lerdau 2002; Koricheva 2002). Stamp (2003b) along with Herms and Mattson (1992) suggested to view CNB as a module of GDB, given some overlap between both theories. However, such a view would ignore that GDB rather than CNB can conceive different physiological interactions to result in similar *C/N*. In particular, GDB mirrors secondary metabolism to obey a parabolic dependency function in relation to resource availability, so that non-linearity in plant response is covered. In addition, GDB is specific in predicting the ways different environmental factors affect trade-offs through altering source/sink relationships, rather than being restricted to immediate effects on the carbon–nutrient balance. On such latter grounds, GDB has been evaluated as advanced in guiding a mechanistic analysis of allocation trade-offs, and in promoting theory development (Stamp 2003b).

1.2.3 Protein Competition Model

PCM (Jones and Hartley 1999) also pursues a mechanistic view on allocation trade-offs in plants and appears to be conceptually related to CNB and GDB because it considers N availability as one driver of allocation amongst nutrients. However, PCM differs from CNB and GDB theories by emphasizing the biochemical process level in the regulation of allocation, specifically, the competition between formation of proteins and phenolic compounds for the common, limiting precursor phenylalanine—which is the core of PCM. As a typical characteristic, plants possess the shikimic acid pathway producing phenyl-propanes, and hence, phenylalanine. Therefore, the trade-off can be claimed to originate from the inversely related allocation into protein *versus* phenolic substances formation, with phenylalanine-ammonia lyase (PAL) as the central metabolic switch. At increasing N availability, protein synthesis should drain, therefore, the carbon pool, creating a sink for growth, while curtailing the capacity for the formation of phenolic metabolites. In essence, and as opposed to GDB, the plant's N rather than C pool is conceived to limit phenolic synthesis, and hence, defence. It is believed that the concentration of phenolic compounds increases with declining sink strength for growth, either through environmental effects constraining N availability, or during advanced stages of organ maturation and ontogeny, or in plant species with inherently low growth rates (Jones and Hartley 1999). PCM was suggested to complement GDB in biochemical terms and to represent a viable alternative to CNB. Doubts arose, however, recently about the validity of PCM under elevated CO_2. According to PCM, the phenylpropanoid pathway should become limited (i.e. with the N pool making proficient use of the high C supply for growth), whereas *sensu* GDB, C availability to the formation of phenylpropanoids should increase relative to the N pool. Respective evidence appears to support GDB rather than PCM (Mattson et al. 2005). In addition, phenylalanine was shown to be regenerated under N limitation through internal recycling of N in amination/desamination reactions (Pankhurst and Jones 1979; Graham 1983; Peng et al. 2007), which questions the regulation of the formation of phenolic substances through N availability. Furthermore, pool sizes of defence metabolites do not necessarily increase with ageing, as high levels in vulnerable young leaves have shown the plants' capability of decoupling defence from ontogeny (Herms and Mattson 1992).

In cross-evaluating growth/defence allocation theories, Koricheva et al. (1998) developed a hierarchical model of carbon allocation to plant secondary compounds. This model implies that resource-based theories as represented by CNB and GDB make valid predictions only about the total amount of carbon that can be allocated to carbon-based secondary compounds (CBSCs). Such theories cannot predict plant responses at lower hierarchical allocation levels, where carbon is shunted into various alternative synthesis pathways to different CBSC classes (e.g. phenolic compounds including lignins, terpenoids, alkaloids, tannins, phenylpropanoids). Instead, Koricheva et al. (1998) suggested that carbon allocation at lower hierarchy levels depends on specific evolutionary plant responses to the various types of biotic and abiotic stresses. Phenylpropanoids are an exception in that they are closely linked—through their precursor phenylalanine—with growth metabolism (see above). Hence, they tend to more distinctly reflect the resource trade-off between growth and defence (Rühmann et al. 2002) than do other compound classes. PCM may attain relevance in view of evolutionary specificities at low hierarchy levels.

1.2.4 *Optimal Defence Theory*

As opposed to GDB, CNB and PCM, OD (Rhoades and Cates 1976) is not related primarily to resource availability and associated phenotypic plasticity (including growth/defence allocation trade-offs). Rather, OD represents a life-history, i.e. adaptation-based explanation model that evaluates defence of plant organs in relation to risk by herbivory and value for plant fitness (Orians and Ward 2010). Economy in the regulation of allocation is presumed, viewing adaptations by their potential of decreasing defence cost, so that sufficient, i.e. "evolutionarily optimized", rather than abundant investment into defence is postulated (cf. Siemens et al. 2010). "Value" is defined by the fitness costs inflicted upon injury or tissue loss and tends to be negatively correlated with the abundance of a tissue or organ within the plant (Zangerl and Bazzaz 1992). "Risk" arises from the probability of vulnerability (by herbivore attack), as incurring in the absence of defence. High priority to defence is implied, if a tissue is exposed to consumers (e.g. peripheral tissue location in the plant), with the chemical quality and fitness costs being determined with the probability of discovery (Feeny 1976; Bustamante et al. 2006). Conversely, low priority is suggested, if consumers are satiated by high tissue/organ abundance. In line are observations that determinate shoot growth, abundantly producing young foliage during short time periods, may not require enhanced defence, whereas leaves continually originating from indeterminate growth by low numbers may do so (Herms and Mattson 1992). OD is complementary to, and consistent with GDB in sharing the fundamental assumption that defence is costly. Both predict that given multiple defensive options of equal efficacy, the least costly is favoured, and that at equal costs, preference is with the most efficacious one. GDB provides a functional foundation to OD, when viewing phenotypic variation in defence as an adaptive trait in the response to nutrient availability (Glynn et al. 2007).

1.2.5 Preliminary Valuation

As the regulatory settings in plants are complex, one cannot *a priori* expect any of the theories introduced above to give a plain answer about the plant's "approach" to resolving the dilemma of resource sequestration *versus* retention. Plants may even invest resources into several different defence traits, being "jack-of-all-trades" *sensu* Koricheva et al. (2004), which may mitigate trade-offs of a particular trait. Haugen et al. (2008) postulate the clarification of allocation conflicts in relation to plant ontogeny and variable site conditions. Phenotypic plasticity in response may pretend plants to—or even makes them—perform in ways unrelated to resource trade-offs and respective theories (Bradshaw 2006; Ballhorn et al. 2008).

1.3 Constraints and Potential of Theory Development

For all theories introduced above, consistency with experimental evidence is limited (Koricheva et al. 1998; Koricheva 2002; Siemens et al. 2010). Scope and state of "theory maturation" attained *prior to* the recent gain in knowledge that will be reported in this book on the plant-internal growth/defence conflict will be outlined, therefore, both in view of science theory and empirical considerations. Constraints on and potential for theory development on plant resource allocation will be highlighted.

1.3.1 Aspects of Science Theory

1.3.1.1 Stages of Theory Maturation

Plants may express high phenotypic plasticity in stress response (Koricheva et al. 1998; Siemens et al. 2010; Cipollini and Heil 2010). The existence of this plasticity has even led to questioning the suitability of the objective, i.e. the regulation of resource allocation between growth and stress defence, for theory building (Berenbaum 1995). The uncertainty relates to the rules of theory development and means of associated hypothesis testing. Theory development goes through early, immature and mature stages (Stamp 2003a). The early stage typically implies the formulation of relatively vague and qualitative hypotheses which cannot be accurately and unambiguously tested. Conclusions are at risk to be caught in conceptual conflicts, because the underlying mechanisms are still unclear. Increasing empirical evidence is one means of promoting sharpness in hypothesis formulation and testing towards approaching the immature stage. If predictions stay vague, competing hypotheses and theories on same, but inadequately understood phenomena tend to prevail (Berenbaum 1995), and intellectual progress may become exhausted in reconciling findings, which typically lack overarching conceptual and experimental frameworks. Novel statistical tools, linking generality of findings with deterministic strength, along with rigorous examination for plausibility and inherent consistency can mitigate seeming conflicts (Chap. 16). Overcoming "immaturity" means posing and falsifying hypotheses on mechanisms that underlie the challenging phenomena, i.e. *sensu* Karl Popper (1989) hypothesis (and theory) elimination through falsification is the pre-requisite for intellectually approaching the actual nature of research objectives. The frontier in research on growth/defence-related allocation has arrived at intensifying this selection process (Stamp 2003a), and it is one aim of this book to explore recent progress in approaching "maturity" in theory development.

The current stage requires the definition of predictive limits of hypotheses under examination rather than the exploration of the entire scope of theories. Findings upon passing comprehensive quality assessment, combining plausibility and consistency evaluation with hypothesis testing, have been adopted, in addition, too rarely into frameworks of mathematical modelling. Such latter abstractions advance hypothesis formulation and testing towards enhanced integration levels to foster further empirical progress. Interaction between experimentation and modelling has been neglected in promoting theory building. Advancements in this field are a subject of this book. In the forefront, none of the introduced theories has reached "maturity", given the shortcoming that most knowledge has not been elaborated in view of theory development, but circumstantially originated from the pursuit of conceptually non-related research questions.

1.3.1.2 Hypothesis Formulation

How to formulate hypotheses and design research in view of theory development? Making research questions operational for testing requires streamlining hypothesis formulation, while preventing simplification prone to misinterpretation. Simulations by mechanistic modelling, representing hypotheses themselves, may provide demanding predictions beyond the scope of experimentation, being testable against empirical evidence. Still, theories can only reflect excerpts from the wide range of biological reality, being defined by plant behaviour and ecological scenario.

Research outcome beyond hypothesis prediction represents falsification, and hence, in terms of science theory, gain in understanding (cf. Popper 1989). The appropriate balance between simplification in hypothesis formulation as an expression of attained intellectual comprehension and aptitude to falsification, therefore, is crucial. In particular, growth/defence resource allocation with its metabolic specificities across stress and plant types (see Sect. 1.2) offers potential for (rather than hinders) theory development, as will be elucidated in this book.

1.3.1.3 Sub-theories

Theories can be viewed as integrative constructs of sub-theories that are insertions for facilitating analysis (Stamp 2004). Such insertions dealing with pre-requisites of the overall construct support focused hypothesis formulation on, e.g., gene regulation, biochemical pathways, biotic interactions—or, often neglected, but fundamentally important—on the roles of genotype, pre-history, plant ontogeny, allometry, phenology or bifurcation in resource allocation. Derived hypotheses are applicable to explicit developmental or ecological settings and can be hedged or interconnected to argumentative chains or statistically ascertainable relationships. At this stage, statistical modelling (Chap. 16) as a tool complementary to mechanistic modelling can foster hypothesis development and testing. Advancement of such kind is constructive, unless intellectually unassimilated findings merely accumulate. Recent evidence reported in this book originated from an approach as suggested here. Only inconclusive stagnation would pretend plant functioning to be governed by exceptions and lead to questioning suitability for theory development.

1.3.2 Biological Aspects

1.3.2.1 Steady-States *versus* Stages of Transition

Theory development is constrained also by the complexity of plant life and related challenges in experimentation that needs to ensure potential for hypothesis testing. One key assumption behind the theories on growth/defence resource allocation typically is steady-state in plant metabolism (Stamp 2004, "steady-state" used here *sensu* pseudo steady-state, as plant are open systems in terms of thermodynamics, cf. Sect. 1.1). However, plant life at field sites is hardly determined by steady-states. Although these are conducive to facilitate and standardize experimentation and hypothesis testing (Koricheva et al. 1998), evolutionary understanding gained about plant performance is limited (Glynn et al. 2007). Complication in conceptually relying on steady-states may arise due to plants' use of reserves from pre-experimental periods (Siemens et al. 2010), while acclimating to experimental treatments (Glynn et al. 2007). Defence costs inherited from the preceding generation through non-genetic parental effects may also bias steady-states (Purrington 2000).

In addition, resource supply *versus* demand can change due to experimentation and/or plant development, which altogether underlines the dynamic rather than steady-state character of resource allocation. In this sense, empirical studies conducted so far tended to neglect the evolutionary dimension of growth/defence allocation which in fact resembles a "moving target" (Glynn et al. 2007). Moreover, plants were often "overloaded" with stress (e.g. parasites) in order to provoke distinct response, which not necessarily was ecologically meaningful (Bolton 2009; cf. van Dam and Heil 2011). Recent assessment by Glynn et al. (2007), however, showed predictions of GDB to stay robust about changes in plant metabolism, as transitioning to new equilibrium upon alteration in resource availability. Hence, the scope of GDB can reach beyond the argument of Stamp (2004) that growth/defence allocation theories are bound to steady-state presumptions. The subsequent account in this book is guided by envisioning dynamics in plant response.

1.3.2.2 Phenotypic Plasticity

In addition to dynamics in allocation, resource availability if severely limiting GPP challenges empirical analyses. Given such a setting to limit both growth and defence, plants apparently obey the principle of "optimal phenotypic plasticity" (Glynn et al. 2007) in assigning high priority to compensatory growth. For example, nutrient or water shortage typically drives below rather than aboveground production (Mooney and Winner 1991). During growth adjustment, growth/defence trade-offs hardly become substantiated. Absence of trade-off may also be explained by high variation in resource acquisition *versus* allocation (van Noordwijk and de Jong 1986). Once adjustment being accomplished, plant performance was found, however, to conform to trade-off predictions by GDB (Glynn et al. 2007). It was proposed, therefore, to extend GDB by including the concept of phenotypic plasticity to enhance the theory's predictive strength. This concept communicates why GPP and allocation to defence and growth can positively correlate while approaching the plant-internal set-point of balanced resource flux *sensu* Mooney and Winner (1991): The constraint on GPP (i.e. production of photosynthate) apparently becomes relieved to a higher degree than on growth (in terms of irreversible biomass increment) so that the eventually resulting surplus in C availability can stimulate defence. Such grounds may explain inconsistencies in theory predictions by ascertaining the parabolic dependency of defence, as postulated by GDB, on the transition from limiting to non-limiting resource availability (cf. Fig. 1.1; Glynn et al. 2007). Complication represents, however, scarcity of knowledge on below-ground plant defence and variable competitive interference during successional progression (Luedemann et al. 2005; Rasmann et al. 2010; Hakes and Cronin 2011).

1.3.2.3 Driving Factors

Misinterpretation of growth/defence allocation theories can arise from confounding driving factors. Exemplifying GDB, these are nutrients and water rather than carbon (Koricheva 2002). The distinction is crucial in understanding growth (biomass increment) rather than GPP (productivity of photosynthate) to be limited by incipient shortage of water and/or nutrients (Kramer and Boyer 1995). The latter drive the sink induction of growth, and by this, are the primary determinants of potential allocation trade-offs. Mitigated sink induction, however, enhances the internal carbon availability which then fosters structural and functional differentiation, including defence. Nevertheless, there is no strict "either-or", regarding the potential usage of resources in growth *versus* defence processes, as some metabolites may serve both plant functions (Riipi et al. 2002; cf. Vannette and Hunter 2011). Phenylpropanoids, however, appear to mostly mirror allocation trade-offs under the fluctuating influence of driving factors (see Sect. 1.2.3 and Chap. 3). Within this compound class, trade-offs may even occur between metabolites, if precursors share same biosynthesis pathways (Koricheva et al. 2004). Trade-offs tended to become conspicuous within the whole-plant phenylpropanoid pool (Koricheva 2002). However, metabolic whole-plant assessments have been rare, despite their high integrative value in theory development.

1.3.2.4 Productivity Parameters

Another challenge is the adequate assessment of photosynthetic productivity and growth parameters in view of growth/defence allocation theories. NAR and RGR (relative growth rate), respectively, have been chosen as substitutes, like in the case of early definitions of GDB (cf. Herms and Mattson 1992). This increases the conceptual uncertainty in addition to ambiguous distinctions between secondary and primary metabolites (cf. Matyssek et al. 2005; Stitt and Schulze 1994; see Sect. 1.2). NAR may be viewed as mirroring the balance between respiratory demand and the remaining portion of carbon, upon the respiratory mitigation of GPP, eventually available to growth (Glynn et al. 2007). The circumstance of NAR and RGR being linked to each other *via* LWR (leaf weight ratio) within the framework of growth analysis (Nobel 1983) is conducive, however, to interpreting predictions of GDB (Glynn et al. 2007), as LWR is an indicator of the plant's phenotypic plasticity (see above). Theory development has neglected, though, the capacity of plants under stress for inducing additional sinks for carbon related to defence. Such sinks require enhanced supply of primary carbon compounds (Schwachtje and Baldwin 2008; Bolton 2009). Photosynthesis can be up-regulated in such cases (Murray and Walters 1992; Williams and Ayres 1981), augmenting the carbon pool available to allocation through enhancing GPP (i.e. the supply of photosynthate). Increasing GPP de-escalates growth/defence trade-offs, i.e. helps the plant to "escape" from the dilemma (in addition to reasons suggested by van Noordwijk and de Jong 1986; see above)—albeit reducing conformity to prevalent theory predictions. Photosynthetic up-regulation may be a trait of defence beyond GDB, unless parasites profit from the enhanced carbon availability (Bolton 2009).

1.3.2.5 Respiratory Costs

Within frameworks of growth/defence allocation theories, the amount of respiratory carbon as a determinant of productivity and substantial component of GPP has largely remained unaccounted for. This deficit is severe as primary carbon compounds do not only feed differentiation and defence, but are crucial also for the associated respiratory demand of both (Bolton 2009). Unlike other aspects of defence, e.g. metabolic pathogen recognition, signalling or biochemical stress responses, the energy recruitment of defence has scarcely been investigated (Schwachtje and Baldwin 2008; Bolton 2009). On such grounds, the concept of GDB was widened by Matyssek et al. (2005) in replacing NAR with GPP (acknowledging respiration) and stressing the potential multi-functionality of metabolites, mediating between rather than separating growth and defence-related metabolism (cf. Fig. 1.1). This considers the actual scope of GDB in a more realistic view, as "full resource costs" of trade-offs must include respiration, and as multi-functionality of metabolites is a basic feature of the plants' biology. Häberle et al. (2009) approximated the substrate disposable for defence to be about 2–5 % of GPP. Estimating full costs of defence and growth is difficult (Gershenzon 1994; Bazzaz 1997), but important for further theory development. Quantification of respiration is relevant also to costs before metabolites deploy their ultimate and specific function (Purrington 2000), as incurred from transport, storage or synthesis and turnover of precursors (Lerdau and Gershenzon 1997). Given the methodological challenges, however, to be overcome, "full-cost" studies hardly have been pursued (Lerdau and Gershenzon 1997).

1.3.2.6 Direct *versus* Indirect Defence Costs

Growth/defence allocation theories typically focus on costs which are directly associated with resource investment to defence *versus* such to growth and other plant functions (allocation costs). Direct costs of *constitutive* defence (being preformed in the absence of stress impact) must not conceptually be confounded with such inducible defences. Costs of these two types of defences may be negatively correlated, representing another kind of trade-off (Koricheva et al. 2004). Conclusions about "inexpensiveness" of inducible defence in relation to plant fitness appear to be premature (Heil, and Baldwin 2002; Koricheva 2002) and perhaps result from limitations in statistical resolution (Purrington 2000). Instead, expenses are likely for sustaining the plant's capacity for inducing defence, if intermittently needed (Purrington 2000; Cipollini and Heil 2010). As this capacity relies on plant properties provided by the growth-related metabolism (e.g. storage structures, defence precursors, multi-functionality of metabolites; Arnold and Targett 2003), and as *inducible* defence tends to be locally restricted in plant tissues, costs may be lower, however, than of constitutive defence (see Chap. 3).

However, indirect costs of defence may also be relevant for plant resource allocation to growth *versus* defence (Koricheva 2002). One type are "opportunity costs" associated with constitutive defence, denoting foregone opportunities (e.g. Stitt and Schulze 1994) upon investing resources into defence instead of growth—*sensu* the additional biomass increment (beyond the equivalent effect of resource partitioning per se) that would have occurred in the absence of defence. Such costs are manifested in the presence of competitors (Baldwin and Hamilton 2000; Hakes and Cronin 2011). Of indirect nature also are "ecological costs" of defence (Koricheva 2002). These denote enhanced defence against one stressor at the expense of other defences (Purrington 2000; Strauss et al. 2002; Haugen et al. 2008). Third-party trade-offs between the different kinds of defences may arise (Koricheva et al. 2004; Ballhorn et al. 2008). The significance of ecological costs has been overlooked in terms of plant evolution and ecology, although indirect and direct costs appear to be equally important (Koricheva 2002). If plants developed multiple stress defences at low ecological costs instead of switching defence within rigid frameworks of allocation trade-offs (Koricheva et al. 2004; Novriyanti et al. 2010), low conformity to current allocation theories would not be surprising (Cipollini and Heil 2010; Siemens et al. 2010). Ecological costs require awareness as incurring under contrasting ecological scenarios, including varying extents of competition or facilitation as modifiers of defence capacity (Baldwin and Hamilton 2000; Siemens et al. 2003; Hakes and Cronin 2011).

1.3.2.7 Mechanistic Perspective

Given the methodological constraints on "full-cost" analyses (cf. Gershenzon 1994; see above), the mechanistic understanding of costs is a pre-requisite for promoting growth/defence allocation theories (Purrington 2000; Stamp 2003b). The cause–effect related "down-link" from phenotypic expression to gene regulation, however, represents a research frontier in plant science. A challenge is posed, in particular, for understanding the genotype determining trade-offs in growth/defence allocation and ecological costs (Purrington 2000), as exemplified for defence–drought interrelationships (Haugen et al. 2008). Gene clusters rather than single genes respond to external stress and/or internal resource demands, hindering proofs of cause–effect relationships (Potters et al. 2009), because both at the gene and process level, straight-forward relationships can be lost and embedded in complex interaction networks (as determined, e.g. by demands for reserve storage, reproduction or by symbionts like mycorrhizal fungi, N_2-fixing bacteria or soil micro-organisms and insects; van Dam and Heil 2011). As a consequence, "third-party trade-offs" can arise, so that the one on growth/defence allocation needs to be viewed as part of the continuum in whole-plant allocation (Fig. 1.2, Matyssek et al. 2005). This continuum is represented by one resource pool (as the sum of primary and secondary metabolites) concurrently serving growth as a means of competitiveness, defence for retaining competitively acquired resources, and (below-ground) symbionts in support of competitiveness and reproductive success

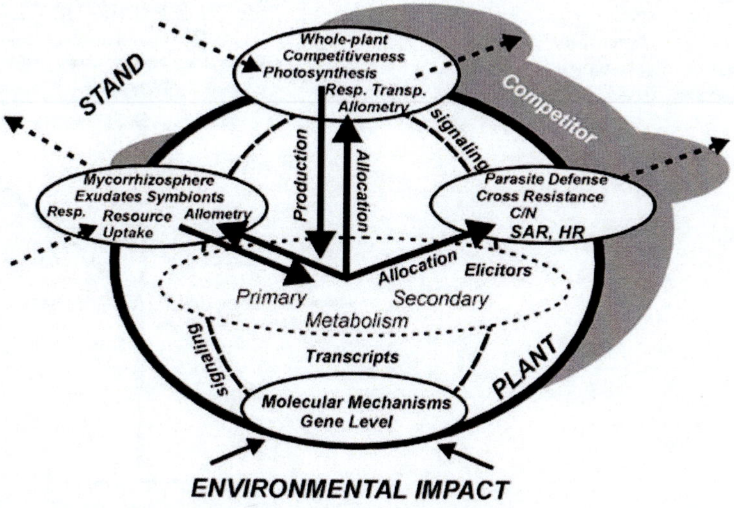

Fig. 1.2 Conceptual model visualizing whole-plant resource allocation. *Arrows* indicate major pathways in resource allocation between physiological demands within the plant and in exchange with the environment, as being under the control of environmental impact and molecular processes (see text for details; from Sandermann and Matyssek 2004)

(van Dam and Heil 2011; Vannette and Hunter 2011). Ecological benefits counteracting trade-offs have been reported, as some kind of facilitation, from mycorrhizae in enhancing defence (Bonello et al. 1993; Ericsson et al. 1996; Bi et al. 2007; Pozo and Azcon-Aguilar 2007), although the mechanistic basis is unravelled only in parts and also mitigation of defence has been reported (Koricheva et al. 2009). However, such kind of factorial and mechanistic complexity represents the actual, biologically relevant stage of growth/defence theories.

1.3.2.8 Spatio-temporal Process Scaling

"Timing" of mechanisms in growth/defence allocation is another aspect of relevancy in related theory development. Gene regulation can proceed instantaneously, whereas the induced metabolic processes may evolve with delay and develop prolonged momentum (regarding growth, up to weeks or months). Conversely, only prolonged factorial impact may end up in gene induction, as well as more than one gene induction event may be necessary for sustaining defence (Purrington 2000). Hence, the temporal coupling between gene expression and physiological response can be weak or even appear to be absent under the influence of factorial networks (Glinski and Weckwerth 2006). The plant's metabolic status may even more readily be assessable through principles of thermodynamics than the pursuit of process pathways (Potters et al. 2009; cf. Sect. 1.1). Process scaling is intrinsically linked with increasing time scales (Sandermann and Matyssek 2004) and

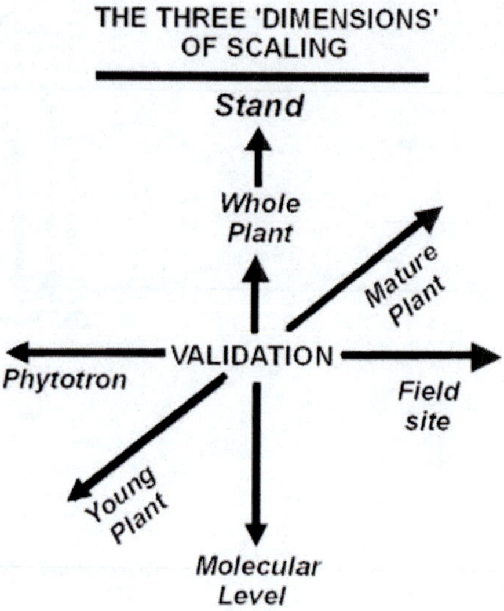

Fig. 1.3 Three spatio-temporal axes ("dimensions") of process scaling (adapted from Matyssek 2001)

decreasing strength of cause/effect relationships in approaching hierarchically high levels of integration (Baldocchi 1993). Mechanistic consistency may be lost when bridging neighbouring scales within the integration hierarchy (see Fig. 3 of the Prologue; Watts 1999). Such scaling characteristics interfere with growth/defence analyses. In addition, ontogeny affects process scaling as one additional driver in resource allocation (see above, Haugen et al. 2008). Also within-plant scaling has prevailed (linking factorial impact with phenotype and gene regulation), although plants in parallel do affect the ecosystem fluxes and site conditions, which in turn set the stage for the plants' performance (Hendriks et al. 2009). Matyssek (2001) stressed, therefore, the "three dimensions of scaling" in developing growth/defence theories on mechanistic grounds (Fig. 1.3), i.e. integrating (1) the gene-to-ecosystem axis and those of (2) ontogeny (in association with plant allometry, see Chap. 13) and (3) growth conditions, i.e. from controlled (phytotrons, glasshouses) and semi-controlled environments (e.g. field lysimeters; Schloter and Matyssek 2009) to actual field sites. Axes (2) and (3) are borne by the experience that juvenile plants physiologically and structurally differ from mature ones, and that plant behaviour displayed under controlled growth conditions does not necessarily conform to field performance. Evidence remains rudimentary, therefore, as long as physiologically and ecologically meaningful validation is missing.

1.3.3 *Theoretical Outset of This Book Project*

Amongst theories on growth and defence-related resource allocation, GDB has been considered as the one with the highest integrative and explanatory potential

(Stamp 2003b). The conceptual strength is based on the advanced framework character in comprehending available knowledge from the cell, tissue, organ and whole-plant levels, across ontogenetic stages, ecological scenarios including stand-level interactions, and beyond steady-state conditions (Glynn et al. 2007). On such grounds, advancement towards "theory maturity" appears to be promising, given the plants' repertoire in answering stress in plentiful and variable ways (Schwachtje and Baldwin 2008), i.e. by means of more than one defence strategy. Challenges will be coped with in this book that originate from dynamics and plasticity, making plant resource allocation and defence "moving targets" both for theoretical treatment and the ecology and evolution of host–parasite relationships (Glynn et al. 2007). In striving for mechanistic understanding, the ecological perspective, neglected in the past, on growth/defence-related allocation will be fostered in the following chapters in combination with and relative to the molecular analysis (cf. Schwachtje and Baldwin 2008).

1.4 Gain in Knowledge Presented in This Book: An Outlook

Recent gain in knowledge, both in empirical and theoretical terms, will be united in this book on one of the plant's and plant research's fundamental challenges, as pointed out above, i.e. the regulation of allocation across physiological demands. Allocation may culminate towards growth (mediating competitive resource sequestration) *versus* defence (enabling for resource retention) so that focus is directed to GDB as a reference, being one leading theory on plant allocation with integrative capacities for mechanistic foundation and extension (Stamp 2003b; Herms and Mattson 1992; Koricheva 2002). Part II will then comprehend advancements across diverse spatio-temporal dimensions of mechanisms in resource allocation, according to the book's sub-title. Relevancy will be examined for herbaceous and woody plants and stands in view of the introductory considerations of the Sects. 1.1–1.3 of this chapter.

Starting point of Part II will be the extent of progress in linking the molecular with the biochemical and physiological level, accounting for plant–pathogen, plant–herbivore and plant–mycorrhizosphere interactions in determining, individually and in combination, allocation control through gene regulation. Resource costs and benefits will be viewed in relation to symbiontic organisms and extended to respiratory costs as further determinants of whole-plant allocation. State-of-the-art methodology will be highlighted for resource tracing so that cost/benefit relationships can be integrated across spatio-temporal scales. Scaling will then be extended by plant ontogeny and growth scenario as driving factors of allocation. Growth/defence-related effects by plant competition will be shown to allow mechanistic understanding only, if resource fluxes of plant–pathogen and plant–mycorrhizosphere interactions are integrated into the analysis. Space-related efficiencies of resource investments into *versus* returns from plant structures will be introduced as determinants of competitiveness and a feature of growth. The extent will be examined to which space may become a "currency" in comparing costs of growth with such of defence when viewing the latter as a means for sustaining the plant's access to space and hence, external resources. Space relates to allometric plasticity in shaping the plants' three-dimensional structure and associated resource flux at the stand level. The extent will be assessed to which stand development mirrors resource allocation within and between plants under abiotic and biotic stress, including such by parasites. Embedding growth/defence allocation in Part II into the different kinds of interactions and scenarios between the molecular and hierarchically higher spatio-temporal scales will convey the issue about resource conflicts *sensu* GDB towards a mechanistically founded and ecologically relevant comprehension.

Part III will examine the capacity of the attained empirical knowledge for theory building on resource allocation in plants. Mechanistic and statistical modelling as complementary components of an integrative analysis will be assessed, regarding compensation for experimental shortcomings and usability for theory development, on the basis of scenario simulations upon parameter variation. Capacities and limitations of empirical and theoretical approaches will be elucidated in promoting clarification and predictability of plant system behaviour. The extent of common underlying mechanisms in resource allocation across plant types, ontogenetic stages and growth scenarios will be examined in conclusion (Part IV). Arguments will be evaluated about a "unifying theory" on plant resource allocation, reconciling the breadth of evidence and knowledge.

Acknowledgements We thank "Deutsche Forschungsgemeinschaft" (DFG) for the support of research covered by the topic of this review through SFB 607 "*Growth and Parasite Defence – Competition for Resources in Economic Plants from Agronomy and Forestry*". Comments on an early version of the manuscript by Dr. D. Herms are acknowledged. Stimulating discussions on the topic with Dr. U. Lüttge and inspiring suggestions by him on the manuscript are highly appreciated.

References

Arnold TM, Targett NM (2003) To grow and defend: lack of tradeoffs for brown algal phlorotannins. Oikos 100:406–408

Baldocchi DD (1993) Scaling water vapor and carbon dioxide exchange from leaves to a canopy: rules and tools. In: Ehleringer JR, Field CB (eds) Scaling physiological processes, leaf to globe. Academic, San Diego, pp 77–114

Baldwin IT, Hamilton W (2000) Jasmonate-induced responses of Nicotiana sylvestris results in fitness costs due to impaired competitive ability for nitrogen. J Chem Ecol 26:915–952

Ballhorn DJ, Kautz S, Lion U, Heil M (2008) Trade-offs between direct and indirect defences of lima bean (*Phaseolus lunatus*). J Ecol 96:971–980

Bazzaz FA (1997) Allocation of resources in plants: state of the science and critical questions. In: Bazzaz FA, Grace J (eds) Plant resource allocation. Academic, San Diego, pp 1–38

Bazzaz FA, Grace J (1997) Plant resource allocation. Academic, San Diego, p 303

Berenbaum MR (1995) The chemistry of defense: theory and practice. Proc Natl Acad Sci USA 92:2–8

Bi HH, Song YY, Zeng RS (2007) Biochemical and molecular responses of host plants to mycorrhizal infection and their role in plant defence. Allelopathy J 20:15–28
Bolton MD (2009) Primary metabolism and plant defense – fuel for the fire. Mol Plant Microbe Interact 22:487–497
Bonello P, Heller W, Sandermann H (1993) Ozone effects on root-disease susceptibility and defence responses in mycorrhizal and non-mycorrhizal seedlings of Scots pine (*Pinus sylvestris* L.). New Phytol 124:653–663
Bradshaw AD (2006) Unravelling phenotypic plasticity – why should we bother? New Phytol 170:644–648
Bryant J, Chapin F III, Klein D (1983) Carbon/nutrient balance of boreal plants in relation to vertebrate herbivory. Oikos 40:357–368
Bustamante RO, Chacón P, Niemeyer HM (2006) Patterns of chemical defences in plants: an analysis of the vascular flora of Chile. Chemoecology 16:145–151
Cipollini D, Heil M (2010) Costs and benefits of induced resistance to herbivores and pathogens in plants. CAB Rev Perspect Agric Vet Sci Nutr Nat Resour 5:1–25
Darwin C (1859) The origin of species by means of natural selection: or, the preservation of favoured races in the struggle for life. John Murray, London, p 502
Ericsson T, Rytter L, Vapaavuori E (1996) Physiology and allocation in trees. Biomass Bioenergy 11:115–127
Feeny P (1976) Plant apparency and chemical defense. In: Wallace JW, Mansell RL (eds) Biochemical interaction between plants and insects. Plenum, New York, pp 1–40
Gershenzon J (1994) The cost of plant chemical defense against herbivory: a biochemical perspective. In: Bernays EA (ed) Insect-plant Interactions. CRC, Boca Raton, pp 105–173
Givnish TJ (ed) (1986) On the economy of plant form and function. Cambridge University Press, Cambridge, p 717
Glinski M, Weckwerth W (2006) The role of mass spectrometry in plant systems biology. Mass Spectrom Rev 25:173–214
Glynn C, Herms DA, Orians CM, Hansen RC, Larsson S (2007) Testing the growth-differentiation balance hypothesis: dynamic responses of willows to nutrient availability. New Phytol 176:623–634
Graham RD (1983) Effects of nutrient stress on susceptibility of plants to disease with particular reference to the trace elements. Adv Bot Res 10:221–276
Grams TEE, Andersen CP (2007) Competition for resources in trees: physiological versus morphological plasticity. In: Esser K, Lüttge U, Beyschlag W, Murata J (eds) Progress in botany, vol 68. Springer, Heidelberg, pp 356–381
Häberle KH, Nunn A, Reiter I, Werner H, Heller W, Bahnweg G, Gayler S, Lütz C, Matyssek R (2009) Variation of defence-related metabolites in the foliage of adult beech and spruce: a conceptual approach to approximating traded-off carbon. Eur J Forest Res 128:99–108
Haeckel E (1870) Über Entwicklungsgang und Aufgabe der Zoologie (Rede vom 12. 1. 1869), Studien über Moneren und andere Protisten. W. Engelmann, Leipzig, pp 3–20
Hakes AS, Cronin JT (2011) Environmental heterogeneity and spatiotemporal variability in plant defense traits. Oikos 120:452–462
Hamilton JG, Zangerl AR, Delucia EH, Berenbaum MR (2001) The carbon-nutrient balance hypothesis: its rise and fall. Ecol Lett 4:86–95
Haugen R, Steffes L, Wolf J, Brown P, Matzner S, Siemens DH (2008) Evolution of drought tolerance and defense: dependence of tradeoffs on mechanism, environment and defense switching. Oikos 117:231–244
Heil M, Baldwin IT (2002) Fitness costs of induced resistance: emerging experimental support for a slippery concept. Trends Plant Sci 7:61–67
Hendriks RJJ, Luijten L, van Groenendael JM (2009) Context-dependent defence in terrestrial plants: the effects of light and nutrient availability on plant resistance against herbivory. Entomol Exp Appl 131:233–242
Herms DA, Mattson WJ (1992) The dilemma of plants: to grow or defend. Q Rev Biol 67:283–335
Jones CG, Hartley SE (1999) A protein competition model of phenolic allocation. Oikos 86:27–44
Koricheva J (2002) Meta-analysis of sources of variation in fitness costs of plant antiherbivore defenses. Ecology 83:176–190

Koricheva J, Larsson S, Haukioja E, Keinänen M (1998) Regulation of woody plant secondary metabolism by resource availability: hypothesis testing by means of meta-analysis. Oikos 83:212–226
Koricheva J, Nykänen H, Gianoli E (2004) Meta-analysis of trade-offs among plant antiherbivore defenses: are plants Jacks-of-all-trades, masters of all? Am Nat 163:E65–E75
Koricheva J, Gange AC, Jones T (2009) Effects of mycorrhizal fungi on insect herbivores: a meta-analysis. Ecology 90:2088–2097
Körner Ch (2003) Carbon limitation in trees. J Ecol 91:4–17
Kramer PJ, Boyer JS (1995) Water relations of plants and soils. Academic, San Diego
Küppers M (1994) Canopy gaps: competitive light interception and economic space filling – a matter of whole-plant allocation. In: Caldwell MM, Pearcy RW (eds) Exploitation of environmental heterogeneity by plants – ecophysiological processes above and below-ground. Academic, San Diego, pp 111–144
Lerdau M (2002) Benefits of the carbon-nutrient balance hypothesis. Oikos 98:534–536
Lerdau M, Gershenzon J (1997) Allocation theory and chemical defense. In: Bazzaz FA, Grace J (eds) Plant resource allocation. Academic, San Diego, pp 265–277
Luedemann G, Matyssek R, Fleischmann F, Grams TEE (2005) Acclimation to ozone affects host/pathogen interaction and competitiveness for nitrogen in juvenile *Fagus sylvatica* and *Picea abies* trees infected with *Phytophthora citricola*. Plant Biol 7:640–649
Loomis WE (1953) Growth and differentiation – an introduction and summary. In: Loomis WE (ed) Growth and differentiation in plants. Iowa State College Press, Ames, pp 1–17
Lorio PL Jr (1988) Growth and differentiation balance relationships in pines affect their resistance to bark beetles (Coleoptera: Scoloytidae). In: Mattson WJ, Levieux J, Bernard-Dagan C (eds) Mechanisms of woody plant defenses against insects: search for pattern. Springer, New York, pp 73–92
Mattson WJ, Julkunen-Tiitto R, Herms DA (2005) CO2 enrichment and carbon partitioning to phenolics: do plants responses accord better with the protein competition model or the growth-differentiation balance models? Oikos 111:337–347
Matyssek R (2001) Trends in forest tree physiological research: biotic and abiotic interactions. In: Huttunen S, Heikkilä H, Bucher J-B, Sundberg B, Jarvis PG, Matyssek R (eds) Trends in European forest tree physiological. Kluwer, Dordrecht, pp 241–246
Matyssek R, Schulze E-D (1987) Heterosis in hybrid larch (*Larix decidua* × *leptolepis*) II. Growth characteristics. Trees 1:225–231
Matyssek R, Schnyder H, Elstner E-F, Munch J-C, Pretzsch H, Sandermann H (2002) Growth and parasite defence in plants: the balance between resource sequestration and retention. Plant Biol 4:133–136
Matyssek R, Agerer R, Ernst D, Munch J-C, Osswald W, Pretzsch H, Priesack E, Schnyder H, Treutter D (2005) The plant's capacity in regulating resource demand. Plant Biol 7:560–580
Mooney HA, Winner WE (1991) Partitioning response of plants to stress. In: Pell E, Mooney HA, Winner WE (eds) Response of plants to multiple stresses. Academic, San Diego, pp 129–141
Murray DC, Walters DR (1992) Increased photosynthesis and resistance to rust infection in upper, uninfected leaves of rusted broad bean (*Vicia faba* L.). New Phytol 120:235–242
Nobel PS (1983) Biophysical plant physiology and ecology. W.H. Freeman, San Francisco, p 608
Novriyanti E, Aoyama C, Watanabe M, Koike T (2010) Plant defense characteristics and hypotheses in birch species. Eurasian J Forest Res 13:77–85
Orians CM, Ward D (2010) Evolution of plant defenses in non-indigenous environments. Annu Rev Entomol 55:439–459
Pankhurst CE, Jones WT (1979) Effectiveness of *Lotus* root nodules III. Effect of combined nitrogen on nodule effectiveness and flavolan synthesis in plant roots. J Exp Bot 30:1109–1118
Peng M, Bi Y-M, Zhu T, Rothstein SJ (2007) Genome-wide analysis of *Arabidopsis* responsive transcriptome to nitrogen limitation and its regulation by the ubiquitin ligase gene NLA. Plant Mol Biol 65:775–797
Popper KR (1989) Logik der Forschung. J.C.B. Mohr (Paul Siebeck), Tübingen, p 477

Potters G, Pasternak TP, Guisez Y, Jansen MAK (2009) Different stresses, similar morphogenic responses: integrating a plethora of pathways. Plant Cell Environ 32:158–169

Pozo MJ, Azcon-Aguilar C (2007) Unraveling mycorrhiza-induced resistance. Curr Opin Plant Biol 10:393–398

Purrington CB (2000) Cost of resistance. Curr Opin Plant Biol 3:305–308

Rasmann S, Bauerle TL, Poveda K, Vannette R (2010) Predicting root defence against herbivores during succession. Funct Ecol. doi:10.1111/j.1365-2435.2010.01811.x

Rhoades DF, Cates RG (1976) A general theory of plant antiherbivore chemistry. In: Wallace JW, Mansell RL (eds) Biochemical interaction between plants and insects, Recent advances in phytochemistry. Plenum, New York, pp 168–213

Riipi M, Ossipov V, Lempa K, Haukioja E, Koricheva J, Ossipova S, Pihlaja K (2002) Seasonal changes in birch leaf chemistry: are there trade-offs between leaf growth and accumulation of phenolics. Oecologia 130:380–390

Rühmann S, Leser C, Bannert M, Treutter D (2002) Relationship between growth, secondary metabolism, and resistance of apple. Plant Biol 4:137–143

Sampedro L, Moreira X, Zas R (2011) Costs of constitutive and herbivore-induced chemical defences in pine trees emerge only under low nutrient availability. J Ecol. doi:10.1111/j.1365-2745.2011.01814.x

Sandermann H, Matyssek R (2004) Scaling up from molecular to ecological processes. In: Sandermann H (ed) Molecular ecotoxicology of plants, vol 170, Ecological studies. Springer, Heidelberg, pp 207–226

Schloter M, Matyssek R (2009) Tuning growth versus defence – belowground interactions and plant resource allocation. Plant Soil 323:1–5

Schulze E-D (1982) Plant life forms and their carbon, water and nutrient relations. In: Lange OL, Nobel PS, Osmond CB, Ziegler H (eds) Physiological plant ecology II, vol 12B. Springer, Berlin, pp 615–676

Schulze E-D, Küppers M, Matyssek R (1986) The roles of carbon balance and branching pattern in the growth of woody species. In: Givnish TJ (ed) On the economy of plant form and function. Cambridge University Press, Cambridge, pp 585–602

Schwachtje J, Baldwin IT (2008) Why does herbivore attack reconfigure primary metabolism? Plant Physiol 146:845–851

Schwinning S (1996) Decomposition analysis of competitive symmetry and size structure dynamics. Ann Bot 77:47–57

Siemens DH, Lischke H, Maggiulli N, Schürch S, Roy BA (2003) Cost of resistance and tolerance under competition: the defence-stress benefit hypothesis. Evol Ecol 17:247–263

Siemens DH, Keck AG, Ziegenbein S (2010) Optimal defense in plants: assessment of resource allocation costs. Evol Ecol 24:1291–1305

Stamp N (2003a) Theory of plant defensive level: example of process and pitfalls in development of ecological theory. Oikos 102:672–678

Stamp N (2003b) Out of the quagmire of plant defenses hypotheses. Q Rev Biol 78:23–55

Stamp N (2004) Can the growth-differentiation balance hypothesis be tested rigorously? Oikos 107:439–448

Stitt M, Schulze E-D (1994) Plant growth, storage, and resource allocation: from flux control in a metabolic chain to the whole-plant level. In: Schulze E-D (ed) Flux control in biological systems. Academic, San Diego, pp 57–118

Strauss SY, Rudgers JA, Lau JA, Irwin RE (2002) Direct and ecological costs of resistance to herbivory. Trends Ecol Evol 17:278–285

van Dam NM, Heil M (2011) Multitrophic interactions below and above ground: en route to the next level. J Ecol 99:77–88

van Noordwijk AJ, de Jong G (1986) Acquisition and allocation of resources: their influence on variation in life history tactics. Am Nat 128:137–142

Vannette RL, Hunter MD (2011) Plant defence theory re-examined: nonlinear expectations based on the costs and benefits of resource mutualisms. J Ecol 99:66–76

Watts DC (1999) Small worlds. The dynamics of networks between order and randomness. Princeton University Press, Princeton

Williams GM, Ayres PG (1981) Effects of powdery mildew and water stress on CO_2 exchange in uninfected leaves of barley. Plant Physiol 68:527–530

Zangerl AB, Bazzaz FA (1992) Theory and pattern in plant defense allocation. In: Fritz RS, Simms EL (eds) Plant resistance to herbicides and pathogens. The University of Chicago Press, Chicago, pp 363–391

Zilber-Rosenberg I, Rosenberg E (2008) Role of microorganisms in the evolution of animals and plants: the hologenome theory of evolution. FEMS Microbiol Rev 32:723–735

Part II
Mechanisms of Resource Allocation in Plants and Stands

Introduction

Part II addresses, in view of the introductory considerations of Part I (Chap. 1), the several spatiotemporal dimensions of mechanisms in resource allocation of plants and stands. Chapter 2 starts up elucidating the molecular regulation of plant resource allocation and explores means for identifying mechanistic links with the biochemical/physiological process level. On such grounds, biotic interactions and their interrelationships are demonstrated as determining allocation control and performance. Chapters 3, 4, and 5 accordingly address plant–pathogen, plant–herbivore, and plant–mycorrhizosphere relationships, respectively. Cost/benefit relations will be introduced in Chap. 6 regarding respiration and carbon-use efficiency as allocation determinants at the whole-plant level. With carbon as a resource that affects allocation at various spatiotemporal scales from cells to ecosystems, Chap. 7 will review state-of-the-art tracing methodologies as a prerequisite for integrating cost/benefit considerations within plants and stands.

Reaching this stage, Part II will focus on the stand-level partitioning of resources driven by competition, exemplifying this crucial variant of biotic interaction for woody-plant systems. Quality and quantity of light as the energy resource, being both consequence and determinant of competition, will be addressed by Chap. 8, given the canopy architecture of a managed beech/spruce forest. The respective structural settings provide the stage for the three-dimensional plant and stand-level partitioning of nutritional elements, as demonstrated by Chap. 9, and Chap. 10 will resume cost/benefit considerations in view of nutrient acquisition and associated efficiencies in belowground space exploration and exploitation through symbiotic soil microorganisms and their interrelationships. A whole-plant perspective will be pursued in Chap. 11 with the about 60 years old and up to 30 m high forest trees, demonstrating the intrinsic link of carbon and water turnover with above and belowground efficiencies of space-related resource use as determinants of individual competitiveness under stand conditions. Responsiveness is examined under natural and experimental abiotic and biotic stress, bridging to Chap. 12 which submits the introduced efficiencies to an analysis of the mechanistic grounds of individual competitiveness. A basis is provided for examining, in Chap. 13, the allometric plasticity of plants in determining structural patterns and associated

R. Matyssek et al. (eds.), *Growth and Defence in Plants*, Ecological Studies 220,
DOI 10.1007/978-3-642-30645-7, © Springer-Verlag Berlin Heidelberg 2012

resource turnover at the stand level. Eventually, the extent will be investigated to which stand development mirrors links to resource allocation within and between neighboring plants (Chap. 14). This assessment at the uppermost spatiotemporal level of biological organization reached in this book will complete the overview on interrelated mechanisms that underlie resource allocation in plants and stands.

Chapter 2
Common Links of Molecular Biology with Biochemistry and Physiology in Plants Under Ozone and Pathogen Attack

D. Ernst, M. Jürgensen, G. Bahnweg, W. Heller, and G. Müller-Starck

2.1 Introduction

Ozone is a ubiquitous phytotoxic air pollutant that severely affects vegetation, including that of forest ecosystems. During the last decade, the effects of ozone and the mechanistic responses of plants at the level of gene expression have become clearer (Kangasjärvi et al. 1994; Langebartels et al. 1997, 2002; Dizengremel 2001 Mahalingam et al. 2003; Jaspers et al. 2005; Ludwikow and Sadowski 2008; Matyssek et al. 2008). Moreover, it has been shown that acute ozone exposure results in a gene expression pattern similar to pathogen attack, and the appearance of ozone-induced cell lesions resembles the hypersensitive response of plants, sharing many of its physiological and molecular features (Sandermann 1996; Pell et al. 1997; Rao et al. 2000; Jaspers et al. 2005; Heath 2008). Ozone exposure and biotic stress seem to have several commonalities. Therefore, ozone has been recognised as an abiotic elicitor of plant defence reactions (Sandermann et al. 1998), and a crosstalk between abiotic and biotic stress responses is evident (Eckey-Kaltenbach et al. 1994; Fujita et al. 2006). Furthermore, the transcriptional regulation of receptor-like protein kinases has also shown strong response similarities between ozone and pathogen stress (Wrzaczek et al. 2010).

Acute ozone exposure (200–300 nl l^{-1} for several hours) results in an up- or down-regulation of well-studied ozone-responsive genes (Matyssek et al. 2008). Microarray analyses have revealed that these transcripts could be categorised into

D. Ernst (✉) • M. Jürgensen • G. Bahnweg • W. Heller
Institute of Biochemical Plant Pathology, Helmholtz Zentrum München-, German Research Center for Environmental Health, Ingolstädter Landstr. 1, 85764 Neuherberg, Germany
e-mail: ernst@helmholtz-muenchen.de

G. Müller-Starck
Section of Forest Genetics, Technische Universität München, Hans-Carl-von-Carlowitz-Platz 2, 85354 Freising, Germany
e-mail: mueller-starck@forst.tu-muenchen.de

R. Matyssek et al. (eds.), *Growth and Defence in Plants*, Ecological Studies 220,
DOI 10.1007/978-3-642-30645-7_2, © Springer-Verlag Berlin Heidelberg 2012

functional groups belonging to different pathways of metabolism, such as primary metabolism, energy, cell growth and division, transcription, protein synthesis and destination, transport, cell structure, signal transduction, disease and defence, secondary metabolism, and non-categorised ones for *Arabidopsis thaliana* (Mahalingam et al. 2003; Tamaoki et al. 2003; Ludwikow et al. 2004; Tosti et al. 2006) and *Medicago truncatula* (Puckette et al. 2008, 2009). In addition, acute ozone exposure of ozone-sensitive *Arabidopsis* mutants showed that the response of plants towards ozone is under complex genetic regulation (Overmyer et al. 2008). Comparing acute and chronic (up to 100 nl l^{-1} for several weeks) ozone exposure of soybean plants it has been shown at the physiological level that changes in the photosynthetic capacity were very similar; however, a different pattern of ozone damage at the leaf level was evident (Chen et al. 2009). Experiments of chronic ozone exposure of up to twice the ambient level over several days, weeks or even years with herbaceous plants and forest trees have been performed under both controlled-chamber conditions (Langebartels et al. 1997; Miyazaki et al. 2004; D'Haese et al. 2006; Chen et al. 2009; Bohler et al. 2010) and open-air conditions, for example, the Aspen Free-Air Carbon Dioxide Enrichment (FACE) Experiment (http://aspenface.mtu.edu/), Kranzberg Ozone Fumigation Experiment (KROFEX; http://www.sfb607.de/) as part of the CASIROZ study (Matyssek et al. 2007; http://www.casiroz.de), the Kuopio open-field exposure system (Wulff et al. 1992; Karnosky et al. 2007), a lysimeter study (Winkler et al. 2009; http://www.helmholtz-muenchen.de/en/lysimeter/home/index.html) or the Soybean Free Air Concentration Enrichment (SoyFACE; http://soyface.illinois.edu/index.htm) experiment. In these free-air field experiments, some genes were similarly up- or down-regulated by chronic ozone fumigation, as compared to acute ozone exposure in the following analysed plants: *A. thaliana* (Miyazaki et al. 2004; Li et al. 2006), *Thellungiella halophila* (Li et al. 2006), trembling aspen (Gupta et al. 2005), European beech (Olbrich et al. 2009), and paper birch (Kontunen-Soppela et al. 2010b). However, differences in transcripts have also been observed that might reflect additional stress factors under field conditions (Miyazaki et al. 2004; Olbrich et al. 2010a).

At the transcriptional level, powerful techniques, such as microarray analysis, quantitative real-time RT-PCR, and, more recently, high-throughput sequencing methods, have been employed that can provide significant information about the transcribed genes that are involved in abiotic/biotic stress responses and the adaptation of plants to changing environments. However, the results have to be considered carefully, as transcript and protein levels are not necessarily and/or clearly correlated (Gygi et al. 1999). In addition, low-abundance mRNAs, often important in regulation processes, are not easy to detect and to quantify, in contrast to transcripts that are abundant. The analysis of proteins and of the proteome in response to different stresses is now possible through sophisticated techniques, such as 2D-DIGE, MALDI-MS/MS, LC-MS/MS or iTRAQ, which can provide information about the proteins encoded by the genes involved in these stress responses (Renaut et al. 2006; Jorrín-Novo et al. 2009). Plant metabolites are the end products of gene and protein expression and can reflect, at least in part, the final step of the plant's response to environmental stress (Tretheway 2001). Therefore, integrated "omic" analyses are important to understand plant responses towards abiotic/biotic stress and to identify the regulatory networks in plant stress responses (Urano et al. 2010). The integration of transcript, protein and metabolite data will reflect the response of the plant to environmental stress at a profound and comprehensive level (Oksman-Caldentey and Saito 2005; Bohnert et al. 2006; Trauger et al. 2008).

2.2 Transcription Analyses upon Abiotic and/or Biotic Stress in Non-woody and Woody Plant Species

2.2.1 Transcription Analysis in Non-woody Plant Species upon Ozone Treatment

Acute ozone-induced transcriptional changes were first observed in an ozone-sensitive tobacco cultivar by analysing pathogenesis-related (PR) protein transcripts (Ernst et al. 1992; Schraudner et al. 1992). Thereafter, many single transcriptional studies have been conducted, reflecting antioxidant enzymes, ethylene biosynthesis, phenylpropanoid metabolism and diverse metabolic pathways, which are known to be affected by biotic stress (Kangasjärvi et al. 1994; Langebartels et al. 2002; Heath 2008). New technologies soon emerged, such as cDNA microarray, which have allowed the expression analyses of thousands of genes; indeed, in the model organism, *A. thaliana*, the whole transcriptome has been analysed. It was found that hundreds of genes were up- or down-regulated and could be grouped into functional categories of whole-plant metabolism (Ludwikow et al. 2004; D'Haese et al. 2006; Li et al. 2006; Tosti et al. 2006). Gene expression profiles of ozone-exposed *A. thaliana* revealed that the three hormones, ethylene, jasmonic acid and salicylic acid, interact in the regulation of ozone-affected genes (Tamaoki et al. 2003). This interaction, as a result of an oxidative burst, has many characteristics in common with the pathogen-caused hypersensitive response, which results in an altered gene expression for genes such as PR proteins, following both ozone and biotic stresses (Overmyer et al. 2003; Jaspers et al. 2005). The hierarchical clustering of ozone stress-regulated genes in pepper plants that were treated either with ozone or *Xanthomonas axonopodis* sp. *glycines* showed an extensive overlap of commonly affected genes (Lee and Yun 2006). In soybean, it was found that mainly defence-related genes, hormone signalling-related genes and genes encoding transcription factors were commonly regulated by elevated ozone and mosaic virus infection, and the highest gene expression changes were observed when both stressors were applied at the same time (Bilgin et al. 2008).

It is known that WRKY (conserved peptide sequence) transcription factors are involved in various plant processes that cope with abiotic and biotic stress and that there is a network of WRKY transcription factors in defence signalling (Eulgem and Somssich 2007; Pandey and Somssich 2009). For binding the WRKY transcription factor, the cognate *cis*-acting W box is essential (Eulgem et al. 2000). The W box motif (TGAC) and a W box-like motif were significantly over-represented in an ozone-induced cDNA library of *A. thaliana* (Mahalingam et al. 2003). Similarly, the promoter of the class I β-1,3-glucanase gene, which contains a GCC box, is induced upon ozone and pathogen stress (Leubner-Metzger and Meins 1999; Grimmig et al. 2003; Ernst and Aarts 2004). Several transcription factors, that are up-regulated by ozone, including WRKY, MYB (myeloblastosis) and the zinc finger protein family, have been described in herbaceous plants (D'Haese et al. 2006; Li et al. 2006; Tosti et al. 2006; Cho et al. 2008), with an over-representation of WRKY transcription factors, indicating again the remarkable similarities in gene expression induced by ozone and pathogen stress.

2.2.2 Transcription Analysis in Woody Plant Species upon Ozone Treatment

Ozone-induced transcriptional changes were first described for conifers and European beech upon acute ozone exposure (Galliano et al. 1993; Schneiderbauer et al. 1995; Langebartels et al. 1997). European beech is the most abundant broadleaf tree species in Europe (Schütt et al. 1992). As it is also of major importance for the European forest industry, it was chosen as an experimental species for our studies to obtain deeper insights, at the molecular level, into the possibly detrimental effects of ozone on European broadleaf forest ecosystems. The transcriptome analyses of long-term ozone-treated beech saplings (Olbrich et al. 2005) resulted in a gene grouping that was similar to that described for herbaceous plants (Tamaoki et al. 2003; Ludwikow et al. 2004; D'Haese et al. 2006; Lee and Yun 2006). Functional classifications revealed genes related to primary metabolism, energy, cell growth/division, transcription, protein synthesis, protein destination and storage, transporters, intracellular traffic, cell structure, signal transduction, disease/defence, transposons, secondary metabolism, unclear categorisation and non-categorised genes (Fig. 2.1; Olbrich et al. 2005). Similarly, following long-term chronic exposure to elevated tropospheric ozone in trembling aspen and paper birch trees, categorised transcripts were detected (Gupta et al. 2005; Kontunen-Soppela et al. 2010b), and the clustering of the normalised data was carried out with self-organising maps (Kontunen-Soppela et al. 2010b). The transcript responses in the leaves of ozone-treated juvenile European beech trees at an outdoor free-air model fumigation site (twice the ambient ozone level) was monitored over two growing seasons, which revealed altered expression patterns of expressed sequence tags (ESTs) that are involved in signal transduction, cell structure, energy and disease/defence. In addition, many unknown ESTs were also affected (Fig. 2.2; Olbrich et al. 2009).

The specific conditions of microarray analyses require careful scrutiny, incorporating data pre-processing, data normalisation and statistical analysis to

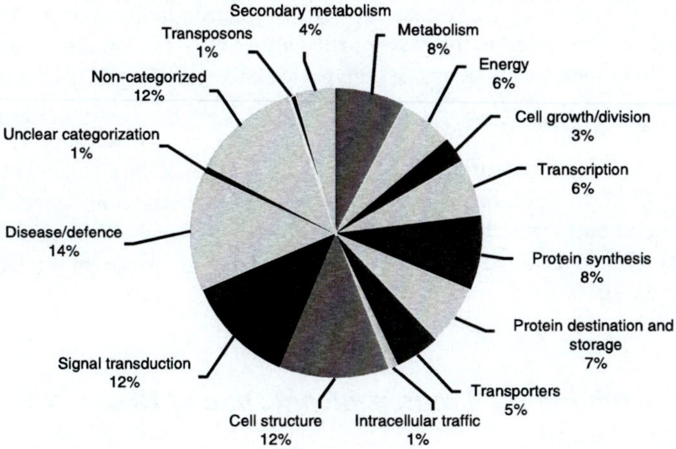

Fig. 2.1 The categorisation of ozone-affected genes in the leaves of European beech saplings (Olbrich et al. 2005)

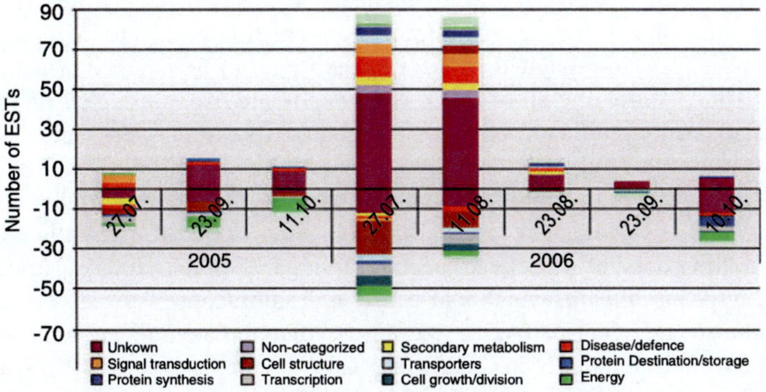

Fig. 2.2 The transcript responses in the leaves of ozone-treated juvenile European beech trees at an outdoor free-air model fumigation site over two growing seasons. The EST are organised into functional categories. The sampling time points are indicated (taken from Olbrich et al. 2009)

select differentially expressed genes, and Loess normalisation, principal component analysis and empirical Bayes methods are necessary in the analysis (Olbrich et al. 2009) (see Chap. 16 for the statistical methods, including support vector machines, for the classification of as yet unknown genes). Direct transcript correlations of woody and herbaceous plants have been performed for genes involved in ethylene biosynthesis (Betz et al. 2009a) and the shikimate pathway (Betz et al. 2009b). Janzik et al. (2005) have demonstrated that ozone had dramatic effects on the regulation of the shikimate pathway in tobacco. Furthermore, in European beech, an ozone-dependent reprogramming of the transcription of all shikimate pathway genes, as a prerequisite for secondary metabolism, has been demonstrated (Betz et al. 2009b).

Cis/trans elements are important for gene regulation, and an ozone-responsive W box has been described in the resveratrol synthase, *Vst1*, promoter of grape vine, which was also induced by *Botrytis cinerea* (Schubert et al. 1997; Grimmig et al. 2003; Ernst and Aarts 2004). Several ozone-induced WRKY transcription factors have been described in herbaceous plants and shrubs (Mahalingam et al. 2003; D'Haese et al. 2006; Heidenreich et al. 2006; Tosti et al. 2006; Paolacci et al. 2007; Cho et al. 2008). In trees, ozone-responsive WRKY transcription factors have also been described and were shown to be up-regulated in beech (Olbrich et al. 2005) and poplar (Rizzo et al. 2007) and down-regulated in paper birch (Kontunen-Soppela et al. 2010b).

2.2.3 Combining the Transcriptional Data of Beech, Apple and Potato

A controlled infection study of beech leaves with the fungal endophyte, *Apiognomonia errabunda*, was performed at the Kranzberger Forst research facility (Olbrich et al. 2010b). It was found that the levels of only a few transcripts were altered upon *A. errabunda* infection. Interestingly, a WRKY transcription factor that is also up-regulated by ozone was found to be induced (Olbrich et al. 2005, 2010b). Comparing all of the *A. errabunda*- and ozone-affected ESTs, it was determined that significant changes in gene expression were more strongly affected by ozone fumigation than by *A. errabunda* infestation (Fig. 2.3; Olbrich et al. 2010b). This is in contrast to a study involving infection with the oomycete, *Phytophthora plurivora*, that resulted in more than 1,000 differentially expressed genes in the roots of European beech saplings (Schlink 2009). However, in contrast to *A. errabunda*, which can cause leaf anthracnose, species of the genus *Phytophthora* are plant-damaging pathogens that threaten forest trees (Oßwald et al. 2004; Fleischmann et al. 2005) (see Chap. 3 for host–parasite interactions). The functional classification of the altered ESTs of ozone-treated leaves and of beech roots infected with *P. plurivora* showed quite a similar distribution of disease/defence (14–15 %) that included PR proteins and transcription (4–7 %) (Olbrich et al. 2005; Schlink 2009). The transcriptome profiling in hybrid poplar following fungal infection identified genes that were predominantly related to primary and secondary metabolism, cell-wall reinforcement and lignification, defence and stress, and signal transduction (Miranda et al. 2007; Azaiez et al. 2009). Moreover, the infection of European beech seedlings with *P. plurivora* resulted in a transient expression of 1-aminocyclopropane-1-carboxylic acid oxidase 1 (*ACO1*) in parallel with an ethylene burst (Portz et al. 2011), which was similar to that in juvenile European beech trees after ozone treatment (Betz et al. 2009a). This clearly points to an overlap of gene expression in European beech after ozone fumigation and pathogen treatment.

The establishment of suppression subtractive hybridisation (SSH) libraries from scab susceptible apple leaves strongly infected with *Venturia inaequalis* produced

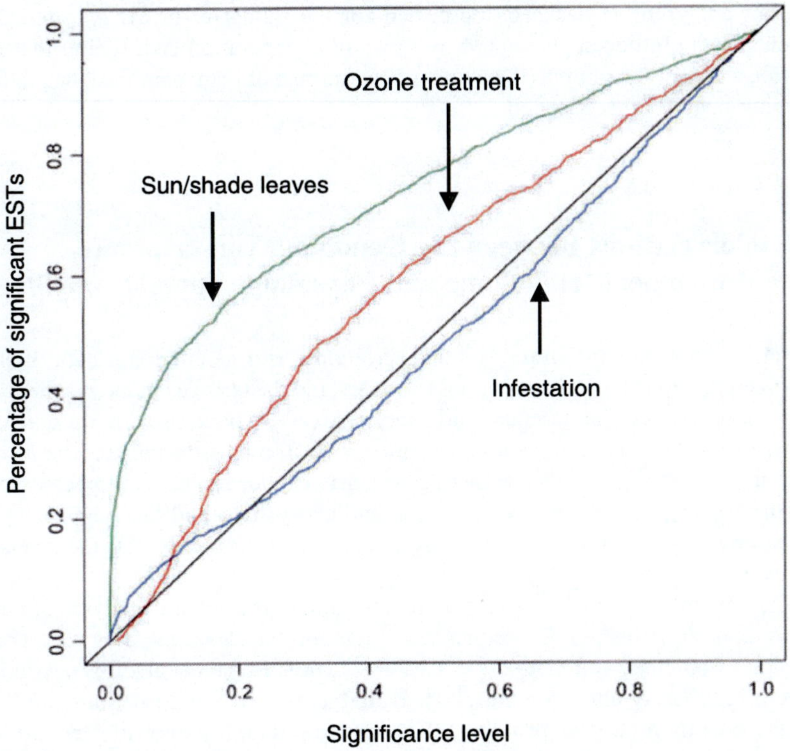

Fig. 2.3 A comparison of the effects of *A. errabunda* and ozone and leaf type on gene expression. The percentage of significant univariate test results (gene expression) with respect to the three factors is shown. A straight line would be expected when no effect of a factor on gene expression was found (taken from Olbrich et al. 2010b)

results that indicated that approximately 10 % of ESTs belonged to the category disease/defence (Zistler 2007), which is in good agreement with the data from ozone-treated European beech (Olbrich et al. 2005). In addition, genes coding for PR proteins and distinct heat shock proteins were also present and were in agreement with the results from European beech (Olbrich et al. 2005; Zistler 2007). Similarly, the sequence analyses of SSH libraries of susceptible and moderately resistant potato cultivars infected with *Phytophthora infestans* resulted in approximately 22 % and 9 %, respectively, of the ESTs belonging to the category of disease/defence (Ros et al. 2004). As was found for European beech and apple, genes encoding PR proteins were also identified as induced in potato (Olbrich et al. 2005, 2010b; Ros et al. 2004; Zistler 2007) (see Chap. 4 for defence genes induced by insects and Chaps. 5 and 10 for the positive interactions of microorganisms and mycorrhizal fungi). In conclusion, most genes belonging to the category disease/defence were up-regulated in the different plant species analysed upon ozone or pathogen treatment. However, it has to be kept in mind that the whole transcriptome was not analysed, as has been described for the model plant, *A. thaliana*. This deficiency might be supplemented by using next-generation sequencing that may provide a deeper insight into the whole transcriptome of plant genomes that, to date, have not been fully sequenced.

2.3 Correlations Between the Genome, Transcriptome, Proteome, Metabolome and Physiology upon Ozone Stress

Correlating the genome, transcriptome, proteome and metabolome data of plant responses to ozone stress is important to understand the changes in the biochemistry of the whole plant. Sandermann and Matyssek (2004) have provided a scaled-up model of the molecular and ecological processes in plants. In this excellent overview, the hierarchical levels and interactions between the molecular processes, cells and tissues, organs, whole-plant systems, and ecosystems and biochemical cycles are outlined, and the influence of environmental factors on each of these levels are discussed.

The integrated "omic" approaches are possible for the holistic analyses of model plants that have been fully sequenced, including *A. thaliana*, rice and poplar (Oksman-Caldentey and Saito 2005; Cho et al. 2008; Renaut et al. 2009). Although custom microarray chips are currently available, proteome/phosphoproteome and metabolome (non-targeted profiling) analyses are still highly sophisticated, difficult to conduct, and expensive. Only a few metabolic studies applied to ozone stress can be found in the literature; the species that have been investigated include white birch (Kontunen-Soppela et al. 2007; Ossipov et al. 2008) and rice (Cho et al. 2008). In white birch, ozone was found to cause increases in phenolic compounds and compounds related to cuticular wax layers, whereas compounds related to carbohydrate metabolism and chloroplast pigments were decreased; interestingly, the metabolite profiles showed differences between different birch genotypes (Kontunen-Soppela et al. 2007). Similarly, the differences within a single tree, between trees within an experimental field, or between different fields was shown to increase the phenotypic variation in silver birch metabolites, which may have led to difficulties in the interpretation of the results (Ossipov et al. 2008).

In contrast, more examples of proteomic data are available in the literature. Gel-based proteomics and immunoblotting techniques have identified bean and maize proteins that were also known to be induced by ozone at the transcriptional level (Torres et al. 2007). Similarly, MS/MS analyses have revealed known ozone-regulated proteins (Torres et al. 2007), and 2D-DIGE analyses have identified a large number of proteins involved in carbon metabolism, electron transport, detoxification, secondary metabolism, protein folding and disease/defence (Bohler et al. 2007, 2010). In soybean, the analyses of total soluble and chloroplast proteins have revealed that proteins involved in photosynthesis and carbon assimilation decreased, and proteins involved in carbon metabolism and antioxidant defence increased, after ozone stress (Ahsan et al. 2010). In rice and wheat, ozone was found to affect the levels of photosynthetic, antioxidative defence, disease/defence and stress-related proteins (Agrawal et al. 2002; Feng et al. 2008; Sarkar et al. 2010), all of which are factors of pathways that are known to be affected at the transcriptomic level. With regards to a holistic approach analysing transcriptomics, proteomics and metabolomics, only a single reference with respect to ozone stress exists in the literature thus far (Cho et al. 2008). It was found that genes categorised into cellular processing and signalling, information processing and storage and metabolism were mainly regulated; the identified proteins were found to be involved in cellular processing and signalling, photosynthesis and defence. On the basis of genes functionally categorised from transcriptome data, Cho et al. (2008) demonstrated an induction of the pentose phosphate pathway, ethylene and jasmonate biosynthesis, transcription of naringenin-related genes, and genes involved in glutamate and γ-aminobutyric acid biosynthesis. This systematic survey showed that ozone triggered a chain reaction of altered gene and protein expression and metabolite accumulation in rice (Cho et al. 2008).

An initial comparison of transcriptional and physiological data of tree growth following ozone treatment was obtained in the Aspen FACE experiment using trembling aspen (Karnosky et al. 2007). The treatment resulted in a pattern of gene expression that was found to fit well with the overall patterns of physiology and growth (Wustman et al. 2001; Karnosky et al. 2003; Gupta et al. 2005). Although protein and metabolite data were not included, these correlations were a breakthrough in the analysis of field-grown trees. Regarding the carbon metabolism of conifers, a review by Dizengremel (2001) showed reduced mRNA levels of the small and large subunits of ribulose 1,5-biphosphate carboxylase/oxygenase (*Rubisco*), a decrease in the protein levels of the large subunit, a decrease in the Rubisco relative activity and a decrease in the relative photosynthetic rate upon ozone treatment, indicating premature senescence. The long-term effects of elevated tropospheric ozone on silver birch have revealed an up-regulation of many senescence-associated genes, in correlation with an earlier abscission of leaves and a decreased chlorophyll content (Kontunen-Soppela et al. 2010a). However, in contrast to the decrease in the gene expression of photosynthesis-related genes and the amount and activity of Rubisco, net photosynthesis was not affected. Therefore, gene expression data do not necessarily reflect biochemical processes or plant physiology (Kontunen-Soppela et al. 2010a).

Stilbenes, a major constitutive of phenolic compounds in Scots pine, have been shown to accumulate in the needles upon abiotic and biotic stresses (Rosemann et al. 1991; Lieutier et al. 1996). In Scots pine seedlings, an ozone-induced increase of stilbene synthase (*STS*) transcripts, STS enzyme activity and stilbene content was demonstrated (Rosemann et al. 1991; Chiron et al. 2000a, b). Lignin, the second most abundant organic material in the biosphere after cellulose, is synthesised via cinnamylalcohol dehydrogenase (CAD). Ozone exposure of poplar leaves resulted in a rapid and strong increase of *CAD* mRNA levels and CAD enzyme activity, independent of the foliar development (Cabané et al. 2004). In addition, the lignin content was substantially increased under ozone exposure, which indicated a regulated correlation between transcripts, enzyme activities and stress-induced lignin content (Cabané et al. 2004).

At the Kranzberger Forst research facility, during the CASIROZ study with adult European beech, treatment with an ozone level that was twice the ambient level revealed an up-regulation of 9-*cis*-epoxycarotenoid dioxygenase, which encodes the key enzyme of abscisic acid (ABA) biosynthesis. This result was in accordance with the observed increase in the level of ABA (Jehnes et al. 2007; Matyssek et al. 2007). Consequently, decreased stomatal conductance and photosynthesis rates were observed in the leaves (Löw et al. 2007; Matyssek et al. 2007). However, an expected reduction of the annual stem growth on the basis of stem diameter measurements was not observed after 3 years of ozone exposure (Wipfler et al. 2005). In contrast, after 8 years of ozone fumigation at twice the ambient level, a drastic decrease in the volume growth was observed on the basis of tree-ring analysis and height-increment measurements (Pretzsch et al. 2010; Matyssek et al. 2010a, b; see Chap. 13 for discussions on allocation and allometry). Again, these results demonstrate that an extrapolation from genes to total-tree growth, even at an ecological level, needs a large number of data for a global perspective.

The shikimate pathway, based on primary metabolism and leading to the secondary pathway and aromatic amino acid metabolism, was shown to be up-regulated in European beech leaves in response to ozone (Fig. 2.4; Betz et al. 2009b, c; see also Chap. 3 for a discussion of the shikimate pathway). A coordinated regulation of all of the genes was evident, and an observed increase in the protein level of 3-deoxy-D-*arabino*-heptulosonate-7-phosphate synthase 3 and 3-dehydroquinate dehydratase/shikimate dehydrogenase was in agreement with the increased transcript levels for these enzymes (Fig. 2.4; Betz et al. 2009b). Furthermore, an increased stress-lignin content was found in severe damaged leaves (Betz et al. 2009c). Although additional genes and enzymes leading to the production of lignin were not analysed, the accumulation of lignin is in accordance with the observed responses in other studies, as described in poplar for enzyme activities of shikimate dehydrogenase, phenylalanine ammonia-lyase and CAD (Cabané et al. 2004). Similarly, phenolics that are derived from phenylalanine have been found to accumulate in many tree species upon ozone fumigation (Langebartels et al. 1997; Peltonen et al. 2005; Kontunen-Soppela et al. 2007; Betz et al. 2009b). Chorismate synthase, the final enzyme step of the shikimate pathway, produces chorismate, which is also one of the basic products of salicylic acid (SA) and gentisic acid (GA) formation. Although chorismate synthase transcripts were only weakly up-regulated in beech upon ozone treatment, its induction occurred in parallel with the appearance of leaf symptoms and with increased levels of conjugated SA and GA (Betz et al. 2009b). SA can be synthesised via isochorismate or, alternatively, via prephenate, phenylalanine, cinnamic acid and benzoic acid. Chorismate mutase, which converts chorismate into prephenate was also up-regulated at the transcriptional level upon ozone fumigation (Fig. 2.4; Betz et al. 2009b). Because both conjugated SA and GA are products of the shikimate pathway, their accumulation is consistent with the increased transcription of shikimate pathway genes (Fig. 2.4; Betz et al. 2009b).

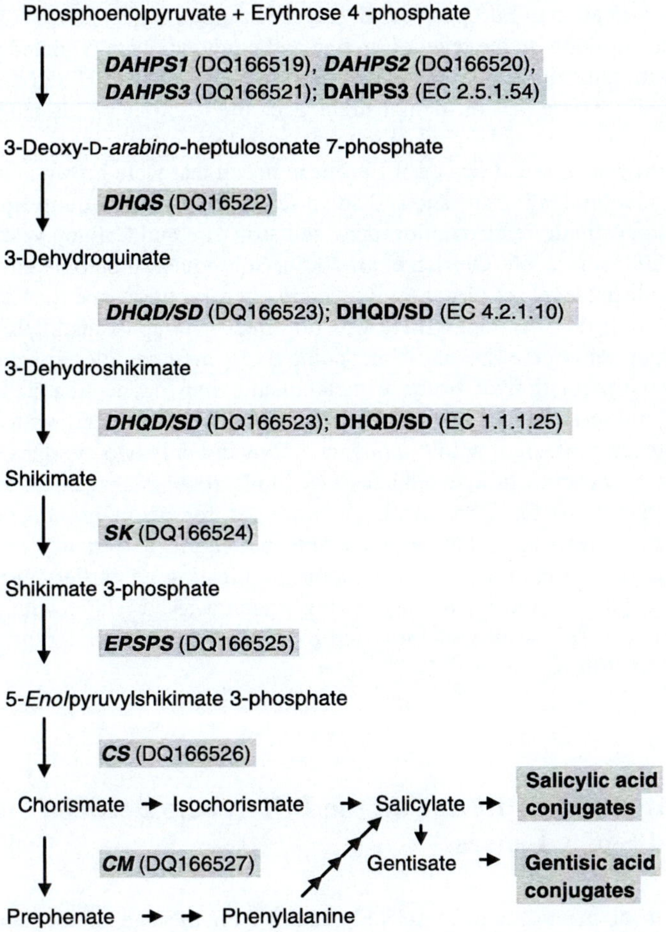

Fig. 2.4 The steps of the shikimate pathway up-regulated by ozone and the related accumulation of conjugated salicylic and gentisic acid in European beech (the related transcripts, proteins and metabolites are highlighted); the enzymes depicted are 3-deoxy-D-*arabino*-heptulosonate 7-phosphate synthase (DAHPS), 3-dehydroquinate synthase (DHQS), 3-dehydroquinate dehydratase/shikimate dehydrogenase (DHQD/SD), shikimate kinase (SK), 5-*enol*pyruvylshikimate 3-phosphate synthase (EPSPS), chorismate synthase (CS), chorismate mutase (CM) (Betz et al. 2009b)

Ozone-induced ethylene biosynthesis has been shown to be the result of 1-aminocyclopropane-1-carboxylic acid (ACC) synthase (ACS) and ACO activity (Kangasjärvi et al. 1997; Moeder et al. 2002). ACC synthesis is catalysed via ACS, and ACO, in turn oxidises, ACC to ethylene. In European beech, it was demonstrated that the emission of ethylene, the levels of its precursor, ACC, and the transcript levels of *ACS2* and *ACO1* showed a persistent increase and preceded cell death upon ozone fumigation (Nunn et al. 2005a). Similarly, it was shown that cell lesion formation in European beech leaves was preceded by persistent increases in ethylene emission, in the level of its free and conjugated malonylated precursor, ACC, and in transcript levels, specifically for *ACS1*, *ACS2*, *ACO1* and *ACO2* (Betz et al. 2009a). These results demonstrate a chain reaction of altered gene and metabolite expression and a change in leaf physiology/morphology.

Microarray analyses of leaves of European beech that were grown at an outdoor free-air model fumigation site showed down-regulated levels for transcripts belonging to photosynthetic-related, chloroplast cell structure and Calvin cycle genes for the years 2005 and 2006 (Olbrich et al. 2009). Subsequent proteome analyses also revealed reduced levels of proteins involved in photosynthesis and the Calvin cycle (Kerner et al. 2011). Although there was no direct overlap of distinctly regulated transcripts/proteins at a specific time point, both methods clearly indicated an overall down-regulation of primary metabolism upon ozone treatment. Similar results of transcriptomic and proteomic analyses were obtained with European beech roots after infection with *P. plurivora*, showing only two overlaps: phosphoglucomutase and cytochrome reductase (Schlink 2009; Valcu et al. 2009) (see Chaps. 3, 5 and 10 for more detail on processes for below-ground organs). In addition, the down-regulation of transcripts belonging to primary metabolism, photosynthesis and chloroplast cell structure, indicating an earlier senescence in ozone-treated trees at the end of the growing season, was correlated with phenological data on leaf senescence, as indicated by an earlier discoloration of the leaves (Pritsch et al. 2008).

2.4 Ontogenetic Effects and the Differences Between Sun and Shade Leaves

The effects of ozone on trees strongly depend on the age of the tree and the differences between shade and sun leaves. Microarray analyses of leaves of 60-year-old European beech trees at the Kranzberger Forst research facility in 2005 and 2006 showed only few ozone-induced transcriptional changes in both the sun and shade leaves of mature trees (Olbrich et al. 2010a) and only weak up-regulation of transcripts for shikimate pathway genes (Betz et al. 2009c). However, no identical transcriptional changes were detected in the 2 years of the study. The few transcriptional changes that were observed were in contrast to the behaviour of juvenile trees, which showed clear transcriptional responses to an ozone level that was twice the ambient level (Olbrich et al. 2009). Comparing the lysimeter experiment (juvenile trees) with the Kranzberger Forst study (sun leaves of adult trees) on a month-to-month basis, it was found that only two genes were similarly regulated, although at different time points. However, as only a few genes were expressed upon ozone fumigation at the Kranzberger Forst research facility, it was difficult to make a comparison (see Chap. 11 for a discussion of further ontogenetic effects). Although the expression profiles of the juvenile trees were not identical between

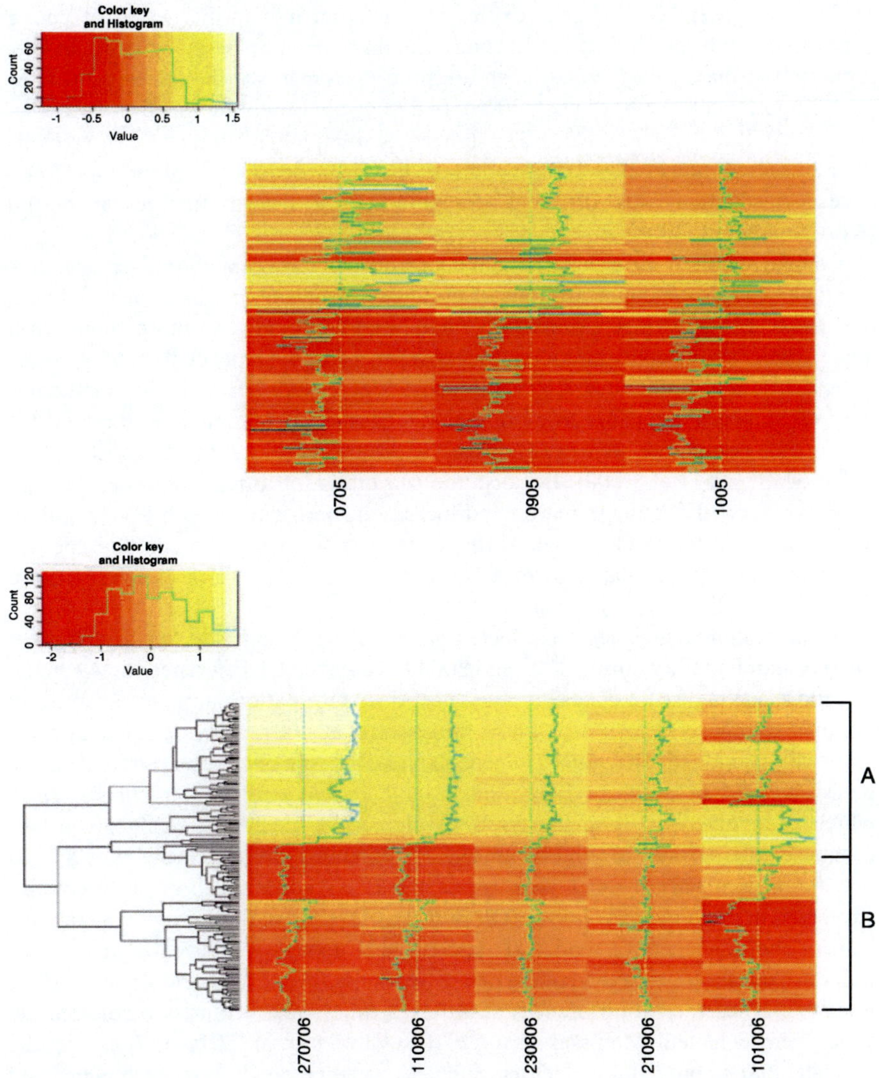

Fig. 2.5 A heat map of identical genes in the years 2005 and 2006. Hierarchical clustering is shown only for data from 2006. The values are $\log_2$ ratios; *white*, *yellow* = up-regulated; *red* = down-regulated (taken from Olbrich et al. 2009)

2005 and 2006, a common trend of gene expression was evident (Fig. 2.5). The contrasting transcript levels of identical genes in both years may have been due to the prominent differences in weather conditions (Olbrich et al. 2009). It is known that juvenile trees have lower concentrations of ascorbate and glutathione, as well as photosynthetic pigments (Wieser et al. 2003; Nunn et al. 2005b; Herbinger et al. 2007). The increased level of antioxidants in the leaves of mature beech trees might explain the lower level of gene expression, as compared to the leaves of young beech trees (Olbrich et al. 2010a). Therefore, an ozone flux-based assessment (i.e. representing the dose of ozone taken up through stomata) at the leaf level was of great importance (Matyssek et al. 2008). In 5-year-old trembling aspen cultivated at Aspen FACE, 238 out of 4,600 ESTs showed qualitatively similar expression levels over 2 years (Gupta et al. 2005). In contrast, the response of wood properties after a 5-year fumigation was different compared to a 3-year fumigation period (Kostiainen et al. 2008).

Comparing all of the transcriptional data from several years of field experiments using chronic ozone concentrations of up to twice the ambient level, it was found that the levels of changed transcripts were weak, compared to those found with acute, higher ozone concentrations (Miyazaki et al. 2004; Gupta et al. 2005; Li et al. 2006; Olbrich et al. 2009, 2010a; Kontunen-Soppela et al. 2010a, b). This indicates the importance of exposure to more realistic ozone concentrations in the field for ozone risk-assessment studies (Matyssek and Sandermann 2003; Karnosky et al. 2007; Matyssek et al. 2008). Because the sun crown of a tree is exposed to high photosynthetically active radiation and increased short-wave length UV radiation, as compared to the shade crown, differences in gene expression can be expected for light- and UV-regulated pathways (Zinser et al. 2007; Götz et al. 2010; see Chap. 8 for a discussion on solar radiation).

Transcriptional analyses for selected genes in beech leaves at the Kranzberger Forst research facility during 2003 and 2004 have indicated differences between the sun and shade leaves for genes involved in ethylene formation, stomatal closure and lignin biosynthesis (Jehnes et al. 2007; Matyssek et al. 2007; Matyssek et al. 2010b). However, significant differences in gene expression were only observed for *ACS2*, a gene involved in ethylene biosynthesis, and *caffeic acid O-methyltransferase,* which is involved in lignin biosynthesis (Jehnes et al. 2007). Using an in-lab-produced microarray (i.e. non-commercial) containing ozone-responsive ESTs, it was found that only 0.1–0.2 % of all of the ESTs appeared to be differentially expressed in sun and shade leaves (Olbrich et al. 2010a, b). Furthermore, a comparison of the ozone effects and leaf type on gene expression revealed a stronger influence of the leaf type than of the ozone treatment (Fig. 2.3; Olbrich et al. 2010b); however, no consistent differences at different time points within two consecutive years were detected (Matyssek et al. 2010b; Olbrich et al. 2010a). These results indicate that unpredictable factors, such as weather conditions, pathogens and nutrient supply, may strongly influence gene expression.

2.5 Concluding Remarks

Ozone acts as an abiotic elicitor in both herbaceous and woody plant species (Sandermann et al. 1998; Matyssek et al. 2005). This has been clearly demonstrated at the level of gene regulation and results in a complex interaction of *cis/trans* elements. W box motifs and WRKY transcription factors are known as crucial regulators of the changes in the transcriptome as a defence response in plants after pathogen attack, and they also act in the response of plants to ozone treatment. Therefore, it is not surprising that after both types of treatment, transcripts have shown a similar up-regulation for distinct biochemical pathways. Combining the transcriptional data of European beech, apple and potato has revealed that most of the genes belonging to disease/defence are up-regulated in these plant species upon ozone or pathogen stress.

Thus far, data of the responses of the entire transcriptome are very rare, and studies may be complemented by the utilisation of next-generation sequencing. A common response of the transcriptome, proteome and metabolome upon ozone treatment was shown as an up-regulation in the shikimate pathway and ethylene biosynthesis. Further insight into this complex regulation is expected by holistic, large-scale "omic" approaches and the application of bioinformatic tools. In addition, phosphoproteomics will also be a valuable tool to unravel plant regulatory mechanisms upon stress (van Bentem et al. 2006). For the first time, DNA-array analyses have shown the distinct differences between sun and shade leaves, as well as between the leaves of juvenile and mature trees grown under free-air ozone exposure. In addition, transcriptional differences between different years were observed, indicating that less predictable factors, given by variations in the environment, may influence gene expression. Moreover, the combined effects of different abiotic stresses under field conditions should be a focus of future research, as combined stress effects may show positive or negative interactions (Mittler 2006; Eastburn et al. 2011).

In recent years, it has been demonstrated that long, non-protein-coding RNAs and small RNAs are major regulators of gene expression (Jacquier 2009; Charon et al. 2010; Chitwood and Timmermans 2010). Small RNAs are important in biotic stress responses, mostly as mediators of repressive gene regulation through RNA silencing (Ruiz-Ferrer and Voinnet 2009). Stress-responsive microRNAs for cold, heat, dehydration and UV-B radiation have been described in *Populus* (Lu et al. 2008; Jia et al. 2009). Novel mechanical stress-responsive microRNAs in poplar, which are absent from *Arabidopsis*, have also been described (Lu et al. 2005). These results have provided new insights into the regulatory networks that modulate the interaction between the plant genome and environmental factors. Ozone effects may, therefore, also be further regulated by such microRNAs, and this will be a challenge for further research. Similarly, gene expression driven by stress factors often depends on histone post-translational modifications and DNA methylation, which accounts for epigenetic gene regulation (Chinnusamy and Zhu 2009; Williams 2011). An epigenetic variation in trees occurring in contrasting environments or affecting climatic adaptation has also been described (Johnson et al. 2009; Lira-Medeiros et al. 2010). Interestingly, memory and carryover effects in the action of ozone on conifers have been portrayed with delayed physiological effects appearing months after the initiation of the treatment, even at a near ambient level of ozone fumigation (Sasek et al. 1991; Langebartels et al. 1998; Oksanen 2003). However, the degree to which extent the memory effects and corresponding imprinting are inherited or acquired remains unknown, even though it has been described in various plants (Ries et al. 2000; Molinier et al. 2006; Lang-Mladek et al. 2010). The resolution of this question will be a major challenge for ongoing research projects.

Acknowledgements We are grateful to our co-workers for their important contribution during the past years. We thank Ana Staninska and Gerhard Welzl for their help with statistical methods. Special thanks go to Pierre Dizengremel (Nancy-Université) and Elina Oksanen (University of Eastern Finland) for critically reading the manuscript and helpful comments. This work was supported by the Deutsche Forschungsgemeinschaft (SFB 607) and, in part, by the European Community (Evoltree, 6th Framework Programme; COST Action E52).

References

Agrawal GK, Rakwal R, Yonekura M, Kubo A, Saji H (2002) Proteome analysis of differentially displayed proteins as a tool for investigating ozone stress in rice (*Oryza sativa* L.) seedlings. Proteomics 2:947–959

Ahsan N, Nanjo Y, Sawada H, Kohno Y, Komatsu S (2010) Ozone stress-induced proteomic changes in leaf total soluble and chloroplast proteins of soybean reveal that carbon allocation is involved in adaptation in the early developmental stage. Proteomics 10:2605–2619

Azaiez A, Boyle B, Levée V, Séguin A (2009) Transcriptome profiling in hybrid poplar following interactions with *Melampsora* rust fungi. Mol Plant Microbe Interact 22:190–200

Betz GA, Gerstner E, Olbrich M, Winkler JB, Langebartels C, Heller W, Sandermann H, Ernst D (2009a) Effects of abiotic stress on gene transcription in European beech: ozone affects ethylene biosynthesis in saplings of *Fagus sylvatica* L. iForest 2:114–118

Betz GA, Gerstner E, Stich S, Winkler B, Welzl G, Kremmer E, Langebartels C, Heller W, Sandermann H, Ernst D (2009b) Ozone affects shikimate pathway genes and secondary metabolites in saplings of European beech (*Fagus sylvatica* L.) grown under greenhouse conditions. Trees 23:539–553

Betz GA, Knappe C, Lapierre C, Olbrich M, Welzl G, Langebartels C, Heller W, Sandermann H, Ernst D (2009c) Ozone affects shikimate pathway transcripts and monomeric lignin composition in European beech (*Fagus sylvatica* L.). Eur J Forest Res 128:109–116

Bilgin DD, Aldea M, O'Neill BF, Benitez M, Li M, Clough SJ, DeLucia EH (2008) Elevated ozone alters soybean-virus interaction. Mol Plant Microbe Interact 21:1297–1308

Bohler S, Bagard M, Oufir M, Planchon S, Hoffmann L, Jolivet Y, Hausman J-F, Dizengremel P, Renaut J (2007) A DIGE analysis of developing poplar leaves subjected to ozone reveals major changes in carbon metabolism. Proteomics 7:1584–1599

Bohler S, Sergeant K, Lefèvre I, Jolivet Y, Hoffmann L, Renaut J, Dizengremel P, Hausman J-F (2010) Differential impact of chronic ozone exposure on expanding and fully expanded poplar leaves. Tree Physiol 30:1415–1432

Bohnert HJ, Gong Q, Li P, Ma S (2006) Unraveling abiotic stress tolerance mechanisms – getting genomics going. Curr Opin Plant Biol 9:180–188

Cabané M, Pireaux J-C, Léger E, Weber E, Dizengremel P, Pollet B, Lapierre C (2004) Condensed lignins are synthesized in poplar leaves exposed to ozone. Plant Physiol 134:586–594

Charon C, Moreno AB, Bardou F, Crespi M (2010) Non-protein-coding RNAs and their interacting RNA-binding proteins in the plant cell nucleus. Mol Plant 3:729–739

Chen CP, Frank TD, Long SP (2009) Is a short, sharp shock equivalent to long-term punishment? Contrasting the spatial pattern of acute and chronic ozone damage to soybean leaves via chlorophyll fluorescence imaging. Plant Cell Environ 32:327–335

Chinnusamy V, Zhu J-K (2009) Epigenetic regulation of stress responses in plants. Curr Opin Plant Biol 12:133–139

Chiron H, Drouet A, Claudot A-C, Eckerskorn C, Trost M, Heller W, Ernst D, Sandermann H (2000a) Molecular cloning and functional expression of a stress-induced multifunctional *O*-methyltransferase with pinosylvin methyltransferase activity from Scots pine (*Pinus sylvestris* L.). Plant Mol Biol 44:733–745

Chiron H, Drouet A, Lieutier F, Payer H-D, Ernst D, Sandermann H (2000b) Gene induction of stilbene biosynthesis in Scots pine in response to ozone treatment, wounding and fungal infection. Plant Physiol 124:865–872

Chitwood DH, Timmermans MCP (2010) Small RNAs are on the move. Nature 467:415–419

Cho K, Shibato J, Agrawal GK, Jung Y-H, Kubo A, Jwa N-S, Tamogami S, Satoh K, Kikuchi S, Higashi T, Kimura S, Saji H, Tanaka Y, Iwahashi H, Masuo Y, Rakwal R (2008) Integrated transcriptomics, proteomics, and metabolomics analyses to survey ozone responses in the leaves of rice seedling. J Proteome Res 7:2980–2998

D'Haese D, Horemans N, De Coen W, Guisez Y (2006) Identification of late O_3-responsive genes in *Arabidopsis thaliana* by cDNA microarray analysis. Physiol Plant 128:70–79

Dizengremel P (2001) Effects of ozone on the carbon metabolism of forest trees. Plant Physiol Biochem 39:729–742

Eastburn DM, McElrone AJ, Bilgin DD (2011) Influence of atmospheric and climate change on plant-pathogen interactions. Plant Pathol 60:54–69

Eckey-Kaltenbach H, Ernst D, Heller W, Sandermann H (1994) Biochemical plant responses to ozone. IV. Cross-induction of defensive pathways in parsley (*Petroselinum crispum* L.) plants. Plant Physiol 104:67–74

Ernst D, Aarts M (2004) *cis* Elements and transcription factors regulating gene promoters in response to environmental stress. In: Sandermann H (ed) Molecular ecotoxicology of plants, vol 170, Ecological studies. Springer, Berlin, pp 151–176

Ernst D, Schraudner M, Langebartels C, Sandermann H (1992) Ozone-induced changes of mRNA levels of *ß*-1,3-glucanase, chitinase and 'pathogenesis-related' protein 1b in tobacco plants. Plant Mol Biol 20:673–682

Eulgem T, Somssich IE (2007) Networks of WRKY transcription factors in defense signaling. Curr Opin Plant Biol 10:366–371

Eulgem T, Rushton PJ, Robatzek S, Somssich IE (2000) The WRKY superfamily of plant transcription factors. Trends Plant Sci 5:199–206

Feng YW, Komatsu S, Furukawa T, Koshiba T, Kohno Y (2008) Proteome analysis of proteins responsive to ambient and elevated ozone in rice seedlings. Agric Ecosyst Environ 125:255–265

Fleischmann F, Koehl J, Portz R, Beltrame AB, Oßwald W (2005) Physiological changes of *Fagus sylvatica* seedlings infected with *Phytophthora citricola* and the contribution of its elictin "citricolin" to pathogenesis. Plant Biol 7:650–658

Fujita M, Fujita Y, Noutoshi Y, Takahashi F, Narusaka Y, Yamaguchi-Shinozaki K, Shinozaki K (2006) Crosstalk between abiotic and biotic stress responses: a current view from the points of convergence in the stress signaling networks. Curr Opin Plant Biol 9:436–442

Galliano H, Cabané M, Eckerskorn C, Lottspeich F, Sandermann H, Ernst D (1993) Molecular cloning, sequence analysis and elicitor-/ozone-induced accumulation of cinnamyl alcohol dehydrogenase from Norway spruce (*Picea abies* L.). Plant Mol Biol 23:145–156

Götz M, Albert A, Stich S, Heller W, Scherb H, Krins A, Langebartels C, Seidlitz HK, Ernst D (2010) PAR modulation of the UV-dependent levels of flavonoid metabolites in *Arabidopsis thaliana* (L.) Heynh. leaf rosettes: cumulative effects after a whole vegetative growth period. Protoplasma 243:95–103

Grimmig B, Gonzalez-Perez MN, Leubner-Metzger G, Vögeli-Lange R, Meins F, Hain R, Penuelas J, Heidenreich B, Langebartels C, Ernst D, Sandermann H (2003) Ozone-induced gene expression occurs via ethylene-dependent and -independent signalling. Plant Mol Biol 51:599–607

Gupta P, Duplessis S, White H, Karnosky DF, Martin F, Podila GK (2005) Gene expression patterns of trembling aspen trees following long-term exposure to interacting elevated CO_2 and tropospheric O_3. New Phytol 167:129–142

Gygi SP, Rochon Y, Franza BR, Aebersold R (1999) Correlation between protein and mRNA abundance in yeast. Mol Cell Biol 19:1720–1730

Heath RL (2008) Modification of the biochemical pathways of plants induced by ozone: what are the varied routes to change? Environ Pollut 155:453–463

Heidenreich B, Bieber E, Sandermann H, Ernst D (2006) Identification of a new member of the WRKY family in tobacco. Involved in ozone-induced gene regulation? Acta Physiol Plant 28:117–125

Herbinger K, Then C, Haberer K, Alexou M, Löw M, Remele K, Rennenberg H, Matyssek R, Grill D, Wieser G, Tausz M (2007) Gas exchange and antioxidative compounds in young beech trees under free-air ozone exposure and comparisons to adult trees. Plant Biol 9:288–297

Jacquier A (2009) The complex eukaryotic transcriptome: unexpected pervasive transcription and novel small RNAs. Nat Rev Genet 10:833–844

Janzik I, Preiskowski S, Kneifel H (2005) Ozone has dramatic effects on the regulation of the prechorismate pathway in tobacco (*Nicotiana tabacum* L. cv. Bel W3). Planta 223:20–27

Jaspers P, Kollist H, Langebartels C, Kangasjärvi J (2005) Plant responses to ozone. In: Smirnoff N (ed) Antioxidative and reactive oxygen species in plants. Blackwell, Oxford, pp 268–292

Jehnes S, Betz G, Bahnweg G, Haberer K, Sandermann H, Rennenberg H (2007) Tree internal signalling and defence reactions under ozone exposure in sun and shade leaves of European beech (*Fagus sylvatica* L.) trees. Plant Biol 9:253–264

Jia X, Ren L, Chen Q-J, Li R, Tang G (2009) UV-B-responsive microRNAs in *Populus tremula*. J Plant Physiol 166:2046–2057

Johnson Ø, Kvaalen H, Yakovlev I, Dæhlen OG, Fossdal CG, Skrøppa T (2009) An epigenetic memory from time of embryo development affects climatic adaptation in Norway spruce. In: Gusta L, Wisniewski M, Tanino K (eds) Plant cold hardiness: from the laboratory to the field. CAB, International, Wallingford, pp 99–107

Jorrín-Novo JV, Maldonado AM, Echevarría-Zomeño S, Valledor L, Castillejo MA, Curto M, Valero J, Sghaier B, Donoso G, Redondo I (2009) Plant proteomics update (2007-2008): second-generation proteomic techniques, an appropriate experimental design, and data analysis to fulfill MIAPE standards, increase plant proteome coverage and expand biological knowledge. J Proteomics 72:285–314

Kangasjärvi J, Talvinen J, Utriainen M, Karjalainen R (1994) Plant defence systems induced by ozone. Plant Cell Environ 17:783–794

Kangasjärvi J, Tuomainen J, Betz C, Ernst D, Langebartels C, Sandermann H (1997) Ethylene synthesis in tomato plants exposed to ozone. The role of ethylene in ozone damage, NATO Science Partnership Sub-Series 3, vol 34. In: Kanellis AK, Chang C, Kende H, Grierson D (eds) Biology and biotechnology of the plant hormone ethylene. Kluwer, Dordrecht, pp 259–265

Karnosky DF, Zak DR, Pregitzer KS, Awmack CS, Bockheim JG, Dickson RE, Hendrey GR, Host GE, King JS, Kopper BJ, Kruger EL, Kubiske ME, Lindroth RL, Mattson WJ, McDonald EP, Noormets A, Oksanen E, Parsons WFJ, Percy KE, Podila GK, Riemenschneider DE, Sharma P, Thakur R, Sôber A, Sôber J, Jones WS, Anttonen S, Vapaavuori E, Mankovska B, Heilman W, Isebrands JG (2003) Tropospheric O_3 moderates responses of temperate hardwood forests to elevated CO_2: a synthesis of molecular to ecosystem results from the Aspen FACE project. Funct Ecol 17:289–304

Karnosky DF, Werner H, Holopainen T, Percy K, Oksanen T, Oksanen E, Heerdt C, Fabian P, Nagy J, Heilman W, Cox R, Nelson N, Matyssek R (2007) Free-air exposure systems to scale up ozone research to mature trees. Plant Biol 9:181–190

Kerner R, Winkler JB, Dupuy JW, Jürgensen M, Lindermayr C, Ernst D, Müller-Starck G (2011) Changes in the proteome of juvenile European beech following three years exposure to free-air elevated ozone. iForest 4:69–76

Kontunen-Soppela S, Ossipov V, Ossipova S, Oksanen E (2007) Shift in birch leaf metabolome and carbon allocation during long-term open-field ozone exposure. Glob Change Biol 13:1053–1067

Kontunen-Soppela S, Riikonen J, Ruhanen H, Brosché M, Somervuo P, Peltonen P, Kangasjärvi J, Auvinen P, Paulin L, Keinänen M, Oksanen E, Vapaavuori E (2010a) Differential gene expression in senescing leaves of two silver birch genotypes in response to elevated CO_2 and tropospheric ozone. Plant Cell Environ 33:1016–1028

Kontunen-Soppela S, Parviainen J, Ruhanen H, Brosché M, Keinänen M, Thakur RC, Kolehmainen M, Kangasjärvi J, Oksanen E, Karnosky DF, Vapaavuori E (2010b) Gene expression responses of paper birch (*Betula papyrifera*) to elevated CO_2 and O_3 during leaf maturation and senescence. Environ Pollut 158:959–968

Kostiainen K, Kaakinen S, Warsta E, Kubiske ME, Nelson ND, Sober J, Karnosky DF, Saranpää P, Vapaavuori E (2008) Wood properties of trembling aspen and paper birch after 5 years of exposure to elevated concentrations of CO_2 and O_3. Tree Physiol 28:805–813

Langebartels C, Ernst D, Heller W, Lütz C, Payer H-D, Sandermann H (1997) Ozone responses of trees: results from controlled chamber exposures at the GSF phytotron. In: Sandermann H, Wellburn AR, Heath Rl (eds) Forest decline and ozone. A comparison of controlled chamber and field experiments, vol 127, Ecological studies. Springer, Berlin, pp 163–200

Langebartels C, Heller W, Führer G, Lippert M, Simons S, Sandermann H (1998) Memory effects in the action of ozone on conifers. Ecotoxicol Environ Saf 41:62–72

Langebartels C, Schraudner M, Heller W, Ernst D, Sandermann H (2002) Oxidative stress and defense reactions in plants exposed to air pollutants and UV-B radiation. In: Inzé D, Van Montagu M (eds) Oxidative stress in plants. Taylor & Francis, London, pp 105–135

Lang-Mladek C, Popova O, Kiok K, Berlinger M, Rakic B, Aufsatz W, Jonak C, Hauser M-T, Luschnig C (2010) Transgenerational inheritance and resetting of stress-induced loss of epigenetic gene silencing in *Arabidopsis*. Mol Plant 3:594–602

Lee S, Yun S-C (2006) The ozone stress transcriptome of pepper (*Capsicum annuum* L.). Mol Cells 21:197–205

Leubner-Metzger G, Meins F (1999) Functions and regulation of plant β-1,3-glucanases (PR-2). In: Datta SK, Muthukrishnan S (eds) Pathogenesis-related proteins in plants. CRC Press, Boca Ratun, pp 49–76

Li P, Mane SP, Sioson AA, Vasquez Robinet C, Heath LS, Bohnert HJ, Grene R (2006) Effects of chronic ozone exposure on gene expression in *Arabidopsis thaliana* ecotypes and in *Thellungiella halophila*. Plant Cell Environ 29:854–868

Lieutier F, Sauvard D, Brignolas F, Picron V, Yart A, Bastien C, Jay-Allemand C (1996) Changes in phenolic metabolites of Scots-pine phloem induced by *Ophiostoma brunneo-ciliatum*, a bark-beetle-associated fungus. Eur J Forest Pathol 26:145–158

Lira-Medeiros CF, Parisod C, Fernandes RA, Mata CS, Cardoso MA, Ferreira PCG (2010) Epigenetic variation in mangrove plants occurring in contrasting natural environment. PLoS One 5:e10326

Löw M, Häberle K-H, Warren CR, Matyssek R (2007) O_3 flux-related responsiveness of photosynthesis, respiration, and stomatal conductance of adult *Fagus sylvatica* to experimentally enhanced free-air O_3 exposure. Plant Biol 9:197–206

Lu S, Sun Y-H, Shi R, Clark C, Li L, Chiang VL (2005) Novel and mechanical stress-responsive microRNAs in *Populus trichocarpa* that are absent from Arabidopsis. Plant Cell 17:2186–2203

Lu S, Sun Y-H, Chiang VL (2008) Stress-responsive microRNAs in *Populus*. Plant J 55:131–151

Ludwikow A, Sadowski J (2008) Gene networks in plant ozone stress response and tolerance. J Integr Plant Biol 50:1256–1267

Ludwikow A, Gallois P, Sadowski J (2004) Ozone-induced oxidative stress response in *Arabidopsis*: transcription profiling by microarray approach. Cell Mol Biol Lett 9:829–842

Mahalingam R, Gomez-Buitrago A, Eckardt N, Shah N, Guevara-Garcia A, Day P, Raina R, Fedoroff NV (2003) Characterizing the stress/defense transcriptome of *Arabidopsis*. Genome Biol 4:R20

Matyssek R, Sandermann H (2003) Impact of ozone on trees: an ecophysiological perspective. In: Esser K, Lüttge U, Beyschlag W, Hellwig F (eds) Progress in botany, vol 64. Springer, Berlin, pp 349–404

Matyssek R, Agerer R, Ernst D, Munch J-C, Oßwald W, Pretzsch H, Priesack E, Schnyder H, Treutter D (2005) The plant's capacity in regulating resource demand. Plant Biol 7:560–580

Matyssek R, Bahnweg G, Ceulemans R, Fabian P, Grill D, Hanke DE, Kraigher H, Oßwald W, Rennenberg H, Sandermann H, Tausz M, Wieser G (2007) Synopsis of the CASIROZ case study: *ca*rbon *si*nk strength of *Fagus sylvatica* L. in a changing environment – experimental *r*isk assessment of mitigation by chronic *oz*one impact. Plant Biol 9:163–180

Matyssek R, Sandermann H, Wieser G, Booker F, Cieslik S, Musselman R, Ernst D (2008) The challenge of making ozone risk assessment for forest trees more mechanistic. Environ Pollut 156:567–582

Matyssek R, Karnosky DF, Wieser G, Percy K, Oksanen E, Grams TEE, Kubiske M, Hanke D, Pretzsch H (2010a) Advances in understanding ozone impact on forest trees: messages from novel phytotron and free-air fumigation studies. Environ Pollut 158:1990–2006

Matyssek R, Wieser G, Ceulemans R, Rennenberg H, Pretzsch H, Haberer K, Löw M, Nunn AJ, Werner H, Wipfler P, Oßwald W, Nikolova P, Hanke DE, Kraigher H, Tausz M, Bahnweg G, Kitao M, Dieler J, Sandermann H, Herbinger K, Grebenc T, Blumenröther M, Deckmyn G, Grams TEE, Heerdt C, Leuchner M, Fabian P, Häberle K-H (2010b) Enhanced ozone strongly reduces carbon sink strength of adult beech (*Fagus sylvatica*) – resume from the free-air fumigation study at Kranzberg Forest. Environ Pollut 158:2527–2532

Miranda M, Ralph SG, Mellway R, White R, Heath MC, Bohlmann J, Constabel CP (2007) The transcriptional response of hybrid poplar (*Populus trichocarpa* x *P. deltoides*) to infection by *Melampsora medusae* leaf rust involves induction of flavonoid pathway genes leading to the accumulation of proanthocyanidins. Mol Plant Microbe Interact 20:816–831

Mittler R (2006) Abiotic stress, the field environment and stress combination. Trends Plant Sci 11:15–19

Miyazaki S, Fredricksen M, Hollis KC, Poroyko V, Shepley D, Galbraith DW, Long SP, Bohnert HJ (2004) Transcript expression profiles of *Arabidopsis thaliana* grown under controlled conditions and open-air elevated concentrations of CO_2 and of O_3. Field Crops Res 90:47–59

Moeder W, Barry CS, Tauriainen AA, Betz C, Tuomainen J, Utriainen M, Grierson D, Sandermann H, Langebartels C, Kangasjärvi J (2002) Ethylene synthesis regulated by biphasic induction of 1-aminocyclopropane-1-carboxlic acid synthase and 1-aminocyclopropane-1-carboxylic acid oxidase genes is required for hydrogen peroxide accumulation and cell death in ozone-exposed tomato. Plant Physiol 130:1918–1926

Molinier J, Ries G, Zipfel C, Hohn B (2006) Transgeneration memory of stress in plants. Nature 442:1046–1049

Nunn AJ, Anegg S, Betz G, Simons S, Kalisch G, Seidlitz HK, Grams TEE, Häberle K-H, Matyssek R, Bahnweg G, Sandermann H, Langebartels C (2005a) Role of ethylene in the regulation of cell death and leaf loss in ozone-exposed European beech. Plant Cell Environ 28:886–897

Nunn AJ, Kozovits AR, Reiter IM, Heerdt C, Leuchner M, Lütz C, Liu X, Löw M, Winkler JB, Grams TEE, Häberle K-H, Werner H, Fabian P, Rennenberg H, Matyssek R (2005b) Comparison of ozone uptake and sensitivity between a phytotron study with young beech and a field experiment with adult beech (*Fagus sylvatica*). Environ Pollut 137:494–506

Oksanen E (2003) Response of selected birch *(Betula pendula* Roth) clones to ozone change over time. Plant Cell Environ 26:875–886

Oksman-Caldentey K-M, Saito K (2005) Integrating genomics and metabolomics for engineering plant metabolic pathways. Curr Opin Biotechnol 16:174–179

Olbrich M, Betz G, Gerstner E, Langebartels C, Sandermann H, Ernst D (2005) Transcriptome analysis of ozone-responsive genes in leaves of European beech (*Fagus sylvatica* L.). Plant Biol 7:670–676

Olbrich M, Gerstner E, Welzl G, Winkler JB, Ernst D (2009) Transcript responses in leaves of ozone-treated beech saplings at an outdoor free air model fumigation site over two growing seasons. Plant Soil 323:61–74

Olbrich M, Gerstner E, Bahnweg G, Häberle K-H, Matyssek R, Welzl G, Heller W, Ernst D (2010a) Transcriptional signatures in leaves of adult European beech trees (*Fagus sylvatica* L.) in an experimentally enhanced free air ozone setting. Environ Pollut 158:977–982

Olbrich M, Knappe C, Wenig M, Gerstner E, Häberle K-H, Kitao M, Matyssek R, Stich S, Leuchner M, Werner H, Schlink K, Müller-Starck G, Welzl G, Scherb H, Ernst D, Heller W, Bahnweg G (2010b) Ozone fumigation (twice ambient) reduces leaf infestation following natural and artificial inoculation by the endophytic fungus *Apiognomonia errabunda* of adult European beech trees. Environ Pollut 158:1043–1050

Ossipov V, Ossipova S, Bykov V, Oksanen E, Koricheva J, Haukioja E (2008) Application of metabolomics to genotype and phenotype discrimination of birch trees grown in a long-term open-field experiment. Metabolomics 4:39–51

Oßwald W, Koehl J, Heiser I, Nechwatal J, Fleischmann F (2004) New insights in the genus *Phytophthora* and current diseases these pathogens cause in their ecosystem. In: Esser K, Lüttge U, Beyschlag W, Murata J (eds) Progress in botany, vol 65. Springer, Berlin, pp 436–466

Overmyer K, Brosché M, Kangasjärvi J (2003) Reactive oxygen species and hormonal control of cell death. Trends Plant Sci 8:335–342

Overmyer K, Kollist H, Tuominen H, Betz C, Langebartels C, Wingsle G, Kangasjärvi S, Brader G, Mullineaux P, Kangasjärvi J (2008) Complex phenotypic profiles leading to ozone sensitivity in *Arabidopsis thaliana* mutants. Plant Cell Environ 31:1237–1249

Pandey SP, Somssich IE (2009) The role of WRKY transcription factors in plant immunity. Plant Physiol 150:1648–1655

Paolacci AR, Miraldi C, Tanzarella OA, Badiani M, Porceddu E, Nali C, Lorenzini G, Ciaffi M (2007) Gene expression induced by chronic ozone in the Mediterranean shrub *Phillyrea latifolia*: analysis by cDNA-AFLP. Tree Physiol 27:1541–1550

Pell EJ, Schlagnhaufer CD, Arteca RN (1997) Ozone-induced oxidative stress: mechanisms of action and reaction. Physiol Plant 100:264–273

Peltonen PA, Vapaavuori E, Julkunen-Tiitto R (2005) Accumulation of phenolic compounds in birch leaves is changed by elevated carbon dioxide and ozone. Glob Change Biol 11:1305–1324

Portz RL, Fleischmann F, Koehl J, Fromm J, Ernst D, Pascholati SF, Osswald WF (2011) Histological, physiological and molecular investigations of *Fagus sylvatica* seedlings infected with *Phytophthora citricola*. Forest Pathol 41:202–211

Pretzsch H, Dieler J, Matyssek R, Wipfler P (2010) Tree and stand growth of mature Norway spruce and European beech under long-term ozone fumigation. Environ Pollut 158:1061–1070

Pritsch K, Ernst D, Fleischmann F, Gayler S, Grams TEE, Göttlein A, Heller W, Koch N, Lang H, Matyssek R, Munch JC, Olbrich M, Scherb H, Stich S, Winkler JB, Schloter M (2008) Plant and soil system responses to ozone after 3 years in a lysimeter study with juvenile beech (*Fagus sylvatica* L.). Water Air Soil Pollut Focus 8:139–154

Puckette MC, Tang Y, Mahalingam R (2008) Transcriptomic changes induced by acute ozone in resistant and sensitive *Medicago truncculata* accessions. BMC Plant Biol 8:46

Puckette M, Peal L, Steele J, Tang Y, Mahalingam R (2009) Ozone responsive genes in *Medicago truncculata*: analysis by suppression subtraction hybridization. J Plant Physiol 166:1284–1295

Rao MV, Koch JR, Davis KR (2000) Ozone: a tool for probing programmed cell death in plants. Plant Mol Biol 44:345–358

Renaut J, Hausman J-F, Wisniewski ME (2006) Proteomics and low-temperature studies: bridging the gap between gene expression and metabolism. Physiol Plant 126:97–109

Renaut J, Bohler S, Hausman J-F, Hoffmann L, Sergeant K, Ahsan N, Jolivet Y, Dizengremel P (2009) The impact of atmospheric composition on plants: a case study of ozone and poplar. Mass Spectrom Rev 28:495–516

Ries G, Heller W, Puchta H, Sandermann H, Seidlitz HK, Hohn B (2000) Elevated UV-B radiation reduces genome stability in plants. Nature 406:98–101

Rizzo M, Bernardi R, Salvini M, Nali C, Lorenzini G, Durante M (2007) Identification of differentially expressed genes induced by ozone stress in sensitive and tolerant poplar hybrids. J Plant Physiol 164:945–949

Ros B, Thümmler F, Wenzel G (2004) Analysis of differentially expressed genes in a susceptible and moderately resistant potato cultivar upon *Phytophthora infestans* infection. Mol Plant Pathol 5:191–201

Rosemann D, Heller W, Sandermann H (1991) Biochemical plant responses to ozone. II. Induction of stilbene biosynthesis in Scots pine (*Pinus sylvestris* L.) seedlings. Plant Physiol 97:1280–1286

Ruiz-Ferrer V, Voinnet O (2009) Roles of plant small RNAs in biotic stress responses. Annu Rev Plant Biol 60:485–510

Sandermann H (1996) Ozone and plant health. Annu Rev Phytopathol 34:347–366

Sandermann H, Matyssek R (2004) Scaling up from molecular to ecological processes. In: Sandermann H (ed) Molecular ecotoxicology of plants, vol 170, Ecological studies. Springer, Berlin, pp 207–226

Sandermann H, Ernst D, Heller W, Langebartels C (1998) Ozone: an abiotic elicitor of plant defence reactions. Trends Plant Sci 3:47–50

Sarkar A, Rakwal R, Agrawal SB, Shibato J, Ogawa Y, Yoshida Y, Agrawal GK, Agrawal M (2010) Investigating the impact of elevated levels of ozone on tropical wheat using integrated phenotypical, physiological, biochemical, and proteomics approaches. J Proteome Res 9:4565–4584

Sasek TW, Richardson CJ, Fendick EA, Bevington SR, Kress LW (1991) Carryover effects of acidic rain and ozone on the physiology of multiple flushes of loblolly pine seedlings. Forest Sci 37:1078–1098

Schlink K (2009) Identification and characterization of differentially expressed genes from *Fagus sylvatica* roots after infection with *Phytophthora citricola*. Plant Cell Rep 28:873–882

Schneiderbauer A, Back E, Sandermann H, Ernst D (1995) Ozone induction of extensin mRNA in Scots pine, Norway spruce and European beech. New Phytol 130:225–230

Schraudner M, Ernst D, Langebartels C, Sandermann H (1992) Biochemical plant responses to ozone. III. Activation of the defense-related proteins β-1,3-glucanase and chitinase in tobacco leaves. Plant Physiol 99:1321–1328

Schubert R, Fischer R, Hain R, Schreier PH, Bahnweg G, Ernst D, Sandermann H (1997) An ozone-responsive region of the grapevine resveratrol synthase promoter differs from the basal pathogen-responsive sequence. Plant Mol Biol 34:417–426

Schütt P, Schuck HJ, Stimm B (1992) Lexikon der Forstbotanik. Ecomed Verlagsgesellschaft, Landsberg

Tamaoki M, Nakajima N, Kubo A, Aono M, Matsuyama T, Saji H (2003) Transcriptome analysis of O_3-exposed Arabidopsis reveals that multiple signal pathways act mutually antagonistically to induce gene expression. Plant Mol Biol 53:443–456

Torres NL, Cho K, Shibato J, Hirano M, Kubo A, Masuo Y, Iwahashi H, Jwa N-S, Agrawal GK, Rakwal R (2007) Gel-based proteomics reveals potential novel protein markers of ozone stress in leaves of cultivated bean and maize species of Panama. Electrophoresis 28:4369–4381

Tosti N, Pasqualini S, Borgogni A, Ederli L, Falistocco E, Crispi S, Paolocci F (2006) Gene expression profiles of O_3-treated *Arabidopsis* plants. Plant Cell Environ 29:1686–1702

Trauger SA, Kalisak E, Kalisiak J, Morita H, Weinberg MV, Menon AL, Poole FL II, Adams MWW, Siuzdak G (2008) Correlating the transcriptome, proteome, and metabolome in the environmental adaptation of a hyperthermophile. J Proteome Res 7:1027–1035

Tretheway RN (2001) Gene discovery via metabolic profiling. Curr Opin Biotechnol 12:135–138

Urano K, Kurihara Y, Seki M, Shinozaki K (2010) 'Omics' analyses of regulatory networks in plant abiotic stress responses. Curr Opin Plant Biol 13:132–138

Valcu C-M, Junqueira M, Shevchenko A, Schlink K (2009) Comparative proteomic analysis of responses to pathogen infection and wounding in *Fagus sylvatica*. J Proteome Res 8:4077–4091 van Bentem SD, Roitinger E, Anrather D, Csaszar E, Hirt H (2006) Phosphoproteomics as a tool to unravel plant regulatory mechanisms. Physiol Plant 126:110–119

Wieser G, Hecke K, Tausz M, Häberle K-H, Grams TEE, Matyssek R (2003) The influence of microclimate and tree age on the defense capacity of European beech (*Fagus sylvatica* L.) against oxidative stress. Ann For Sci 60:131–135

Williams ME (2011) Introduction to epigenetic gene regulation. In: Plant Cell. Teaching tools in plant biology. http:www.plantcell.org/site/teachingtools/TTPB4LectureNotes.pdf Cited 31 March 2011

Winkler JB, Lang H, Graf W, Reth S, Munch JC (2009) Experimental setup of field lysimeters for studying effects of elevated ozone and below-ground pathogen infection on a plant-soil-system of juvenile beech (*Fagus sylvatica* L.). Plant Soil 323:7–19

Wipfler P, Seifert T, Heerdt C, Werner H, Pretzsch H (2005) Growth of adult Norway spruce (*Picea abies* [L.] Karst.) and European beech (*Fagus sylvatica* L.) under free-air ozone fumigation. Plant Biol 7:611–618

Wrzaczek M, Brosché M, Salojärvi J, Kangasjärvi S, Idänheimo N, Mersmann S, Robatzek S, Karpiński S, Karpińska B, Kangasjärvi J (2010) Transcriptional regulation of the CRK/DUF26 group of receptor-like protein kinases by ozone and plant hormones in Arabidopsis. BMC Plant Biol 10:95

Wulff A, Hänninen O, Tuomainen A, Kärenlampi L (1992) A method for open-air exposure of plants to ozone. Ann Bot Fennici 29:253–262

Wustman BA, Oksanen E, Karnosky DF, Noormets A, Isebrands JG, Pregitzer KS, Hendrey GR, Sober J, Podila GK (2001) Effects of elevated CO_2 and O_3 on aspen clones varying in O_3 sensitivity: can CO_2 ameliorate the harmful effects of O_3? Environ Pollut 115:473–481

Zinser C, Seidlitz HK, Welzl G, Sandermann H, Heller W, Ernst D, Rau W (2007) Transcriptional profiling of summer wheat, grown under different realistic UV-B irradiation regimes. J Plant Physiol 164:913–922

Zistler C (2007) Studien zur Genexpression in Apfelblättern (*Malus domestica*) nach Inokulation mit *Venturia inaequalis*. PhD thesis, Technische Universität München bitte. http://nbn-resolving.de/urn/resolver.pl?urn:de:bvb:91-diss-20070905-627157-1-4

Chapter 3
Host–Parasite Interactions and Trade-offs Between Growth- and Defence-Related Metabolism Under Changing Environments

W. Oßwald, F. Fleischmann, and D. Treutter

3.1 Introduction

The growth differentiation balance (GDB) theory by Herms and Mattson (1992) attempts to predict the allocation of resources between differentiation-related and growth-related processes under different environmental conditions (Chap. 1). Among the differentiation-related processes, all those metabolic events that increase the resistance of plants against herbivores, parasites, and pathogens are of particular importance. Several structural features such as trichomes or a thick cuticle as well as the chemical armada of secondary metabolites account for defence. It is generally assumed that the main function of secondary metabolites is their role in defence. This, however, may not be true for individual plant–pathogen interactions where the pathogen has learnt to cope with the plant's defence equipment. In this case, a given trade-off between growth and secondary metabolism will not influence the susceptibility of the host against that particular pathogen. Varying environments influence the secondary metabolism in different ways. According to the prediction of the GDB enhancement of concentrations of secondary metabolites will generally improve the defensive power of plants. Resource availability is regarded as the modulator of the trade-off between the mainly growth-related primary metabolism and its defence-related counterpart, the secondary metabolism. However, agricultural crops have been selected for high yield which stands synonymous for growth and reproduction. At the same time, many of these crop plants have also been selected for resistance against environmental stress and/or pathogens. Thus, both growth-related processes as well as

W. Oßwald (✉) • F. Fleischmann
Pathology of Woody Plants, Technische Universität München, Hans-Carl-von-Carlowitz-Platz 2, 85354 Freising, Germany
e-mail: osswald@wzw.tum.de

D. Treutter
Fruit Science, Technische Universität München, Dürnast 2, 85354 Freising, Germany

R. Matyssek et al. (eds.), *Growth and Defence in Plants*, Ecological Studies 220,
DOI 10.1007/978-3-642-30645-7_3, © Springer-Verlag Berlin Heidelberg 2012

defence-related biosyntheses are both active at a high level and at the same time. Particular regulation mechanisms must be available to manage the trade-offs. Furthermore, the secondary metabolism is characterized by a wide diversity. Only a few metabolites show effective defensive activity against highly specialized and adapted pathogens or invaders.

In this chapter, we discuss trade-offs between growth and secondary metabolism and the impact of a changing environment. Our reflection is opened by presenting defence mechanisms on a general base as well as on the level of defence-related metabolites. A special focus is set on phenolic compounds and their role in plant defence against various adverse environments. Before a discussion on trade-offs between growth-related processes and defence-related metabolism is started, the respective metabolic pathways are defined and primary and secondary metabolites are separated and possible switches between the respective biosynthetic pathways are shown. A synoptic section (Sect. 3.3) aims in describing trade-offs between growth and defence with special regards to the competing primary and secondary metabolism.

3.2 General Aspects of Plant Defence

Interactions of host plants with different pathogens such as virus, bacteria, oomycetes or fungi result in resistance or susceptibility. In general, most plants show resistance towards all genetic variants of parasites, what is called non-host or horizontal resistance, and such plants are called non-host plants (Nürnberger and Lipka 2005). In consequence resistance is the rule and susceptibility the exception. Non-host resistance is the most robust and durable plant resistance known in nature. This kind of resistance is expressed by preformed barriers (constitutive defence) and by the induction of specifically induced defence mechanisms (Fig. 3.1). In this case pathogen-associated molecular patterns (PAMPs) are recognized by pattern recognition receptors (PRRs) of the host (Dodds and Rathjen 2010; Hückelhoven 2007; Boller and Felix 2009; Pieterse et al. 2009). PAMPs that fulfil these criteria are flagellin from bacteria or chitin, β-glucans and ergosterol from fungi (Nürnberger and Lipka 2005). In consequence to this interaction, a range of basal defence mechanisms are expressed, causing PAMP-triggered immunity (PTI), what is also called innate immunity. In order to overcome PTI, virulent pathogens have developed effector proteins which target signalling components downstream PAMP recognition and suppress basal defence responses (Schornack et al. 2009). This interaction is called effector triggered susceptibility (ETS). During ETS basal defence mechanisms are also induced in host plants but they are ineffective or insufficient to control the growth of the pathogen. As an evolutionary result, cultivars of host plants have developed resistance genes that code for resistance (R) proteins which specifically recognize pathogen effectors what leads to the second defence system of plants, the so-called race-cultivar specific resistance (Ingle et al. 2006). This kind of resistance is also termed vertical resistance or

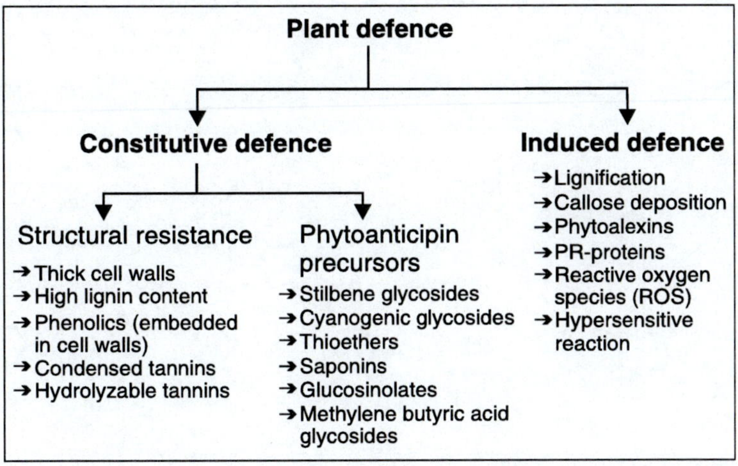

Fig. 3.1 The multilayered defence arsenal expressed in resistant plants

effector-triggered immunity (ETI) and follows the gene-for-gene model of Flor (1971). This recognition event causes the hypersensitive response (HR), characterized by rapid apoptotic cell death and local necrosis (Martin et al. 2003). Some of the induced defence mechanisms, found during PTI, are also expressed during vertical resistance.

Constitutive defence factors are already expressed in healthy plants in parallel to plant growth at strategic sites and consist either of constitutive barriers (structural resistance) or of preformed antimicrobial secondary metabolites such as the phytoanticipins (Figs. 3.1, 3.2, and 3.3 and Table 3.1). By definition "*phytoanticipins are low molecular weight, antimicrobial compounds that are present in plants before challenge by microorganisms, or are produced after infection solely from preexisting constituents*" (Van Etten et al. 1994). For example, oat roots synthesize the saponin avenacin A-1 (triterpene glycoside) which is mainly localized in the root epidermis where it forms a protective barrier towards saponin-sensitive fungi, such as *Gaeumannomyces graminis* (Papadopoulou et al. 1999; Carter et al. 1999). With the exception of saponins all other precursors have to be enzymatically converted into the corresponding toxic aglycones.

For many host–pathogen interactions this defence arsenal is sufficient to successfully stop the penetration of the pathogen or the growth of the invader within host tissue. In case the pathogen has overcome the constitutive defence it may be recognized by the interaction of PAMPs and PRRs, the latter being localized on plasma membranes. In consequence, the whole or parts of the inducible defence arsenal is activated that consists of reactive oxygen species (ROS) (Hückelhoven and Kogel 2003; Baker and Orlandi 1995; Thordal-Christensen et al. 1997), lignification (Cahill and McComb 1992), cell wall reinforcement, the formation of pathogenesis-related (PR) proteins (van Loon et al. 2006) as well as the synthesis of phytoalexins and other inducible phenolics. This entire defence is highly localized at the penetration site or the adjacent tissue.

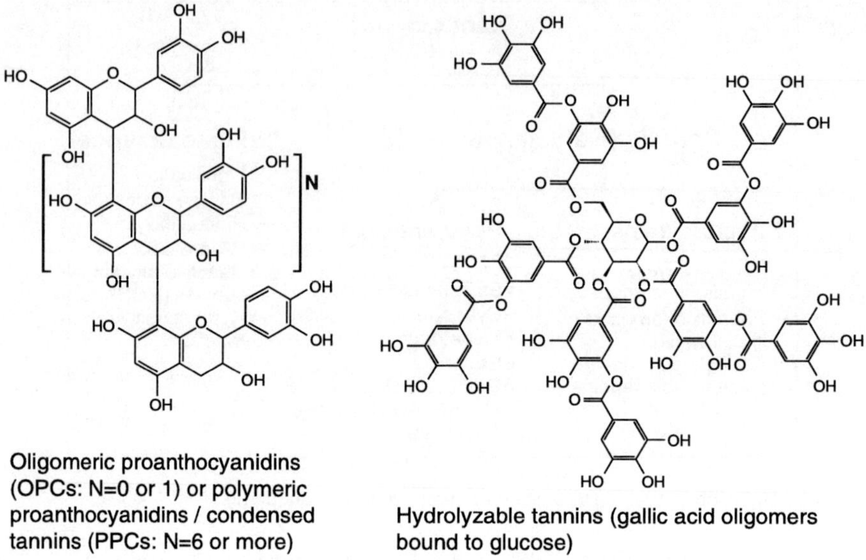

Fig. 3.2 Tannins as examples for constitutive defence compounds of plants

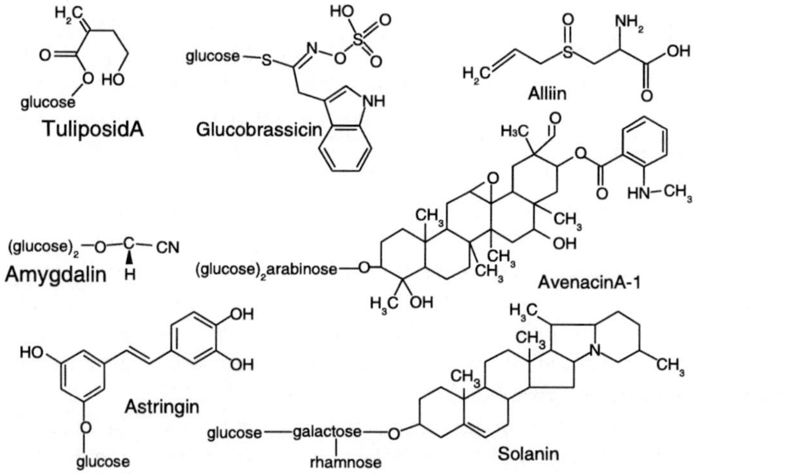

Fig. 3.3 Structure of some prominent phytoanticipin precursors as pre-existing defence compounds of plants

Pathogenesis-related proteins are induced in many plant families upon infection with various types of pathogens and have been classified up to now into 17 families (van Loon et al. 2006). Many of inducible pathogenesis-related proteins show antimicrobial activity against fungi and bacteria in vitro such as chitinases and β-1,3-glucanases. Therefore, they are discussed to act as defence proteins in resistant

Table 3.1 Prominent phytoanticipin precursors (with the exception of saponins) of plants

Chemical characterization				
Class of compounds	Metabolite	Family	Species	Organ
Stilbene glycoside	Astringin and rhaponticin (precursor for astringenin and isorhapontigenin)	Pinaceae	*Picea sitchensis*	Bark
Saponin	α-Solanin	*Solanaceae*	*Solanum tuberosum*	Green potato peel
	α-Tomatin	*Solanaceae*	*Lycopersicon esculentum*	Green fruit
	Avenacin A	*Poaceae*	*Avena sativa*	Roots
Glucosinolate	Glucobrassicin (precursor for isothiocyanate)	*Brassicaceae*	*Brassica oleraceae, Brassica alba*	Fruit
Glycosides of hydroxy carbonic acid	Tuliposid A (precursor for tulipalin A; butyrolacton)	*Liliaceae*	*Tulipa* spec.	Stalk, bulb
Cyanogenic glycosides	Amygdalin (precursor for HCN)	*Rosaceae*	*Prunus dulcis, P. armeniaca, P. serotina*	Leaf, seed
Thioether	Alliin (precursor for Allicin)	*Allioideae*	*Allium cepa*	Bulb

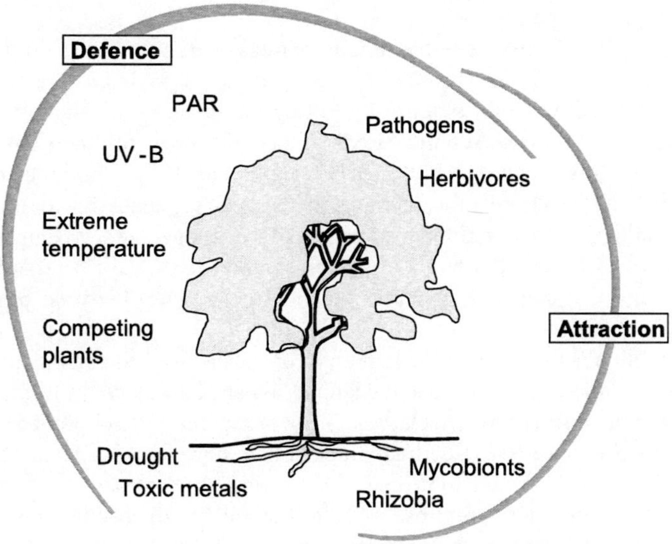

Fig. 3.4 Functions of phenolic compounds at the interface between plant and environment (*PAR* photosynthetic active radiation)

host–pathogen interactions. However, it was proved that only some few of many potential PR-genes are specifically expressed during defence. Most of other PR-genes are up-regulated constitutively in roots and floral tissue of plants, possibly indicating that they are additionally involved in plant development.

Phenolic compounds occur widespread in plants and are a biologically important and chemically diverse group of secondary metabolites (Lattanzio et al. 2008) that often accumulate in cells at the interface to the environment, where they function as defensive compounds or as attractants (Fig. 3.4). In addition to their preformed status, their biosynthesis is often stimulated/induced by several biotic and abiotic stressors (Dixon and Paiva 1995). In contrast to former times, these compounds are no longer judged as waste products or as evolutionary remnants without current function, nor as mere metabolic end products that are toxic to the plant and are, therefore, stored away in vacuoles. Obviously, it was first recognized by entomologists that these natural chemical products are involved in the interaction between plants and herbivores acting both as defence-related metabolites and as attractants (Hartmann 2007). Antifungal, antibacterial and antiviral effects have been described for several phenolic classes (Wink 1988; Treutter 2005). In this context, they occur in plants either as constitutive or as inducible defence compounds including the so-called phytoalexins. The belonging to one of these groups is independent of the chemical nature. Meanwhile it is generally accepted that these so-called secondary metabolites possess a wide range of biological activities and show multifunctional roles. They are beneficial for plants (Lattanzio et al. 2008) as antioxidants, as UV-protectants, as growth regulating compounds, as signals modulating gene expression and as phytoalexins (Bednarek and Osbourn 2009). By definition "*phytoalexins are low molecular weight, antimicrobial compounds that are both synthesized by and accumulated in plants after exposure to microorganisms*" (Van Etten et al. 1994). A summary of important phytoalexins is given in Table 3.2, and Fig. 3.5.

The multifunctionality of phenolic compounds is mainly based on their structural diversity. Due to their antimicrobial properties, phytoalexins are discussed to be important in active plant defence by killing the invading pathogen (Mert-Türk 2002; Kuc 1995; Hammerschmidt 1999). Good evidence for their role in plant defence exists (Snyder and Nicholson 1990; Hain et al. 1993; Thomma et al. 1999; Graham et al. 2007). Despite the chemical diversity of phytoalexins found in plants, the type of antimicrobial compounds formed after pathogen attack is mainly family specific (Table 3.2). The overview of key biosynthetic pathways for their synthesis (given in Figs. 3.6 and 3.7) indicates the interrelationship between primary and secondary metabolism.

All phenolic classes directly or indirectly originate from the shikimate pathway with most of them passing phenylalanine as an intermediate leading to the phenylpropanoids and derived classes. This particular phenyl-propane aromatic amino acid is one of the links to primary metabolism since it is also needed for protein biosynthesis. Important links to primary metabolism are represented by the pentose phosphate cycle providing erythrose 4-phosphate and by glycolysis supplying phosphoenolpyruvate (PEP) for the shikimate pathway. Furthermore the high-energy key metabolite acetyl-CoA which is oxidized within the Krebs-cycle to finally release ATP after passing through the respiratory chain is on one hand used for the synthesis of

Table 3.2 Prominent phytoalexins of gymnosperms and dicotyledons

Plant family	Chemical type	Example: Name	Host	Organ	Pathogen
Phytoalexins of gymnosperms					
Pinaceae	Stilbene	Pinosylvin	*Pinus resinosa*, *P. sylvestris*, *P. taeda*	Sapwood	*Fomes annosus*; *Peridermium pini*; *Ceratocystis minor*
Pinaceae	Flavanone	Pinocembrin	*P. taeda*; *P. radiata*	Sapwood	*Fomes annosus*
Pinaceae	Aromatic carboxylic acid	Benzoic acid	*P. radiata*	Needles	*Dothistroma septospora*
Pinaceae	Lignan	Hydroxymatairesinol	*Picea avies*	Sapwood	*F. annosus*
Pinaceae	Dihydroflavonol	Taxifolin	*Pseudotsuga menziesii*	Roots	*Poria weirrii*
Cupressaceae	Tropolone glycoside	Cupressotropolone A	*Cupressus sempervirens*	Bark	*Diplodia pinea*
Phytoalexins of dicotyledones					
Brassicaceae	Indol alkaloid	Camalexin	*Arabidopsis thaliana*	Leaves	*Pseudomonas syringae*; *different fungal pathogens*
Brassicaceae	Indol alkaloid	Brassilexin	*Brassica carinata*, *B. juncea*, *B. nigra*	Leaves	*Phoma lingam*
Asteraceae	Coumarin	Scopoletin	*Helianthus annusus*	Stem	*Helminthosporium carbonum*
Platanaceae			*Platanus acerifolia*	Stem	*Ceratocystis fimbriata*
Fabaceae	Pterocarpane	Pisatin	*Pisum sativum*	Leaf, stem	*Mycosphaerella pinodes*
Fabaceae	Pterocarpane	Glyceollin I	*Glycine max*	Leaf, root	*Phytophthora sojae*
Fabaceae	Pterocarpane	Phaseollin	*Phaseolus vulgaris*	Leaf	
Fabaceae	Isoflavone	Kieviton	*Phaseolus vulgaris*	Leaf	
Fabaceae	Pterocarpane	Medicarpin	*Cicer arietinum*	Leaf	*Ascochyta rabiei*
Fabaceae	Pterocarpane	Maackiain	*Cicer arietinum*	Leaf	*Ascochyta rabiei*
Solanaceae	Sesquiterpenoid	Rishitin, lubimin, phytuberin	*Solanum tuberosum*	Tuber	*Phytophthora infestans*
Solanaceae	Sesquiterpenoid	Rishitin, phytuberin	*Nicotiana tabaccum*	Leaf	
Solanaceae	Sesquiterpenoid	Rishitin	*Lycopersicon esculentum*	Leaf, fruit	*Phytophthora infestans*

(continued)

Table 3.2 (continued)

Plant family	Chemical type	Example: Name	Host	Organ	Pathogen
Solanaceae	Phenylpropanoid	Chlorogenic acid	*Nicotiana tabaccum*	Leaf	*TMV-virus*
Apiaceae	Furanocoumarin	Marmesin	*Apium graveolens*	Stalk	*Botrytis cinerea*
Apiaceae	Furanocoumarin	Psoralen, bergapten	*Petroselinum crispum*	Roots	*Phytophthora sojae*
Apiaceae	Furanocoumarin	Psoralen, bergapten, xanthotoxin	*Glehnia littoralis*	Roots	*Pseudomonas cichorii*
Vitaceae	Stilbene	Resveratrol	*Vitis vinifera*	Leaf, fruits	*Botrytis cinerea*
Poaceae	3-Deoxyanthocyanidin	Apigeninidin/ luteolinidin	*Sorghum*	Leaf	*Colletotrichum*

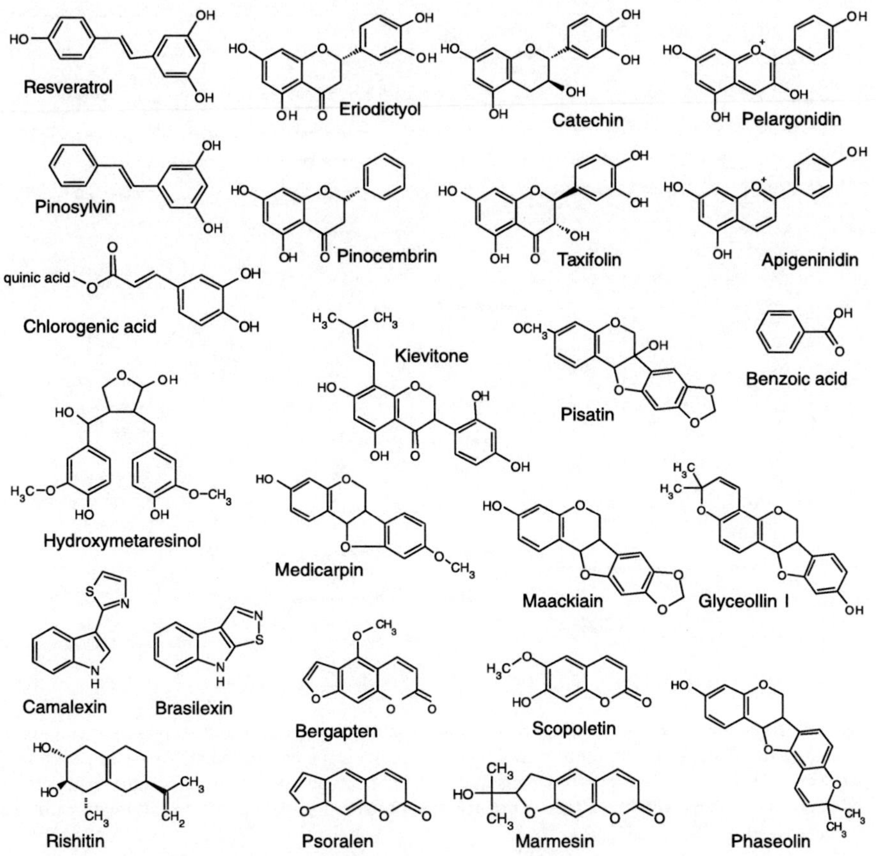

Fig. 3.5 Structure of phytoalexins as induced defence compounds in plants

malonate which is needed as a CoA-ester by chalcone synthase (CHS) and stilbene synthase (STS) to form the flavonoid and stilbene structures, respectively (Fig. 3.7). On the other hand acetyl-CoA feeds the mevalonate pathway from which terpenoid phytoalexins such as rishitin and lubimin are derived. This shows that the stimulation of the secondary pathway causes a competitive situation for ATP, what in consequence might result in a trade-off between growth and defence, one of the basic assumptions of the GDB (Chap. 1).

3.3 Switching from Primary to Secondary Metabolism

Under resource limiting conditions, such as carbon limitation which is supposed to be induced by high N-nutrition, carbon-based secondary metabolism is down-regulated. The reverse effect is often described for low N-nutrition. But it is not

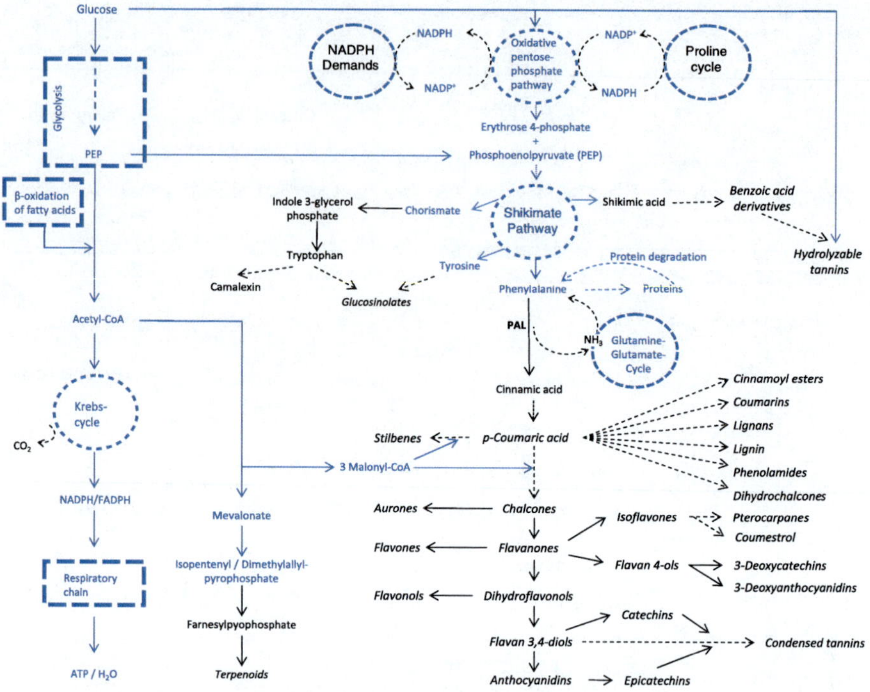

Fig. 3.6 Biosynthetic routes leading to different classes of defence-related secondary metabolites indicating the general link to the primary metabolism with the central positions of the shikimate pathway and the oxidative pentose phosphate cycle; *PAL* phenylalanine ammonia lyase. *Blue* indicates the primary and black the secondary metabolism. The details of the shikimate pathway are visualized in Chap. 2, Fig. 2.4

understood, if the enhanced polyphenol biosynthesis under these conditions is just an overflow reaction utilizing excessive carbon or if other mechanisms are involved or even prevalent. In experiments studying competition and resource-mediated conflicts between growth and defensive chemistry in trembling aspen, Donaldson et al. (2006) confirmed that resource availability influences the realization of defence costs. However, they also stated that the increased production of condensed tannins under limited resource availability was not solely a consequence of "excess" carbon. Under such conditions the growth rate was lower than would be expected. Growth reduction may therefore not only have been based on the construction costs of condensed tannins but additional factors may have played a role (Donaldson et al. 2006).

Under N-limiting conditions, N-recycling may be one of the mechanisms behind the often observed increase of phenylpropanoids (Fig. 3.6; Lewis and Yamamoto 1989). A further trigger for switching from primary to secondary metabolism was recently discussed by Lattanzio et al. (2008) which couples the accumulation of the stress metabolite proline with an activated phenylpropanoid biosynthesis via the oxidative pentose phosphate pathway (Shetty 2004). Under several conditions

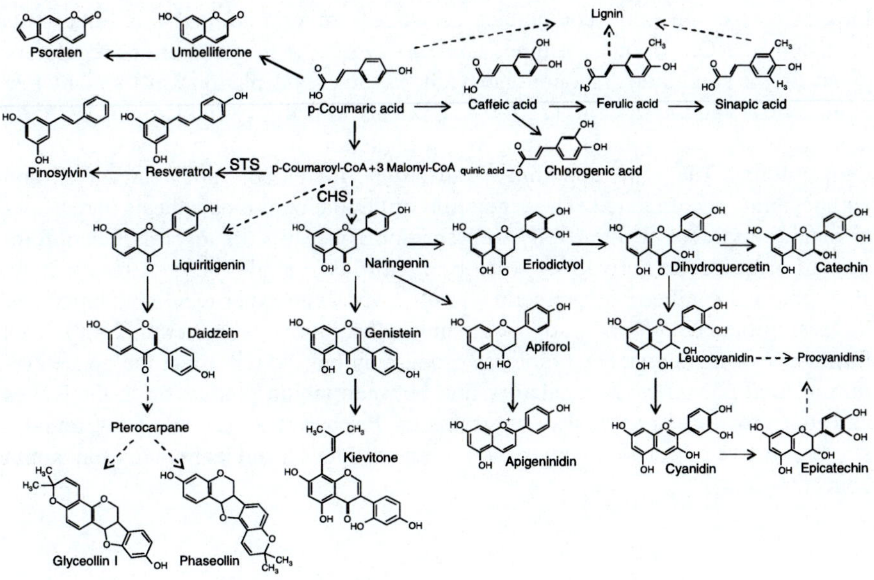

Fig. 3.7 Biosynthetic pathway of prominent phenolic phytoalexins, catechins and lignin; *CHS* chalcone synthase, *STS* stilbene synthase

of stress (pathogen infection, nutrient and water deficiency), the plant is forced to accumulate a large quantity of free proline. Its synthesis is accompanied by the oxidation of NADPH. An increased $NADP^+$/NADPH ratio is likely to enhance the activity of the oxidative pentose phosphate pathway providing precursors for phenolic biosynthesis via the shikimic acid pathway. An enhancement of phenylpropanoid biosynthesis may also be driven by a deficiency in reduced pyridine nucleotides (NADPH). In that case, it is supposed that the pentose phosphate pathway is stimulated for providing NADPH thus releasing substrates for shikimate and phenylpropanoid biosynthesis (Shetty 2004).

Fritz et al. (2006) stated that nitrogen deficiency activates the phenylpropanoid metabolism via a mechanism that is independent of changes in the precursor supply. They provide evidence that nitrate contributes to the transcriptional regulation of phenylpropanoid metabolism in tobacco. They showed that a nitrate reductase-deficient tobacco mutant grown under high N supply produced as much chlorogenic acid and rutin as a wild-type tobacco did under low N conditions. The switch from growth-related primary metabolism towards polyphenol biosynthesis as reported for high N fertilization may, therefore, be related to regulatory mechanisms other than resource limitation. This idea is supported by Palumbo et al. (2007) who also found that the carbon/nitrogen ratios alone were insufficient for predicting resource allocation to secondary metabolites in *Ilex vomitoria*.

Under N-deficiency conditions, recently nitric oxide (NO) was described as a signal involved in the up-regulation of PAL activity (Kováčik et al. 2009). In the roots of nitrogen-deficient *Matricaria chamomilla* plants internal NO stimulated the biosynthesis of phenolic compounds as well as of proline, and this effect was mediated by ROS. In model studies on *Arabidopsis* the involvement of regulatory genes for the stimulated flavonoid biosynthesis under N depletion was found by Lea et al. (2007) and Olsen et al. (2009). The flavonoid pathway activators *PAP1/GL3* and *PAP2/Myb12* responded strongly to a depletion of nitrogen as well as to low temperature. The corresponding transcripts increased, and flavanols and anthocyanins accumulated. The regulation of the flavonol biosynthesis under low-N conditions seems to be even more specific, since a shift towards kaempferol instead of quercetin derivatives was observed (Olsen et al. 2009). A biosynthetic shift from isorhamnetin to quercetin was observed in an experiment with onions by increasing the NO_3^-/NH_4^+-ratio of the nutrient solution (Perner et al. 2008). This fertilizer treatment enhanced biomass production but did not affect the concentration of total phenolics. A regulatory link between tannin production in the leaves and root growth of poplar was suggested by Fischer et al. (2006). They found a strong positive correlation between fine-root production and leaf condensed tannin concentration.

3.4 Defence-Related Metabolism and Trade-off with Growth

As shown before many metabolic precursors have to be provided by primary metabolism that flow into secondary metabolism to guarantee the synthesis of the multilayered defence arsenal of plants. This raises the question of whether a trade-off between growth-related processes and defence-related metabolism exists. These aspects were recently reviewed by Treutter (2010) in the context of enrichment of health-related secondary metabolites in agricultural crops, the so-called "chemical farming". The Growth differentiation balance describes how plants may regulate between growth and differentiation processes responsible also to defence under different environmental conditions (see Chap. 1). Such differentiation processes include e.g. the thickening of leaf cuticle or the secondary metabolism in general.

All these processes will in total effect the constitutive resistance of plants. GDB attempts to explain the influence of a resource gradient on the trade-off between growth- and defence-related processes of plants and in consequence tries to give information of costs mainly for constitutive rather than for induced plant defence. If a plant has to allocate a substantial amount of resources to growth and or reproduction due to a change in resource availability, then defence should become limited; in consequence, such a plant will become more susceptible to pathogens. It must be noticed that the experimental approaches for testing GDB as well as interpretational attempts of observations mostly do not cover the complete range of resource availability. Le Bot et al. (2009) for instance indicate that the domain of N deficiency "*is rarely described in research papers on agricultural crops*". A general distinction has to be made also between constitutive resistance which permanently requires activation of the defence-related metabolism and, conversely, induced resistance. Figure 3.8 outlines the relationship of defensive compounds to the biochemical pools (Gayler et al. 2004).

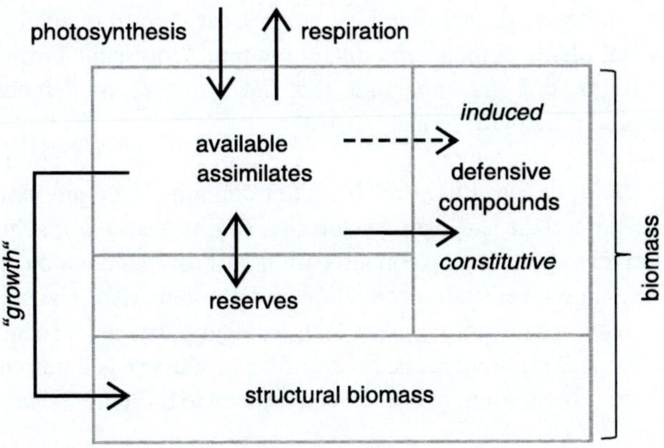

Fig. 3.8 The aggregated biochemical pools (modified after Gayler et al. 2004)

Bergelson and Purrington (1996) in their review paper analyzed data sets of 88 publications on trade-offs between resistance and fitness traits. Only 50 % of the resistant plants showed a cost of resistance, 5 % of the comparisons exhibited benefits for the resistance trait and 45 % of the studies showed no differences regarding costs between resistant and susceptible plants. They also figured out that costs were highest for herbicide resistance, intermediate for pathogen and smallest for herbivore resistance. Finally they could not find evidence that costs for resistance were more pronounced when plants were grown under stressful conditions such as intraspecific competition.

3.4.1 Trade-off and Defence Against Herbivores

Many of the studies trying to test Growth differentiation balance were done, correlating the amount of secondary compounds, e.g. in leaves with the attack of herbivores. Coley (1986) investigated the costs and benefits of defence by tannins for the tree *Cecropia peltata*. He showed that leaves of *C. peltata* plants with high tannin concentration had lower damage in herbivory experiments as compared to those with low tannin concentrations. He proved that the total number of leaves produced was negatively correlated with the tannin concentration. Since the ontogenetic and environmental effects on growth were minimized in the experiment the author concluded that there is a significant cost of tannin production, which becomes apparent in a reduced leaf formation.

Baldwin et al. (1990) investigated the consequences of induced alkaloid production of wild tobacco plants on their reproduction. Nicotine and nornicotine concentrations are known to increase in tobacco leaves after actual or simulated herbivory. In consequence, the growth of tobacco herbivores was reduced of protected leaves (Baldwin 1988a, b; Baldwin 1989). The removal of about 20 % of total shoot mass, simulating herbivory, caused a significant reduction of seed mass per plant. However, when nicotine synthesis was inhibited by auxin treatment, damaged plants did not differ from control plants in their reproductive output. Unless the hormone treatment had any side effects, one may conclude that the synthesis of defence compounds induced by simulated herbivory must be costly.

Han and Lincoln (1994) investigated the costs of carbon allocation to secondary metabolites in *Diplacus aurantiacus*. This plant contains large amounts of phenolic resin in leaves to defend the leaf-eating caterpillar *Euphydryas chalcedona*. They were able to quantitatively assess the carbon allocation cost for leaf resin by applying linear negative regression between resin production and plant growth. The surprising result was that 1 mg increase in resin per gram leaf dry weight caused a reduction of either 2.1 mg dry shoot or 1.3 mg stem mass. Since 2.52 g of glucose is theoretically required to form 1 g of resin, this amount of carbon can be used to build up twice as much shoot biomass as resin.

One main conclusion of GDB is that resource limitation causes slower growth which, however, favours large investments into defence. This statement is convincingly corroborated by the review paper of Coley et al. (1985). They compared the degree of defence in slow and fast-growing plants with their anti-herbivore characteristics and showed that the concentration of defence compounds in leaves of slow-growing plants at resource-limited sites was higher as compared to those in leaves of fast-growing plants at resource-rich sites. In consequence, the rates of herbivory were the highest on leaves of fast-growing plants.

Agrawal (2000) investigated the effect of defence induction of *Lepidium virginicum* on total biomass production, allowing *Pieris rapae* larvae to feed on one leaf. Similar to *A. thaliana*, induced leaves of *L. virginicum* showed a higher number of trichomes and a higher concentration of glucosinolates. When growing under low density in an enemy-free environment, there was no difference in total biomass of induced compared to control plants. However, under high-density induced plants were characterized by lower root biomass and increased aboveground growth. These results prove that induced resistance towards herbivores can cause a change in allocation rather than a loss of total biomass.

Glynn et al. (2007) tested GDB by measuring relative growth rate in parallel to net assimilation and phenylpropanoid concentrations of two willow species along five fertility levels. With this kind of experimental setup, they fulfilled the demands of Stamp (2004) and Wilkens (1997) that at least five concentrations along a resource gradient are necessary to detect a curvilinear pattern of plant defence. Glynn et al. (2007) analyzed their data on day 23, 40 and 85 of growth and concluded that both willow species in general responded as predicted by GDB. However, their results also showed that no general prediction for components of the secondary metabolism can be made. For instance, they found a quadratic trend for condensed tannins and a linear effect for total phenylpropanoids across the five fertility levels on day 40 for mature foliage. The same was reported for 85 days of old immature foliage.

Glynn et al. (2003) were the first who quantified the effects of two nitrogen levels on constitutive as well as on induced resistance in poplar, in the presence of two different herbivores. They discovered that high nutrient treatment compared to the low N fertilization had no effect on photosynthesis, whereas there was a strong increase for all growth parameters, as the total concentration of leaf phenolics decreased. This negative correlation between plant growth and leaf phenolic concentrations is in agreement with GDB according to Herms and Mattson (1992) saying that increased nutrient availability increases biomass without effecting net assimilation rate. Thus, gypsy moth (*Lymantria dispar*) larvae grew much faster on high-fertility plants. Surprisingly, increase in nutrient availability had no effect on constitutive resistance towards the white-marked tussock moth (*Orgyia leucostigma*). Herbivore-induced systemic resistance (HISR) on younger leaves was obtained, allowing the gypsy moth to feed for 72 h on older leaves. HISR to gypsy moth was most expressed for high-fertility plants, whereas induced resistance to the white-marked tussock moth was highest in the low-fertility treatment. These results prove that total leaf phenolics or tannins cannot be taken as reliable indicators predicting defence capacity of plants towards different pathogens.

Recently Häring et al. (2008) tried to check GDB for *Onobrychis viciifolia*, comparing constitutive and induced tannin concentration of leaves after wounding or treatment with different elicitors. They used four different phosphorus concentrations as nutrient solutions ranging from 0.0027 up to 2 mM. Regarding constitutive tannin concentrations and nutrient availability, they did not find the pattern postulated by Herms and Mattson (1992) (Chap. 1, Fig. 1.1). In addition, there was no significant correlation between growth and tannin concentration of leaves at any level of nutrient availability. Therefore, the authors conclude that the postulated inevitable trade-off between growth and defence predicted by GDB not necessarily is compelling.

Summarizing these results, one can conclude that the expression of defence traits regarding herbivory are costly for many plants. However, there are also host plants for which defence traits and growth parameters are not negatively correlated. Unfortunately, defence costs, if detected, are expressed in most cases qualitatively and referred to constitutive resistance traits in the absence of any herbivore.

3.4.2 Resistance Inducers and Trade-off Studies

Many trade-off studies were carried out with chemicals, such as BION (benzothiadiazole) that induce resistance in plants. Heil et al. (2000) showed that wheat plants sprayed with the resistance inducer BION produced lower biomass, developed fewer shoots and ears and, therefore, achieved fewer seeds. This effect became most apparent when treated plants suffered from a shortage of nitrogen, which was explained by the need of the plants to allocate nitrogen to the synthesis of pathogenesis-related (PR) proteins that are known as defence enzymes (van Loon et al. 2006).

Application of jasmonic acid (JA), methyl-jasmonate (MeJA) or salicylic acid (SA) can induce systemic responses in plants known as induced systemic resistance (ISR) and systemic acquired resistance (SAR), respectively (see Chap. 5, Fig. 5.1). These plant hormones were often used to study costs of induced resistance.

However, the results are inconsistent as shown by the following examples. Van Dam and Baldwin (2001) treated *Nicotiana attenuate* plants with MeJA which were grown in competition with untreated plants in either high or low-N soils. ^{15}N-labelled KNO_3 was used to quantify N acquisition and allocation. MeJA-treated plants competing with un-induced plants produced significantly fewer seeds, acquired less ^{15}N per unit of total biomass and allocated less ^{15}N to seeds. Costs for induced resistance were much more expressed in plants growing under high nitrogen. Similar results were reported by Cipollini (2002) who manipulated the expression of SAR of *Arabidopsis thaliana* with SA and of ISR with JA. He proved that the application of both wound-related hormones significantly stimulated levels of chemical defences and, in parallel, reduced seed production on average by 15 % in both studies. However, Thaler (1999) did not detect costs for induced resistance of tomato plants treated with JA. He did not measure any difference in yield between induced and control plants, whereas treated plants showed increased levels of characteristic defence enzymes and reflected 60 % less leaf damage. Häring et al. (2008) investigated the effect of insect elicitors, which are components of the saliva of the lepidopteron *Spodoptera littoralis*, on foliar condensed tannins (CT) of *Onobrychis viciifolia*. There was a significant induction of CT of elicitor-treated leaves as compared to controls. However, the authors did not find any evidence for a trade-off between growth and defence in this specific interaction.

3.4.3 Trade-off Studies on Fungal and Bacterial Infections as well as Mutants and Transgenic Plants

As shown above, many experiments were carried out in the past mainly with herbivores in order to calculate costs for resistance under different environmental constraints. However, experiments with fungi are rare. In most cases susceptible and resistant genotypes were compared regarding growth and/or fitness costs in a disease-free environment. Burdon and Müller (1987) tried to measure the fitness costs of 15 lines of *Avena fatua* susceptible or resistant to *Puccina coronata* grown at different temperatures under controlled conditions in the glasshouse or in the field, but their results were not consistent. All resistant lines showed a slower germination rate at 20 and 25 °C as compared to the susceptible ones, whereas there was no longer any significant difference at 15 °C. However, when growing in the glasshouse at high temperatures in a disease-free environment, susceptible genotypes performed better than did resistant lines regarding fecundity, whereas the opposite was found under field conditions. According to Brown (2002) the only reliable way to study any costs of resistance or of fitness is the use of mutants or transgenic plants. Mutants that either constitutively express resistance or have lesions in genes responsible for signalling cascades are often used to prove a trade-off between growth and resistance traits. One mutant named cev1 constitutively expressing several defence-related genes had enhanced resistance to powdery mildew diseases and showed reduced growth, stunted roots and accumulated anthocyanin (Ellis and Turner 2001). Royo et al. (1999) proved that potato plants antisense for a lipoxygenase showed reduced induction of proteinase inhibitors and increased susceptibility to Colorado potato beetles. These plants exhibiting reduced defence traits were characterized by enhanced tuber yields. Magg et al. (2001) compared several maize genotypes expressing the *Bacillus thuringensis* crystal protein with their non-transgenic controls for resistance to the European corn borer (*Ostrinia nubilalis*) and for important agronomic traits. They showed in two independent sets of experiments that transgenic plants, constitutively expressing the insecticidal toxin from *Bacillus thuringensis*, did not differ for grain yield in the absence of larvae of *Ostrinia nubilalis*. Tian et al. (2003) calculated the costs for the maintenance of an agent resistance (R) gene in *Arabidopsis thaliana*. According to the gene-for-gene model, the *RPM1* resistance gene in *A. thaliana* codes for a plasma membrane protein that specifically recognizes *Pseudomonas syringae* pathogens carrying the corresponding *AvrRpml* or *AvrB* gene (Bisgrove et al. 1994). In order to test for fitness costs of RPM1 expression the authors inserted the whole gene together with its promotor and terminator region into a susceptible ecotype. The four generated independent transgenic lines were grown to maturity in the field in the absence of detectable disease. The experiments proved that RPM1 plants had significantly fewer siliques and seed per silique as well as lower shoot biomass as compared to *A. thaliana* plants without the resistance gene. In total, transgenic plants suffered a 9 % decrease in seed production relative to their counterparts. The authors do not believe that the decline in fitness can be explained just by metabolic costs of RPM1 synthesis. They speculated that the relatively high fitness costs of 9 % in their field trial might be due to the over-expression of the R-gene by environmental signals that could lead to constitutive defence response in the absence of any enemies.

Recently Zeller et al. (2010) compared various plant performance traits of genetically modified (GM) wheat *Triticum aestivum* with control plants under glasshouse and field conditions. The GM lines expressed the *Pm3b* gene which confers race-specific resistance towards the fungus powdery mildew *Blumeria graminis* f.sp. *tritici*. Resistant plants undergo local cell death as a result of the hypersensitive reaction (HR) upon a penetration attempt by a germinating fungal spore (Yahiaoui et al. 2004). In a first greenhouse experiment the authors showed that in the presence of a fungicide all GM wheat lines produced about 25 % less seeds, expressed as tonnes ha^{-1}, as compared to control plants. This indicates that the costs for constitutive resistance of GM wheat might be high if the pathogen is absent. However, costs for all defence reactions induced after pathogen attack of resistant plants appear to be negligible, because seed yield of GM plants attacked by the pathogen did not differ without additional fertilizer, or was even higher for fertilized wheat as compared to controls. A respective conclusion can be drawn for resistance costs and total vegetative mass. Surprisingly, the opposite was found for costs of resistance referring to various plant performance traits when the experiment was run under field conditions. Under such conditions resistance appears to be costly, because all GM lines, successfully expressing mildew resistance,

Table 3.3 Host–pathogen interactions investigated within the SFB607

Host plant	Pathogen	Target organ	Environmental factor
Potato (*Solanum tuberosum*)	*Phytophthora infestans*	Leaf	CO_2, O_3
	Alternaria solani	Leaf	CO_2
			CO_2, N
Barley (*Hordeum vulgare*)	*Drechslera teres*	Leaf	CO_2, O_3
Apple (*Malus* spp.)	*Venturia inaequalis*	Leaf	N
	Phytophthora cactorum	Stem	CO_2, N
European beech (*Fagus sylvatica*)	*Phytophthora plurivora*[a]	Roots	CO_2, N

[a]Please note that *Phytophthora citricola* isolates of deciduous European trees were recently transferred to a new species *Phytophthora plurivora* (Jung and Burgess 2009). Therefore citations regarding *P. citricola* actually address to *Phytophthora plurivora*

show the lowest seed yield. In contrast, resistance costs appear to be low, even under field conditions, when costs are related to total vegetative mass.

The experiments of Zeller et al. (2010) prove for the first time that a trade-off might not only exist between defence reactions and plant performance traits but also is found for different defence reactions towards variable pathogens. Under field conditions, three out of four GM wheat lines, which were resistant towards *Blumeria graminis*, had lost their resistance towards *Claviceps purpurea*, and the opposite was true for all control lines.

3.4.4 *Effect of a Changing Environment on Trade-offs Between Growth and Defence*

Within the collaborative research centre SFB607 the rationale of which was outlined by Matyssek et al. (2005), trade-off studies with several host–pathogen interactions were investigated under different environmental regimes. Four economically important host plants from agriculture and forestry, such as potato (*Solanum tuberosum*), barley (*Hordeum vulgare*), apple (*Malus domestica*) and European beech (*Fagus sylvatica*) were chosen in combination with their corresponding pathogens (Table 3.3). Four different leaf pathogens were object of the investigations: late blight and early blight of potato foliage, caused by *Phytophthora infestans* and *A. solani*, respectively (Fry 2007), the two most important potato diseases worldwide. *Drechslera teres* which causes net blotch, the major foliar disease of barley (McLean et al. 2009) and *Venturia inaequalis* which induces apple scab (Gessler et al. 2006). While *Phytophthora infestans*, *D. teres* and *Venturia inaequalis* are biotrophic pathogens, which predominantly grow and sporulate only on living plant tissue, *A. solani* is necrotrophic, i.e. kills host tissue on which it grows.

Phytophthora cactorum and *Phytophthora plurivora* are root and stem pathogens of apple (Harris 1991) and European beech (Jung et al. 2005; Jung 2009) respectively. *Phytophthora plurivora* was recently described as a distinct species formerly classified as *P. citricola* (Jung and Burgess 2009). For potato and apple several cultivars with known differences in susceptibility towards the corresponding pathogens were

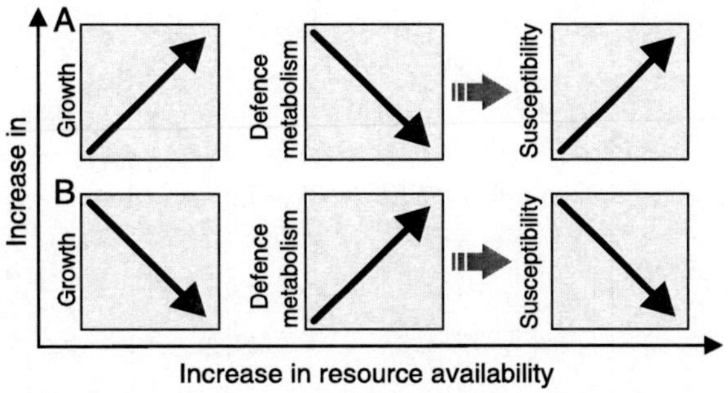

Fig. 3.9 Simplified scheme to test for trade-offs between growth and pathogen defence on the basis of GDB (Herms and Mattson 1992)

used, while research on beech was done with a specified provenience, representing the natural genetical diversity.

In terms of growth differentiation balance, the different host–pathogen interactions of potato, barley, apple and beech were investigated for possible trade-offs between growth- and defence-related metabolism. In order to achieve this, growth promoting and growth constraining environmental conditions were selected. Atmospheric carbon dioxide concentration was elevated from ambient to ambient + 300 μL L^{-1} CO_2 to increase the carbon availability of host plants. Nitrogen availability was modified by increasing or decreasing the standard nitrogen fertilization of the respective plant species by a factor of two. In some experiments chronically high ozone concentrations were used to decrease carbon availability within the plants, as it is known that this air pollutant decreases net photosynthesis (Matyssek and Sandermann 2003). Usually twofold ambient ozone concentrations with a restriction to 150 nL L^{-1} O_3 were applied, to prevent the plants from acute ozone damage. Additionally, for potato a reduction of ozone to one-fifth of the ambient ozone concentration was applied, which should result in an increase of carbon availability similar to an increase in CO_2.

As the prediction of GDB mainly refers to the synthesis of carbon-based secondary metabolites, soluble and cell wall bound phenolic compounds were selected as the main surrogates of plant defence. Simplified, a trade-off between growth- and defence-related metabolisms should appear as an antipodal behaviour of growth parameters and phenolic compounds of a specified plant organ under changing resource availability. This may turn out as an increase in growth and a decrease in the concentration of secondary metabolites resulting in an increase in susceptibility (Fig. 3.9a), or as a decrease in growth in combination with an increase in the concentration of secondary metabolites, and hence a decrease in susceptibility (Fig. 3.9b). The latter correlation would indicate that resistance is costly. Trade-off studies were mainly carried out for healthy plants showing constitutive defence mechanisms. However, trade-offs should also be present for infected plants expressing induced resistance mechanisms in addition to constitutive defence (Fig. 3.1). Such a kind of correlation between growth, defence and susceptibility was found for the apple variety "Golden Delicious", which is susceptible towards the apple scab (Fig. 3.10). With increasing nitrogen fertilization, shoot growth

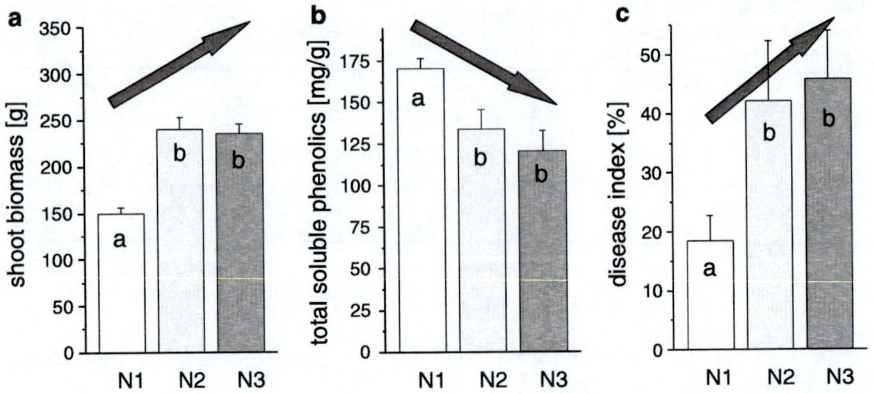

Fig. 3.10 Shoot biomass (**a**), leaf total soluble phenolics (**b**), and *Venturia inaequalis* infestation (**c**) of the susceptible apple cv. "Golden Delicious", in dependence of the nitrogen supply (3.6 g (N1), 8.2 g (N2), 13 g (N3) N fertilization per tree and year). Modified after Leser and Treutter (2005)

increased, and the concentration of phenolic compounds in leaves decreased. In accordance with GDB susceptibility towards the fungus increased.

The basal resistance of scab-susceptible apple cultivars is based on the accumulation of flavan-3-ols (catechins and procyanidins) surrounding the infection zones of leaves and fruits (Mayr and Treutter 1998). These phenolic compounds are efficient in inhibiting pectinases and cellulases of the fungus (Golba et al. 2012). Flavan-3-ols occur also constitutively in apple tissues. It is assumed that above a threshold concentration defence against *Venturia inaequalis* is successful (Mayr et al. 1997). This constitutive biosynthesis of phenolics is sensitive to environmental conditions such as high N fertilization, which reduces the activity of phenylalanine ammonia lyase (PAL; Strissel et al. 2005), the key enzyme of the phenylpropanoid pathway, and thus results in a general decrease of phenolic compounds (Leser and Treutter 2005; Rühmann et al. 2002).

Besides the basal resistance to scab, vertical resistance (or ETI) is mediated by several resistance genes in apple (Gessler et al. 2006). The scab resistance of the cultivar "Rewena" can be traced back to the specific Vf resistance gene (Leser and Treutter 2005; Gessler et al. 2006). In similar trade-off studies as for "Golden Delicious", the cultivar "Rewena" showed a similar relation between shoot growth and soluble phenolics in leaves. However, resistance towards *Venturia* was not mitigated by nitrogen fertilization to any extent (Fig. 3.11). The defence mechanisms involved in ETI are also related to phenylpropanoids. When PAL was inhibited in leaves of the resistant cultivar "Sir Prize", resistance got completely lost (Mayr et al. 1997). In contrast to basal resistance, where flavanols are active compounds, lignification seems to be involved in ETI. Recently, it was found that the resistant variety "Rewena" exhibits higher levels of transcripts corresponding both to lignification and to primary metabolism, when compared

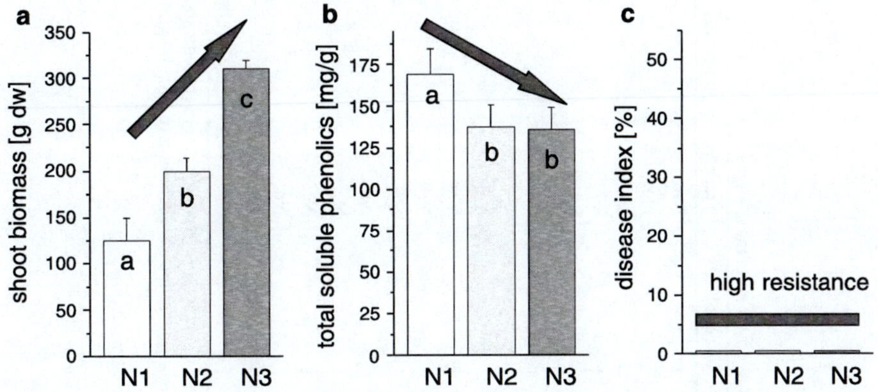

Fig. 3.11 Shoot biomass (**a**), leaf total soluble phenolics (**b**), and *Venturia inaequalis* infestation (**c**) of the resistant apple cv. "Rewena", in dependence of the nitrogen supply (3.6 g (N1), 8.2 g (N2), 13 g (N3) N fertilization per tree and year). Modified after Leser and Treutter (2005)

Table 3.4 Significant differences in gene expression ($p < 0.05$) between the cultivars Rewena (Re) and Golden Delicious (GD) occurring in two independent experiments (Holzapfel et al. 2012)

	Re/GD Exp.1		Re/GD Exp. 2	
Homology	Ratio	*p*	Ratio	*p*
Oxygen evolving enhancer protein 2	1.3	0.0350	1.5	0.0033
NAPH-dependent hydroxypyruvate reductase	1.2	0.0097	1.6	<0.0001
Polypeptide of the oxygen-evolving complex of photosystem II	1.3	0.0344	1.6	<0.0001
Cytosolic ascorbate peroxidase (APX)	1.4	0.0024	1.2	0.0211
Cinnamyl alcohol dehydrogenase (CAD)	1.2	0.0011	1.2	0.0016
Photosystem I reaction centre subunit X psaK	1.4	<0.0001	1.6	<0.0001
Cysteine protease 1	1.3	0.0046	1.3	0.0022

to the susceptible variety "Golden Delicious" (Holzapfel et al. 2012) (Table 3.4). However, reduced PAL activity due to increased nitrogen fertilization was not sufficient to overcome the Vf-mediated resistance in "Rewena".

Similar to the host–pathogen system apple-*Venturia inaequalis*, the other host–pathogen associations were evaluated regarding trade-offs and possible costs for resistance. The results are summarized in Table 3.5. A trade-off between growth and phenolic compounds was found for 10 out of 15 investigated plant–pathogen–environment combinations, thus verifying the predictions of GDB. In other cases (e.g. beech and barley grown under elevated CO_2 or ozone) no trade-off was confirmed. In Chap. 14 (mechanistic modelling) it is shown that this finding is not necessarily a falsification of GDB. Crucial for the occurrence of a trade-off is the plant-internal resource availability, but not the amount of external resources availability. Thus, different experiments with identical environmental conditions

Table 3.5 Responses of growth, phenolic compounds and susceptibility of different host–pathogen interactions to environmental changes

Host	Pathogen	Environm. variable	Growth	Phenolics	*Trade-off*	Susceptibility	Trade-off explains susc.
Beech	*Phytophthora plurivora*	N	↑	↓	Yes	↓	No
		CO_2	(↑)	–	No	↑	No
		O_3	↓	–	No	–	–
Apple cv. Gol. Del.	*Venturia inaequalis*	N	↑	↓	Yes	↑	Yes
Apple cv. Rewena	*Venturia inaequalis*	N	↑	↓	Yes	–	No
Apple cv. M9, GD Rewena	*P. cactorum* (*P. citricola*)	N	↑	↓	Yes	↑	Yes
		CO_2, low N	–	↑	(Yes)	↓	Yes
		CO_2, high N	↑	↓	Yes	↑	Yes
Potato cv. Indira	*Phytophthora infestans*	N	↑	↓	Yes	↑	Yes
		CO_2	(↑)	↓	(Yes)	(↓)	No
		O_3	–	↑	No	↑	No
Potato cv. Bettina	*A. solani*	CO_2	↑	↓	Yes	↓	No
		N	↑	↓	Yes	↓	No
Barley	*D. teres*	CO_2	↑	–	No	–	–
		O_3	↓	–	No	↓	No

Arrows indicate the direction of a significant plant response ($p < 0.05$), when the specified environmental variable is increased. Arrows in brackets indicate trends in plant responses ($p < 0.1$), while "–" indicates no change ($p > 0.1$). Trade-offs between growth and phenolic compounds according to Fig. 3.9 and their relation to susceptibility of the host are indicated can result in different plant-internal resource availabilities, when the physiological states of the plants were different.

A remarkable result of the above meta-study was that only in 5 of 15 cases, the responses of phenolic compounds were able to explain changes in susceptibility, e.g. increasing susceptibility in parallel with decreasing phenolic contents or vice versa. It is important to emphasize that also in cases where growth differentiation ballance was applicable, a negative correlation between phenolic compounds and susceptibility does not prove any causality between plant phenolics and resistance (see Sect. 1.2.3). As shown in Sect. 3.2 many other defence mechanisms, besides phenolics are involved in pathogen defence. Some of them, such as the pathogenesis-related proteins depend on nitrogen availability and might respond differently to environmental changes than do phenolics.

For the analysis of other defence reactions besides phenolic compounds cDNA-arrays for potato, apple and beech were developed based on subtractive cDNA-libraries enriched for pathogen-induced ESTs. It turned out that *Phytophthora plurivora* was able to successfully suppress defence reactions in beech roots successfully (Valcu et al. 2009; Schlink 2010). Although many beech saplings died shortly after infection, some manage to tolerate the infection for a long time, even under elevated CO_2 where *Phytophthora plurivora* infestation of beech saplings was strongly increased compared to ambient CO_2 levels (Fleischmann et al. 2010). There was no indication of a trade-off between growth and defence in surviving plants. However, these plants were characterized by higher net assimilation rates per unit leaf area, by a smaller root system, but unchanged total biomass, and showed an increased number of root tips per unit of root biomass as compared to healthy control trees (Fig. 3.12; Fleischmann et al. 2010; see also Sect. 16.3). Thus, the additional costs for pathogen defence and for the putatively enhanced turnover of fine roots were compensated by an increased carbon gain. In parallel, the size of the attacked root system was reduced, while its efficiency increased at the same time.

Potato and apple showed comparable responses at the transcript level upon pathogen attack. Exemplified transcript levels of selected defence ESTs of potato infected with *Phytophthora infestans* at 72 hpi are shown in Table 3.6 (Ros et al. 2005). At the leaf level, induction of defensive genes was higher in the late blight-susceptible cultivar "Indira" as compared to the moderately resistant cultivar "Bettina". Similar results were found in scab-infected apple leaves of the susceptible cultivar "Golden Delicious" and in the resistant cultivar "Rewena" (data not shown). With the assumption that defence reactions start locally surrounding the infection site of the pathogen, above transcript levels were calculated on the basis of the infected leaf area. In this case transcript levels of the moderately resistant cultivar "Bettina" exceeded those in "Indira" in most cases (Ros et al. 2008). Such results show that there might be trade-offs between growth-related metabolism and pathogen defence at a micro-scale level, which do not extend to higher levels such as a plant organ or even the whole plant. In such cases, the validation or falsification of the GDB strongly depends on the scaling level of the examinations.

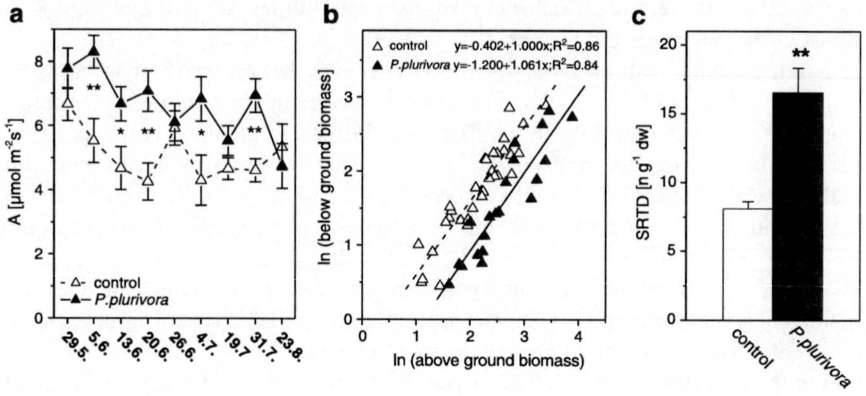

Fig. 3.12 Net photosynthesis rates (**a**), above-ground vs. below-ground biomass (**b**) and specific root tip density (**c**) of beech seedlings grown under elevated CO_2 and infected with *Phytophthora plurivora* for two growing seasons in comparison to non-inoculated controls (*SRTD* specific root tip density). Modified after Fleischmann et al. (2010)

Table 3.6 Gene expression of selected defence genes in the potato cultivars "Indira" (susceptible) and "Bettina" (moderate resistant) 72 h post inoculation with *Phytophthora infestans*

	Total leaf		Symptomatic leaf area	
Calculation basis	Indira	Bettina	Indira	Bettina
pr-1a	**44.7**	3.6	**112**	72
pr-1b	**34.5**	3.4	**86**	68
pr-2	**35.8**	8.6	89	**172**
Glucanase	**30.5**	13.2	76	**264**
Thaumatin-like	14.0	14.4	35	**288**
Proteinase inhibitor	**8.7**	3.4	22	**68**
Peroxidase	**17.1**	5.6	43	**112**
Cell-death ass. protein	**16.7**	5.3	42	**106**

Gene expression is either calculated on the basis of total leaves (left), and on the basis of symptomatic leaf area (right; cultivar "Indira" 40 %, cultivar "Bettina" 5 % symptomatic leaf area), respectively. For accentuation the higher values of each calculation are printed in bold. Original data was taken from Ros et al. (2005)

3.5 Critical Remarks on Trade-off-Studies

Trade-offs between growth and secondary metabolism were described in this review for many plants and under different environmental conditions, which modified availabilities of resources. The mechanisms underlying these observations are in a strict sense not well understood. A limitation of substrates for competing metabolic pathways, i.e. primary and secondary metabolism, may be assumed. Conversely, plants may have other options to channel resources towards occasionally favoured growth and differentiation activities. It might be, as calculated for apple with the generic plant model PLATHO (Chap. 17) that an investment in growth results in a higher benefit than investment in defence (Gayler et al. 2004). Regarding the rather low costs for biosynthesis of defence-related polyphenols it is hard to believe that an active growing shoot with photosynthetically active leaves is unable to provide some carbon for secondary metabolites. Findings on soybean (Kretzschmar et al. 2009) corroborate that idea. Soybean seedlings grown in a CO_2-enriched atmosphere showed stimulated growth and an altered *C*/*N* ratio. When elicitated by NO or by a glucan (PAMP) elicitor from *Phytophthora sojae* the phenylpropanoid pathway was strongly activated resulting in the accumulation of flavonoids and isoflavonoids and of the glyceollin phytoalexin, respectively.

Thus regulation mechanisms beyond a limitation of metabolites may be assumed.

Assessing trade-offs between growth and defence one has to take into account the individual host–herbivore or host–pathogen interaction as well as the type of defence, i.e. constitutive or induced. Constitutive resistance based on a permanent presence of defensive compounds appears to be more costly than the induced supplement. However, also preformed secondary metabolites never occur in the whole plant or in all cells within a tissue. They are rather localized only on strategically important sites for plant defence, such as epidermal cells. This exactly is the defence strategy of resistant plants: They tend to commit "partial suicide" instead of a broad induction of defence compounds all over the plant. On the other hand, the widespread existence of induced resistance may suggest a selective advantage over constitutive defence which is pointed out by Walters et al. (2005). These authors also postulate that induced resistance is more targeted to the invading organism and they showed that the kind of elicitor determines the particular response of the host plants.

In many experiments, artificial resistance inducers have been used to increase our understanding of plant–pathogen interactions and to elucidate trade-offs between growth and defence. However, Heil and Baldwin (2002) doubt in their review paper that a negative correlation between fitness parameters and induced resistance by JA, MeJA or SA can always be interpreted as a trade-off, and that induced defence inevitably causes fitness costs. Their disbelief is based on the knowledge that all these compounds directly affect fitness parameters besides their capacity to induce defence reactions (Creelman and Mullet 1997). Jasmonates are known to reduce seed germination, root growth and photosynthesis. Chaudhry et al. (1994) showed that jasmonate induced a 60 kDa ribosome-inactivating protein in barley. Reinbothe et al. (1993) proved that exogenously applied MeJA affected plastid gene expression at the mRNA and protein level. For instance the translation of the large subunit of ribulose-1,5-bisphosphate carboxylase/oxygenase as well as of the 65 and 68 kDa proteins of photosystem I was reduced in barley leaves. This dilemma between induction of resistance traits and additional side effects on plant growth was recently demonstrated by Gould et al. (2008). MeJA treated *Pinus radiata* seedlings exhibited induced resistance towards *Diplodia pinea* during the first 2 weeks of treatment, afterwards this effect declined. In parallel with elevated disease resistance a significant reduction in seedling growth rate was monitored after the second week of MeJA exposure. Due to the fact that MeJA treatment strongly reduced CO_2 uptake rate shortly after application the authors stated that this side effect of the resistance inducer can account for growth reduction.

Therefore all the studies with resistance inducers have to be discussed carefully. It cannot be generally hypothesized that increased disease resistance is associated with reduced growth due to a change in resource allocation from growth to defence (Herms and Mattson 1992; Stamp 2003). The results of Gould et al. (2008) clearly show that simply correlating growth or fitness parameters with resistance traits is not adequate to deduce trade-offs or costs for resistance. The work of Zeller et al. (2010) also proves that it is crucial which physiological traits are used to calculate cost for resistance and under which conditions the experiments are carried out. For example they got the opposite result regarding cost for resistance for glasshouse and field experiments.

A general problem in interpretation of literature with regards to the trade-off between growth and defence is an analytical one. In many studies only information about total phenolics is given and the particular defence-related compounds are not separately examined. The methods for estimating total phenolics detect only the predominating compounds whereas minor components are overseen.

Moreover, secondary plant metabolites may act as developmental regulators. This was recently reviewed for flavonoids (Taylor and Grotewold 2005). Therefore, a feedback from defence-related secondary metabolites towards the growth-related metabolism has also to be taken into account.

3.6 Conclusions and Outlook

In this chapter many examples of possible trade-offs between growth-related processes and defence-related metabolism are described. It provides a range of environmentally affected plant–pathogen interactions which could be used as models for further studies aiming in describing the consequences of changes in the concentrations of secondary metabolites for plant–parasite/pathogen interactions. General assumptions and open questions are outlined briefly:

- It is generally accepted that plant defence is very often based on its secondary metabolism. Thus, metabolic precursors have to be shared and several enzymatic switches compete for the same carbon skeletons as substrates.
- A prominent deficiency of many studies in this respect is the use of inappropriate methods for measuring defence-related compounds. The widely used estimation of i.e. total phenolics or tannins ignores the structural diversity of secondary metabolites and different physiological activities of the individual metabolites. The application of crude analytical methods gives no insight into the balance among different branches of the secondary metabolism.
- For further understanding of the trade-off between growth and defence-related compounds more intense metabolomic studies should be performed and attempts should be made in elucidating the regulatory mechanisms of the related switches

between the pathways. Their embedding in developmental processes, thus influenced by plant growth regulators has also to be taken into account.

- Pathogens, herbivores and abiotic elicitors (Chap. 2) often affect secondary metabolism as well as susceptibility or resistance of plants. In most cases, however, clear relationships between these observations or even causalities cannot be deduced. In case of a trade-off between growth and secondary metabolism and an increase in resistance, one can only conclude costs for resistance when it is proven that the secondary metabolites investigated are casually linked with defence. This exactly is the dilemma of many of the studies discussed above.

References

Agrawal AA (2000) Benefits and costs of induced plant defence for *Lepidium virginicum* (Brassicaceae). Ecology 81:1804–1813

Baker JC, Orlandi EW (1995) Active oxygen in plant pathogenesis. Ann Rev Phytopathol 33:299–321

Baldwin IT (1988a) Short-term damage-induced alkaloids protected plants. Oecologia 75:367–370

Baldwin IT (1988b) The alkaloidal responses of wild tobacco to real and simulated herbivory. Oecologia 77:378–381

Baldwin IT (1989) The mechanism of damage-induced of alkaloids in wild tobacco. J Chem Ecol 15:1661–1680

Baldwin IT, Cynthia L, Sims L, Kean SE (1990) The reproductive consequences associated with inducible alkaloidal responses in wild tobacco. Ecology 71:252–262

Bednarek P, Osbourn A (2009) Plant-microbe interactions: chemical diversity in plant defence. Science 324:746–748

Bergelson J, Purrington CB (1996) Surveying patterns in the cost of resistance in plants. Am Nat 148:536–558

Bisgrove SR, Simonich MT, Smith NM, Sattler A, Innes RW (1994) A disease resistance gene in *Arabidopsis* with specificity for two different pathogen avirulence genes. Plant Cell 6:927–933

Boller T, Felix G (2009) A renaissance of elicitors: perception of microbe-associated molecular patterns and danger signals by pattern-recognition receptors. Annu Rev Plant Biol 60:379–406

Brown JKM (2002) Yield penalties of disease resistance in crops. Curr Opin Plant Biol 5:1–6

Burdon JJ, Müller WJ (1987) Measuring the cost of resistance to *Puccinia coronata cda* in *Avena fatua*. J Appl Ecol 24:191–202

Cahill DM, McComb JA (1992) A comparison of changes in phenylalanine ammonia-lyase activity, lignin and phenolic synthesis in the roots of *Eucalyptus calophylla* (field resistant) and *E. marginata* (susceptible) when infected with *Phytophthora-cinnamomi*. Physiol Mol Plant Pathol 40:315–332

Carter JP, Spink J, Cannon PF, Daniels MJ, Osbourn AE (1999) Isolation, characterization, and avenacin sensitivity of a diverse collection of cereal-root-colonizing fungi. Appl Environ Microbiol 65:3364–3372

Chaudhry B, Müller-Uri F, Cameron-Mills V, Gough S, Simpson D, Skriver K, Mundy J (1994) The barley 60 kDa jasmonate-induced protein (JIP60) is a novel ribosome-inactivating protein. Plant J 6:815–824

Cipollini DF (2002) Does competition magnify the fitness costs of induced responses in *Arabidopsis thaliana*? A manipulative approach. Oecologia 13:514–520

Coley PD (1986) Costs and benefits of defense by tannins in a neotropical tree. Oecologia 70:238–241

Coley PD, Bryant JP, Chapin FS (1985) Resource availability and plant antiherbivore defense. Science 230:895–899

Creelman RA, Mullet JE (1997) Biosynthesis and action of jasmonates in plants. Annu Rev Plant Physiol Plant Mol Biol 48:355–381

Dixon RA, Paiva NL (1995) Stress-induced phenylpropanoid metabolism. Plant Cell 7:1085–1097

Dodds PN, Rathjen JP (2010) Plant immunity: towards an integrated view of plant-pathogen interactions. Nat Rev Biol 11:539–548

Donaldson JR, Kruger EL, Lindroth RL (2006) Competition- and resource-mediated tradeoffs between growth and defensive chemistry in trembling aspen (*Populus tremuloides*). New Phytol 169:561–570

Ellis C, Turner JG (2001) The Arabidopsis mutant cev1 has constitutively active jasmonate and ethylene signalling pathways and enhanced resistance to pathogens. Plant Cell 13:1025–1033

Fischer DG, Hart SC, Rehill BJ, Lindroth RL, Keim P, Whitham TG (2006) Do high-tannin leaves require more roots? Oecologia 149:668–675

Fleischmann F, Raidl S, Oßwald W (2010) Changes in susceptibility of beech (*Fagus sylvatica*) seedlings towards *Phytophthora citricola* under the influence of elevated atmospheric CO_2 and nitrogen fertilization. Environ Pollut (Special Issue) 158:1051–1060

Flor HH (1971) Current status of the gene-for-gene concept. Annu Rev Phytopathol 9:275–296

Fritz C, Palacios-Rojas N, Feil R, Stitt M (2006) Regulation of secondary metabolism by the carbon–nitrogen status in tobacco: nitrate inhibits large sectors of phenylpropanoid metabolism. Plant J 46:533–548

Fry WE (2007) The canon of potato science: 10. Late blight and early blight. Potato Res 50:243–245

Gayler S, Leser C, Priesack E, Treutter D (2004) Modelling the effect of environmental factors on the "trade-off" between growth and defensive compounds in young apple trees. Trees 18:363–371

Gessler C, Patocchi A, Sansavini S, Tartarini S, Gianfranceschi L (2006) *Venturia inaequalis* resistance in apple. Crit Rev Plant Sci 25:473–503

Glynn C, Herms DA, Egawa M, Hansen R, Mattson WJ (2003) Effects of nutrient availability on biomass allocation as well as constitutive and rapid induced herbivore resistance in poplar. Oikos 101:385–397

Glynn C, Herms DA, Orians CM, Hansen RC, Larsson S (2007) Testing the growth-differentiation balance hypothesis: dynamic responses of willows to nutrient availability. New Phytol 176:623–634

Golba B, Treutter D, Kollar A (2012) Effects of apple (*Malus x domestica* Borkh.) phenolic compounds on proteins and cell wall-degrading enzymes of *Venturia inaequalis*. Trees 26:131–139

Gould N, Reglinski T, Spiers M, Taylor T (2008) Physiological trade-offs associated with methyl jasmonate – induced resistance in Pinus radiate. Can J For Res 38:677–684

Graham TL, Graham MY, Subramanian S, Yu O (2007) RNAi silencing of genes for elicitation or biosynthesis of 5-deoxyisoflavonoids suppresses race-specific resistance and hypersensitive cell death in *Phytophthora sojae* infected tissues. Plant Physiol 144:728–740

Hain R, Reif H-J, Krause E, Langebartels R, Kindl H, Vornam B, Wiese W, Schmelzer E, Schreier PH, Stöcker RH, Stenzel K (1993) Disease resistance results from foreign phytoalexin expression in a novel plant. Nature 361:153–156

Hammerschmidt R (1999) Phytoalexins: what have we learned after 60 years? Annu Rev Phytopathol 37:285–306

Han K, Lincoln DE (1994) The evolution of carbon allocation to plant secondary metabolites: a genetic analysis of cost in *Diplacus aurantiacus*. Evolution 48:1550–1563

Häring DA, Huber MJ, Suter D, Edwards PJ, Lüscher A (2008) Plant enemy-derived elicitors increase the foliar tannin concentration of *Onobrychis viciifolia* without a trade-off to growth. Ann Bot 102:979–987

Harris DC (1991) The phytophthora diseases of apple. J Hortic Sci 66:513–544

Hartmann T (2007) From waste products to ecochemicals: fifty years research of plant secondary metabolism. Phytochemistry 68:2831–2846

Heil M, Baldwin IT (2002) Fitness costs of induced resistance: emerging experimental support for a slippery concept. Trends Plant Sci 7:61–67

Heil M, Hilpert A, Kaiser W, Linsenmair KE (2000) Reduced growth and seed set following chemical induction of pathogen defence: does systemic acquired resistance (SAR) incur allocation costs? J Ecol 88:645–654

Herms DA, Mattson WJ (1992) The dilemma of plants: to grow or defend. Q Rev Biol 67:283–335

Holzapfel C, Meisel B, Thümmler F, Leser C, Treutter D (2012) Differential gene expression in leaves of a scab susceptible and a resistant apple cultivar upon *Venturia inaequalis* inoculation. Trees 26:121–129

Hückelhoven R (2007) Cell wall-associated mechanisms of disease resistance and susceptibility. Annu Rev Phytopathol 45:101–127

Hückelhoven R, Kogel K-H (2003) Reactive oxygen intermediates in plant-microbe interactions: who is who in powdery mildew resistance? Planta 216:891–902

Ingle RA, Carstens M, Denby KJ (2006) PAMP recognition and the plant-pathogen arms race. Bioessays 28:880–889

Jung T (2009) Beech decline in Central Europe driven by the interaction between *Phytophthora* infections and climatic extremes. Forest Pathol 39:73–94

Jung T, Burgess TI (2009) Re-evaluation of *Phytophthora citricola* isolates from multiple woody hosts in Europe and North America reveals a new species, *Phytophthora plurivora* sp nov. Persoonia 22:95–110

Jung T, Hudler G, Jensen-Tracy S, Griffiths H, Fleischmann F, Oßwald W (2005) Involvement of *Phytophthora* species in the decline of European beech in Europe and the USA. Mycologist 19:159–166

Kováčik J, Klejdus B, Bačkor M (2009) Nitric oxide signals ROS scavenger-mediated enhancement of PAL activity in nitrogen-deficient *Matricaria chamomilla* roots: side effects of scavengers. Free Radic Biol Med 46:1686–1693

Kretzschmar FDS, Aidara MPM, Salgado I, Braga MR (2009) Elevated CO_2 atmosphere enhances production of defence-related flavonoids in soybean elicited by NO and a fungal elicitor. Environ Expl Bot 65:319–329

Kuc J (1995) Phytoalexins, stress metabolism, and disease resistance in plants. Annu Rev Phytopathol 33:275–297

Lattanzio V, Kroon PA, Quideau S, Treutter D (2008) Introduction: plant phenolics – secondary metabolites with diverse functions. In: Daayf F, Lattanzio V (eds) Recent advances in polyphenol research, vol 1. Wiley-Blackwell, Chichester, pp 1–35

Le Bot J, Bénard C, Robin C, Bourgaud F, Adamowicz S (2009) The 'trade-off' between synthesis of primary and secondary compounds in young tomato leaves is altered by nitrate nutrition: experimental evidence and model consistency. J Exp Bot 60:4301–4314

Lea US, Slimestad R, Smedvig P, Lillo C (2007) Nitrogen deficiency enhances expression of specifc MYB and bHLH transcription factors and accumulation of end products in the flavonoid pathway. Planta 225:1245–1253

Leser C, Treutter D (2005) Effects of nitrogen supply on growth, contents of phenolic compounds and pathogen (scab) resistance of apple trees. Physiol Plant 123:49–56

Lewis NG, Yamamoto E (1989) Tannins – their place in plant metabolism. In: Hemingway RW, Karchesy JJ (eds) Chemistry and significance of condensed tannins. Plenum, New York, pp 23–43

Magg T, Melchinger AE, Klein D, Bohn M (2001) Comparison of *Bt* maize hybrids with their non-transgenic counterparts and commercial varieties for resistance to European corn borer and for agronomic traits. Plant Breed 120:397–403

Martin GB, Bogdanove AJ, Sessa G (2003) Understanding the functions of plant disease resistance proteins. Annu Rev Plant Biol 54:23–61

Matyssek R, Agerer R, Ernst D, Munch JC, Oßwald W, Pretzsch H, Priesack E, Schnyder H, Treutter D (2005) The plant's capacity in regulating resource demand. Plant Biol 7:560–580
Matyssek R, Sandermann H (2003) Impact of ozone on trees: an ecophysiological perspective, Bot, vol 64, Progress in botany. Springer verlag, Heidelberg, pp 349–404
Mayr U, Treutter D (1998) Flavanols as defence barriers in apple leaves against the apple scab fungus (Venturia inaequalis). Acta Hortic 456:79–82
Mayr U, Michalek S, Treutter D, Feucht W (1997) Phenolic compounds of apple and their relationship to scab resistance. J Phytopathol 145:69–75
McLean MS, Howlett BJ, Hollaway GJ (2009) Epidemiology and control of spot form of net blotch (*Pyrenophora teres* f. *maculata*) of barley: a review. Crop Pasture Sci 60:303–315
Mert-Türk F (2002) Phytoalexins: defence or just a response to stress? J Cell Mol Biol 1:1–6
Nürnberger T, Lipka V (2005) Non-host resistance in plants: new insights into an old phenomenon. Mol Plant Pathol 6:335–345
Olsen KM, Slimestad R, Lea US, Brede C, Løvdal T, Ruoff P, Verheul M, Lillo C (2009) Temperature and nitrogen effects on regulators and products of the flavonoid pathway: experimental and kinetic model studies. Plant Cell Environ 32:286–299
Palumbo MJ, Putz FE, Talcott ST (2007) Nitrogen fertilizer and gender effects on the secondary metabolism of yaupon, a caffeine-containing North American holly. Oecologia 151:1–9
Papadopoulou K, Melton RE, Leggett M, Daniels MJ, Osbourn AE (1999) Compromised disease resistance in saponin-deficient plants. Proc Natl Acad Sci USA 96:12923–12928
Perner H, Rohn S, Driemel G, Batt N, Schwarz D, Kroh LW, George E (2008) Effect of nitrogen species supply and mycorrhizal colonization on organosulfur and phenolic compounds in onions. J Agric Food Chem 56:3538–3545
Pieterse MJ, Leon-Reyes A, Van der Ent S, Van Wees SCM (2009) Networking by small-molecular hormones in plant immunity. Nat Chem Biol 5:308–316
Reinbothe S, Reinbothe C, Heintzen C, Seidenbecher C, Parthier B (1993) A methyljasmonate-induced shift in the length of the 5′ untranslated region impairs translation of the plastid *rbcL* transcript in barley. EMBO J 12:1505–1512
Ros B, Thümmler F, Wenzel G (2005) Comparative analysis of *Phytophthora infestans* induced gene expression in potato cultivars with different levels of resistance. Plant Biol 7:686–693
Ros B, Mohler V, Wenzel G, Thümmler F (2008) *Phytophthora infestans*-triggered response of growth- and defense-related genes in potato cultivars with different levels of resistance under the influence of nitrogen availability. Physiol Plant 133:386–396
Royo RJ, Leon J, Vancanneyt G, Albar JP, Rosahl S, Ortego F, Castanera P, Sanchez-Serrano JJ (1999) Antisense-mediated depletion of a potato lipoxygenase reduces wound induction of proteinase inhibitors and increases weight gain of insect pests. Proc Natl Acad Sci USA 96:1146–1151
Rühmann S, Leser C, Bannert M, Treutter D (2002) Relationship between growth, secondary metabolism, and resistance of apple. Plant Biol 4:137–143
Schlink K (2010) Down-regulation of defence genes and resource allocation into infected roots as factors for compatibility between Fagus sylvatica and Phytophthora citricola. Funct Integr Genomics 10:253–264
Schornack S, Huitema E, Cano LM, Bozkurt TO, Oliva R, Van Damme M, Schwizer S, Raffaele S, Chaparro-Garcia A, Farrer R, Segretin ME, Bos J, Haas BJ, Zody MC, Nusbaum C, Win J, Thines M, Kamoun S (2009) Ten things to know about oomycete effectors. Mol Plant Pathol 10:795–803
Shetty K (2004) Role of proline-linked pentose phosphate pathway in biosynthesis of plant phenolics for functional food and environmental applications: a review. Process Biochem 9:789–803
Snyder BA, Nicholson RL (1990) Synthesis of phytoalexins in sorghum as a site specific response to fungal ingress. Science 248:1637–1639
Stamp N (2003) Out of the quagmire of plant defense hypotheses. Q Rev Biol 78:23–55

Stamp N (2004) Can the growth-differentiation balance hypothesis be tested rigorously? Oikos 107:439–448

Strissel T, Halbwirth H, Hoyer U, Zistler C, Stich K, Treutter D (2005) Growth promoting nitrogen nutrition affects flavonoid biosynthesis in young apple (*Malus domestica* Borkh.) leaves. Plant Biol 7:677–685

Taylor LP, Grotewold E (2005) Flavonoids as developmental regulators. Curr Opin Plant Biol 8:317–323

Thaler J (1999) Induced resistance in agricultural crops: effects of jasmonic acid on herbivory and yield in tomato plants. Popul Ecol 28:30–37

Thomma BPHJ, Nelissen I, Eggemont K, Broekaert WF (1999) Deficiency in phytoalexin production causes enhanced susceptibility of *Arabidopsis thaliana* to the fungus *Alternaria brassicicola*. Plant J 19:163–171

Thordal-Christensen H, Zhang Z, Wie Y, Collinge DB (1997) Subcellular localization of H_2O_2 in plants. H_2O_2 accumulation in papillae and hypersensitive response during the barley-powdery mildew interaction. Plant J 11:1187–1194

Tian D, Traw MB, Chen JQ, Kreitman M, Bergelson J (2003) Fitness costs of R-gene-mediated resistance in *Arabidopsis thaliana*. Nature 423:74–77

Treutter D (2005) Significance of flavonoids in plant resistance and enhancement of their biosynthesis. Plant Biol 7:581–591

Treutter D (2010) Managing phenol contents in crop plants by phytochemical farming and breeding – visions and constraints. Int J Mol Sci 10:1–x. doi:10.3390/ijms100x000x

Valcu CM, Junqueira M, Shevchenko A, Schlink K (2009) Comparative proteomic analysis of responses to pathogen infection and wounding in Fagus sylvatica. J Proteome Res 8:4077–4091

Van Dam NM, Baldwin IT (2001) Competition mediates costs of jasmonate-induced defences, nitrogen acquisition and transgenerational plasticity in *Nicotiana attenuate*. Funct Ecol 15:406–415

Van Etten HD, Mansfield JW, Bailey JA, Farmer EE (1994) Two classes of plant antibiotics: phytoalexins versus "phytoanticipins". Plant Cell 6:1191–1192

Van Loon LC, Rep M, Pieterse CMJ (2006) Significance of inducible defense-related proteins in infected plants. Annu Rev Phytopathol 44:135–162

Walters D, Walsh D, Newton A, Lyon G (2005) Induced resistance for plant disease control: maximizing the efficacy of resistance elicitors. Phytopathology 95:1368–1373

Wilkens RT (1997) Limitations of evaluating the growth-differentiation balance hypothesis with only two levels of light and water. Ecoscience 4:319–326

Wink M (1988) Plant breeding: importance of plant secondary metabolites for protection against pathogens and herbivores. Theor Appl Genet 75:225–233

Yahiaoui N, Srichumpa P, Dudler R, Keller B (2004) Genome analysis at different ploidy levels allows cloning of the powdery mildew resistance gene Pm3b from hexaploid wheat. Plant J 37:528–538

Zeller SL, Kalinina O, Brunner S, Keller B, Schmid B (2010) Transgene × environment interactions in genetically modified wheat. PLoS One 5:e11405

Chapter 4
Conifer Defense Against Insects and Fungal Pathogens

N. Kolosova and J. Bohlmann

4.1 Introduction

During their long life times conifer trees are exposed to a large number of herbivorous insect species and pathogenic fungi. For example, specialized insects such as bark beetles, weevils, and budworms are causing substantial losses to both natural and plantation forests, in particular during outbreaks that affect vast forest areas. Most recently, the mountain pine beetle (*Dendroctonus ponderosae*) epidemic has affected more than 15 million hectares of western Canada's forests and is spreading further east breaching geographical and climatic barriers and affecting potentially new host species (Raffa et al. 2008). Bark beetles, associated with pathogenic blue-stain fungi, can kill healthy and weakened trees by mass attack (Paine et al. 1997). Among weevils, one of the most destructive forest pests in North America is the white pine weevil or spruce shoot weevil (*Pissodes strobi*) (Alfaro et al. 2002). Among defoliators, the western spruce budworm (*Choristoneura occidentalis*) is one of the most destructive conifer pests in North America affecting several host species including species of spruce (Alfaro et al. 1982; Nealis et al. 2009). In addition, conifers are exposed to an array of pathogenic fungi. Among fungi transferred by bark beetles, some of the most pathogenic and widely spread are the ophistomatoid fungi such as the mountain pine beetle-associate *Grosmannia clavigera*, which can rapidly colonize the phloem and sapwood (Yamaoka et al. 1995). Among fungi that infect conifer species independent of insect vectors, *Heterobasidion annosum* is one of the most destructive pathogens affecting several conifer species (Asiegbu et al. 2005a, b). *Heterobasidion annosum* infects tree roots and spreads up into stem tissues. Pitch canker disease caused by *Fusarium circinatum* infects pine species throughout the world (Gordon 2006).

N. Kolosova • J. Bohlmann (✉)
Michael Smith Laboratories, University of British Columbia, 2185 East Mall, Vancouver, BC, Canada V6T 1Z4,
e-mail: bohlmann@msl.ubc.ca

R. Matyssek et al. (eds.), *Growth and Defence in Plants*, Ecological Studies 220,
DOI 10.1007/978-3-642-30645-7_4, © Springer-Verlag Berlin Heidelberg 2012

Conifers are often attacked simultaneously by insects and fungal pathogens. Association of blue-stain fungi with bark beetles provides the fungi with the opportunity not only to move from tree to tree, but also to breach host defense barriers such as thick bark tissues that protect conifers from pathogen invasion. The fungi also assist bark beetles in colonizing the tree. Pathogenic blue-stain fungi weaken tree defenses, cause tree dehydration, and accelerate tree death (Paine et al. 1997). The presence of blue-stain fungi is necessary for successful reproduction of the mountain pine beetle (Six and Paine 1998). Despite the sometimes severe effects from attack of specialized pests and pathogens, conifer trees have a range of defenses that protect them against many potential pests and pathogens. These defenses include the production of chemicals such as oleoresin terpenoids and phenolics that can be toxic to insects and fungi, anatomical structures to store and transport these chemicals, and the activation of pathogenesis-related genes (Franceschi et al. 2005; Keeling and Bohlmann 2006a; Bohlmann 2008). This massive enhancement of defense compounds and transcripts clearly points to a trade-off between growth and defense according to GDB, and the production of anatomical structures for storage of secondary compounds is not conducive for growth. Please note that topics dealing with phenolics and pathogenesis-related proteins in plant defense will also be addressed in Chap. 3.

4.2 Constitutive and Induced Anatomical Defenses

The primary defensive anatomy of conifers includes the outer bark, which consists mostly of lignified and suberized cells, and serves as a first physical barrier against insect and fungal pathogens. The inner bark and secondary phloem contain a variety of defense-related cell types such as heavily lignified sclerenchyma and sclereids, cells containing calcium oxalate crystals, polyphenolic parenchyma, which may be involved in the production and storage of toxic phenolics, and resin ducts or resin cells, which store terpenoids (Franceschi et al. 2005; Bohlmann 2008). Upon wounding, fungal infection, or insect attack, conifer stem tissues respond with lesion formation, which is commonly referred to as the hypersensitive response involving cell death and the accumulation of constitutive and induced phenolics and terpenoids in the affected areas (Lieutier 2002). The hypersensitive response and the release of toxic chemicals may restrict and possibly kill insects and fungal pathogens. Formation of the wound periderm localizes the damage. The induced activation of polyphenolic parenchyma and traumatic resin ducts, which are formed in response to the attacks, further enhance conifer defense capacity against the current threat and additional attacks (Franceschi et al. 2005; Bohlmann 2008).

4.2.1 Resin Ducts

Species of spruce and pine have constitutive, axial resin ducts in the bark and xylem which are connected via radial resin ducts forming a three-dimensional network

(Franceschi et al. 2005). Once disrupted by wounding or insect damage, resin ducts release a terpenoid rich oleoresin that may repel insects and pathogens. In response to wounding, pathogen invasion, herbivore attack, or upon treatment with the defense elicitor methyl jasmonate, the formation of new resin ducts occurs in the xylem of several conifer species (Franceschi et al. 2000; Martin et al. 2002; Hudgins et al. 2003, 2004; Huber et al. 2005; Miller et al. 2005; Zulak and Bohlmann 2010). Induced formation of traumatic resin ducts has been studied in much detail in Norway spruce (*Picea abies*). Traumatic resin ducts in Norway spruce xylem are formed in response to mechanical wounding (Nagy et al. 2000), inoculation with blue-stain fungus (Franceschi et al. 2000), inoculation with root rot fungus (Krekling et al. 2004), and in response to treatment with methyl jasmonate (Franceschi et al. 2002; Martin et al. 2002). Formation of traumatic resin ducts was also reported for wound-induced and fungus-inoculated Austrian pine (*Pinus nigra*) (Luchi et al. 2005) and for fungus-infected western white pine (*Pinus monticola*) (Hudgins et al. 2005). It was shown that wounding and fungal infection lead to the formation of traumatic resin ducts in spruce and pine, not only at the site of infection but to some degree also in distant tissues (Christiansen et al. 1999; Nagy et al. 2000; Krekling et al. 2004; Luchi et al. 2005). Traumatic resin ducts contribute to increased resin production and the newly formed resin can be different in chemical composition compared to constitutive resin (Martin et al. 2002; Faldt et al. 2003; Miller et al. 2005; Zulak et al. 2009).

4.2.2 Polyphenolic Parenchyma, Sclerenchyma, and Sclereids

Phloem polyphenolic parenchyma (PP) cells tare localized in the bark and are thought to be involved in the production of toxic phenolics (Franceschi et al. 1998). Wounding, fungal infection, and methyl jasmonate induce the formation of additional layers of PP cells (Franceschi et al. 2000, 2002; Krokene et al. 2003; Hudgins et al. 2004; Krekling et al. 2004). Comparison of two clones of Norway spruce selected for their resistance to blue-stain fungi demonstrated a correlation between resistance to blue-stain fungi and the density of phenolic parenchyma cells (Franceschi et al. 1998). Phenolics in the form of lignin also contribute to the formation of sclerenchyma and sclereids, which contribute to defense due to their mechanical strength. Analysis of phenolics present in individual sclereid cells revealed that, in addition to lignin, these cells also accumulate low-molecular weight phenolics such as stilbenes and flavonoids (Li et al. 2007). This observation suggests that sclereids may be involved in chemical and physical defense. In spruce bark, increased sclereid cell density is associated with higher resistance to weevils (King and Alfaro 2009).

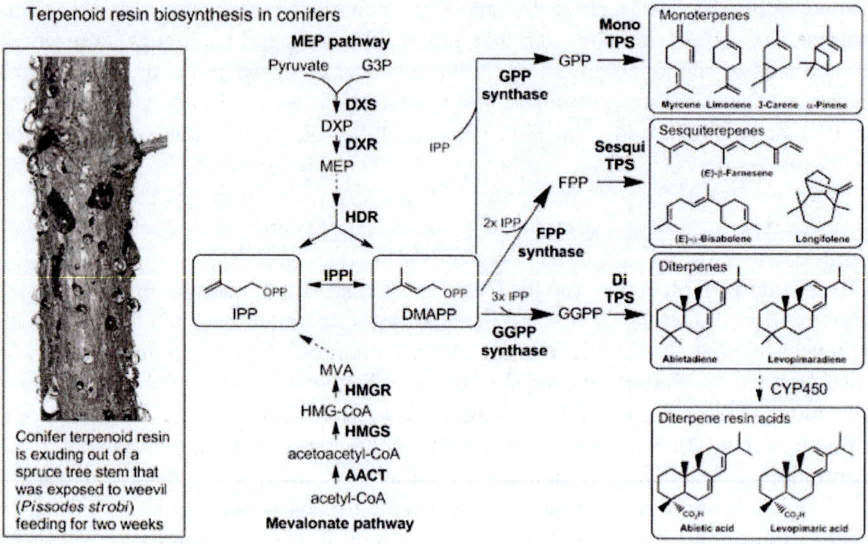

Fig. 4.1 Terpenoid resin biosynthesis in conifers. Monoterpenes, sesquiterpenes, and diterpenes are synthesized from IPP and DMAPP by the activity of prenyltransferases and terpene synthases. Diterpenes are further converted to diterpene resin acids through cytochrome P450 enzymes. Abbreviations: *MEP* 2-C-methyl-D-erythritol 4-phosphate, *G3P* glyceraldehyde 3-phosphate, *DXP* 1-deoxy-D-xylulose 5-phosphate, *DXS* 1-deoxy-D-xylulose 5-phosphate synthase, *DXR* 1-deoxy-D-xylylose 5-phosphate reductoisomerase, *HDR* 4-hydroxy-3-methylbut-2-enyl diphosphate reductase, *IPP* isopentenyl diphosphate, *IPPI* isopentenyl diphosphate isomerase, *DMAPP* dimethylallyl diphosphate, *AACT* acetoacetyl-CoA thiolase, *HMG-CoA* hydroxymethylglutaryl CoA, *HMGS* hydroxymethylglutaryl CoA synthase, *HMGR* hydroxymethylglutaryl CoA reductase, *MVA* mevalonic acid, *GPP* geranyl diphosphate, *FPP* farnesyl diphosphate, *GGPP* geranylgeranyl diphosphate, *TPS* terpene synthase, *CYP450* cytochrome P450

4.3 Constitutive and Induced Chemical Defenses

4.3.1 Terpenoid Oleoresin

A prominent group of chemicals involved in conifer defense are oleoresin terpenoids (Fig. 4.1). Constitutive terpenoid resin production is considered to be one of the most important defenses of conifers against the initial invasion of pests and fungi (Paine et al. 1997; Langenheim 2003; Raffa et al. 2005; Bohlmann 2008). When bark integrity is breached by wounding or insect attack, resin, which is usually contained under pressure, is released and can trap invading insects or pathogens and seals the wound. Conifer resin typically contains large amounts of monoterpenes and diterpene resin acids and smaller quantities of sesquiterpenes. The oleoresin of spruce species contains more than 30 different known terpenoid chemicals (Martin et al. 2002; Miller et al. 2005; Zeneli et al. 2006) and similar chemical diversity of terpenoids was detected in other conifer species (Huber et al. 2005; Manninen et al. 2002).

Monoterpenes and sesquiterpenes are volatile compounds that evaporate from resin exposed at wound sites, whereas the less volatile diterpene resin acids accumulate and crystallize, ultimately forming a solid wound seal (Phillips and Croteau 1999; Trapp and Croteau 2001; Huber et al. 2004; Keeling and Bohlmann 2006a). Several oleoresin terpenoids have been shown to have antibacterial properties (Himejima et al. 1992), antifungal properties, including toxicity against blue-stain fungi (Delorme and Lieutier 1990; Paine and Hanlon 1994; Kopper et al. 2005) and toxicity and repellent properties towards bark beetles (Paine et al. 1997) and weevils (Nordlander 1990; Tomlin et al. 1996; Alfaro et al. 2002).

4.3.2 Induced Terpenoid Defenses

The production of oleoresin in a number of conifer species is induced by wounding, insect damage, pathogen infection, or treatment with methyl jasmonate (Keeling and Bohlmann 2006a). Induced resin can have a different terpenoid composition, as compared to constitutive resin (Tomlin et al. 2000; Martin et al. 2002; Miller et al. 2005; Zulak et al. 2009). Bark beetle-associated blue-stain fungus induced a nearly 100-fold increase in total mono- and sesquiterpene levels at the inoculation site in Norway spruce (Viiri et al. 2001). The blue-stain fungus *Grosmannia clavigera* (then known as *Ophiostoma clavigerum*) induced the formation of monoterpenes and diterpenes in 2-year-old lodgepole pine (*Pinus contorta*) saplings (Croteau et al. 1987) and monoterpene formation in mature (about 80 years old) lodgepole pines (Miller et al. 1986). Induced resin production in Norway spruce was correlated with resistance to blue-stain fungus *Ceratocystis polonica* (Zeneli et al. 2006). The importance of induced terpenoid resin production against bark beetle attack was demonstrated in a study which correlated induced resin production in lodgepole pine inoculated with *Grosmannia clavigera* with the survival of these trees during mass mountain pine beetle attack (Raffa and Berryman 1982).

4.3.3 Phenolics

Phenolics, including stilbenes, lignans, flavonoids, proanthocyanidins, and tannins, are abundant in the bark of conifers (Pan and Lundgren 1996; Viiri et al. 2001; Franceschi et al. 2005). Stilbenes and flavonoids may have antifeedant effect on bark beetles (Faccoli and Schlyter 2007) and antifungal properties (Woodward and Pearce 1988; Evensen et al. 2000; Celimene et al. 2001; Vargas-Arispuro et al. 2005). In particular, the stilbene resveratrol inhibited the growth of beetle-associated ophistomatoid fungi (Salle et al. 2005). A higher constitutive concentration of phenolics in conifer bark was associated with conifer resistance to fungal pathogens (Brignolas et al. 1995; Bois and Lieutier 1997; Brignolas et al. 1998). In addition to soluble phenolics, highly lignified cells are considered to be one of the key defense-related structures of tree bark (Pearce 1996; King and Alfaro 2009; Wainhouse et al. 1990, 1997, 1998; Bonello et al. 2003). Wounding and fungal infection induces the production of phenolics in conifer trees (Franceschi et al. 2005). Norway spruce trees that were wounded and inoculated with blue-stain fungus had higher concentrations of (+)-catechin, the flavonoid taxifolin, and the stilbene *trans*-resveratrol when compared to controls (Evensen et al. 2000). Fungal infection induced the accumulation of higher quantities of phenolics than wounding alone. Infection of Austrian pine with *Sphaeropsis sapinea*, a fungus that causes shoot blight and cankers, induced the production of soluble and cell wall bound phenolics, including increased lignin deposition (Bonello and Blodgett 2003). The ability to accumulate phenolics in response to wounding and fungal inoculation, including the accumulation of flavonoids in Norway spruce, was correlated with fungal resistance in conifer species (Brignolas et al. 1995, 1998), the accumulation of stilbenes and flavonoids in Scots pine (*Pinus sylvestris*) (Bois and Lieutier 1997), and the accumulation of stilbenes and lignin in Austrian pine (Wallis et al. 2008).

4.4 Molecular Mechanisms of Conifer Defense Response

4.4.1 Genes Involved in the Formation of Oleoresin Terpenoids

Molecular mechanisms involved in the formation of oleoresin terpenoids have been the topic of recent review articles (Fig. 4.1; e.g., Keeling and Bohlmann 2006a, b; Zulak and Bohlmann 2010). In brief, terpenoids are synthesized from isopentenyl diphosphate (IPP) and dimethylallyl diphosphate (DMAPP). IPP and DMAPP are used by prenyltransferases to form geranyl diphosphate (GPP), farnesyl diphosphate (FPP), and geranylgeranyl diphosphate (GGPP). GPP, FPP, and GGPP are the substrates for monoterpene synthases, sesquiterpene synthases, and diterpene synthases, respectively. IPP and DMAPP are produced through the mevalonic acid (MEV) and methylerythritol phosphate (MEP) pathways. The MEP pathway is thought to be the primary route for biosynthesis of monoterpenoids and diterpenoids of conifer resin (Cordoba et al. 2009). Several conifer genes involved in the MEP pathway were recently characterized in Norway spruce (Phillips et al. 2007) and Japanese red pine (*Pinus densiflora*) (Kim et al. 2009) including members of the 1-deoxyxylulose 5-phosphate synthase (DXS), 1-deoxyxylulose 5-phosphate reductoisomerase (DXR), and hydroxymethylbutenyl 4-diphosphate reductase (HDR) gene families. Increased expression of MEP pathway genes was associated with increased resin production in conifers (Phillips et al. 2007; Kim et al. 2009; Zulak et al. 2009). IPP and DMAPP are the substrates for GPP synthase, FPP synthase, and GGPP synthase. Several prenyltransferases have been characterized in conifers (Keeling and Bohlmann 2006a; Schmidt and Gershenzon 2007, 2008), including a bifunctional GGPP synthase that produces GGPP and GPP (Schmidt et al. 2009). Terpene synthases (TPSs) use DMAPP, GPP, FPP, and GGPP as substrates and are represented by large gene families in angiosperms and gymnosperms. TPSs have been functionally characterized in several conifer species with the majority of the enzymes being characterized from grand fir (*Abies grandis*), Norway spruce, Sitka spruce (*Picea sitchensis*), white spruce (*Picea glauca*), and loblolly pine (*Pinus taeda*) (Keeling and Bohlmann 2006a, b). Many of the conifer monoterpene synthases, sesquiterpene synthases, and diterpene synthases produce multiple products and thus play a central role in generating the structural diversity of terpenoids in conifer defense.

The production of terpenoids is metabolically costly (Gershenzon 1994). However, conifers produce these compounds in very large amounts. The production of oleoresin terpenoids in some pine species may exceed perhaps the production of any other group of defense chemicals in any other plant species. Resource allocation can be controlled at the level of induced response of part of the defensive oleoresin biosynthesis. The TPS genes of conifers play a key role in the regulation of oleoresin terpenoid biosynthesis. Expression of TPS genes is upregulated in response to wounding, insect attack, and methyl jasmonate treatment (Keeling and Bohlmann 2006a). Wounding of grand fir induced the expression of all three classes of TPSs (Steele et al. 1998). Monoterpene synthases were induced within 2 h of wounding but the induction of sesqui- and diterpene synthases was only detectable 3 days after wounding. Additionally, wounding and weevil attack of Sitka spruce resulted in increased transcript levels of monoterpene synthases (McKay et al. 2003; Miller et al. 2005). The level of transcript induction was comparable between wounding and weevil attack, although insect attack resulted in a more rapid transcript upregulation. A detailed study of the effect of weevil attack on Sitka spruce revealed a slower induction of diterpene synthases as compared to the induction of monoterpene synthases, with only weak induction of sesquiterpene synthases (Miller et al. 2005). Blue-stain fungal inoculation also resulted in the increased enzyme activity of monoterpene and diterpene synthases in lodgepole pine (Croteau et al. 1987). Methyl jasmonate treatment induced gene expression and enzyme activities of mono-, sesqui-, and diterpene synthases in Sitka spruce and Norway spruce (Martin et al. 2002, 2003; Miller et al. 2005; Zulak et al. 2009; Hall et al. 2011). Increased transcript and protein levels of TPSs correlate well with the increased production of resin terpenoids in Norway spruce (Zulak et al. 2009). Diterpene resin acids of conifer defense are formed from diterpenes by cytochrome P450 dependent oxidations (Ro et al. 2005; Hamberger and Bohlmann 2006; Ro and Bohlmann 2006). Cytochrome P450 enzymes involved in oxidation of resin diterpenoids have been cloned and functionally characterized in loblolly pine (Ro et al. 2005) and in Sitka spruce (Hamberger and Bohlmann 2006).

4.4.2 Genes Involved in the Production of Phenolics

A large variety of phenylpropanoids are synthesized from phenylalanine through a complex grid of biosynthetic pathways. Genes involved in the production of lignin, flavonoids, and other phenylpropanoid phytoalexins have been characterized in numerous angiosperm species (Dixon et al. 2002; Boerjan et al. 2003). Several genes involved in the phenylpropanoid pathway of conifers and their involvement in defense mechanisms have also been characterized. The first committed step in phenylpropanoid biosynthesis is the conversion of phenylalanine to cinnamic acid by phenylalanine ammonia lyase (PAL). PAL was characterized in loblolly pine (Whetten and Sederoff 1992) and a family of *PAL* genes has been characterized in jack pine, *Pinus banksiana* (Butland et al. 1998). Expression of *PAL* genes was upregulated in jack pine cell cultures treated with a fungal elicitor extracted from the ectomycorrhizal fungus, *Thelephora terrestris* (Butland et al. 1998). In Scots pine seedlings, the expression of *PAL* transcripts was upregulated by endophytic and pathogenic *Rhizoctonia* fungi with higher levels of *PAL* transcript being induced by the pathogenic fungi 2 days post-inoculation (Gronberg et al. 2009). In Norway spruce, PAL protein was localized to phenolic parenchyma cells, supporting a role in the biosynthesis of defense-associated phenolics (Franceschi et al. 1998).

Chalcone synthase catalyzes the first pathway-specific step in flavonoid biosynthesis. Chalcone synthase mRNA accumulated in white spruce (*Picea glauca*) needles treated with wounding both locally and systemically (Richard et al. 2000). In another study, an increase in chalcone synthase transcript level was correlated with resistance of Norway spruce clones to blue-stain fungi (Nagy et al. 2004). Chalcone synthase transcript levels peaked several days earlier in a Norway spruce clone that had higher resistance to blue-stain fungus as compared to the less resistant clone. A rapid induction of chalcone synthase expression in Norway spruce correlated with the increased resistance to bark beetle-associated blue-stain fungus suggests that the increased activation of flavonoid biosynthesis confers improved resistance of conifers to fungal pathogens (Brignolas et al. 1995; Nagy et al. 2004).

One of the most common stilbenes in conifers is pinosylvin. Accumulation of pinosylvin and an increase of pinosylvin synthase activity were observed in seedlings of Scots pine treated with the fungus *Botrytis cinerea* (Gehlert et al. 1990). Several stilbene synthases were cloned (Fliegmann et al. 1992; Schwekendiek et al. 1992) and the multigene family of elicitor-responsive stilbene synthases involved in pinosylvin production has been characterized in Scots pine (Preisig-Muller et al. 1999). Analysis of Scots pine pinosylvin synthase promoters using a tobacco system revealed inducibility in response to fungal treatment (Preisig-Muller et al. 1999). Other studies of pinosylvin synthase and pinosylvin methyltransferase revealed that expression of these genes is upregulated in wounded and fungus-inoculated Scots pine trees (Chiron et al. 2000). Inoculation of Scots pine seedlings with endophytic and pathogenic *Rhizoctonia* fungi species resulted in increased stilbene synthase transcript levels, with increased induction occurring 2 days after inoculation with pathogenic fungus suggesting a defense-related role for the gene (Gronberg et al. 2009). Biochemical characterization of three stilbene synthases and a chalcone synthase from Japanese red pine, *Pinus densiflora*, indicated the potential interaction of stilbenoid and flavonoid biosynthesis through inhibition of chalcone synthase by pinosylvin (Kodan et al. 2002).

Lignans and lignin are synthesized from monolignols which undergo oxidation and random or directed radical coupling. Lignin is synthesized from *p*-coumaryl, coniferyl, and sinapyl alcohols. These monolignols are incorporated into hydroxyphenyl (H), guaiacyl (G), and syringyl (S) lignin, respectively (Boerjan et al. 2003). Gymnosperm lignins are mostly composed of G units with a minor amount of H units. S units are considered to be unique to angiosperms, although S lignin was detected in few gymnosperm species as an exception (Weng and Chapple 2010). Laccases and peroxidases are involved in the production of phenoxy radicals (Boerjan et al. 2003). Laccases are represented in conifers by large gene families (Sato et al. 2001; Ralph et al. 2006a; Koutaniemi et al. 2007) and laccase gene expression was induced in the bark of Sitka spruce by wounding, weevil feeding, and budworm herbivory (Ralph et al. 2006a). Peroxidases are also present in conifers as multigene families (Koutaniemi et al. 2007) and likely have multiple roles in lignification and the defense-related production of reactive oxygen species (Passardi et al. 2005). A conifer peroxidase was first cloned from Norway spruce (Fossdal et al. 2001), and the activity of this peroxidase was upregulated in Norway spruce infected with the blue-stain fungus, *Ceratocystis polonica*. The speed of peroxidase activation was correlated with increased resistance to fungal inoculation (Nagy et al. 2004). Two peroxidase transcripts, which were associated with lignification, were induced in Norway spruce bark by *Heterobasidion annosum* (Koutaniemi et al. 2007), and similarly, peroxidase transcript levels were induced in the roots of Scots pine infected with *Heterobasidion annosum* (Adomas et al. 2007).

Dirigent proteins are involved in the directed stereospecific coupling involved in the biosynthesis of lignans (Davin et al. 1997; Davin and Lewis 2000). A large family of dirigent proteins has been recently characterized in Sitka spruce (Ralph et al. 2006b, 2007a). Several dirigent protein genes were induced by wounding, weevil herbivory, and budworm herbivory in Sitka spruce (Ralph et al. 2006b; Ralph et al. 2007a).

4.4.3 Antimicrobial Proteins in Conifers

In addition to specialized metabolites (e.g., terpenes, phenolics) that serve as chemical barriers to insects and pathogens, plants produce a large number of antimicrobial proteins that may have additional roles in defense (De Lucca et al. 2005). Many of these proteins are also classified as pathogenesis-related (PR) proteins based on their induction by pathogen attack and their lack of constitutive expression. Seventeen families of PR proteins have been described in angiosperm plants (Van Loon and Van Strien 1999; Sels et al. 2008). Several antimicrobial proteins have been characterized in conifer species, including chitinases (Wu et al. 1997; Hietala et al. 2004; Liu et al. 2005), β-1,3-glucanases (Sharma et al. 1993; Asiegbu et al. 1995), peroxidases (Fossdal et al. 2001; Adomas et al. 2007), defensins (Sharma and Lonneborg 1996; Kovalyova and Gout 2008), thaumatin-like proteins (Piggott et al. 2004), PR-10 proteins (Liu et al. 2003), and antimicrobial peptides (AMPs) (Asiegbu et al. 2003, 2005b).

Chitinases are able to digest chitin and degrade the fungal cell wall, resulting in the direct inhibition of fungal growth, and the release of fungal cell wall elicitors that contribute to the induction of plant defense (Collinge et al. 1993). In addition to antifungal properties, chitinases can affect insects by damaging the chitin containing peritrophic matrix (Kramer and Muthukrishnan 1997). Chitinases may also play a role in programmed cell death in angiosperms and conifers (Kasprzewska 2003; Wiweger et al. 2003). Plant chitinases are represented by multigene families that are divided into seven classes based on their gene structure (Meins et al. 1994; Neuhaus 1999). Several class I, II, and IV chitinases were cloned from conifer species (Wu et al. 1997; Davis et al. 2002; Wiweger et al. 2003; Liu et al. 2005; Schmidt et al. 2005). In conifers, chitinase expression is induced by wounding, fungal inoculation, herbivory, methyl jasmonate, salicylic acid, or chitosan (Wu et al. 1997; Davis et al. 2002; Schmidt et al. 2005; Ralph et al. 2006a). In Douglas fir (*Pseudotsuga menziesii*) infected with the root rot fungus *Phellinus suphurascens*, chitinase was localized to the fungal hyphae, consistent with a role in the degradation of fungal cell walls (Islam et al. 2009). Increased expression of chitinases was associated with Norway spruce resistance to *Heterobasidion annosum* (Fossdal et al. 2006), and different patterns of chitinase protein expression was observed in resistant and susceptible western white pine (*Pinus monticola*) infected with *Cronartium ribicola* (Liu et al. 2005).

Beta-1,3-glucanases are often induced along with chitinases and contribute to antifungal activity by digesting β-1,3-glucan fibers that are part of the fungal cell wall (Mauch et al. 1988; Collinge et al. 1993). Induction of β-1,3-glucanases and chitinases was observed in Norway spruce roots infected with the parasitic oomycete *Pythium* sp. (Sharma et al. 1993). Further studies revealed the localization of spruce chitinase and β-1,3-glucanase proteins on the hyphal walls of *Heterobasidion annosum* in infected Norway spruce roots supporting a role for these enzymes in the digestion of fungal cell walls (Asiegbu et al. 1995).

In addition to the antimicrobial proteins with known biochemical function there are a number of antimicrobial plant proteins with unknown biochemical function such as thaumatin-like proteins, defensins, PR-10 proteins, and AMPs. Expression of the thaumatin-like protein gene from western white pine was induced by wounding and blister rust pathogen (*Cronartium ribicola*) infection (Piggott et al. 2004) and transcripts annotated as thaumatin were among the most strongly upregulated in Scots pine inoculated with *Heterobasidion annosum* (Adomas et al. 2007). The function of this protein is unknown, although it may participate in the degradation of the fungal cell wall based on the function of a related thaumatin-like protein in barley (Zareie et al. 2002; Piggott et al. 2004). It is also suggested that thaumatin-like proteins may disrupt hyphal and spore membranes by forming transmembrane pores (De Lucca et al. 2005). A thaumatin-like protein accumulated in Douglas fir root infected with root rot fungus and was localized extracellularly on the host cell membrane, supporting a role in affecting fungal cell wall permeability (Islam et al. 2009).

Several defensin genes have been characterized from conifer species (Sharma and Lonneborg 1996; Kovalyova and Gout 2008). Defensin gene expression was induced by wounding, methyl jasmonate, and fungal treatment in white spruce and the antifungal activity of selected conifer defensins was established in vitro (Pervieux et al. 2004; Gout and Kovalyova 2008; Kovalyova and Gout 2008). Transgenic tobacco and Norway spruce overexpressing a Norway spruce defensin-like gene were more resistant to bacterial and fungal pathogens respectively (Elfstrand et al. 2001).

PR-10 proteins were cloned from several conifer species including maritime pine, western white pine, white spruce, and Douglas fir. Several PR-10 genes are induced by wounding, fungal inoculation, and methyl jasmonate in western white pine (Ekramoddoullah et al. 2000; Dubos and Plomion 2001; Liu et al. 2003; Mattheus et al. 2003). Angiosperm PR-10 proteins are known to have ribonuclease properties and antifungal activity (De Lucca et al. 2005) and PR-10 proteins were localized on fungal cell walls in western white pine infected with blister rust fungus, supporting involvement in antifungal defense (Liu et al. 2003).

A variety of additional unique AMP groups were characterized in angiosperm plants (De Lucca et al. 2005). A family of AMPs (Sp-AMP) was characterized in Scots pine (Asiegbu et al. 2003). Based on homology with plant and yeast proteins it was suggested that these AMPs may inhibit fungal and bacterial cell wall biosynthesis. Expression of an AMP transcript was induced by *Heterobasidium annosum* in Scots pine and the AMP was accumulated on the host cell surface, indicating its direct participation in antifungal defense (Asiegbu et al. 2003, 2005a, b; Adomas et al. 2007).

4.5 Elicitors of Conifer Defense

Research into the signaling mechanisms involved in plant defense in angiosperms revealed three common types of low-molecular weight signaling molecules that affect the downstream expression of defense-related genes: octadecanoids, including jasmonates; ethylene; and salicylic acid (Koornneef and Pieterse 2008). Pathways activated by these signal molecules may interact with each other to produce various patterns of disease and herbivore-induced responses. While defense-associated signaling has been extensively studied in angiosperms using *Arabidopsis thaliana* and other model species (Thomma et al. 2001; Kunkel and Brooks 2002; Koornneef and Pieterse 2008; Wu and Baldwin 2009), very little is known about defense signaling in conifers.

4.5.1 Jasmonate-Induced Defense Responses in Conifers

In conifers, application of methyl jasmonate induces anatomical and biochemical changes similar to those induced by wounding, insect herbivore attack, and pathogen invasion (e.g., Huber et al. 2004; Miller et al. 2005; Zulak et al. 2009). At the anatomical level, polyphenolic parenchyma activation and traumatic resin duct development is induced by methyl jasmonate treatment (Franceschi et al. 2002;

Martin et al. 2002; Hudgins et al. 2003). At the biochemical and molecular levels, methyl jasmonate induces accumulation and biosynthesis of terpenoids (Martin et al. 2002, 2003; Phillips et al. 2007; Faldt et al. 2003; Miller et al. 2005). A comparison of methyl jasmonate induction and white pine weevil (*Pissodes strobi*) feeding on Sitka spruce showed that both treatments produced remarkably similar anatomical, biochemical, and molecular responses (Miller et al. 2005). Both treatments induced the formation of traumatic resin ducts, accumulation of terpenoids, and strong upregulation of TPS expression.

The application of methyl jasmonate induces pathogenesis-related protein gene expression, including increased chitinase transcript levels in slash pine (*Pinus elliottii*) seedlings (Davis et al. 2002), and enhanced expression of PR-10 transcripts in wound-induced western white pine (Liu et al. 2003). In addition, methyl jasmonate application induced chalcone synthase in white spruce (Richard et al. 2000).

Gene expression analysis of Sitka spruce in response to white pine weevil feeding revealed an increased expression of genes putatively involved in the octadecanoid pathway (Miller et al. 2005; Ralph et al. 2006a). The ability of methyl jasmonate to induce defense responses similar to those elicited by wounding, insect feeding, and fungal inoculations is consistent with a role for octadecanoid signaling in induced conifer defense.

4.5.2 *Ethylene Is Involved in the Wound- and Jasmonate-Induced Defense Responses in Conifers*

The involvement of ethylene in the induction of anatomical defenses and correlation between ethylene and jasmonate signaling has recently been demonstrated in Douglas fir (Hudgins and Franceschi 2004). The application of methyl jasmonate and ethylene induced the formation of traumatic resin ducts and activated of polyphenolic parenchyma cells. Inhibition of ethylene signaling resulted in a decreased response to wounding and methyl jasmonate treatment indicating that the methyl jasmonate response is mediated by ethylene. Ethylene production was also correlated with monoterpene biosynthesis following fungal inoculation of slash pine and loblolly pine (Popp et al. 1995), and application of an ethylene releaser, ethrel, increased monoterpene synthase activity in grand fir (Katoh and Croteau 1998).

Several ACC (1-aminocyclopropane-1 carboxylic acid) synthases, enzymes involved in ethylene biosynthesis, were cloned and characterized in white spruce, interior spruce (*Picea glauca* × *Picea engelmannii*), and Douglas fir (Ralph et al. 2007b). Gene expression of selected ACC synthases was induced by wounding and white pine weevil feeding in Sitka spruce and ACC synthase protein accumulated in Douglas fir induced by wounding (Ralph et al. 2007b). Another enzyme involved in ethylene biosynthesis, ACC oxidase, was cloned from Sitka spruce, white spruce, and Douglas fir (Hudgins et al. 2006). Both methyl jasmonate and wounding treatments increased the accumulation of ACC oxidase protein in Douglas fir

(Hudgins and Franceschi 2004; Hudgins et al. 2006). In Douglas fir bark, both ACC synthase and ACC oxidase were localized to resin duct epithelial cells, polyphenolic parenchyma cell, which are involved in conifer defense, and ray parenchyma cells, which connect phloem, cambium, and xylem and may serve in spreading of defense-related signaling (Hudgins et al. 2006; Ralph et al. 2007b). The effect of ethylene in inducing conifer defense responses in addition to defense-related increased expression and localization of the enzymes involved in ethylene biosynthesis supports the role of ethylene in conifer defense signaling.

4.5.3 Salicylic Acid and Conifer Defense Responses

The role of salicylic acid as defense elicitor in conifers is not well established. Salicylic acid accumulated in Norway spruce seedlings treated with the pathogen *Pythium irregulare* and methyl jasmonate (Kozlowski and Metraux 1998; Kozlowski et al. 1999). Treatment of Norway spruce with methyl jasmonate resulted in the induced emission of methyl salicylate (Martin et al. 2003). Application of salicylic acid induced the expression of chitinases in slash pine seedlings (Davis et al. 2002). However salicylic acid failed to induce any anatomical defenses in Douglas fir while MeJa and ethylene treatment induced the activation of phenolic parenchyma and the formation of traumatic resin ducts (Hudgins and Franceschi 2004). Wound-induced expression of PR-10 protein in western white pine was enhanced by methyl jasmonate but was suppressed by salicylic acid (Liu et al. 2003), suggesting different roles of these elicitors in conifer defense.

4.6 Acquired Resistance in Conifers

Acquired resistance can develop in plants after they have been exposed to pathogens, insect attack or wounding, and may result in higher resistance to subsequent infection or insect attack (Kiraly et al. 2007). Local acquired resistance develops in tissues surrounding the infection or damage site and systemic acquired resistance develops in distant tissues. These responses have been extensively studied in angiosperms (Kiraly et al. 2007; Walters 2009). Successful induction of Norway spruce resistance against the beetle-associated blue-stain fungus *Ceratocystis polonica* is possible by prior inoculation of the tree with a sublethal dose of the same fungus (Christiansen et al. 1999; Krokene et al. 1999, 2003). Pretreatment with a sublethal dose of fungus induced the formation of traumatic resin ducts and increased the number of polyphenolic parenchyma cells (Krokene et al. 2003). Pretreatment with fungus was significantly more effective in inducing resistance when compared to wounding alone (Christiansen et al. 1999; Krokene et al. 1999). In addition, the timing of pretreatment is important for successful subsequent resistance to fungi. Trees that were pretreated with fungus 1 week prior to mass inoculation were not significantly more resistant to subsequent fungal inoculations, whereas trees pretreated 3–9 weeks prior to mass inoculation had reduced disease symptoms when inoculated in the pretreated stem areas (Krokene et al. 2003). In these studies, pretreatment and the subsequent inoculations were done on the same part of the stem, supporting the involvement of local acquired resistance. However only small areas of the stems were pretreated (Christiansen et al. 1999) and increased resistance was observed in the entire stem section indicating the spread of resistance from the inoculation sites. Systemic induction of traumatic resin ducts was observed in spruce (Christiansen et al. 1999; Krekling et al. 2004) and pine (Luchi et al. 2005). Systemically acquired resistance was observed in Monterey pine (*Pinus radiata*) pretreated with the pitch canker pathogen *Fusarium circinatum*, and resulted in the increased resistance against the pathogen for at least a year following initial inoculation (Bonello et al. 2001, 2006). A similar increase in resistance was observed in Austrian pine inoculated with canker pathogens, and a resistance increase was observed above and below the inoculation sites, indicating bidirectional signaling (Blodgett et al. 2007). Systemically induced resistance was associated with increased lignin content and may involve the accumulation of soluble phenolics (Bonello and Blodgett 2003; Blodgett et al. 2007; Wallis et al. 2008). In addition, cross induction of systemic acquired resistance by a canker pathogen and insect *Neodiprion sertifer* (European pine sawfly) was observed in Austrian pine (Eyles et al. 2007). Locally and systemically acquired resistance may be an important factor of conifer defense (Bonello et al. 2006).

Although not likely practical for large forest areas, increased resistance of pretreated conifers provides the potential to use pretreatment techniques as protective measures at least for individual trees. The effect of methyl jasmonate treatment without wounding of mature Norway spruce was similar to that of the fungal pretreatment, in that the formation of traumatic resin ducts was induced, resin flow was increased, and phenolic parenchyma were activated resulting in the increased resistance of Norway spruce to the blue-stain fungi *Ceratocystis polonica* (Franceschi et al. 2002; Zeneli et al. 2006; Krokene et al. 2008) and resistance to colonization by the bark beetle *Ips typographus* (Erbilgin et al. 2006). Treatment of Norway spruce seedlings with methyl jasmonate also increased seedling resistance to *Pythium ultimum* (Kozlowski et al. 1999), substantially reducing the mortality rate of spruce seedlings. Treatment of Maritime pine (*Pinus pinaster*) seedlings with methyl jasmonate increased their resistance to large pine weevil (*Hylobious abietis*) (Moreira et al. 2009). Acquired resistance develops in methyl jasmonate pretreated trees within 3 weeks to a month and is likely to be sustained for several months because of the formation of stable anatomical structures (Erbilgin et al. 2006). Methyl jasmonate treatment does not seem to have a phytotoxic effect but may result in moderate reduction of conifer sapwood growth (Krokene et al. 2008); however, the reduced growth may be transient due to the temporal nature of methyl jasmonate action (Gould et al. 2008).

4.7 Transcriptome and Proteome Analysis of Conifer Defense

Substantial genomics resources have been developed for species of spruce and pine with the development of extensive EST databases and their applications, for example for microarray transcriptome profiling (Allona et al. 1998; Kirst et al. 2003; Stasolla et al. 2003; Ralph et al. 2006a, 2008). Transcriptome profiling of the Sitka spruce defense response induced by mechanical wounding, spruce budworm (*Choristoneura occidentalis*) and white pine weevil (*Pissodes strobi*) using a cDNA microarray containing 9.7K cDNA elements revealed a large reorganization of the transcriptome, with 25–36% of the studied transcriptome being differentially expressed in Sitka spruce induced by these treatments (Ralph et al. 2006a). Similarly, a smaller scale defense-related pine transcriptome profiling revealed altered expression of about 10 % of transcripts in Scots pine induced by the pathogenic fungus *Heterobasidion annosum* using a 2.1K loblolly pine microarray (Adomas et al. 2007). These studies revealed, on the transcriptome level, strong induction of several branches of the phenylpropanoid pathway, antimicrobial proteins such as chitinases and thaumatins by wounding, herbivory, and pathogen treatment in conifers. In addition, microarray studies of Sitka spruce induced by wounding and insect feeding showed strong induction of the terpenoid pathway and the induction of transcripts annotated to octadecanoid and ethylene signaling pathways (Ralph et al. 2006a). The defense response in conifers included the reorganization of primary metabolic processes such as the downregulation of photosynthesis by wounding, insect feeding, and fungal inoculation (Ralph et al. 2006a; Adomas et al. 2007). These microarray studies provided a variety of new candidate genes, including a number of transcripts annotated to the phenylpropanoid pathway, antimicrobial proteins, transcription factors, and transcripts involved in signaling as targets for further study into their involvement in conifer defense. The development of EST resources also allowed the application of proteomics to evaluate the Sitka spruce bark proteome response to wounding and weevil feeding (Lippert et al. 2007). Among over a hundred proteins, which were differentially expressed in induced Sitka spruce, only a small number were also identified as similarly differentially expressed by the spruce microarray, supporting the necessity of complementary genomics and proteomics approaches to the study of conifer defense.

4.8 Conclusions and Outlook

Conifer defense involves several different mechanisms that allow conifer trees to survive for a long time in the environment in which they are exposed to potentially faster evolving insect pests and pathogens. The variety of anatomical, chemical, and molecular defense mechanisms, which includes multiple cell types and tissue structures, production of a large number of chemically diverse terpenoids and phenolics, as well as antimicrobial proteins, affords conifer trees the capacity to adversely affect pathogens and insect pests in a number of ways simultaneously.

Constitutive conifer defense strategy is enhanced by the capacity for induction of most of the defense mechanisms in response to attack from pathogens or herbivores.

Current knowledge of conifer defense mechanisms offers a substantial foundation and a variety of directions for future research, much of which will employ the integration of genomics, proteomics, and metabolomics approaches (Hall et al. 2011). The development of conifer EST resources, full-length cDNA collections (Ralph et al. 2008) and new conifer genome sequencing will allow for further gene discovery, functional characterization, and improvement in conifer gene annotation. The available transcriptome profiling of conifer defense can be further extended by including new conifer/insect and conifer/pathogen systems, by comparing defense responses of resistant and susceptible trees and by evaluating systemic defense mechanisms. In addition, whole transcriptome profiling will also include primary metabolism and growth. Such efforts will strongly contribute to elucidate the trade-off between growth and defense on the basis of GDB.

The systematic application of metabolite analysis, targeted proteomics, and transcript analysis of target defense-related genes, which was demonstrated in the study of genes involved in conifer terpenoid biosynthesis (Zulak et al. 2009; Hall et al. 2011), can be used in future studies of the involvement of phenylpropanoid pathway and pathogenesis-related proteins in conifer defense. Further analysis of the localization of metabolites, proteins, and transcripts in different cell types using advanced methods such as laser-microdissection (Li et al. 2007; Abbott et al. 2010) may clarify spatial distribution and the role of these gene products in conifer defense.

A large variety of candidate genes potentially important in the conifer defense response suggested by targeted and genomics studies requires functional characterization to clarify the role of these genes in conifer defense. Biological functions of these genes need to be assessed by correlating metabolite accumulation and gene expression with the resistance level of conifer trees, and by evaluation of the effect of these metabolites and proteins on pathogens and insects. Furthermore, biological function of conifer defense-related genes can be evaluated in vivo in transgenic angiosperms or conifer systems, which have been successfully applied in conifer research (Davis et al. 2002; Chatthai et al. 2004; Wadenback et al. 2008; Bedon et al. 2009; Wagner et al. 2009). In addition, the potential of synergistic interaction of different defense mechanisms needs to be explored. Interactions between primary (growth) and secondary metabolism (defense) need also to be studied for understanding a trade-off as predicted by GDB.

Understanding of conifer defense mechanisms can provide valuable insights in the dynamic of forest infestation through integration of the effects of conifer defense responses with the ecology and biology of complex interactions of conifer species, insect herbivores, and fungal pathogens (Raffa et al. 2008). Such multi-scale evaluation of conifer defense capacity ranging from cellular level in individual tree to forest stand and landscape level events will allow a better prediction of forest infestation progression and will contribute to the development of improved forest management strategies.

Acknowledgment Research in the laboratory of JB has been generously supported with funding from the Natural Science and Engineering Research Council of Canada in the form of NSERC Discovery and Strategic Grants; and with funding from Genome Canada and Genome British Columbia in support of the Treenomix Project (http://www.treenomix.ca), the smart forest project and the Tria Project (http://www.thetriaproject.ca). JB is Distinguished University Scholar of the University of British Columbia, Vancouver, Canada.

References

Abbott E, Hall D, Hamberger B, Bohlmann J (2010) Laser microdissection of conifer stem tissues: Isolation and analysis of high quality RNA, terpene synthase enzyme activity and terpenoid metabolites from resin ducts and cambial zone tissue of white spruce (Picea glauca). BMC Plant Biol 10(106):1–16

Adomas A, Heller G, Li GS, Olson A, Chu TM, Osborne J, Craig D, Van Zyl L, Wolfinger R, Sederoff R, Dean RA, Stenlid J, Finlay R, Asiegbu FO (2007) Transcript profiling of a conifer pathosystem: response of *Pinus sylvestris* root tissues to pathogen (*Heterobasidion annosum*) invasion. Tree Physiol 27:1441–1458

Alfaro RI, Vansickle GA, Thomson AJ, Wegwitz E (1982) Tree mortality and radial growth losses caused by the western spruce budworm in a Douglas-fir stand in British-Columbia. Can J Forest Res 12:780–787

Alfaro RI, Borden JH, King JN, Tomlin ES, McIntosh RL, Bohlmann J (2002) Mechanisms of resistance in conifers against shoot infesting insects – The case of the white pine weevil *Pissodes strobi* (Peck): (Coleoptera: Curculionidae). In: Wagner MR, Clancy KM, Lieutier F, Paine TD (eds) 21st International congress of entomology. Springer, Iguassu Falls, Brazil, pp 105–130

Allona I, Quinn M, Shoop E, Swope K, St Cyr S, Carlis J, Riedl J, Retzel E, Campbell MM, Sederoff R, Whetten RW (1998) Analysis of xylem formation in pine by cDNA sequencing. Proc Natl Acad Sci USA 95:9693–9698

Asiegbu FO, Denekamp M, Daniel G, Johansson M (1995) Immune cytochemical-localization of pathogenesis-related proteins in roots of Norway Spruce infected with *Heterobasidion-Annosum*. Eur J Forest Pathol 25:169–178

Asiegbu FO, Choi WB, Li GS, Nahalkova J, Dean RA (2003) Isolation of a novel antimicrobial peptide gene (Sp-AMP) homologue from *Pinus sylvestris* (Scots pine) following infection with the root rot fungus Heterobasidion annosum. FEMS Microbiol Lett 228:27–31

Asiegbu FO, Adomas A, Stenlid J (2005a) Conifer root and butt rot caused by Heterobasidion annosum (Fr.) Bref. s.l. Mol Plant Pathol 6:395–409

Asiegbu FO, Nahalkova J, Li GS (2005b) Pathogen-inducible cDNAs from the interaction of the root rot fungus *Heterobasidion annosum* with Scots pine (Pinus sylvestris L.). Plant Sci 168:365–372

Bedon F, Levasseur C, Grima-Pettenati J, Seguin A, MacKay J (2009) Sequence analysis and functional characterization of the promoter of the *Picea glauca* cinnamyl alcohol dehydrogenase gene in transgenic white spruce plants. Plant Cell Rep 28:787–800

Blodgett JT, Eyles A, Bonello P (2007) Organ-dependent induction of systemic resistance and systemic susceptibility in Pinus nigra inoculated with *Sphaeropsis sapinea* and *Diplodia scrobiculata*. Tree Physiol 27:511–517

Boerjan W, Ralph J, Baucher M (2003) Lignin biosynthesis. Annu Rev Plant Biol 54:519–546

Bohlmann J (2008) Insect-induced terpenoid defenses in spruce. In: Schaller A (ed) Induced plant resistance to herbivory. Springer, New York, pp 173–187

Bois E, Lieutier F (1997) Phenolic response of Scots pine clones to inoculation with Leptographium wingfieldii, a fungus associated with Tomicus piniperda. Plant Physiol Biochem 35:819–825

Bonello P, Blodgett JT (2003) *Pinus nigra-Sphaeropsis sapinea* as a model pathosystem to investigate local and systemic effects of fungal infection of pines. Physiol Mol Plant Pathol 63:249–261

Bonello P, Gordon TR, Storer AJ (2001) Systemic induced resistance in Monterey pine. Forest Pathol 31:99–106

Bonello P, Storer AJ, Gordon TR, Wood DL (2003) Systemic effects of Heterobasidion annosum on ferulic acid glucoside and lignin of presymptomatic ponderosa pine phloem, and potential effects on bark-beetle-associated fungi. J Chem Ecol 29:1167–1182

Bonello P, Gordon TR, Herms DA, Wood DL, Erbilgin N (2006) Nature and ecological implications of pathogen-induced systemic resistance in conifers: A novel hypothesis. Physiol Mol Plant Pathol 68:95–104

Brignolas F, Lacroix B, Lieutier F, Sauvard D, Drouet A, Claudot AC, Yart A, Berryman AA, Christiansen E (1995) Induced responses in phenolic metabolism in 2 Norway spruce clones after wounding and inoculations with *Ophiostoma-polonicum*, a bark beetle-associated fungus. Plant Physiol 109:821–827

Brignolas F, Lieutier F, Sauvard D, Christiansen E, Berryman AA (1998) Phenolic predictors for Norway spruce resistance to the bark beetle *Ips typographus* (Coleoptera: Scolytidae) and an associated fungus, Ceratocystis polonica. Can J Forest Res 28:720–728

Butland SL, Chow ML, Ellis BE (1998) A diverse family of phenylalanine ammonia-lyase genes expressed in pine trees and cell cultures. Plant Mol Biol 37:15–24

Celimene CC, Smith DR, Young RA, Stanosz GR (2001) In vitro inhibition of Sphaeropsis sapinea by natural stilbenes. Phytochemistry 56:161–165

Chatthai M, Osusky M, Osuska L, Yevtushenko D, Misra S (2004) Functional analysis of a Douglas-fir metallothionein-like gene promoter: transient assays in zygotic and somatic embryos and stable transformation in transgenic tobacco. Planta 220:118–128

Chiron H, Drouet A, Lieutier F, Payer HD, Ernst D, Sandermann H (2000) Gene induction of stilbene biosynthesis in Scots pine in response to ozone treatment, wounding, and fungal infection. Plant Physiol 124:865–872

Christiansen E, Krokene P, Berryman AA, Franceschi VR, Krekling T, Lieutier F, Lonneborg A, Solheim H (1999) Mechanical injury and fungal infection induce acquired resistance in Norway spruce. Tree Physiol 19:399–403

Collinge DB, Kragh KM, Mikkelsen JD, Nielsen KK, Rasmussen U, Vad K (1993) Plant chitinases. Plant J 3:31–40

Cordoba E, Salmi M, Leon P (2009) Unravelling the regulatory mechanisms that modulate the MEP pathway in higher plants. J Exp Bot 60:2933–2943

Croteau R, Gurkewitz S, Johnson MA, Fisk HJ (1987) Biochemistry of oleoresinosis 1 – monoterpene and diterpene biosynthesis in lodgepole pine saplings infected with *Ceratocystis-clavigera* or treated with carbohydrate elicitors. Plant Physiol 85:1123–1128

Davin LB, Lewis NG (2000) Dirigent proteins and dirigent sites explain the mystery of specificity of radical precursor coupling in lignan and lignin biosynthesis. Plant Physiol 123:453–461

Davin LB, Wang HB, Crowell AL, Bedgar DL, Martin DM, Sarkanen S, Lewis NG (1997) Stereoselective bimolecular phenoxy radical coupling by an auxiliary (dirigent) protein without an active center. Science 275:362–366

Davis JM, Wu HG, Cooke JEK, Reed JM, Luce KS, Michler CH (2002) Pathogen challenge, salicylic acid, and jasmonic acid regulate expression of chitinase gene homologs in pine. Mol Plant Microbe Interact 15:380–387

De Lucca AJ, Cleveland TE, Wedge DE (2005) Plant-derived antifungal proteins and peptides. Can J Microbiol 51:1001–1014

Delorme L, Lieutier F (1990) Monoterpene composition of the preformed and induced resins of Scots pine, and their effect on bark beetles and associated fungi. Eur J Forest Pathol 20:304–316

Dixon RA, Achnine L, Kota P, Liu CJ, Reddy MSS, Wang LJ (2002) The phenylpropanoid pathway and plant defence - a genomics perspective. Mol Plant Pathol 3:371–390

Dubos C, Plomion C (2001) Drought differentially affects expression of a PR-10 protein, in needles of maritime pine (Pinus pinaster Ait.) seedlings. J Exp Bot 52:1143–1144

Ekramoddoullah AKM, Yu XS, Sturrock R, Zamani A, Taylor D (2000) Detection and seasonal expression pattern of a pathogenesis-related protein (PR-10) in Douglas-fir (*Pseudotsuga menziesii*) tissues. Physiol Plant 110:240–247

Elfstrand M, Fossdal CG, Swedjemark G, Clapham D, Olsson O, Sitbon F, Sharma P, Lonneborg A, von Arnold S (2001) Identification of candidate genes for use in molecular breeding - A case study with the Norway spruce defensin-like gene, Spi 1. Silvae Genet 50:75–81

Erbilgin N, Krokene P, Christiansen E, Zeneli G, Gershenzon J (2006) Exogenous application of methyl jasmonate elicits defenses in Norway spruce (*Picea abies*) and reduces host colonization by the bark beetle *Ips typographus*. Oecologia 148:426–436

Evensen PC, Solheim H, Hoiland K, Stenersen J (2000) Induced resistance of Norway spruce, variation of phenolic compounds and their effects on fungal pathogens. Forest Pathol 30:97–108

Eyles A, Chorbadjian R, Wallis C, Hansen R, Cipollini D, Herms D, Bonello P (2007) Cross-induction of systemic induced resistance between an insect and a fungal pathogen in Austrian pine over a fertility gradient. Oecologia 153:365–374

Faccoli M, Schlyter F (2007) Conifer phenolic resistance markers are bark beetle antifeedant semiochemicals. Agric Forest Entomol 9:237–245

Faldt J, Martin D, Miller B, Rawat S, Bohlmann J (2003) Traumatic resin defense in Norway spruce (*Picea abies*): Methyl jasmonate-induced terpene synthase gene expression, and cDNA cloning and functional characterization of (+)-3-carene synthase. Plant Mol Biol 51:119–133

Fliegmann J, Schroder G, Schanz S, Britsch L, Schroder J (1992) Molecular analysis of chalcone and dihydropinosylvin synthase from Scots pine (*Pinus-sylvestris*), and differential regulation of these and related enzyme-activities in stressed plants. Plant Mol Biol 18:489–503

Fossdal CG, Sharma P, Lonneborg A (2001) Isolation of the first putative peroxidase cDNA from a conifer and the local and systemic accumulation of related proteins upon pathogen infection. Plant Mol Biol 47:423–435

Fossdal CG, Hietala AM, Kvaalen H, Solheim H (2006) Changes in host chitinase isoforms in relation to wounding and colonization by Heterobasidion annosum: early and strong defense response in 33-year-old resistant Norway spruce clone. Tree Physiol 26:169–177

Franceschi VR, Krekling T, Berryman AA, Christiansen E (1998) Specialized phloem parenchyma cells in Norway spruce (Pinaceae) bark are an important site of defense reactions. Am J Bot 85:601–615

Franceschi VR, Krokene P, Krekling T, Christiansen E (2000) Phloem parenchyma cells are involved in local and distant defense responses to fungal inoculation or bark-beetle attack in Norway spruce (Pinaceae). Am J Bot 87:314–326

Franceschi VR, Krekling T, Christiansen E (2002) Application of methyl jasmonate on Picea abies (Pinaceae) stems induces defense-related responses in phloem and xylem. Am J Bot 89:578–586

Franceschi VR, Krokene P, Christiansen E, Krekling T (2005) Anatomical and chemical defenses of conifer bark against bark beetles and other pests. New Phytol 167:353–375

Gehlert R, Schoppner A, Kindl H (1990) Stilbene synthase from seedlings of *Pinus-Sylvestris* - purification and induction in response to fungal infection. Mol Plant Microbe Interact 3:444–449

Gershenzon J (1994) Metabolic costs of terpenoid accumulation in higher-plants. J Chem Ecol 20:1281–1328

Gordon TR (2006) Pitch canker disease of pines. Phytopathology 96:657–659

Gould N, Reglinski T, Spiers M, Taylor JT (2008) Physiological trade-offs associated with methyl jasmonate – induced resistance in *Pinus radiata.* Can J Forest Res 38:677–684

Gout R, Kovalyova V (2008) Inhibition of growth of the phytopathogenic organisms by Scots pine (*Pinus sylvestris L.*) defensin. Sylwan 152:54–58

Gronberg H, Hietala AM, Haahtela K (2009) Analysing scots pine defence-related transcripts and fungal DNA levels in seedlings single- or dual-inoculated with endophytic and pathogenic Rhizoctonia species. Forest Pathol 39:377–389

Hall DE, Robert JA, Keeling CI, Domanski D, Qesada AL, Jancsik S, Kuzyk M, Br H, Borchers CH, Bohlmann J (2011) An integrated genomic, proteomic, and biochemical analysis of (+)-3-carene biosynthesis in Sitka spruce (*Picea sitchensis*) genotypes which are resistant or susceptible to white pine weevil. Plant J 65:936–948

Hamberger B, Bohlmann J (2006) Cytochrome P450 mono-oxygenases in conifer genomes: discovery of members of the terpenoid oxygenase superfamily in spruce and pine. Biochem Soc Trans 34:1209–1214

Hietala AM, Kvaalen H, Schmidt A, Johnk N, Solheim H, Fossdal CG (2004) Temporal and spatial profiles of chitinase expression by Norway spruce in response to bark colonization by *Heterobasidion annosum.* Appl Environ Microbiol 70:3948–3953

Himejima M, Hobson KR, Otsuka T, Wood DL, Kubo I (1992) Antimicrobial terpenes from oleoresin of Ponderosa pine Tree *Pinus-ponderosa* - a defense-mechanism against microbial invasion. J Chem Ecol 18:1809–1818

Huber DPW, Ralph S, Bohlmann J (2004) Genomic hardwiring and phenotypic plasticity of terpenoid-based defenses in conifers. J Chem Ecol 30:2399–2418

Huber DPW, Philippe RN, Madilao LL, Sturrock RN, Bohlmann J (2005) Changes in anatomy and terpene chemistry in roots of Douglas-fir seedlings following treatment with methyl jasmonate. Tree Physiol 25:1075–1083

Hudgins JW, Franceschi VR (2004) Methyl jasmonate-induced ethylene production is responsible for conifer phloem defense responses and reprogramming of stem cambial zone for traumatic resin duct formation. Plant Physiol 135:2134–2149

Hudgins JW, Christiansen E, Franceschi VR (2003) Methyl jasmonate induces changes mimicking anatomical defenses in diverse members of the Pinaceae. Tree Physiol 23:361–371

Hudgins JW, Christiansen E, Franceschi VR (2004) Induction of anatomically based defense responses in stems of diverse conifers by methyl jasmonate: a phylogenetic perspective. Tree Physiol 24:251–264

Hudgins JW, McDonald GI, Zambino PJ, Klopfenstein NB, Franceschi VR (2005) Anatomical and cellular responses of *Pinus monticola* stem tissues to invasion by Cronartium ribicola. Forest Pathol 35:423–443

Hudgins JW, Ralph SG, Franceschi VR, Bohlmann J (2006) Ethylene in induced conifer defense: cDNA cloning, protein expression, and cellular and subcellular localization of 1-aminocyclopropane-1-carboxylate oxidase in resin duct and phenolic parenchyma cells. Planta 224:865–877

Islam MA, Sturrock RN, Holmes TA, Ekramoddoullah AKM (2009) Ultrastructural studies of *Phellinus sulphurascens* infection of Douglas-fir roots and immunolocalization of host pathogenesis-related proteins. Mycol Res 113:700–712

Kasprzewska A (2003) Plant chitinases - regulation and function. Cell Mol Biol Lett 8:809–824

Katoh S, Croteau R (1998) Individual variation in constitutive and induced monoterpene biosynthesis in grand fir. Phytochemistry 47:577–582

Keeling CI, Bohlmann J (2006a) Genes, enzymes and chemicals of terpenoid diversity in the constitutive and induced defence of conifers against insects and pathogens. New Phytol 170:657–675

Keeling CI, Bohlmann J (2006b) Diterpene resin acids in conifers. Phytochemistry 67:2415–2423

Kim YB, Kim SM, Kang MK, Kuzuyama T, Lee JK, Park SC, Shin SC, Kim SU (2009) Regulation of resin acid synthesis in Pinus densiflora by differential transcription of genes encoding multiple 1-deoxy-D-xylulose 5-phosphate synthase and 1-hydroxy-2-methyl-2-(E)-butenyl 4-diphosphate reductase genes. Tree Physiol 29:737–749

King JN, Alfaro RI (2009) Developing Sitka spruce populations for resistance to the white pine weevil: summary of research and breeding program. Technical Report 50, Ministry of Forests and Range, Forest Science Program, Victoria, BC. http://www.for.gov.bc.ca/hfd/pubs/Docs/Tr/Tr050.htm

Kiraly L, Barnaz B, Kiralyz Z (2007) Plant resistance to pathogen infection: Forms and mechanisms of innate and acquired resistance. J Phytopathol 155:385–396

Kirst M, Johnson AF, Baucom C, Ulrich E, Hubbard K, Staggs R, Paule C, Retzel E, Whetten R, Sederoff R (2003) Apparent homology of expressed genes from wood-forming tissues of loblolly pine (*Pinus taeda L.)* with *Arabidopsis thaliana*. Proc Natl Acad Sci USA 100:7383–7388

Kodan A, Kuroda H, Sakai F (2002) A stilbene synthase from Japanese red pine (Pinus densiflora): Implications for phytoalexin accumulation and down-regulation of flavonoid biosynthesis. Proc Natl Acad Sci USA 99:3335–3339

Koornneef A, Pieterse CMJ (2008) Cross talk in defense signaling. Plant Physiol 146:839–844

Kopper BJ, Illman BL, Kersten PJ, Klepzig KD, Raffa KF (2005) Effects of diterpene acids on components of a conifer bark beetle-fungal interaction: Tolerance by Ips pini and sensitivity by its associate Ophiostoma ips. Environ Entomol 34:486–493

Koutaniemi S, Warinowski T, Karkonen A, Alatalo E, Fossdal CG, Saranpaa P, Laakso T, Fagerstedt KV, Simola LK, Paulin L, Rudd S, Teeri TH (2007) Expression profiling of the lignin biosynthetic pathway in Norway spruce using EST sequencing and real-time RT-PCR. Plant Mol Biol 65:311–328

Kovalyova VA, Gout RT (2008) Molecular cloning and characterization of Scotch pine defensin 2. Cytol Genet 42:408–412

Kozlowski G, Metraux JP (1998) Infection of Norway spruce (*Picea abies* (L) Karst) seedlings with Pythium irregulare Buism and *Pythium ultimum* Trow: histological and biochemical responses. Eur J Plant Pathol 104:225–234

Kozlowski G, Buchala A, Metraux JP (1999) Methyl jasmonate protects Norway spruce [*Picea abies* (L.) Karst.] seedlings against *Pythium ultimum* Trow. Physiol Mol Plant Pathol 55:53–58

Kramer KJ, Muthukrishnan S (1997) Insect chitinases: Molecular biology and potential use as biopesticides. Insect Biochem Mol Biol 27:887–900

Krekling T, Franceschi VR, Krokene P, Solheim H (2004) Differential anatomical response of Norway spruce stem tissues to sterile and fungus infected inoculations. Trees 18:1–9

Krokene P, Christiansen E, Solheim H, Franceschi VR, Berryman AA (1999) Induced resistance to pathogenic fungi in Norway spruce. Plant Physiol 121:565–569

Krokene P, Solheim H, Krekling T, Christiansen E (2003) Inducible anatomical defense responses in Norway spruce stems and their possible role in induced resistance. Tree Physiol 23:191–197

Krokene P, Nagy NE, Solheim H (2008) Methyl jasmonate and oxalic acid treatment of Norway spruce: anatomically based defense responses and increased resistance against fungal infection. Tree Physiol 28:29–35

Kunkel BN, Brooks DM (2002) Cross talk between signaling pathways in pathogen defense. Curr Opin Plant Biol 5:325–331

Langenheim JH (2003) Plant resins: chemistry, evolution, ecology, and ethnobotany. Timber, Portland, OR, p 586

Li SH, Schneider B, Gershenzon J (2007) Microchemical analysis of laser-microdissected stone cells of Norway spruce by cryogenic nuclear magnetic resonance spectroscopy. Planta 225:771–779

Lieutier F (2002) Mechanisms of resistance in conifers and bark beetle attack strategies. In: Wagner MR, Clancy KM, Lieutier F, Paine TD (eds) 21st International congress of entomology. Springer, Iguassu Falls, Brazil, pp 105–130

Lippert D, Chowrira S, Ralph SG, Zhuang J, Aeschliman D, Ritland C, Ritland K, Bohlmann J (2007) Conifer defense against insects: Proteome analysis of Sitka spruce (*Picea sitchensis*) bark induced by mechanical wounding or feeding by white pine weevils (*Pissodes strobi*). Proteomics 7:248–270

Liu JJ, Ekramoddoullah AKM, Yu XS (2003) Differential expression of multiple PR10 proteins in western white pine following wounding, fungal infection and cold-hardening. Physiol Plant 119:544–553

Liu RJ, Ekramoddoullah AKM, Zamani A (2005) A class IV chitinase is up-regulated by fungal infection and abiotic stresses and associated with slow-canker-growth resistance to Cronartium ribicola in Western white pine (*Pinus monticola*). Phytopathology 95:284–291

Luchi N, Ma R, Capretti P, Bonello P (2005) Systemic induction of traumatic resin ducts and resin flow in Austrian pine by wounding and inoculation with Sphaeropsis sapinea and Diplodia scrobiculata. Planta 221:75–84

Manninen AM, Utriainen J, Holopainen T, Kainulainen P (2002) Terpenoids in the wood of Scots pine and Norway spruce seedlings exposed to ozone at different nitrogen availability. Can J Forest Res 32:2140–2145

Martin D, Tholl D, Gershenzon J, Bohlmann J (2002) Methyl jasmonate induces traumatic resin ducts, terpenoid resin biosynthesis, and terpenoid accumulation in developing xylem of Norway spruce stems. Plant Physiol 129:1003–1018

Martin DM, Gershenzon J, Bohlmann J (2003) Induction of volatile terpene biosynthesis and diurnal emission by methyl jasmonate in foliage of Norway spruce. Plant Physiol 132:1586–1599

Mattheus N, Ekramoddoullah AKM, Lee SP (2003) Isolation of high-quality RNA from white spruce tissue using a three-stage purification method and subsequent cloning of a transcript from the PR-10 gene family. Phytochem Anal 14:209–215

Mauch F, Mauchmani B, Boller T (1988) Antifungal hydrolases in pea tissue.2. inhibition of fungal growth by combinations of chitinase and beta-1,3-Glucanase. Plant Physiol 88:936–942

McKay SAB, Hunter WL, Godard KA, Wang SX, Martin DM, Bohlmann J, Plant AL (2003) Insect attack and wounding induce traumatic resin duct development and gene expression of (-)-pinene synthase in Sitka spruce. Plant Physiol 133:368–378

Meins F, Fritig B, Linthorst H, Mikkelsen J, Neuhaus J, Ryals J (1994) Plant chitinase genes. Plant Mol Biol Rep 12:S22–S28

Miller RH, Berryman AA, Ryan CA (1986) Biotic elicitors of defense reactions in lodgepole pine. Phytochemistry 25:611–612

Miller B, Madilao LL, Ralph S, Bohlmann J (2005) Insect-induced conifer defense. White pine weevil and methyl jasmonate induce traumatic resinosis, de novo formed volatile emissions, and accumulation of terpenoid synthase and putative octadecanoid pathway transcripts in Sitka spruce. Plant Physiol 137:369–382

Moreira X, Sampedro L, Zas R (2009) Defensive responses of *Pinus pinaster* seedlings to exogenous application of methyl jasmonate: Concentration effect and systemic response. Environ Exp Bot 67:94–100

Nagy NE, Franceschi VR, Solheim H, Krekling T, Christiansen E (2000) Wound-induced traumatic resin duct development in stems of Norway spruce (Pinaceae): Anatomy and cytochemical traits. Am J Bot 87:302–313

Nagy NE, Fossdal CG, Krokene P, Krekling T, Lonneborg A, Solheim H (2004) Induced responses to pathogen infection in Norway spruce phloem: changes in polyphenolic parenchyma cells, chalcone synthase transcript levels and peroxidase activity. Tree Physiol 24:505–515

Nealis VG, Noseworthy MK, Turnquist R, Waring VR (2009) Balancing risks of disturbance from mountain pine beetle and western spruce budworm. Can J Forest Res 39:839–848

Neuhaus JM (1999) Plant chitinases (PR-3, PR-4, PR-8, PR-11). In: Datta SK, Muthukrishnan S (eds) Pathogenesis-related proteins in plants. CRC, Boca Ranton, pp 77–105

Nordlander G (1990) Limonene inhibits attraction to alpha-pinene in the pine weevils *Hylobius-abietis* and *H-pinastri*. J Chem Ecol 16:1307–1320

Paine TD, Hanlon CC (1994) Influence of oleoresin constituents from *Pinus-Ponderosa* and *Pinus-Jeffreyi* on growth of mycangial fungi from *Dendroctonus-Ponderosae* and *Dendroctonus-Jeffreyi*. J Chem Ecol 20:2551–2563

Paine TD, Raffa KF, Harrington TC (1997) Interactions among scolytid bark beetles, their associated fungi, and live host conifers. Annu Rev Entomol 42:179–206

Pan HF, Lundgren LN (1996) Phenolics from inner bark of *Pinus sylvestris*. Phytochemistry 42:1185–1189

Passardi F, Cosio C, Penel C, Dunand C (2005) Peroxidases have more functions than a Swiss army knife. Plant Cell Rep 24:255–265

Pearce RB (1996) Antimicrobial defences in the wood of living trees. New Phytol 132:203–233

Pervieux I, Bourassa M, Laurans F, Hamelin R, Seguin A (2004) A spruce defensin showing strong antifungal activity and increased transcript accumulation after wounding and jasmonate treatments. Physiol Mol Plant Pathol 64:331–341

Phillips MA, Croteau RB (1999) Resin-based defenses in conifers. Trends Plant Sci 4:184–190

Phillips MA, Walter MH, Ralph SG, Dabrowska P, Luck K, Uros EM, Boland W, Strack D, Rodriguez-Concepcion M, Bohlmann J, Gershenzon J (2007) Functional identification and differential expression of 1-deoxy-D-xylulose 5-phosphate synthase in induced terpenoid resin formation of Norway spruce (*Picea abies*). Plant Mol Biol 65:243–257

Piggott N, Ekramoddoullah AKM, Liu JJ, Yu XS (2004) Gene cloning of a thaumatin-like (PR-5) protein of western white pine (*Pinus monticola* D. Don) and expression studies of members of the PR-5 group. Physiol Mol Plant Pathol 64:1–8

Popp MP, Johnson JD, Lesney MS (1995) Changes in ethylene production and monoterpene concentration in slash pine and loblolly-pine following inoculation with bark beetle vectored fungi. Tree Physiol 15:807–812

Preisig-Muller R, Schwekendiek A, Brehm I, Reif HJ, Kindl H (1999) Characterization of a pine multigene family containing elicitor-responsive stilbene synthase genes. Plant Mol Biol 39:221–229

Raffa KF, Berryman AA (1982) Physiological differences between lodgepole pines resistant and susceptible to the mountain pine-beetle (Coleoptera, Scolytidae) and associated microorganisms. Environ Entomol 11:486–492

Raffa K, Aukema B, Erbilgin N, Klepzig K, Wallin K (2005) Interactions among conifer terpenoids and bark beetles across multiple levels of scale: an attempt to understand links between population patterns and physiological processes. Recent Adv Phytochem 39:80–118

Raffa KF, Aukema BH, Bentz BJ, Carroll AL, Hicke JA, Turner MG, Romme WH (2008) Cross-scale drivers of natural disturbances prone to anthropogenic amplification: The dynamics of bark beetle eruptions. Bioscience 58:501–517

Ralph SG, Yueh H, Friedmann M, Aeschliman D, Zeznik JA, Nelson CC, Butterfield YSN, Kirkpatrick R, Liu J, Jones SJM, Marra MA, Douglas CJ, Ritland K, Bohlmann J (2006a) Conifer defence against insects: microarray gene expression profiling of Sitka spruce (*Picea sitchensis*) induced by mechanical wounding or feeding by spruce budworms (*Choristoneura occidentalis*) or white pine weevils (*Pissodes strobi*) reveals large-scale changes of the host transcriptome. Plant Cell Environ 29:1545–1570

Ralph S, Park JY, Bohlmann J, Mansfield SD (2006b) Dirigent proteins in conifer defense: gene discovery, phylogeny, and differential wound- and insect-induced expression of a family of DIR and DIR-like genes in spruce (*Picea spp.*). Plant Mol Biol 60:21–40

Ralph SG, Jancsik S, Bohlmann J (2007a) Dirigent proteins in conifer defense II: Extended gene discovery, phylogeny, and constitutive and stress-induced gene expression in spruce (*Picea spp.*). Phytochemistry 68:1975–1991

Ralph SG, Hudgins JW, Jancsik S, Franceschi VR, Bohlmann J (2007b) Aminocyclopropane carboxylic acid synthase is a regulated step in ethylene-dependent induced conifer defense. Full-length cDNA cloning of a multigene family, differential constitutive, and wound- and insect-induced expression, and cellular and subcellular localization in spruce and Douglas fir. Plant Physiol 143:410–424

Ralph SG, Chun HJE, Kolosova N, Cooper D, Oddy C, Ritland CE, Kirkpatrick R, Moore R, Barber S, Holt RA, Jones SJM, Marra MA, Douglas CJ, Ritland K, Bohlmann J (2008) A conifer genomics resource of 200,000 spruce (*Picea spp.*) ESTs and 6,464 high-quality, sequence-finished full-length cDNAs for Sitka spruce (*Picea sitchensis*). BMC Genomics 9:484

Richard S, Lapointe G, Rutledge RG, Seguin A (2000) Induction of chalcone synthase expression in white spruce by wounding and jasmonate. Plant Cell Physiol 41:982–987

Ro DK, Bohlmann J (2006) Diterpene resin acid biosynthesis in loblolly pine (*Pinus taeda*): Functional characterization of abietadiene/levopimaradiene synthase (PtTPS-LAS) cDNA and subcellular targeting of PtTPS-LAS and abietadienol/abietadienal oxidase (PtAO, CYP720B1). Phytochemistry 67:1572–1578

Ro DK, Arimura GL, Lau SYW, Piers E, Bohlmann J (2005) Loblolly pine abietadienol/abietadienal oxidase PtAO (CYP720B1) is a multifunctional, multisubstrate cytochrome P450 monooxygenase. Proc Natl Acad Sci USA 102:8060–8065

Salle A, Monclus R, Yart A, Lieutier F (2005) Effect of phenolic compounds on the in vitro growth of two fungi associated with Ips typographus. Forest Pathol 35:298–304

Sato Y, Bao WL, Sederoff R, Whetten R (2001) Molecular cloning and expression of eight laccase cDNAs in loblolly pine (*Pinus taeda*). J Plant Res 114:147–155

Schmidt A, Gershenzon J (2007) Cloning and characterization of isoprenyl diphosphate synthases with farnesyl diphosphate and geranylgeranyl diphosphate synthase activity from Norway spruce (*Picea abies*) and their relation to induced oleoresin formation. Phytochemistry 68:2649–2659

Schmidt A, Gershenzon J (2008) Cloning and characterization of two different types of geranyl diphosphate synthases from Norway spruce (*Picea abies*). Phytochemistry 69:49–57

Schmidt A, Zeneli G, Hietala AM, Fossdal CG, Krokene P, Christiansen E, Gershenzon J (2005) Induced chemical defences in conifers: Biochemical and molecular approaches to studying their function. In: Romeo JT (ed) Chemical ecology and phytochemistry in forest ecosystems. Elsevier, Amsterdam, pp 1–28

Schmidt A, Wachtler B, Temp U, Krekling T, Seguin A, Gershenzon J (2009) A bifunctional geranyl and geranylgeranyl diphosphate synthase is involved in terpene oleoresin formation in Norway spruce (*Picea abies*). Plant Physiol 152(2):639–655

Schwekendiek A, Pfeffer G, Kindl H (1992) Pine stilbene synthase cDNA, a tool for probing environmental-stress. FEBS Lett 301:41–44

Sels J, Mathys J, De Coninck BMA, Cammue BPA, De Bolle MFC (2008) Plant pathogenesis-related (PR) proteins: A focus on PR peptides. Plant Physiol Biochem 46:941–950

Sharma P, Lonneborg A (1996) Isolation and characterization of a cDNA encoding a plant defensin-like protein from roots of Norway spruce. Plant Mol Biol 31:707–712

Sharma P, Borja D, Stougaard P, Lonneborg A (1993) Pr-proteins accumulating in spruce roots infected with a pathogenic *Pythium sp* isolate include chitinases, chitosanases and beta-1,3-glucanases. Physiol Mol Plant Pathol 43:57–67

Six DL, Paine TD (1998) Effects of mycangial fungi and host tree species on progeny survival and emergence of Dendroctonus ponderosae (Coleoptera: Scolytidae). Environ Entomol 27:1393–1401

Stasolla C, van Zyl L, Egertsdotter U, Craig D, Liu WB, Sederoff RR (2003) The effects of polyethylene glycol on gene expression of developing white spruce somatic embryos. Plant Physiol 131:49–60

Steele CL, Katoh S, Bohlmann J, Croteau R (1998) Regulation of oleoresinosis in grand fir (*Abies grandis*) - Differential transcriptional control of monoterpene, sesquiterpene, and diterpene synthase genes in response to wounding. Plant Physiol 116:1497–1504

Thomma B, Penninckx I, Broekaert WF, Cammue BPA (2001) The complexity of disease signaling in Arabidopsis. Curr Opin Immunol 13:63–68

Tomlin ES, Borden JH, Pierce HD (1996) Relationship between cortical resin acids and resistance of Sitka spruce to the white pine weevil. Can J Bot 74:599–606

Tomlin ES, Antonejevic E, Alfaro RI, Borden JH (2000) Changes in volatile terpene and diterpene resin acid composition of resistant and susceptible white spruce leaders exposed to simulated white pine weevil damage. Tree Physiol 20:1087–1095

Trapp S, Croteau R (2001) Defensive resin biosynthesis in conifers. Annu Rev Plant Physiol Plant Mol Biol 52:689–724

Van Loon LC, Van Strien EA (1999) The families of pathogenesis-related proteins, their activities, and comparative analysis of PR-1 type proteins. Physiol Mol Plant Pathol 55:85–97

Vargas-Arispuro I, Reyes-Baez R, Rivera-Castaneda G, Martinez-Tellez MA, Rivero-Espejel I (2005) Antifungal lignans from the creosotebush (Larrea tridentata). Ind Crops Prod 22:101–107

Viiri H, Annila E, Kitunen V, Niemela P (2001) Induced responses in stilbenes and terpenes in fertilized Norway spruce after inoculation with blue-stain fungus, Ceratocystis polonica. Trees 15:112–122

Wadenback J, von Arnold S, Egertsdotter U, Walter MH, Grima-Pettenati J, Goffner D, Gellerstedt G, Gullion T, Clapham D (2008) Lignin biosynthesis in transgenic Norway spruce plants harboring an antisense construct for cinnamoyl CoA reductase (CCR). Transgenic Res 17:379–392

Wagner A, Donaldson L, Kim H, Phillips L, Flint H, Steward D, Torr K, Koch G, Schmitt U, Ralph J (2009) Suppression of 4-coumarate-CoA ligase in the coniferous gymnosperm *Pinus radiata*. Plant Physiol 149:370–383

Wainhouse D, Cross DJ, Howell RS (1990) The role of lignin as a defense against the spruce bark beetle *Dendroctonus-Micans* - effect on larvae and adults. Oecologia 85:257–265

Wainhouse D, Rose DR, Peace AJ (1997) The influence of preformed defences on the dynamic wound response in Spruce bark. Funct Ecol 11:564–572

Wainhouse D, Ashburner R, Ward E, Boswell R (1998) The effect of lignin and bark wounding on susceptibility of spruce trees to Dendroctonus micans. J Chem Ecol 24:1551–1561

Wallis C, Eyles A, Chorbadjian R, Gardener BM, Hansen R, Cipollini D, Herms DA, Bonello P (2008) Systemic induction of phloem secondary metabolism and its relationship to resistance to a canker pathogen in Austrian pine. New Phytol 177:767–778

Walters DR (2009) Are plants in the field already induced? Implications for practical disease control. Crop Prot 28:459–465

Weng J, Chapple C (2010) The origin and evolution of lignin biosynthesis. New Phytol 187:273–285

Whetten RW, Sederoff RR (1992) Phenylalanine ammonia-lyase from loblolly-pine - purification of the enzyme and isolation of complementary-DNA clones. Plant Physiol 98:380–386

Wiweger M, Farbos I, Ingouff M, Lagercrantz U, von Arnold S (2003) Expression of Chia4-Pa chitinase genes during somatic and zygotic embryo development in Norway spruce (*Picea abies*): similarities and differences between gymnosperm and angiosperm class IV chitinases. J Exp Bot 54:2691–2699

Woodward S, Pearce RB (1988) The role of stilbenes in resistance of Sitka Spruce (Picea-Sitchensis (Bong) Carr) to entry of fungal pathogens. Physiol Mol Plant Pathol 33:127–149

Wu JQ, Baldwin IT (2009) Herbivory-induced signalling in plants: perception and action. Plant Cell Environ 32:1161–1174

Wu HG, Echt CS, Popp MP, Davis JM (1997) Molecular cloning, structure and expression of an elicitor-inducible chitinase gene from pine trees. Plant Mol Biol 33:979–987

Yamaoka Y, Hiratsuka Y, Maruyama PJ (1995) The ability of *Grosmannia clavigera* to kill mature lodgepole pine trees. Eur J Forest Pathol 25:401–404

Zareie R, Melanson DL, Murphy PJ (2002) Isolation of fungal cell wall degrading proteins from barley (Hordeum vulgare L.) leaves infected with *Rhynchosporium secalis*. Mol Plant Microbe Interact 15:1031–1039

Zeneli G, Krokene P, Christiansen E, Krekling T, Gershenzon J (2006) Methyl jasmonate treatment of mature Norway spruce (*Picea abies*) trees increases the accumulation of terpenoid resin components and protects against infection by Ceratocystis polonica, a bark beetle-associated fungus. Tree Physiol 26:977–988

Zulak KG, Bohlmann J (2010) Terpenoid biosynthesis and specialized vascular cells of conifer defense. J Integr Plant Biol 52:86–97

Zulak KG, Lippert DN, Kuzyk MA, Domanski D, Chou T, Borchers CH, Bohlmann J (2009) Targeted proteomics using selected reaction monitoring reveals the induction of specific terpene synthases in a multi-level study of methyl jasmonate-treated Norway spruce (*Picea abies*). Plant J 60:1015–1030

Chapter 5
The Rhizosphere: Molecular Interactions Between Microorganisms and Roots

R. Hampp, A. Hartmann, and U. Nehls

5.1 Introduction

Roots constitute important plant organs for water and nutrient uptake. They, however, also release a wide range of carbon compounds of low molecular weight. This release can amount up to 30% of total net fixed carbon (Smith and Read 2008; Rovira 1991) and forms the basis for an environment inhabited by a highly diverse and active microbial community, the rhizosphere (Hiltner 1904; Hartmann et al. 2008), which is defined as soil compartment influenced by living roots.

The release of assimilates by plant roots results in a greater microbial density and activity in the rhizosphere than in the bulk soil. The specific conditions lead to the selection of distinct microbial communities, where fungi play an important role (Frey-Klett et al. 2005). Especially, symbiotic fungi (ecto/arbuscular mycorrhiza) release a substantial amount of plant-derived carbon to the soil, creating another sphere, the mycorrhizosphere. These organic carbon enriched spheres are highly attractive for other microorganisms. For example, the rhizosphere/bulk soil ratio for Gram-negative bacteria reaches from 2 to 20, for actinomycetes from 5 to 10, and for fungi from 10 to 20 (Morgan et al. 2005). The diversity and structure of bacterial communities are plant specific and vary over time (Smalla et al. 2001; Barriuso et al. 2005; Berg and Smalla 2009; Hartmann et al. 2009). Bacteria can have a negative, neutral, or beneficial effect to plant fitness. Detrimental effects are caused by

R. Hampp (✉)
Physiological Ecology of Plants, IMIT, University of Tübingen, Auf der Morgenstelle 1, 72076 Tübingen, Germany
e-mail: ruediger.hampp@uni-tuebingen.de

A. Hartmann
Department Microbe-Plant Interactions, German Research Center for Environmental Health, Helmholtz Zentrum München, Ingolstädter Landstrasse 1, 85764 Neuherberg, Germany

U. Nehls
Botany, University of Bremen, Leobener Str. 2, 28359 Bremen, Germany

R. Matyssek et al. (eds.), *Growth and Defence in Plants*, Ecological Studies 220,
DOI 10.1007/978-3-642-30645-7_5,

bacterial pathogens and parasites, and bacteria that produce phytotoxic substances. The occurrence of pathogenic bacteria is, however, low in healthy plant populations. This is due to plant defence systems, which are selective and could cause the enrichment of plant beneficial microbes within the rhizosphere (Frey-Klett et al. 2005). The plant beneficial bacteria include saprotrophs that degrade the organic litter, antagonists of plant root pathogens, and plant growth promoting rhizobacteria (PGPR) (Barea et al. 2005).

5.2 Bacteria of the Rhizosphere and Effective Metabolites

5.2.1 Plant Growth Promoting Rhizobacteria

Plant growth promoting rhizobacteria (PGPR) are usually in contact with the root surface as well as the hyphal cell walls of symbiotic fungi (Bonfante and Anca 2009), and increase plant growth (Weller 1988; Lucy et al. 2004; Haas and Défago 2005). PGPR must be able to colonise the root and have to be present in sufficient numbers to exert their functions. *Pseudomonas* spp. and *Bacillus* spp. are the most commonly investigated PGPR, and are often the dominating bacterial groups in the rhizosphere of herbs and grasses (Marilley et al. 1999; Morgan et al. 2005). Diverse PGPR strains have been used successfully for crop inoculations, including members of *Azospirillum*, *Azotobacter*, *Bacillus*, *Enterobacter*, *Pseudomonas*, *Serratia*, and *Xanthomonas* (see Lucy et al. 2004 for a comprehensive list). In addition there is a considerable diversity of endophytic microbes (bacteria and fungi) which are commonly colonising plants systemically without harming the plant, but rather exert plant growth promoting activity in some cases (Rosenblueth and Martinez-Romero 2006; de Almeida et al. 2009). For example, diazotrophic plant growth promoting bacteria belonging to the species *Herbaspirillum seropedica*, *Burkholderia tropica*, *Gluconacetobacter diazotrophicus*, *Azoarcus* sp., or *Azospirillum brasilense* occur as endophytic symbiotic bacteria in Gramineae and other plants (Rothballer et al. 2009), although their exact mechanism of symbiotic interaction is not well understood.

Two groups of PGPR exist: those that are involved in nutrient cycling and plant growth stimulation (biofertilizers), and those that are involved in the biological control of plant pathogens (biopesticides). Bacteria may support plant growth by the mobilisation of inorganic nutrients, by nitrogen fixation, by the production of phytohormones including auxins, cytokinins as well as gibberellins (Dobbelaere et al. 2001). Most interestingly, a widely distributed activity and gene cluster for the degradation of indole acetic acid was discovered recently in diverse bacteria (Leveau and Gerards 2008), which could be another means to interfere with plant growth. Volatile substances, such as acetoine or 2,3-butanediol, were also shown to stimulate plant growth substantially (Ryu et al. 2004; Barea et al. 2005). Some root-associated bacteria are able to stimulate plant growth by reducing inhibitory levels of ethylene in the rhizosphere through the hydrolysis of the ethylene precursor 1-aminocyclopropane-1-carboxylic acid (ACC) (Blaha et al. 2006; Glick et al. 1998; Grichko and Glick 2001; Prigent-Combaret et al. 2008) and rhizobitoxin (Sugarawa et al. 2006). Root-associated and endophytic bacteria with ACC-deaminase activity have a good potential for practical applications (Mayak et al. 2004). In soils with low phosphate (P), P-solubilising bacteria release phosphate ions from low-soluble inorganic P-containing minerals and from organic phosphate sources. Although many P-solubilising bacteria have been characterised, their relative importance in the PGPR effect is uncertain. However, if the phosphate ions are released in an area rich in mycorrhizal fungal hyphae, the hyphae may transport the P to the plants and the PGPR effect is detectable (Artursson et al. 2005; Barea et al. 2005).

The inoculation of roots with *Azospirillum* spp. often promotes plant growth, probably not primarily through biological nitrogen fixation, but mostly due to the ability of the bacteria to produce phytohormones and thus stimulate root development, increase the volume of explored soil space, and finally nutrient uptake efficiency (Steenhoudt and Vanderleyden 2000; Dobbelaere et al. 2001). PGPR often enhance plant growth through the production of plant growth regulators (Lucy et al. 2004). The auxin-type phytohormones produced by *Azospirillum* spp. induce root branching and thus improve plant nutrient uptake from the soil (Dobbelaere and Okon 2007), an example for a possible trade-off of altered resource allocation (GDB). Recently, a salt-tolerant *Azospirillum brasilense* strain NH was characterised, which produced the auxin indole acetic acid at 250 mM NaCl and conferred salt tolerance to wheat upon inoculation (Nabti et al. 2007). Interestingly, an efficient endophytic colonisation of the apoplastic space in the root epidermis by *A. brasilense* NH was observed (Nabti et al. 2010). The growth of the plants may also be stimulated by bacterial cytokinin production (Lucy et al. 2004).

5.2.2 *Plant Disease Suppressing Rhizobacteria*

The second major PGPR mechanism is to reduce the incidence of plant disease. Infectious diseases are often caused by soil-borne organisms including both bacteria and fungi. The soils, where soil-borne diseases are infrequent, are called suppressive soils, and it has been shown that the disease suppression is often caused by specific bacterial and fungal populations (Weller et al. 2002). Recent studies highlight three major mechanisms for the disease suppression: antagonism, direct pathogen–agonist (plant growth promoting microorganism) interactions, and induced systemic resistance in the plant (Compant et al. 2005).

Antagonists are naturally occurring organisms that express traits which enable them to interfere with pathogen growth, survival, and infection. Bacteria, antagonistic to plant pathogens, represent an important part of the rhizosphere communities, and antagonistic strains amount up to 35% of the culturable bacteria (Opelt and Berg 2004). The Gram-negative rods of *Stenotrophomonas maltophilia* (earlier

Xanthomonas maltophilia) are also typical rhizosphere inhabitants, and of scientific interest due to their potential for biological control (Nakayama et al. 1999).

The most thoroughly investigated group of PGPR agonists are still the fluorescent pseudomonads (Haas and Défago 2005). These bacteria produce diverse antagonistic secondary metabolites that suppress the growth of other organisms. As an example, the extracellular pigment pyoverdin is an efficient siderophore (iron carrier), and the production of pyoverdin by pseudomonads in iron-poor soils is an effective way to suppress the growth of non-producers by depriving the pathogens from iron (Kloepper et al. 1980). Pseudomonads also produce metal-chelating agents with proposed properties other than iron scavenging. Pyochelin, e.g., binds effectively copper and zinc, and possesses strong antimicrobial activity (Cornelis and Matthijs 2002). However, the antimicrobial effect of pyochelin, and of some other siderophores, can be explained by their effective metal-chelating activity (Haas and Défago 2005). Gram-positive PGPR antagonists, like *Bacillus subtilis* GB03 (Kloepper et al. 2004) and *B. amyloliquefaciens* FZB42 (Koumoutsi et al. 2004; Chen et al. 2009a) are also very efficient PGPR strains with biocontrol activity, also having effects on systemic resistance in plants. Since their spores withstand adverse conditions, they have wide acceptance for practical applications, because of easier handling and excellent stability of inoculant preparations.

Direct antibiosis is used by several PGPR as a mechanism for biocontrol. *Antibiosis* by PGPR pseudomonads is often caused by the production of several antimicrobial substances. These chemicals not only suppress fungi, but are often also toxic against bacteria (Compant et al. 2005). From antimicrobial compounds produced by pseudomonads, the mode of action has been partly determined for six classes of substances. These include the electron transport inhibitors phenazines, phloroglucinols (causing membrane damage in *Pythium* spp. and being phytotoxic at higher concentrations), pyrrolnitrin (acting as a fungicide), cyclic lipopeptides (surfactant properties against fungi and plants, chelation of cations), and HCN (potent inhibitor of metalloenzymes). A comprehensive list of the antibiotics, producer strains, target organisms, and effects on the host plants has been covered in a review by Raaijmakers et al. (2002). Production of siderophores, lipopeptides, and antibiotics production has been observed in other PGPR isolates as well, including *Bacillus amyloliquefaciens* (Chen et al. 2009b), *Stenotrophomonas* spp. (Compant et al. 2005), and *Streptomyces* spp.

Another group of antagonistic compounds are lytic enzymes, such as cell wall hydrolases that attack pathogens. The ability to degrade fungal cell walls by chitinases is shared by many biocontrol PGPR including *Pseudomonas*, *Serratia*, and *Streptomyces* spp. (Whipps 2001). In addition to chitinases, some bacterial strains produce β-glucanases and proteases (Dunne et al. 2000). Synergism between the action of cell wall degrading agents and antibiotics was observed by Woo et al. (2002). The authors showed that the pre-treatment of plant pathogenic fungi with cell wall degrading enzymes rendered them more susceptible to the antifungal substance, syringomycin.

Inoculation of plants with some PGPR elicits a phenomenon known as induced *systemic resistance* (ISR; Bakker et al. 2007; see Fig. 5.1). ISR allows the plants to

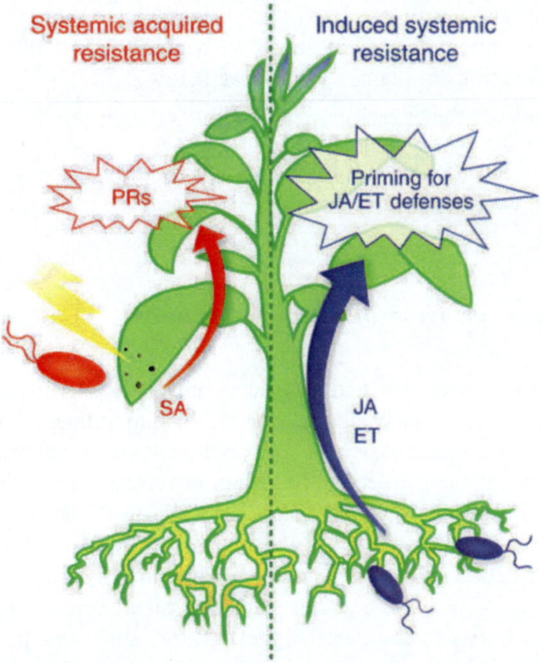

Fig. 5.1 Schematic representation of systemically induced immune responses. Systemic acquired resistance starts with a local infection and can induce resistance in yet not affected distant tissues. Transport of salicylic acid (SA) is essential for this response. Induced systemic resistance can result from root colonisation by non-pathogenic microorganisms and, by long-distance signalling, induces resistance in the shoot. Ethylene (ET) and jasmonic acid (JA) are involved in the regulation of the respective pathways. Depending on the pathogen, JA/ET can also be involved in SAR. They induce pathogenesis-related genes different from those induced by SA (taken from Pieterse et al. 2009, with permission)

endure pathogen attacks that, without bacterial pre-inoculation, could be lethal. The effect is systemic, e.g., root inoculation with the biocontrol PGPR yields the whole plant non-susceptible (Haas and Défago 2005). Thus far, *Pseudomonas*, *Burkholderia*, and *Bacillus* spp. have been shown to elicit ISR (e.g., Ryu et al. 2004), and the search for effective substances is in progress (see also Sect. 3.4.2). Root treatment of *Phaseolus vulgaris* with *Pseudomonas putida* BTP1 leads to a significant reduction of the disease caused by the pathogen *Botrytis cinerea* on leaves. Ongena et al. (2005) isolated the molecular determinant of *P. putida* mainly responsible for the induced systemic resistance and identified it as a polyalkylated benzylamine structure. Exposure to butanediol, the volatile that induces the growth of *Arabidopsis* seedlings (Ryu et al. 2003), decreased disease severity by the bacterial pathogen *Erwinia carotovora* in the same plant (Ryu et al. 2004). Transgenic lines of *Bacillus subtilis*, that emitted reduced levels of 2,3-butanediol, decreased *Arabidopsis thaliana* protection against pathogen infection compared with seedlings exposed to volatiles from wild-type bacterial lines. Furthermore, bacterial signalling molecules of the *N*-acyl homoserine lactone type which are quorum sensing compounds of Gram-negative bacteria (Eberl 1999) were found to exert systemic functions in plants as well (see below).

The multiple mechanisms as to how plant-beneficial bacteria promote plant fitness are only beginning to be resolved. It is obvious that the use of *Pseudomonas* and *Bacillus* spp. has yielded a mass of relevant results, but much remains to be learned from the bacteria in other taxa.

5.2.3 The Special Role of Actinomycetales

Ample evidence indicates that actinomycetes are quantitatively and qualitatively important in the rhizospheres of plants, where they may influence plant growth and protect plant roots against invasion by root pathogens (for a review see Tarkka and Hampp 2008). Within the order of the Streptomycetales, members of the genus *Streptomyces* are traditionally considered as soil dwelling organisms (Janssen 2006), and have been reported to be the most prolific producers of a variety of antibiotics (Berdy 2005). There is now abundant evidence that some *Streptomyces* species colonise the root surface and even plant tissues (Sardi et al. 1992; Coombs and Franco 2003), and it has been suggested that antibiotic production by the streptomycete may protect the host plants against phytopathogens (Challis and Hopwood 2003; Tarkka and Hampp 2008). Streptomycetes causing plant disease were covered recently in Loria et al. (2006).

Streptomycetes can inhibit diverse plant pathogens, including Gram-positive and Gram-negative bacteria, fungi, and nematodes (Crawford et al. 1993; Chamberlain and Crawford 1999; El-Abyad et al. 1993; Samac and Kinkel 2001). Biocontrol strains from *Streptomyces* spp. have often been isolated from suppressive soils (Weller et al. 2002). For example, the common scab disease of potato, caused by *Streptomyces scabies* (Loria et al. 1997), is effectively suppressed by streptomycete isolates from *S. scabies* suppressive soils. The disease suppression effect can be long lasting. Liu et al. (1995) added streptomycete strains to a soil that was non-suppressive against potato scab. Even by the fourth year after the inoculation, the disease reduction due to these strains was at 63–73%. The biocontrol activity of these *Streptomyces* spp. can often be explained by direct inhibition of the growth of the pathogen (Eckwall and Schottel 1997), but it has been observed that stronger biocontrol agents also have a capacity for resource competition (Neeno-Eckwall et al. 2001; Schottel et al. 2001). Disease suppression by streptomycetes owes partially to their exudation of various antimicrobials, helminticides, and enzymes degrading fungal cell walls and insect exoskeletons (Weller et al. 2002).

To date, approximately 17% of biologically active secondary metabolites (7,600 out of 43,000; Berdy 2005) have been characterised from the filamentous streptomycetes. These commonly produce two to three dominant secondary metabolites, along with approximately ten minor compounds. For such mixtures to become effective, the bacteria often produce synergistically acting compounds, and the cost of their production is greatly decreased due to the so-called combinatory biosynthesis. Acquisition and retention of the substance diversity are enabled due to horizontal gene transfer between streptomycete isolates, together with strong microbial competition (Firn and Jones 2000; Challis and Hopwood 2003; Davelos et al. 2004; Weissman and Leadlay 2005).

When applied as a single substance, most of the chemicals produced by streptomycetes do not possess any activity against specific target microbes unless tested at high concentrations (Firn and Jones 2000). Streptomycetes, however, simultaneously produce several bioactive secondary metabolites, which, in combination, possess a strong biological activity (Challis and Hopwood 2003). Such mixtures, often consisting of antibiotics, metal chelators, and growth regulators, may act in a synergistic way (Cocito et al. 1997; Liras 1999). Other sorts of streptomycete chemicals may also act together. In their search for secondary metabolites from streptomycetes, Fiedler et al. (2001) detected an uncommon iron-chelating substance from the culture extracts of two streptomycete strains. This was identified as enterobactin, a characteristic siderophore of *Enterobacteriaceae* spp. The production of enterobactin in addition to the common siderophores, desferri-ferrioxamine B and E, could be an important fitness factor in an iron-poor soil substrate (Challis and Hopwood 2003).

As the frequency of microbial encounters increases with increased population density, the extent to which streptomycetes release chemicals to the soil is also strongly affected. For example, populations of high density have a stronger relative benefit from antibiotics production than low-density populations (Wiener 2000). Davelos et al. (2004) showed that frequency and intensity of antibiotics production by streptomycetes are related to the location of the streptomycetes in the soil. Where high population densities were achieved, the isolates produced more frequently antibiotic substances (Davelos et al. 2004). In the rhizosphere, the microbial population densities are extremely high, and the data from Davelos et al. (2004) suggest that the rhizosphere could be a hotspot for antibiosis. Indeed, Frey-Klett et al. (2005) were able to show that the rhizosphere effect leads to the enrichment of bacteria that suppress plant pathogenic fungi.

In addition, the presence of biocontrol activity determinants other than direct antagonism has been suggested by several studies. There is evidence that bacterial influence on plant growth is also an important determinant in biocontrol. Eight antagonistic *Streptomyces* isolates were tested for their ability to control *Phytophthora* root rots on alfalfa and soybean (Xiao et al. 2002). The strongest indicator of disease suppression in alfalfa by the antagonist was an increase in alfalfa biomass following inoculation with the bacterial isolate. In this case, direct enhancement of alfalfa growth by *Streptomyces* may be one of the key mechanisms by which *Streptomyces* antagonists enhance plant health. In our own experiments, we have observed a similar dependence between plant growth promotion and increased disease resistance, while screening for streptomycete strains that are able to suppress *Brassica* dark leaf spot development in *A. thaliana* (Herold M, Schrey S, Hampp R, Tarkka M, unpublished).

Upon infection by certain rhizobacteria, plants acquire an increased resistance to pathogen attack. This phenomenon has been classified as priming (Conrath et al. 2002). Rhizosphere and endophytic streptomycetes have been recently identified as such disease resistance inducing species (Lehr et al. 2008; Conn et al. 2008). We have studied the mechanisms of disease suppression by the streptomycete GB 4-2 against *Heterobasidion* root and butt rot in Norway spruce seedlings. Unexpectedly, GB 4-2 promoted the physiology of the pathogenic fungus: mycelial growth, germination rate of fungal spores, extension of fungal germ tubes, and even early colonisation of outer cortical layer of the plant root were all enhanced in the presence of the bacteria. However, later, disease development was blocked by the bacterium, since the port of fungal entry into the vascular tissue, the root cortex, was blocked by the formation of cell wall thickenings in co-inoculated plants. In addition, the vascular tissue was rendered inaccessible due to increased xylem formation and strong lignification (Lehr et al. 2008). Together, these data indicate that the inoculation with GB 4-2 sensitised the plant to enhance its responsiveness to the root pathogen. This is another example for a trade-off of increased carbon allocation to the root system (see GDB).

An important finding was associated with this latter result. The infection of needles by *Botrytis cinerea* was also reduced in Norway spruce due to pre-treatment seedlings with *Streptomyces* GB 4-2, suggesting increased systemic defence reactions. To further analyse the underlying mechanism, the accumulation of defence-related transcripts in *A. thaliana* during its interaction with GB 4-2 in the roots and/or with the phytopathogenic fungus *Alternaria brassicicola* in the leaves was investigated (Schrey, unpublished). The aim of these gene expression analyses is to unravel if GB 4-2 provokes a plant immune response similar to ISR or systemic acquired resistance (SAR). In both ISR and SAR, prior treatment results in an enhanced defence response against a subsequent challenge by a pathogen (reviewed by Conrath 2006). In general, ISR is commonly induced upon the challenge of plants by non-pathogenic root-colonising bacteria, while SAR is mediated by phytopathogens. But leaves also can be the site of ISR (Fig. 5.1). Using a set of *A. thaliana* genes related to plant immunity (von Rad et al. 2005), we have observed that the response of *A. thaliana* to GB 4-2 involves changes in the expression of genes associated not only with SAR but also with ISR. Interestingly, ISR-related changes occur in the absence of a challenge by the pathogenic organism, indicating a novel, possibly GB 4-2 specific response pattern in *A. thaliana* (Schrey, unpublished).

Conn et al. (2008) investigated the impact of plant protecting actinomycetes on disease resistance related gene expression in *A. thaliana*. The bacterial inoculation promoted *A. thaliana* growth and endophytic colonisation in the plant tissues. Suppression of *Erwinia carotovora* soft rot as well as of *Fusarium oxysporum* wilt disease was also observed. Gene expression responses to streptomycetes were specific to the bacterial isolate, e.g., inoculation of *A. thaliana* seeds with *Streptomyces* sp. EN27 resulted in a 19-fold induction of the *PR-1* transcript, whereas the closely related strain, *Streptomyces* sp. EN28, was able to induce the defence gene *PDF1.2* by 23-fold. In dual inoculations, the bacteria were able to prime both the SAR and the ISR pathways of *A. thaliana*, upregulating genes in either pathway depending on the infecting pathogen. The use of defence-compromised mutants of *Arabidopsis* showed that *Streptomyces* sp. EN27 induced resistance to *E. carotovora* by a NPR1 (nonexpression of PR proteins)-independent and to *F. oxysporum* by a NPR1-dependent pathway. In conclusion, the gene expression responses to streptomycetes indicate novel patterns of priming by these bacteria, sharing features of both previously characterised pathways, ISR and SAR.

Plant defence responses can be suppressed by specialised organisms. These organisms can produce physiologically active levels of metabolites such as enzymes that act on plant toxins, or exude their own toxins *in planta* that interfere with plant metabolism in ways that benefit the attacker (Bruce and Pickett 2007). They can also produce signals that disrupt the plant's own defence signalling pathways (Cui et al. 2005). There is evidence for the attenuation of plant by the streptomycete strain *Streptomyces* AcH 505.

AcH 505 is a so-called mycorrhization helper bacterium, i.e., a bacterium that facilitates the formation of ectomycorrhizal symbiosis (Frey-Klett et al. 2007). Both water soluble as well as volatile bacterial substances are involved in AcH 505–fungus–plant interactions. First of all, in ectomycorrhiza fungal mycelial growth is promoted by auxofuran, an auxin-related compound produced by AcH 505. In addition, AcH 505 produces an inhibitor of mycelial growth, the antibiotic WS-5995 B (Riedlinger et al. 2006). As the influence of WS-5995 B dominates over that of auxofuran, only the WS-5995 B tolerant fungal strains are able to colonise the host plant.

In addition, the growth of a WS-5995 B sensitive isolate of an important plant pathogen, *Heterobasidion* sp., was inhibited by AcH 505, and we envisaged a potential application for AcH 505: simultaneous growth promotion of mycorrhizal as well as growth suppression of pathogenic fungi (Maier et al. 2004; Schrey et al. 2005). To determine if AcH 505 could serve as a biocontrol agent against *Heterobasidion* root and butt rot, the bacterial influence on mycelial growth of *Heterobasidion* sp. isolates, cultured on wood discs and roots of Norway spruce, was determined. It had been previously suggested (Schottel et al. 2001) that single pathogen–antagonist strain studies may contribute only limited insights into the dynamics of antagonist–pathogen interactions. Our data agreed well with that suggestion: whereas 11 tested *Heterobasidion* strains were sensitive against the antibiotic and suppressed by the bacterium, the growth of another strain, *Heterobasidion annosum* 331, was unaffected by AcH 505. Hazardous in the light of biocontrol applications, root colonisation by *Heterobasidion annosum* 331 was promoted by the bacterium due to a decrease in defence-related gene expression in the host, Norway spruce, which is of advantage only for symbiotic fungi (Lehr et al. 2007). Using a culture system where the bacterium was separated from plant roots and fungal mycelium, we have been able to show that increased fungal colonisation is due to volatiles produced by AcH 505 (Störk M, Lehr N, Hampp R, Tarkka M, unpublished). In conclusion, metabolite production by bacteria like AcH 505 can lead to the inhibition of some and to the benefit of other microorganisms. Depending on the species spectrum in the habitat of the host plant, the development of symbiosis and/or disease may be promoted by such streptomycetes.

5.3 Quorum Sensing Within Bacterial Communities and Trans-kingdom Interactions of *N*-Acyl Homoserine Lactones with Plants

Many environmental and interactive important traits in bacteria, such as antibiotic, siderophore, or exoenzyme (like cellulose, pectinase) production, are regulated in a population density dependent manner using small signalling molecules. This phenomenon, called quorum sensing, is more widespread amongst all bacteria than originally thought. Many different bacterial species are communicating or "speaking" through various secreted molecules. The production is often sophistically regulated as an auto-inducing mechanism, such as the synthesis of *N*-acyl homoserine lactones (AHL), which occur in many variations of molecular structure in a wide variety of Gram-negative bacteria (Eberl 1999). In Gram-positive bacteria, other compounds, such as peptides, AI-2, and quinolone (PQS), regulate cellular activity and behaviour through sensing the cell density. However, it is probably not just the cell density but an integrated measure of the quality of the cellular surrounding (diffusion space, etc.) and the colonisation density, which finally provides the information about crucial habitat conditions leading to higher precision of adaptive gene expression and physiological efficiency (Hense et al. 2007), which has a direct positive impact on evolutionary selection. In addition, some of these signal molecules have also non-signalling roles for important processes, such as nutrient scavenging, ultrastructure modification, and competition between bacteria (Schertzer et al. 2009). In particular, iron siderophores, like pyochelin and quinolone (see above) which interfere with cellular iron stores, are small signal molecules with central functions in the iron homoeostasis. Most interestingly, evidence is accumulating that quorum sensing compounds produced by rhizosphere bacteria are also recognised by plants, and specific responses are induced upon specific molecular interactions. *Serratia liquefaciens*, e.g., strain MG1 is known as a producer of C4- and C6-homoserine lactones and in situ AHL production, occur in the rhizoplane (Gantner et al. 2006). When *Serratia liquefaciens* was inoculated to roots of tomato (Microtom^{R}) plants, the systemic resistance was clearly increased (Schuhegger et al. 2006). The induction of genes related to systemic pathogen response in tomato, like chitinase and PR1, by C4- and C6-homoserine lactones in axenic test systems is contrasted by the response of *A. thaliana* towards AHL compounds with short side chains, because systemic resistance responses are not elevated (von Rad et al. 2008). However, most recently, the induction of systemic resistance responses towards the leaf pathogen *Pseudomonas syringae* DS3000 by oxo-C14-homoserine lactone treatment in the rhizosphere of *A. thaliana* was demonstrated (Zuccharo and Kogel, personal communication). This indicates that there may be different cellular receptors and signalling pathways of bacterial signalling compounds in plants. In addition, AHL compounds with acyl side chain lengths smaller than C8 were found to enter the roots more readily and are transported up to the shoots (Götz et al. 2007). Additionally, it could be shown that the ^{3}H-labelled C6- and C8-AHLs are taken up into the central cylinder and that their transport within the roots is an active process, which is further accelerated by the transpiration flow (Riedel, personal communication). AHL compounds with long side chains (e. g. larger than C10) stick to the root surface and are not transported substantially within, e.g., barley, maize, or *A. thaliana* (Götz et al. 2007; von Rad et al. 2008). In some other plants, like many legumes, the plant's AHL-hydrolyzing activities are efficiently degrading AHLs and thus prohibit a substantial uptake of the AHL compounds into the plants (Götz et al. 2007; Delalande et al. 2005).

Apart from plants, the effects of quorum sensing molecules on fungi were already found, like the induced morphological changes of *Candida albicans* under the influence of *Pseudomonas aeruginosa* (Hogan et al. 2004). Long side chain as well as short side chain *N*-acyl-homoserine lactones were found to modulate also the host immune response and inflammatory signalling pathways of invertebrates (for review see Cooley et al. 2008).

5.4 Mycorrhiza-Associated Bacteria and Their Effects on Symbiosis Development

5.4.1 Endo- and Ectomycorrhizal Fungi

There are many examples for close bacteria–mycorrhiza interactions. In the case of arbuscular mycorrhizae, a group of Gram-negative bacteria closely related to *Burkholderia*, described as *Candidatus Glomeribacter margarita*, was identified as endofungal bacteria by the group of Bonfante (Lumini et al. 2007). This bacterium could not be cultivated until now. Using fluorescent staining techniques and confocal laser scanning microscopy, these bacteria could be located mostly in spores of *Gigaspora margarita* (Bianciotto et al. 2004). There is recent evidence that these endofungal bacteria exert beneficial functions to their mycorrhizal host. When the lines of *G. margarita* containing the endosymbiotic bacteria were compared with the lines which had been cured of the bacteria, it became clear that the endosymbiotic bacteria strongly improved the presymbiotic growth of the arbuscular mycorrhiza-forming fungus, as shown by increased hyphal elongation and branching following treatment with root exudates. Therefore, these bacteria support the growth of such fungi and could thus possibly support mycorrhiza establishment under soil conditions which are unfavorable to the fungi (Frey-Klett et al. 2007). It was demonstrated by Brulé et al. (2001) that in the case of the MHB, *P. fluorescens* BBc6R8 ectomycorrhizal *Douglas-fir–Laccaria bicolor* symbiosis, a similar stimulation of presymbiotic fungal growth may occur. Furthermore, the *Paenibacillus* sp. isolate of *L. bicolor*, which was found as endofungal bacterium within hyphae of *L. bicolor* (Bertaux et al. 2003), was shown to significantly promote the growth of *L. bicolor* in vitro (Deveau et al. 2007). As suggested by Frey-Klett et al. (2007), the two mycorrhizal fungi, *G. margarita* (AM-forming) and *L. bicolor* (ectomycorrhiza (EcM)-forming) illustrate two different evolutionary processes of bacterial colonisation of fungal cells. *G. margarita* is an example of long-lasting co-evolution between the fungus and its endobacterium, *Glomeribacter gigasporarum*, through fungal spore generations (Bianciotto et al. 2004). The small genome size of the endobacteria and the difficulties in cultivating these bacteria in vivo (Jargeat et al. 2004) support the hypothesis of co-evolution.

In contrast, *L. bicolor* harbors fluctuating endobacterial communities that appear to be environmentally acquired, because in EcM collected from forest soils, a high variety of α-proteobacteria was found to colonise hyphae of *L. bicolor* internally (Bertaux et al. 2005). It can be speculated that the intracellular colonisation of *L. bicolor* by soil bacteria would support the fungal host to adapt to changing environmental challenges, especially during the presymbiotic life in soil. This would provide examples for the hologenome theory, which states that the combined metabolic potential and activity of symbiotic associations of microorganisms with plants and animals provide a better basis for the evolution of holobionts with improved fitness in view of environmental challenges (Zilber-Rosenberg and Rosenberg 2008).

5.4.2 *Sebacinales: The Plant Growth Promoting Fungus* Piriformospora indica

More recently, endofungal bacteria were also described in fungi of the order *Sebacinales* (Basidiomycota; Sharma et al. 2008), which constitute a special group of fungi, being mostly involved in ericoid and orchid mycorrhizae (Selosse et al. 2007). *Piriformospora indica* and other members of the *Sebacinales* are wide-host range root-colonising, mostly symbiotic, fungi which allow the plants to grow better under physical and nutrient stress conditions. The big advantage in practical use over arbuscular mycorrhizae is the possibility to cultivate some of these fungi on complex and even minimal substrates on agar plates or in submerged culture. *P. indica* in particular was originally isolated from the roots of different xerophytes in the Indian Thar desert (Verma et al. 1998). Since it can be grown easily without the plant and is amenable to molecular genetic techniques, it is most interesting not only for basic research, but also for biotechnological applications (for review see Oelmüller et al. 2009). It has plant growth promoting and biofertilizer properties especially in nutrient-deficient soils and can act as a biocontrol agent against biotic and abiotic stresses including root and leaf pathogens and insect invaders (Badge et al. 2010). Furthermore, it is a bioregulator for plant growth development, early flowering, and enhanced seed production and it is a bio-agent for the hardening of tissue-culture-raised plants. The *P. indica*/*A. thaliana* as well as the *P. indica*/barley systems were successfully used in the identification of important target compounds of the fungus in the course of interaction and molecular colonisation (Oelmüller et al. 2009).

It turned out that in *P. indica* as well as in several *Sebacina vermifera* isolates, endofungal bacteria could be demonstrated by FISH analysis and confocal laser scanning microscopy or at least by 16S PCR analysis (Sharma et al. 2008). While *Rhizobium radiobacter* (formerly *Agrobacterium tumefaciens*) was identified as endofungal bacterium, different isolates of *S. vermifera* had *Paenibacillus* sp., *Acinetobacter* sp., or *Rhodococcus* sp. as bacterial associate. With the exception of *R. radiobacter*, the other bacteria could not be cultivated. Since all attempts to cure the fungi from their endobacteria were unsuccessful so far, they seem to have important functions for the fungus. In the case of *R. radiobacter*, it could be shown that upon inoculation to barley, this bacterium produced a similar shoot weight increase and decrease in powdery mildew pustules formation as compared to inoculation with the *P. indica* (plus its endofungal *R. radiobacter*) (Sharma et al. 2008). Looking for properties of plant growth stimulation in the bacterium, not only the production of indole acetic acid, but also the biosynthesis of several *N*-acyl homoserine lactones was noticed (Li and Fekete, unpublished results). Interestingly, *R. radiobacter* also was found to use *p*-coumaric acid as a precursor for *N*-coumaryl homoserine lactone (Schaefer et al. 2008). The production of this variety of homoserine lactones may contribute to the plant growth promotion ability in the tripartite symbiotic system barley (or *A. thaliana*), *P. indica*, and the bacterium *R. radiobacter*.

5.5 Ectomycorrhiza-Forming Fungi

Among other types of mutualistic interactions, the formation of ectomycorrhizas (EcM), a symbiosis between fine roots of trees and certain soil fungi, is a way to overcome limitations in mineral nutrients and (easily degradable) carbohydrates, typical for many forest ecosystems. Due to the tree canopy, soil temperature rarely exceeds 15°C and degradation of organic substances often containing high amounts of phenolic substances results in relatively low pH values (below 5). As a consequence, bacterial turnover and mineralisation of organic matter is reduced and soil fungi have a large impact on nutrient recycling and plant mineral nutrition. EcM formation is thus typical for trees in boreal and temperate forests of the northern hemisphere, and alpine regions world-wide. The exchange of fungus-derived nutrients for plant-derived carbohydrates enables the colonisation of mineral nutrient-poor environments by the trees (Smith and Read 2008).

EcM fungi can both live in association with plant roots as symbionts and, possibly also as facultative saprotrophs. EcM fungal colonies remain functionally interconnected, revealing intense nutrient and carbohydrate exchange over variable distances. Vegetative mycelia can differentiate into distinct hyphal networks with different functions, differently densely growing solitary hyphae or organised as rhizomorphs (Agerer 2001) and hyphal mantle/intercellularly growing hyphae (Hartig net). When fungal hyphae recognise an emerging fine root of a compatible plant partner, they direct their growth towards it and colonise the root surface, forming a mantle of hyphae, which encloses the root, and isolates it physically from the surrounding soil (Blasius et al. 1986). Root hairs, which are normally formed by rhizodermal cells, are suppressed by EcM formation. After or parallel to mantle formation, fungal hyphae grow inside the colonised fine root, forming highly branched structures in the apoplast of the rhizodermis (also in the root cortex in gymnosperms). This so-called Hartig net creates a large surface area between both the partners (Kottke and Oberwinkler 1987). EcM are usually composed of two fungal networks with different functions (Harley and Smith 1983; Kottke and Oberwinkler 1987; Smith and Read 2008): the Hartig net as plant/fungus interface, adapted to the exchange of plant-derived carbohydrates for fungus-derived nutrients and the fungal mantle for intermediate storage of nutrients that are delivered by soil growing hyphae and further directed to the Hartig net, as well as carbohydrates that are taken up by hyphae of the Hartig net and are then supplied to the mycelium growing within the soil.

5.5.1 Progress with Regard to Genetic Information from Ectomycorrhiza-Forming Fungi

The *Laccaria bicolor* genome is the first of a fungal symbiont to be sequenced and its release is taking mycorrhizal genomics one large step further (Martin et al. 2008). This, together with the available genomes of saprotrophic and pathogenic fungi (~50; e.g. Galagan et al. 2005; Soanes et al. 2008), offers the opportunity to decipher key components of functions of ectomycorrhiza-forming fungi.

The nuclear genome of *L. bicolor* is estimated to contain 60 million bases, spread out over 12 chromosomes and is thus bigger than that of most previously published fungal genomes. About 19,000 predicted coding regions are found of which almost 25% still lack characterised orthologues in other systems. The larger size is partly explained by an unusually high number of transposable elements that constitute more than 20% of the genome. The even higher density of transposable elements which have been found in *T. melanosporum* and the poplar rust, *Melampsora laricipopulina* (Martin et al. unpublished), suggests a possible relationship between biotrophy and transposable elements richness. Compared with other fungal genomes, the *L. bicolor* genome contains both more and larger gene families which evolved from common ancestors found in other fungi (Lucic et al. 2008). These include several fungal multigene families not only coding for membrane, cell wall, and secreted proteins, but also up to 1,000 lineage-specific families (Martin et al. 2008).

5.5.2 The Saprotrophic Face of ECM Fungi

Through the net carbon input (sequestration of host carbohydrates within fungal networks) and loss (decomposition of soil organic matter, for reviews see (Smith and Read 2008; Talbot et al. 2008), mycorrhizal fungi significantly influence the CO_2 sink efficiency of forest ecosystems.

In order to understand the role played by the many ECM fungal partners in complex forest ecosystems, direct methods to assess a range of activities relevant to tree nutrition and metabolic activity on single excised ectomycorrhizal root tips were introduced. To assess their physiological activity, Jany et al. (2003) compared the [^{14}C] glucose respiration of two ECM species and found a high influence of the respective soil conditions (e.g. drought) and species-specific differences. Using sensitive microplate assays for the detection of phosphatase, chitinase, dehydrogenases, glucosidases, and other hydrolytic enzymes, impacts of season, temperature, and soil moisture effects on enzyme activities were found (Pritsch et al. 2004; Buee et al. 2005; Courty et al. 2005). It can thus be concluded that the ability of degrading litter polymers and assimilating absorbed nutrients widely varies between ECM species, that is enhanced upon carbon and nitrogen starvation, and that it fluctuates seasonally (Buee et al. 2005, 2007; Courty et al. 2007).

Laccaria bicolor possesses expanded glycosyl hydrolase families and a large set of secreted proteases, chitinases, and glucanases associated presumably with the hydrolysis of organic matter from dead organisms (Martin and Tunlid 2009; Nuutinen and Timonen 2008; Martin and Nehls 2009). The thereby indicated strong saprotrophic capability of the fungus enables a symbiosis-driven access of trees to nutrients held up in complex molecules in the soil that are outside of the symbiosis barely available to the plant partner. However, even when capable of catabolising chitin-, glucan-, and protein-complexes from decaying litter, *L. bicolor* is not able to degrade plant cell wall polysaccharides (cellulose, pectins, and pectates; Martin et al. 2008) due to massive gene loss. This prevents ECM fungi from degrading their host cells and, as a consequence, triggering plant defence reactions and could be a prerequisite of adaptation to symbiotic lifestyle. This result has to be taken, however, with care as it may reflect the evolution of one specific clade of mycorrhizal taxon and not the lifestyle of all ectomycorrhizal species. The by-products such as amino acids or monosaccharides resulting from degradation of organic matter are efficiently taken up (Fajardo Lopez et al. 2008; Lucic et al. 2008; Morel et al. 2008), and metabolised (Fajardo López et al. 2007; Deveau et al. 2008; Nuutinen and Timonen 2008; see below). Due to its importance for plant and fungal nutrition, research was mainly focused on nitrogen and only to a lesser extent on phosphate (Tatry et al. 2008), sulphur (Mansouri-Bauly et al. 2006), and potassium (Corratge et al. 2007).

The whole range of inorganic and organic nitrogen sources found in forest soils can be used by *L. bicolor*. Genes encoding putative importers for nitrate, ammonium, urea, amino acids, peptides, nucleotides, allantoin, and polyamines were found in the genome (Lucic et al. 2008). While a proof of function is currently under progress for selected *L. bicolor* gene families, individual nitrogen import proteins of other ECM fungi were previously characterised, e.g. (Willmann et al. 2007; Morel et al. 2008). Like in other organisms with the capability to exude hydrolases, the expression of genes encoding high-affinity importers is frequently induced/enhanced by nitrogen starvation while the transcript levels of low-affinity importers are only marginally affected (for reviews, see Tarkka et al. 2005; Müller et al. 2007). However, as mobilisation capability of a given nitrogen source differs between ECM fungi (Nygren et al. 2008), additional models have to be investigated in future research to take into account fungal adaptation to various forest ecosystems.

In contrast to high levels of inorganic phosphate (Pi) within living systems (in the mM range), free concentrations of Pi in soil are very low, ranging from 1 to 10 μM (Bieleski 1973; Vance et al. 2003). The low availability of Pi in soil is due to its negative charges, resulting in rapid sequestration by cations (Vance et al. 2003) and renders Pi highly immobile (Hinsinger 2001). Uptake of phosphate by plant roots quickly generates a depletion zone, making this element mostly limiting for plant growth. In the association between *Hebeloma cylindrosporum* and its natural host plant *Pinus pinaster*, accumulation of Pi in whole plants was significantly correlated with soil exploration by external hyphae, suggesting that plant Pi mainly originates from fungal Pi uptake (Aquino and Plassard 2004). Genes, revealing significant similarities to phosphate transporters, were identified for *Amanita muscaria* (Barbosa 2004) and *H. cylindrosporum* (*HcPT1* and *HcPT2*; van Aarle et al. 2007; Tatry et al. 2008). Complementation of a yeast mutant (Δpho84), defective in phosphate transport, confirmed both corresponding *H. cylindrosporum* proteins (HcPT1 and HcPT2) to mediate Pi:H^+ symport, exhibiting K_M values of 55 and 4 μM, respectively. Fluorescent in situ RT-PCR (van Aarle et al. 2007) and real-time RT-PCR (Tatry et al. 2008) showed that Pi starvation increased *HcPT1* expression, while transcript levels of *HcPT2* were less dependent on Pi availability, rendering both genes good candidates for a role in fungal Pi uptake from the soil solution when the host plant is grown in soil with low Pi availability.

The macronutrient sulphur is assimilated in the inorganic form by plants, fungi, and most bacteria. Starvation results in an increased uptake by plants (Lappartient et al. 1999), bacteria (Kredich 1993), and fungi (Ono et al. 1999; Van de Kamp et al. 2000), while it is strongly repressed in the presence of reduced sulphur compounds (e.g. glutathione, cysteine, H_2S; Herschbach and Rennenberg 1994; Lappartient et al. 1999; Van de Kamp et al. 2000; Westerman et al. 2001). This indicates a tight regulation by sulphur homeostasis. Contrasting the enhanced sulfate uptake by host plants (clover and maize) with arbuscular mycorrhiza (Gray and Gerdemann 1972; Banerjee et al. 1999), the ectomycorrhizal association had no impact on the rate of sulfate uptake but influenced sulfate loading into the tree xylem (Seegmüller et al. 1996; Kreuzwieser and Rennenberg 1998). Ectomycorrhizal beech roots did not reveal a de-repression of sulfate import by sulphur deficiency as non-mycorrhizal roots do, indicating a significant contribution of the fungal partner in sulfate uptake (Kreuzwieser et al. 1996; Kreuzwieser and Rennenberg 1998). As in other organisms, sulfate uptake was increased by sulphur starvation in the ECM fungus, *L. bicolor* (Mansouri-Bauly et al. 2006). In contrast to bacteria, yeast, and plants, sulfate uptake in *L. bicolor* was not inhibited by glutathione. However, the regulation of sulfate assimilation in *L. bicolor* (indicated by the reduction in activity of 3′-phosphoadenosine 5′-phosphosulfate reductase after treatment with glutathione) is again similar to that of other organisms. Together these data clearly indicate a regulatory uncoupling of sulfate uptake and its reduction in the presence of reduced sulphur compounds in *L. bicolor*, a response which has not yet been reported for other organisms. A potential explanation is that *L. bicolor* improves sulfate uptake and export towards the plant in symbiosis, and receives reduced sulphur back from the plant partner (Mansouri-Bauly et al. 2006).

5.5.3 The Symbiotic Interface of ECM Fungi

In contrast to the uptake by growing soil hyphae, nutrient export at the plant/fungus interface is still poorly understood. Two major potential nitrogen sources, ammonium and amino acids, are thought to be released at the symbiotic interface (for a review see Chalot et al. 2006), and homologs of yeast proteins involved in their excretion are found in different ECM fungi (Selle et al. 2005; Chalot et al. 2006; Lucic et al. 2008). Regarding the plant host, no data on the impact of symbiosis on amino acid import are available yet. However, high-affinity ammonium importers are clearly upregulated upon ectomycorrhiza formation (Selle et al. 2005; Couturier et al. 2007). A prerequisite for an efficient export of any nitrogen source is that the corresponding fungal import is repressed at the plant/fungus interface. In agreement with this, transcript levels of a high-affinity ammonium importer (Willmann et al. 2007), but not that of an amino acid importer (Nehls, unpublished; amino acids are organically bound nitrogen resource preferentially taken up by EcM fungi) were strongly reduced in *Amanita muscaria* ECM. In contrast, transcript levels of putative high-affinity plant ammonium importers were induced upon EcM formation in *L. bicolor* (Lucic et al. 2008). Furthermore, and similar to *A. muscaria* (Nehls, unpublished), urease transcript levels increased in *L. bicolor* during symbiosis. It has been suggested that the release of ammonium through urease activity may be involved in fungal nitrogen export in mycorrhizal symbiosis towards the host plant (Morel et al. 2005; Cruz et al. 2007; Lucic et al. 2008). As a consequence, two scenarios for nitrogen export into the plant tissue at the plant/fungus interface can be suggested: (1) Cytoplasmic ammonium is assimilated into amino acids which are transferred to the host, and re-import of leaked ammonium is enhanced to avoid unattended loss. (2) Ammonium is released by the hyphae at the symbiotic interface (here re-import is repressed by a combination of transcriptional and post-transcriptional control), while the fungal sheath may be active in ammonium re-import to avoid its loss into the soil due to leakage. To get a better understanding of the fungal nitrogen export mechanism, microdissection (distinction of metabolite content and protein activity within hyphal networks) together with the use of fungal mutants (Kemppainen et al. 2008), defective in nitrogen export, will be necessary.

5.5.4 Carbohydrate Allocation

Essential for mycorrhizal symbiosis is a continuous fungal carbohydrate nutrition by the host plant (Saravesi et al. 2008). As a consequence, ECM reveal an elevated sink strength compared to non-mycorrhizal fine roots (for a review, see Nehls 2008). The *L. bicolor* genome (Martin et al. 2008) enables for the first time the exploration of ECM fungal capacity to take up carbohydrates. Due to the lack of fungal invertase and sucrose importer genes (Fajardo Lopez et al. 2008; Martin et al. 2008), apoplastic sucrose (the supposed major carbon source) can be hydrolyzed by host-derived enzymes only, making *L. bicolor* also locally dependent on its host.

Fifteen genes encoding putative hexose importers were found, of which five are functional hexose importers as shown by heterologous expression in yeast (Fajardo Lopez et al. 2008). Transcript profiling showed an enhanced transcript level for half of these genes, indicating a strong increase in hexose uptake capacity at the plant/fungus interface, similar to what was observed for *Amanita muscaria*. However, the comparison of the two ECM fungi revealed large differences in their control of gene expression. While the expression of *A. muscaria* genes was regulated by the apoplastic hexose concentration, development-dependent control was observed for *L. bicolor* (Fajardo Lopez et al. 2008). Further contrasting the situation in *A. muscaria* (but also *H. cylindrosporum*), where glucose and fructose are taken up simultaneously by fungal hyphae, *L. bicolor* does not use fructose until the external glucose concentration is below the K_M value of its high-affinity hexose importer. This behaviour might indicate a less efficient carbohydrate exploitation by *L. bicolor* in symbiosis, reflecting different capabilities of EM fungi.

To maintain a strong carbohydrate sink in symbiosis, increased carbon fluxes through glycolysis (Kowallik et al. 1998) and into storage compounds were observed in ECM of different fungi, in addition to an enhanced hexose uptake (Martin et al. 1985; Fajardo López et al. 2007; Wiemken 2007). This view was confirmed by whole genome transcript analysis in *L. bicolor* (Deveau et al. 2008). By physical separation of ectomycorrhizal networks and monitoring of transcript levels and activities of key enzymes of fungal trehalose biosynthesis, Fajardo López et al. (2007) could localise the enhanced flux into these storage carbohydrates in *Amanita muscaria* hyphae of the plant/fungus interface. However, further biochemical analysis and generation of mutants defective in biosynthesis of trehalose and other storage compounds will be necessary to confirm their function as integral components of fungal sink generation in symbiosis.

5.5.5 Changing the Transcriptome: A First Step Towards Understanding Ectomycorrhiza Formation

For expression profiling of ECM development and function, comprehensive microarray-based gene expression analyses of different developmental stages (pre-symbiotic, fully developed), using different ectomycorrhizal systems (*Laccaria bicolor*/*Pisolithus microcarpus*/*Tuber borchii* + *Tilia americana*, *Paxillus involutus* + *Betula pendula*, *Laccaria bicolor* + *Populus trichocarpa* or *Pseudotsuga menziesii*, *Amanita muscaria* + *Populus tremula* × *tremuloides*), have been performed (Peter et al. 2003; Johansson et al. 2004; Menotta et al. 2004; Duplessis et al. 2005; Wright et al. 2005; Küster et al. 2007; Nehls et al. 2007; Martin et al. 2008). They provide a quantitative assessment of transcript abundance and can be used to predict gene function based on the hypothesis that functionally related genes are mainly transcriptionally regulated. With the exception of those from *L. bicolor* (Martin et al. 2008; see below), these arrays represent a maximum of about 10% of the gene repertoire of a given fungus, and conclusions, that can be drawn, are thus limited.

The number of symbiosis-regulated and symbiosis-specific *L. bicolor* transcripts is very low (<3%; Martin et al. 2008). Transcript profiling of *L. bicolor* interacting with either *Populus trichocarpa* or *Pseudotsuga menziesii* roots has revealed several genes with a striking upregulation in symbiotic tissues (>100-fold; Martin et al. 2008). Most of them are coding for proteins belonging to expanding and lineage-specific gene families (Martin and Tunlid 2009). A large number of the mycorrhiza-upregulated transcripts are encoding small, putatively secreted, cysteine-rich proteins, which may play, together with RGD-motive (arginine–glycine–aspartic acid)-containing acidic proteins (Le Quere et al. 2005) and hydrophobins (Laurent et al. 1999), a role in the construction of the novel symbiotic apoplastic interface. Several cysteine proteinase inhibitors (mycocypin gene family) are amongst the most highly upregulated transcripts suggesting that *L. bicolor* may use proteinase inhibitors for counteracting plant secreted protease activities during its apoplastic growth.

5.6 Conclusions and Relevance with Respect to SFB Data

Light, carbon dioxide, water, and nutrients are the basis for plant growth, at least under controlled conditions, e.g. greenhouses, etc. In the field, a plant has to cope with many more parameters such as competing plants, herbivores, pathogens, or pollutants.

Within the rhizosphere, multi-level organismic interactions become increasingly visible. We know now that large groups of plant-growth promoting bacteria exist within the rhizosphere and mycorrhizosphere, which release products of their secondary metabolism. These can selectively interact with plant symbiotic and plant pathogenic bacteria and fungi, thereby establishing new balances between such organisms and thus affecting plant viability. In addition, by systemic signal propagation, compounds released can also render plants more resistant to shoot pathogens. Such support is not for free, and the plant has to invest a considerable amount of its photo assimilates into the support of the rhizosphere community. This is done by allocation of carbohydrates to the rhizo-/mycorrhizosphere which affects shoot growth, delivering a good example for the GDB theory. Air pollutants such as ozone have been shown to interfere with carbon allocation, and can thus severely affect such fine-tuned organismic interactions (comp. Chap. 10). Altered fungal communities, together with less carbon allocation will also have consequences with respect to the attraction of bacterial communities. For these, twice-ambient ozone caused a shift from Gram-negative to Gram-positive bacteria. The latter include streptomycetes, which are important producers of antibiotics. These can affect pathogenic microorganisms (see Sect. 5.2.2). Surprisingly, elevated ozone resulted in both a higher microbial biomass and abundance of microbes utilising plant C (comp. Chap. 4). This could be due to the starch-enriched litter as carbon allocation into fungal symbionts of the soil is reduced. In summary, the findings show that environmental changes can considerably interact with the microbial rhizosphere communities. This should have an impact on plant performance, and could counteract pollutant-related forms of damage (see marginal effects of elevated ozone on the investigated beech and spruce trees). Rhizospheric interactions certainly will have an impact on allocation of carbon to either growth or protection against pathogens or herbivores, and will make discussion about possible trade-offs not easier. This becomes clear from Chap. 3.

Acknowledgements As far as own research has been addressed, financial support by the Deutsche Forschungsgemeinschaft is gratefully acknowledged. We are indebted to Nina Lehr, Silvia Schrey, and Mika Tarkka for their input, and to Reinhard Agerer for critical reading and many helpful suggestions. GB 4-2 was isolated by K. Poralla (University of Tübingen).

References

Agerer R (2001) Exploration types of ectomycorrhizae. Mycorrhiza 11:107–114

Aquino TM, Plassard C (2004) Dynamics of ectomycorrhizal growth and P transfer to the host plant in response to low and high soil P availability. FEMS Microbiol Ecol 48:149–156

Artursson V, Finlay RD, Jansson JK (2005) Combined bromodeoxyuridine immunocapture and terminal-restriction fragment length polymorphism analysis highlights differences in the active soil bacterial metagenome due to *Glomus mosseae* inoculation or plant species. Environ Microbiol 17:1952–1966

Badge US, Prasad R, Varma A (2010) Interaction of mycobiont: *Piriformospora indica* with medicinal plants and plants of economic importance. Afr J Biotechnol 9:9214–9226

Bakker PA, Pieterse CM, van Loon LC (2007) Induced systemic resistance by fluorescent *Pseudomonas* spp. Phytopathology 97:239–243

Banerjee MR, Chapman SJ, Killham K (1999) Uptake of fertilizer sulfur by maize from soils of low sulfur status as affected by vesicular-arbuscular mycorrhizae. Can J Soil Sci 79:557–559

Barbosa PG (2004) Identifizierung und molekulare Charakterisierung von Phosphataufnahmesystemen des Ektomykorrhizapilzes Amanita muscaria. Dissertation, Eberhard-Karls University, Tuebingen

Barea JM, Pozo MJ, Azcón R, Azcón-Aguilar C (2005) Microbial co-operation in the rhizosphere. J Exp Bot 56:1761–1778

Barriuso J, Pereyra MT, Lucas García JA, Megías M, Gutiérrez Mañero FJ, Ramos B (2005) Screening for putative PGPR to improve establishment of the symbiosis *Lactarius deliciosus–Pinus* sp. Microb Ecol 50:82–89

Berdy J (2005) Bioactive microbial metabolites: a personal view. J Antibiot 58:1–26

Berg G, Smalla K (2009) Plant species and soil type cooperatively shape the structure and function of microbial communities in the rhizosphere. FEMS Microbiol Ecol 68:1–13

Bertaux J, Schmid M, Chemidlin Prevost-Boure N, Churin JL, Hartmann A, Garbaye J, Frey-Klett P (2003) In situ identification of intracellular bacteria related to Paenibacillus sp. in the mycelium of the ectomycorrhizal fungus Laccaria bicolor S238N. Appl Environ Microbiol 69:4243–4248

Bertaux J, Schmid M, Hutzler P, Hartmann A, Garbaye J, Frey-Klett P (2005) Occurrence and distribution of endobacteria in the plant-associated mycelium of the ectomycorrhizal fungus *Laccaria bicolor* S238N. Environ Microbiol 7:1786–1795

Bianciotto V, Genre A, Jargeat P, Lumini E, Bécard G, Bonfante P (2004) Vertical transmission of endobacteria in the arbuscular mycorrhizal fungus *Gigaspora margarita* through vegetative spore generations. Appl Environ Microbiol 70:3600–3608

Bieleski RL (1973) Phosphate pools, phosphate transport and phosphate availability. Ann Rev Plant Physiol 24:225–252

Blaha D, Prigent-Combaret C, Mirza MS, Moenne-Loccoz Y (2006) Phylogeny of the 1-aminocyclopropane-1-carboxylic acid deaminase-encoding gene *acdS* in phytobeneficial and pathogenic Proteobacteria and relation with strain biogeography. FEMS Microbiol Ecol 56:455–470
Blasius D, Feil W, Kottke I, Oberwinkler F (1986) Hartig net structure and formation of fully ensheated ectomycorrhizas. Nord J Bot 6:837–842
Bonfante P, Anca J-A (2009) Plants, mycorrhizal fungi, and bacteria: a network of interactions. Annu Rev Microbiol 63:363–383
Bruce TJ, Pickett JA (2007) Plant defence signalling induced by biotic attacks. Curr Opin Plant Biol 10:387–92
Brulé C, Frey-Klett P, Pierrat JC, Courrier S, Gérard F, Lemoine MC, Rousselet JL, Sommer J, Garbaye J (2001) Survival in the soil of the ectomycorrhizal fungus *Laccaria bicolor* and effect of a mycorrhiza helper *Pseudomonas fluorescens*. Soil Biol Biochem 33:1683–1694
Buee M, Courty PE, Mignot D, Garbaye J (2007) Soil niche effect on species diversity and catabolic activities in an ectomycorrhizal fungal community. Soil Biol Biochem 39:1947–1955
Buee M, Vairelles D, Garbaye J (2005) Year-round monitoring of diversity and potential metabolic activity of the ectomycorrhizal community in a beech (*Fagus silvatica*) forest subjected to two thinning regimes. Mycorrhiza 15:235–245
Challis GL, Hopwood DA (2003) Synergy and contingency as driving forces for the evolution of multiple secondary metabolite production by *Streptomyces* species. Proc Natl Acad Sci USA 100(Suppl 2):14555–14561
Chalot M, Blaudez D, Brun A (2006) Ammonia: a candidate for nitrogen transfer at the mycorrhizal interface. Trends Plant Sci 11:263–266
Chamberlain K, Crawford DL (1999) In vitro and vivo antagonism of pathogenic turfgrass fungi by *Streptomyces hygroscopicus* strains YCED9 and WYE53. J Ind Microbiol Biotechnol 23:641–646
Chen XH, Koumoutsi A, Scholz R, Schneider K, Vater J, Süssmuth R, Piel J, Borriss R (2009a) Genome analysis of *Bacillus amyloliquefaciens* FZB42 reveals its potential for biocontrol of plant pathogens. J Biotechnol 140:27–37
Chen XH, Scholz R, Borriss M, Junge H, Mögel G, Kunz S, Borriss R (2009b) Difficidin and bacilysin produced by plant-associated *Bacillus amyloliquefaciens* are efficient in controlling fire blight disease. J Biotechnol 140:38–44
Cocito C, Di Giambattista M, Nyssen E, Vannuffel P (1997) Inhibition of protein synthesis by streptogramins and related antibiotics. J Antimicrob Chemother 39:7–13
Compant S, Duffy B, Nowak J, Clément C, Barka EA (2005) Use of plant growth-promoting bacteria for biocontrol of plant diseases: principles, mechanisms of action, and future prospects. Appl Environ Microbiol 71:4951–4959
Conn VM, Walker AR, Franco CM (2008) Endophytic actinobacteria induce defense pathways in *Arabidopsis thaliana*. Mol Plant Microbe Interact 21:208–18
Conrath U (2006) Systemic acquired resistance. Plant Signal Behav 1:179–184
Conrath U, Pieterse CM, Mauch-Mani B (2002) Priming in plant-pathogen interactions. Trends Plant Sci 7:210–16
Cooley M, Chhabra SR, Williams P (2008) *N*-Acylhomoserine lactone-mediated quorum sensing: a twist in the tail and a blow for host immunity. Chem Biol 15:1141–1147
Coombs JT, Franco CM (2003) Isolation and identification of actinobacteria from surface-sterilized wheat roots. Appl Environ Microbiol 69:5603–5608
Cornelis P, Matthijs S (2002) Diversity of siderophore-mediated iron uptake systems in fluorescent pseudomonads: not only pyoverdins. Environ Microbiol 4:787–798
Corratge C, Zimmermann S, Lambilliotte R, Plassard C, Marmeisse R, Thibaud JB, Lacombe B, Sentenac H (2007) Molecular and functional characterization of a Na^+-K^+ transporter from the Trk family in the ectomycorrhizal fungus *Hebeloma cylindrosporum*. J Biol Chem 282:26057–26066

Courty P-E, Breda N, Garbaye J (2007) Relation between oak tree phenology and the secretion of organic matter degrading enzymes by *Lactarius quietus* ectomycorrhizas before and during bud break. Soil Biol Biochem 39:1655–1663

Courty P-E, Pritsch K, Schloter M, Hartmann A, Garbaye J (2005) Activity profiling of ectomycorrhiza communities in two forest soils using multiple enzymatic tests. New Phytol 167:309–319

Couturier J, Montanini B, Martin F, Brun A, Blaudez D, Chalot M (2007) The expanded family of ammonium transporters in the perennial poplar plant. New Phytol 174:137–150

Crawford DL, Lynch JM, Whipps JM, Ousley MA (1993) Isolation and characterization of actinomycete antagonists of a fungal root pathogen. Appl Environ Microbiol 59:3899–3905

Cruz C, Egsgaard H, Trujillo C, Ambus P, Requena N, Martins-Loucao MA, Jakobsen I (2007) Enzymatic evidence for the key role of arginine in nitrogen translocation by arbuscular mycorrhizal fungi. Plant Physiol 144:782–792

Cui J, Bahrami AK, Pringle EG, Hernandez-Guzman G, Bender CL, Pierce NE, Ausübel FM (2005) *Pseudomonas syringae* manipulates systemic plant defenses against pathogens and herbivores. Proc Natl Acad Sci USA 102:1791–96

Davelos AL, Kinkel LL, Samac DA (2004) Spatial variation in frequency and intensity of antibiotic interactions among *Streptomycetes* from prairie soil. Appl Environ Microbiol 70:1051–58

De Almeida CV, Andreote FD, Yara R, Tanaka FAO, Azevedo JL, de Almeida M (2009) Bacteriosomes in axenic plants: endophytes as stable endosymbionts. World J Microbiol Biotechnol. doi:10.1007/s11274-009-0073-8

Delalande L, Faure D, Raffoux A, Uroz S, D'Angelo-Picard C, Elasri M, Carlier A, Berruyer R, Petit A, Williams P, Dessaux Y (2005) *N*-hexanoyl-L-homoserine lactone, a mediator of bacterial quorum-sensing regulation, exhibits plant-dependent stability and may be inactivated by germinating *Lotus corniculatus* seedlings. FEMS Microbiol Ecol 52:13–20

Deveau A, Kohler A, Frey-Klett P, Martin F (2008) The major pathways of carbohydrate metabolism in the ectomycorrhizal basidiomycete *Laccaria bicolor* S238N. New Phytol 180:379–390

Deveau A, Palin B, Delaruelle C, Peter M, Kohler A, Pierrat JC, Sarniguet A, Garbaye J, Martin F, Frey-Klett P (2007) The mycorrhizal helper *Pseudomonas fluorescens* BBc6R8 has a specific priming effect on the growth, morphology and gene expression of the ectomycorrhizal fungus *Laccaria bicolor* S238N. New Phytol 175:743–755

Dobbelaere S, Croonenborghs A, Amber T, Ptacek D, Vanderleyden J, Dutto P, Labandera-Gonzalez C, Caballero-Mellado J, Aguiire JF, Kapulnik Y, Shimon B, Burdman S, Kadouri D, Sarig S, Okon Y (2001) Responses of agronomically important crops to inoculation with *Azospirillum*. Aust J Plant Physiol 28:871–879

Dobbelaere S, Okon Y (2007) The plant growth-promoting effect and plant responses. In: Elmerich C, Newton WE (eds) Associative and endophytic nitrogen-fixing bacteria and cyanonbacterial associations. Springer, Dordrecht, pp 145–170

Dunne C, Moenne-Loccoz Y, de Bruijn F, O'Gara F (2000) Overproduction of an inducible extracellular serine protease improves biological control of *Pythium ultimum* by *Stenotrophomonas maltophilia* strain W81. Microbiology 146:2069–2078

Duplessis S, Courty PE, Tagu D, Martin F (2005) Transcript patterns associated with ectomycorrhiza development in *Eucalyptus globulus* and *Pisolithus microcarpus*. New Phytol 165:599–611

Eberl L (1999) N-Acyl homoserine lactone-mediated gene regulation in Gram-negative bacteria. Syst Appl Microbiol 22:493–506

Eckwall EC, Schottel JL (1997) Isolation and characterization of an antibiotic produced by the scab disease-suppressive *Streptomyces diastatochromogenes* strain PonSSII. J Ind Microbiol Biotechnol 19:220–225

El-Abyad MS, El-Sayed MA, El-Shanshoury AR, El-Sabbagh SM (1993) Towards the biological control of fungal and bacterial diseases of tomato using antagonistic *Streptomyces* spp. Plant Soil 149:185–195

Fajardo Lopez M, Dietz S, Grunze N, Bloschies J, Weiss M, Nehls U (2008) The sugar porter gene family of *Laccaria bicolor:* function in ectomycorrhizal symbiosis and soil-growing hyphae. New Phytol 180:365–378

Fajardo López M, Männer P, Willmann A, Hampp R, Nehls U (2007) Increased trehalose biosynthesis in Hartig net hyphae of ectomycorrhizas. New Phytol 74:389–398

Fiedler HP, Krastel P, Muller J, Gebhardt K, Zeeck A (2001) Enterobactin: the characteristic catecholate siderophore of *Enterobacteriaceae* is produced by *Streptomyces* species. FEMS Microbiol Lett 196:147–151

Firn RD, Jones CG (2000) The evolution of secondary metabolism – a unifying model. Mol Microbiol 37:989–994

Frey-Klett P, Chavarte M, Clausse ML, Courrier S, Le Roux C, Raaijmakers J, Martinotti M, Pierrat JC, Garbaye J (2005) Ectomycorrhizal symbiosis affects functional diversity of rhizosphere fluorescent pseudomonads. New Phytol 165:317–328

Frey-Klett P, Garbaye J, Tarkka M (2007) The mycorrhiza helper bacteria revisited. New Phytol 176:22–36

Galagan JE, Henn MR, Ma L-J, Cuomo CA, Birren B (2005) Genomics of the fungal kingdom: insights into eukaryotic biology. Genome Res 15:1620–1631

Gantner S, Schmid M, Dürr C, Schuhegger R, Steidle A, Hutzler P, Langebartels C, Eberl L, Hartmann A, Dazzo FB (2006) In situ spatial scale of calling distances and population density-independent *N*-Acylhomoserine lactone mediated communication by rhizobacteria colonized on plant roots. FEMS Microbiol Ecol 56:188–194

Glick BR, Penrose DM, Li J (1998) A model for the lowering of plant ethylene concentrations by plant growth promoting bacteria. J Theor Biol 190:63–68

Götz C, Fekete A, Gebefuegi I, Forczek S, Fuksová K, Li X, Englmann M, Gryndler M, Hartmann A, Matucha M, Schmitt-Kopplin P, Schröder P (2007) Uptake, degradation and chiral discrimination of N -acyl-D/L-homoserine lactones by barley (*Hordeum vulgare*) and yam bean (*Pachyrhizus erosus*) plants. Anal Bioanal Chem 389:1447–1457

Gray LE, Gerdemann JW (1972) Uptake of sulphur^{-35} by vesicular-arbuscular mycorrhizae. Plant Soil 39:687–698

Grichko VP, Glick BR (2001) Ameloriation of flooding stress by ACC deaminase-containing plant growth production. Plant Physiol Biochem 39:11–18

Haas D, Défago G (2005) Biological control of soil-borne pathogens by fluorescent pseudomonads. Nat Rev Microbiol 3:307–319

Harley JL, Smith SE (1983) Mycorrhizal symbiosis. Academic, London

Hartmann A, Rothballer M, Schmid M (2008) Lorenz Hiltner, a pioneer in rhizosphere microbial ecology and soil bacteriology research. Plant Soil 312:7–14

Hartmann A, Schmid M, van Tuinen D, Berg G (2009) Plant-driven selection of microbes. Plant Soil 321:235–255

Hense BA, Kuttler C, Mueller J, Rothballer M, Hartmann A, Kreft J-U (2007) Does efficiency sensing unify diffusion and quorum sensing? Nat Rev Microbiol 5:230–239

Herschbach C, Rennenberg H (1994) Influence of glutathione (gsh) on net uptake of sulphate and sulphate transport in tobacco plants. J Exp Bot 45:1069–1076

Hiltner L (1904) Über neuere Erfahrungen und Probleme auf dem Gebiete der Bodenbakteriologie unter besonderer Berücksichtigung der Gründüngung und Brache. Arb Dtsch Landw Ges 98:59–78

Hinsinger P (2001) Bioavailability of soil inorganic P in the rhizosphere as affected by root-induced chemical changes. Plant Soil 237:173–195

Hogan DA, Vik A, Kolter R (2004) A *Pseudomonas aeruginosa* quorum-sensing molecule influences *Candida albicans* morphology. Mol Microbiol 54:1212–1223

Janssen PH (2006) Identifying the dominant soil bacterial taxa in libraries of 16S rRNA and 16S rRNA genes. Appl Env Microbiol 72:1719–28

Jany JL, Martin F, Garbaye J (2003) Respiration activity of ectomycorrhizas from *Cenococcum geophilum* and *Lactarius* sp. in relation to soil water potential in five beech forests. Plant Soil 255:487–494

Jargeat P, Cosseau C, Ola'h B, Jauneau A, Bonfante P, Batut J, Bécard G (2004) Isolation, free-living capacities and genome structure of Candidatus *Glomeribacter gigasporarum*, the endocellular bacteria of the mycorrhizal fungus *Gigaspora margarita*. J Bacteriol 186:6876–6884

Johansson T, Le Quèrè A, Ahren D, Söderström B, Erlandsson R, Lundeberg J, Uhlen M, Tunlid A (2004) Transcriptional responses of *Paxillus involutus* and *Betula pendula* during formation of ectomycorrhizal root tissue. Mol Plant Microb Interact 17:202–215

Kemppainen M, Duplessis S, Martin F, Pardo AG (2008) T-DNA insertion, plasmid rescue and integration analysis in the model mycorrhizal fungus *Laccaria bicolor*. Microb Biotechnol 1:258–269

Kloepper JW, Leong J, Teintze M, Schroth MN (1980) Enhanced plant growth by siderophores produced by plant growth-promoting rhizobacteria. Nature 286:885–886

Kloepper JW, Ryu CM, Zhang S (2004) Induced systemic resistance and promotion of plant growth by Bacillus spp. Phytopathology 94:1259–1266

Kottke I, Oberwinkler F (1987) The cellular structure of the Hartig net: coenocytic and transfer cell-like organization. Nord J Bot 7:85–95

Koumoutsi A, Chen X-H, Henne A, Liesegang H, Hitzeroth G, Franke P, Vater J, Borriss R (2004) Structural and functional characterization of gene clusters directing nonribosomal synthesis of bioactive cyclic lipopeptides in *Bacillus amyloliquefaciens* strain FZB42. J Bacteriol 186:1084–1096

Kowallik W, Thiemann M, Huang Y, Mutumba G, Beermann L, Broer D, Grotjohann N (1998) Complete sequence of glycolytic enzymes in the mycorrhizal basidiomycete, *Suillus bovinus*. Z Naturforsch 53:818–827

Kredich NM (1993) Gene regulation of sulfur assimilation. In: DeKok LJ, Stulen I, Rennenberg H, Brunold C, Rauser WE (eds) Sulfur nutrition and sulfur assimilation in higher plants. SPB, The Hague, pp 61–76

Kreuzwieser J, Herschbach C, Rennenberg H (1996) Sulphate uptake and xylem loading of non-mycorrhizal excised roots of young beech (*Fagus sylvatica*) trees. Plant Physiol Biochem 34:409–416

Kreuzwieser J, Rennenberg H (1998) Sulphate uptake and xylem loading of mycorrhizal beech roots. New Phytol 140:319–329

Küster H, Becker A, Firnhaber C, Hohnjec N, Manthey K, Perlick AM, Bekel T, Dondrup M, Henckel K, Goesmann A, Meyer F, Wipf D, Requena N, Hildebrandt U, Hampp R, Nehls U, Krajinski F, Franken P, Pühler A (2007) Development of bioinformatic tools to support EST-sequencing, in silico- and microarray-based transcriptome profiling in mycorrhizal symbioses. Phytochemistry 68:19–32

Lappartient AG, Vidmar JJ, Leustek T, Glass ADM, Touraine B (1999) Inter-organ signaling in plants: regulation of ATP sulfurylase and sulfate transporter genes expression in roots mediated by phloem-translocated compound. Plant J 18:89–95

Laurent P, Voiblet C, Tagu D, de Carvalho D, Nehls U, De Bellis R, Balestrini R, Bauw G, Bonfante P, Martin F (1999) A novel class of ectomycorrhiza-regulated cell wall polypeptides in *Pisolithus tinctorius*. Mol Plant Microb Interact 12:862–871

Le Quere A, Wright D, Soederstroem B, Tunlid A, Johansson T (2005) Global patterns of gene regulation associated with the development of ectomycorrhiza between birch (*Betula pendula* roth.) and *Paxillus involutus* (batsch) Fr. Mol Plant Microb Interact 18:659–673

Lehr NA, Schrey SD, Bauer R, Hampp R, Tarkka MT (2007) Suppression of plant defence response by a mycorrhiza helper bacterium. New Phytol 174:892–903

Lehr NA, Schrey SD, Hampp R, Tarkka MT (2008) Root inoculation with a forest soil streptomycete leads to locally and systemically increased resistance against phytopathogens in Norway spruce. New Phytol 177:965–76

Leveau JHJ, Gerards S (2008) Discovery of a bacterial gene cluster for catabolism of the plant hormone indole-3-acetic acid. FEMS Microbiol Ecol 65:238–250

Liras P (1999) Biosynthesis and molecular genetics of cephamycins. Cephamycins produced by actinomycetes. Antonie Leeuwenhoek 75:109–124

Liu D, Anderson NA, Kinkel LL (1995) Biological control of potato scab in the field with antagonistic *Streptomyces scabies*. Phytopathology 85:827–831

Loria R, Bukhalid RA, Fry BA, King RR (1997) Plant pathogenicity in the genus *Streptomyces*. Plant Dis 81:836–846

Loria R, Kers J, Joshi M (2006) Evolution of plant pathogenicity in *Streptomyces*. Annu Rev Phytopathol 44:469–87

Lucic E, Fourrey C, Kohler A, Martin F, Chalot M, Brun-Jacob A (2008) A gene repertoire for nitrogen transporters in *Laccaria bicolor*. New Phytol 180:343–364

Lucy M, Reed E, Glick BR (2004) Application of free living plant growth-promoting rhizobacteria. Antonie van Leeuwenhoek 86:1–25

Lumini E, Ghignone S, Bianciotto V, Bonfante P (2007) Endobacteria or bacterial symbionts? To be or not to be. New Phytol 170:205–208

Maier A, Riedlinger J, Fiedler H-P, Hampp R (2004) Actinomycetales bacteria from a spruce stand: characterization and effects on growth of root symbiotic, and plant parasitic soil fungi in dual culture. Mycol Progr 3:129–36

Mansouri-Bauly H, Sýkorová Z, Scheerer U, Kopriva S (2006) Sulfur uptake in the ectomycorrhizal fungus *Laccaria bicolor* S238N. Mycorrhiza 16:421–427

Marilley L, Hartwig UA, Aragno M (1999) Influence of an elevated atmospheric CO_2 content on soil and rhizosphere bacterial communities beneath *Lolium perenne* and *Trifolium repens* under field conditions. Microb Ecol 38:39–49

Martin F, Aerts A, Ahrén D, Brun A, Duchaussoy F, Kohler A, Lindquist E, Salamov A, Shapiro HJ, Wuyts J, Blaudez D, Buée M, Brokstein P, Canbäck B, Cohen D, Courty PE, Coutinho PM, Danchin EGJ, Delaruelle C, Detter JC, Deveau A, DiFazio S, Duplessis S, Fraissinet-Tachet L, Lucic E, Frey-Klett P, Fourrey C, Feussner I, Gay G, Gibon J, Grimwood J, Hoegger P, Jain P, Kilaru S, Labbé J, Lin Y, Le Tacon F, Marmeisse R, Melayah D, Montanini B, Muratet M, Nehls U, Niculita-Hirzel H, Oudot-Le Secq MP, Pereda V, Peter M, Quesneville H, Rajashekar B, Reich M, Rouhier N, Schmutz J, Yin T, Chalot M, Henrissat B, Kües U, Lucas S, Van de Peer Y, Podila G, Polle A, Pukkila PJ, Richardson PM, Rouzé P, Sanders I, Stajich JE, Tunlid A, Tuskan G, Grigoriev I (2008) The genome sequence of the basidiomycete fungus *Laccaria bicolor* provides insights into the mycorrhizal symbiosis. Nature 452:88–92

Martin F, Canet D, Marchal JP (1985) ^{13}C nuclear magnetic resonance study of mannitol cycle and trehalose synthesis during glucose utilization by the ectomycorrhizal ascomycete *Cenococcum geophilum*. Plant Physiol 77:499–502

Martin F, Nehls U (2009) Harnessing ectomycorrhizal genomics for ecological insights. Curr Opin Plant Biol 12:508–515

Martin F, Tunlid A (2009) The ectomycorrhizal symbiosis: a marriage of convenience. In: Deising HB (ed) The mycota, vol V, 2nd edn, Plant relationships. Springer, Heidelberg, pp 237–257

Mayak S, Tirosh T, Glick BR (2004) Plant growth-promoting bacteria that confer resistance to water stress in tomatoes and peppers. Plant Sci 166:525–530

Menotta M, Amicucci A, Sisti D, Gioacchini AM, Stocchi V (2004) Differential gene expression during pre-symbiotic interaction between *Tuber borchii* vittad. and *Tilia americana* l. Curr Genet 46:158–165

Morel M, Jacob C, Fitz M, Wipf D, Chalot M, Brun A (2008) Characterization and regulation of PiDur3, a permease involved in the acquisition of urea by the ectomycorrhizal fungus *Paxillus involutus*. Fungal Genet Biol 45:912–921

Morel M, Jacob J, Kohler A, Johansson A, Martin F, Chalot M, Brun A (2005) Identification of genes differentially expressed in extraradical mycelium and ectomycorrhizal roots during *Paxillus involutus-Betula pendula* ectomycorrhizal symbiosis. Appl Environ Microbiol 71:382–391

Morgan JAW, Morgan G, Bending D, White PJ (2005) Biological costs and benefits to plant–microbe interactions in the rhizosphere. J Exp Bot 56:1729–1739

Müller T, Avolio M, Olivi M, Benjdia M, Rikirsch E, Kasaras A, Fitz M, Chalot M, Wipf D (2007) Nitrogen transport in the ectomycorrhiza association: the *Hebeloma cylindrosporum-Pinus pinaster* model. Phytochemistry 68:41–51

Nabti E, Sahnoune M, Adjrad S, Van Dommelen A, Ghoul M, Schmid M, Hartmann A (2007) A halophilic and osmotolerant *Azospirillum brasilense* strain from Algerian soil restores wheat growth under saline conditions. Eng Life Sci 7:354–360

Nabti E, Sahnoune M, Ghoul M, Fischer D, Hofmann A, Rothballer M, Schmid M, Hartmann A (2010) Restoration of growth of durum wheat (*Triticum durum* var. waha) under saline conditions due to inoculation with the rhizosphere bacterium *Azospirillum brasilense* NH and extracts of the marine alga *Ulva lactuca*. J Plant Growth Regul 29:6–22

Nakayama T, Homma Y, Hashidoko Y, Mitzutani J, Tahara S (1999) Possible role of xanthobaccins produced by *Stenotrophomonas* sp. strain SB-K88 in suppression of sugar beet damping-off disease. Appl Environ Microbiol 65:4334–4339

Neeno-Eckwall EC, Kinkel LL, Schottel JL (2001) Competition and antibiosis in the biological control of potato scab. Can J Microbiol 47:332–340

Nehls U (2008) Mastering ectomycorrhizal symbiosis: the impact of carbohydrates. J Exp Bot 59:1097–1108

Nehls U, Grunze N, Willmann M, Reich M, Küster H (2007) Sugar for my honey: carbohydrate partitioning in ectomycorrhizal symbiosis. Phytochemistry 68:82–91

Nuutinen JT, Timonen S (2008) Identification of nitrogen mineralization enzymes, l-amino acid oxidases, from the ectomycorrhizal fungi *Hebeloma* spp. and *Laccaria bicolor*. Mycol Res 112:1453–1464

Nygren CMR, Eberhardt U, Karlsson M, Parrent JL, Lindahl BD, Taylor AFS (2008) Growth on nitrate and occurrence of nitrate reductase-encoding genes in a phylogenetically diverse range of ectomycorrhizal fungi. New Phytol 180:875–889

Oelmüller R, Sherameti I, Tripathi S, Varma A (2009) *Piriformopsora indica*, a cultivable root endophyte with multiple biotechnological applications. Symbiosis 49:1–18

Ongena M, Jourdan E, Schäfer M, Kech C, Budzikiewicz H, Luxen A, Thonart P (2005) Isolation of an N-alkylated benzylamine derivative from *Pseudomonas putida* BTP1 as elicitor of induced systemic resistance in bean. Mol Plant Microbe Interact 18:562–569

Ono BI, Hazu T, Yoshida S, Kawato T, Shinoda S, Brzvwczy J, Paszewski A (1999) Cysteine biosynthesis in *Saccharomyces cerevisiae*: a new outlook on pathway and regulation. Yeast 15:1365–1375

Opelt K, Berg G (2004) Diversity and antagonistic potential of bacteria associated with bryophytes from nutrient-poor habitats of the baltic sea coast. Appl Environ Microbiol 70:6569–6579

Peter M, Courty P, Kohler A, Delaruelle C, Martin D, Tagu D, Frey-Klett P, Duplessis S, Chalot M, Podila G, Martin F (2003) Analysis of expressed sequence tags from the ectomycorrhizal basidiomycetes *Laccaria bicolor* and *Pisolithus microcarpus*. New Phytol 159:117–129

Pieterse MJC, Leon-Reyes A, Van der Ent S, Van Wees SCM (2009) Networking by small-molecule hormones in plant immunity. Nat Chem Biol 5:309–316

Prigent-Combaret C, Blaha D, Pothier JF, Vial L, Poirier M-A, Wisniewski-Dyé F, Moenne-Loccoz Y (2008) Physical organization and phylogenetic analysis of *acdR* as leucine-responsive regulator of the 1-aminocyclooropane-1-carboxylate deaminase gene *acdS* in phytobeneficial *Azospirillum lipoferum* 4B and other Proteobacteria. FEMS Microbiol Ecol 65:202–219

Pritsch K, Raidl S, Marksteiner E, Blaschke H, Agerer R, Schloter M, Hartmann A (2004) A rapid and highly sensitive method for measuring enzyme activities in single mycorrhizal tips using

4-methylumbelliferonelabelled fluorogenic substrates in a microplate system. J Microbiol Methods 58:233–241

Raaijmakers JM, Vlami M, de Souza JT (2002) Antibiotic production by bacterial biocontrol agents. Antonie van Leeuwenhoek 81:537–547

Riedlinger J, Schrey SD, Tarkka MT, Hampp R, Kapur M, Fiedler H-P (2006) Auxofuran, a novel metabolite stimulating growth of fly agaric, produced by the mycorrhiza helper bacterium *Streptomyces* AcH 505. Appl Environ Microbiol 72:3550–3557

Rosenblueth M, Martinez-Romero E (2006) Bacterial endophytes and their interactions with hosts. Mol Plant Microbe Interact 19:827–837

Rothballer M, Schmid M, Hartmann A (2009) Diazotrophic bacterial endophytes in Gramineae and other plants. In: Pawlowski K (ed) Prokaryotic symbionts in plants, vol 8, Microbiol monographs. Springer, Heidelberg, pp 273–302

Rovira AD (1991) Rhizosphere research – 85 years of progress and frustration. In: Kleister DL, Cregan PB (eds) The rhizosphere and plant growth. Kluwer, Amsterdam, pp 3–13

Ryu C-M, Farag MA, Hu C-H, Reddy MS, Kloepper JW, Pare PW (2004) Bacterial volatiles induce systemic resistance in *Arabidopsis*. Plant Physiol 134:1017–1026

Ryu CM, Farag MA, Hu CH, Reddy MS, Wei HX, Paré PW, Kloepper JW (2003) Bacterial volatiles promote growth in *Arabidopsis*. Proc Natl Sci Acad USA 100:4927–4932

Samac DA, Kinkel LL (2001) Suppression of the root-lesion nematode (*Pratylenchus penetrans*) in alfalfa (*Medicago sativa*) by *Streptomyces* spp. Plant Soil 235:35–44

Saravesi K, Markkola A, Rautio P, Roitto M, Tuomi J (2008) Defoliation causes parallel temporal responses in a host tree and its fungal symbionts. Oecologia 156:117–123

Sardi P, Saracchi M, Quaroni S, Petrolini B, Borgonovi GE, Merli S (1992) Isolation of endophytic *Streptomyces* strains from surface-sterilized roots. Appl Environ Microbiol 58:2691–93

Schaefer AL, Greenberg EP, Oliver CM, Oda Y, Huang JJ, Bittan-Banin G, Peres CM, Schmidt S, Juhaszova K, Sufrin JR, Harwood CS (2008) A new class of homoserine lactone quorum-sensing signals. Nature 454:595–599

Schertzer JW, Boulette ML, Whiteley M (2009) More than a signal: non-signaling properties of quorum sensing molecules. Trends Microbiol 17:189–195

Schottel JL, Shimizu K, Kinkel LL (2001) Relationships of *in vitro* pathogen inhibition and soil colonization to potato scab biocontrol by antagonistic *Streptomyces* spp. Biol Control 20:102–112

Schrey SD, Schellhammer M, Ecke M, Hampp R, Tarkka MT (2005) Mycorrhiza helper bacterium *Streptomyces* AcH 505 induces differential gene expression in the ectomycorrhizal fungus *Amanita muscaria*. New Phytol 168:205–216

Schuhegger R, Ihring A, Gantner S, Bahnweg G, Knappe C, Vogg G, Hutzler P, Schmid M, van Breusegem F, Eberl L, Hartmann A, Langebartels C (2006) Induction of systemic resistance in tomato by N-acylhomoserine lactone-producing rhizosphere bacteria. Plant Cell Environ 29:909–918

Seegmüller S, Schulte M, Herschbach C, Rennenberg H (1996) Interactive effects of mycorrhization and elevated atmospheric CO_2 on sulphur nutrition of young pedunculate oak (*Quercus robur* l.) trees. Plant Cell Environ 19:418–426

Selle A, Willmann M, Grunze N, Geßler A, Weiß M, Nehls U (2005) The high-affinity poplar ammonium importer PttAmt1.2 and its role in ectomycorrhizal symbiosis. New Phytol 168:697–706

Selosse MA, Setaro S, Glatard F, Richard F, Urcelay C, Weiss M (2007) Sebacinales are common mycorrhizal associates of Ericaceae. New Phytol 174:864–878

Sharma M, Schmid M, Rothballer M, Hause G, Zuccaro A, Imani J, Schäfer P, Hartmann A, Kogel K-H (2008) Detection and identification of mycorrhiza helper bacteria intimately associated with representatives of the order Sebacinales. Cell Microbiol 10:2235–2246

Smalla K, Wieland G, Buchner A, Zock A, Parzy J, Roskot N, Heuer H, Berg B (2001) Bulk and rhizosphere bacterial communities studied by denaturing gradient gel electrophoresis: plant-dependent enrichment and seasonal shifts revealed. Appl Environ Microbiol 67:4742–4751

Smith SE, Read DJ (2008) Mycorrhizal symbiosis, 3rd edn. Academic, Oxford, UK
Soanes DM, Alam I, Cornell M, Wong HM, Hedeler C, Paton NW, Rattray M, Hubbard SJ, Oliver SG, Talbot NJ (2008) Comparative genome analysis of filamentous fungi reveals gene family expansions associated with fungal pathogenesis. PLoS One 3(6):e2300. doi:10.1371/journal.pone.0002300
Steenhoudt O, Vanderleyden J (2000) Azospirillum, a free-living nitrogen-fixing bacterium closely associated with grasses: genetic, biochemical and ecological aspects. FEMS Microbiol Rev 24:487–506
Sugarawa M, Okazaki S, Nukui N, Ezura H, Mitsui H, Minamisawa K (2006) Rhizobitoxin modulates plant microbe interactions by ethylene inhibition. Biotechnol Adv 24:382–388
Talbot JM, Allison SD, Treseder KK (2008) Decomposers in disguise: mycorrhizal fungi as regulators of soil C dynamics in ecosystems under global change. Funct Ecol 22:955–963
Tarkka M, Nehls U, Hampp R (2005) Physiology of ectomycorrhiza (ECM). In: Lüttge U (ed) Progress in botany. Springer, Berlin, pp 247–276
Tarkka MT, Hampp R (2008) Secondary metabolites of soil streptomycetes in biotic interactions. In: Karlovsky P (ed) Secondary metabolites in soil ecology, vol 14, Soil biology series. Springer, Berlin, pp 107–28
Tatry MV, Kassis EE, Lambilliotte R, Corratgé C, van Aarle I, Amenc LK, Alary R, Zimmermann S, Sentenac H, Plassard C (2008) Two differentially regulated phosphate transporters from the symbiotic fungus *Hebeloma cylindrosporum* and phosphorus acquisition by ectomycorrhizal *Pinus pinaste*r. Plant J 57:1092–1102
van Aarle I, Viennois G, Amenc LK, Tatry MV, Luu D, Plassard C (2007) Fluorescent in situ rt-PCR to visualise the expression of a phosphate transporter gene from an ectomycorrhizal fungus. Mycorrhiza 17:487–494
Van de Kamp M, Schuurs TA, Vos A, Van der Lende TR, Konings WN, Driessen AJM (2000) Sulfur regulation of the sulfate transporter genes *Suta* and *Sutb* in *Penicillium chrysogenum.* Appl Env Microbiol 66:4536–4538
Vance C, Uhde-Stone C, Allan DL (2003) Phosphorus acquisition and use: critical adaptations by plants for securing a nonrenewable resource. New Phytol 157:423–447
Verma SA, Varma A, Rexer KH, Hassel A, Kost G, Sabhoy A, Bisen P, Bütehorn B, Franken P (1998) *Piriformospora indica*, gen. et spec. nov., a new root-colonizing fungus. Mycologia 90:898–905
von Rad U, Klein I, Dobrev PI, Kottova J, Zazimalova E, Fekete A, Hartmann A, Schmitt-Kopplin P, Durner J (2008) The response of *Arabidopsis thaliana* to *N*-hexanoyl-DL-homoserine-lactone, a bacterial quorum sensing molecule produced in the rhizosphere. Planta 229:73–85
von Rad U, Mueller MJ, Durner J (2005) Evaluation of natural and synthetic stimulants of plant immunity by microarray technology. New Phytol 165:191–202
Weissman KJ, Leadlay PF (2005) Combinatorial biosynthesis of reduced polyketides. Nat Rev Microbiol 3:925–936
Weller DM (1988) Biological control of soilborne pathogens in the rhizosphere with bacteria. Annu Rev Phytopathol 26:379–407
Weller DM, Raaijmakers JM, Gardener BB, Thomashow LS (2002) Microbial populations responsible for specific soil suppressiveness to plant pathogens. Annu Rev Phytopathol 40:309–348
Westerman S, Stulen I, Suter M, Brunold C, De Kok LJ (2001) Atmospheric H_2S as sulphur source for *Brassica oleracea*: consequences for the activity of the enzymes of the assimilatory sulphate reduction pathway. Plant Physiol Biochem 39:425–432
Whipps J (2001) Microbial interactions and biocontrol in the rhizosphere. J Exp Bot 52:487–511
Wiemken V (2007) Trehalose synthesis in ectomycorrhizas – a driving force of carbon gain for fungi? New Phytol 174:228–230
Wiener P (2000) Antibiotic production in a spatially structured environment. Ecol Lett 3:122–133
Willmann A, Weiss M, Nehls U (2007) Ectomycorrhiza-mediated repression of the high-affinity ammonium importer gene Am Amt2 in Amanita muscaria. Curr Genet 51:71–78

Woo JH, Kitamura E, Myuoga H, Yuto K (2002) An antifungal protein from the marine bacterium Streptomyces sp. strain AP77 is specific for *Pythium porphyrae*, a causative agent of red rot disease in *Porphyra* spp. Appl Environ Microbiol 68:2666–2675

Wright DP, Johansson T, Le Quere A, Soderstrom B, Tunlid A (2005) Spatial patterns of gene expression in the extramatrical mycelium and mycorrhizal root tips formed by the ectomycorrhizal fungus *Paxillus involutus* in association with birch (*Betula pendula*) seedlings in soil microcosms. New Phytol 167:579–596

Xiao K, Samac DA, Kinkel LL (2002) Biological control of *Phytophthora* root rots on alfalfa and soybean with *Streptomyces*. Biol Control 23:285–95

Zilber-Rosenberg I, Rosenberg E (2008) Role of microorganisms in the evolution of animals and plants: the hologenome theory of evolution. FEMS Microbiol Rev 32:723–735

Chapter 6
Stores as Substrate Sources of Respiration: Effects of Nitrogen Stress and Day Length

C.A. Lehmeier, F.A. Lattanzi, and H. Schnyder

6.1 Introduction

6.1.1 Unknowns on the Role of Stores as Sources of Respiratory Substrate

This chapter is concerned with the substrate supply system of dark respiration. This system is an integral component of the total pool of non-structural biomass that is available to meet the substrate demands of sinks. Plants face the challenge to allocate their limited resources gained in photosynthesis towards numerous sinks, including respiration, the increase in structural biomass for above- and below-ground space occupation (Chaps. 13 and 14), the transfer of assimilate to root bacteria and mycorrhiza (Chaps. 5 and 10) or the synthesis of carbon-based secondary defence compounds (Chaps. 2, 4 and 17). The allocation of assimilate today has important implications for the plants' fitness during the course of plant development (Chaps. 1 and 11).

Dark respiration is a major sink of assimilate, consuming between 30 and 80 % of the gross primary production of plants and providing energy, reductants and carbon skeletons for growth and maintenance processes (Amthor 1989, 2000; Gifford 2003; Van Iersel 2003; Lötscher et al. 2004; Plaxton and Podestá 2006; Chap. 2). There are, in principle, two types of substrate for respiration: substrate that is available through current photosynthetic activity and stored substrate that was assimilated previously.

It has been commonly held for a long time that stores provide substrate for growth and maintenance demands of plants whenever the supply from current assimilation is

C.A. Lehmeier • F.A. Lattanzi • H. Schnyder (✉)
Lehrstuhl für Grünlandlehre, Technische Universität München, Alte Akademie 12, 85350 Freising, Germany
e-mail: schnyder@wzw.tum.de

R. Matyssek et al. (eds.), *Growth and Defence in Plants*, Ecological Studies 220,
DOI 10.1007/978-3-642-30645-7_6,

insufficient to satisfy the current demand (Graber 1931; Sullivan and Sprague 1943; Davidson and Milthorpe 1966a, b; Kozlowski 1992; Schnyder 1993; Zeeman et al. 2007). In this sense, stores form the basis of a homeostatic mechanism that underlies plant individual fitness. Empirical observations about how plants achieve the balance between resource acquisition, storage and growth are important to understand and model plant growth dynamics and yield in a changing environment (Smith and Stitt 2007; Graf and Smith 2011; Chaps. 15 and 17).

However, there have been only a few investigations on the actual role of stores as supplies for respiration (Ryle et al. 1976; Prosser and Farrar 1981; Kouchi et al. 1985, 1986; Avice et al. 1996; Dilkes et al. 2004; Lötscher and Gayler 2005), although provision of substrate for respiration is perhaps the most vital function of stores. Plants will eventually die, if the substrate needs of maintenance respiration are not served. Still, surprisingly little is known about the quantitative characteristics of the substrate supply system of respiration in terms of the size and turnover of the component substrate pools and their chemical identity. This is true, except for a study of Prosser and Farrar (1981) and recent experiments of Lehmeier et al. (2008, 2010a, b). The latter provided a quantitative characterisation of the substrate supply system of plant respiration in terms of the size and kinetic properties of respiratory substrate pools and the localisation of these pools. Furthermore, it determined the effects of nitrogen stress and day/night cycles on these pools. This chapter provides a synthesis of that work.

6.1.2 Which Stores Could Potentially Supply Substrate for Respiration?

Carbohydrates are considered as the main storage form of respiratory substrate in most plants (ap Rees 1980; Tcherkez et al. 2003). Most plants store carbohydrates as starch in plastids (Zeeman et al. 2007), such as transitory starch in chloroplasts, or as sucrose and fructans (fructose-based oligo- and polysaccharides) in vacuoles (Wagner et al. 1983; Sicher et al. 1984; Gerhardt et al. 1987; Pollock and Cairns 1991). Proteins constitute another fraction which turns over continuously and is closely connected with respiratory pathways (Penning de Vries 1975; Lea and Ireland 1999). Given the complexity of plant metabolic networks, one should suspect that any organic compound that can be degraded/mobilised by the plant could serve as substrate for respiration. Yet, it is generally believed that compounds other than carbohydrates and protein generally play no important storage role in vegetative plant parts (Kozlowski 1992; Schnyder 1993; Volenec et al. 1996; Muntz 1998; but see also Hoch 2007).

The size of carbohydrate stores is strongly affected by stress conditions: for instance, nitrogen limitation usually causes an increase in reserve/storage carbohydrate concentration (Gebbing et al. 1999; Morvan-Bertrand et al. 1999; Stitt and Krapp 1999). Yet, it is unknown whether a relationship exists between the size of stores and their actual importance (share) in providing substrates for respiration.

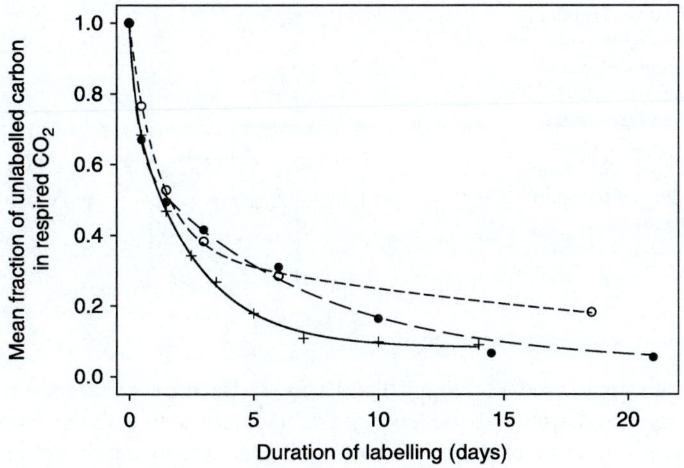

Fig. 6.1 Evolution of the fraction of unlabelled carbon ($f_{unlabelled}$) in respiratory CO_2 of plants grown in different environmental conditions: 16-h light/8-h dark cycles with high nitrogen supply (*plus symbols*), and continuous light with high nitrogen (*closed circles*) and low nitrogen supply (*open circles*). The daily sum of photosynthetically active radiation was the same in all treatments (for details see Lehmeier et al. 2008, 2010a, b). *Lines* give fits of two-pool models to the data (compare Fig. 6.2)

Also, it is unknown if plants growing in day/night cycles are relying more on stores for respiration than plants in continuous light.

6.1.3 *How Can We Discern Contributions of Stores and Current Assimilate to Respiration?*

Proof for the participation of stores in supplying respiration can be obtained by analysing the residence time of substrate carbon in the plant. If carbon fixed in photosynthesis is transferred directly to centres of respiration, then the residence time in the plant is short (seconds to minutes for carbon respired within the photosynthetically active cell to a few hours for carbon transferred to distant plant parts). In contrast, if carbon is first deposited in long-lived molecules (such as proteins or storage carbohydrates) then the residence time is long (days to months).

Such studies are best conducted with quantitative (pulse- or continuous/dynamic-) tracer techniques (e.g. Geiger et al. 1969; Ryle et al. 1976; Kouchi et al. 1985; Schnyder 1992; see also Chap. 7) in combination with monitoring of tracer appearance in respiratory CO_2 (e.g. Schnyder et al. 2003; Klumpp et al. 2005; Gamnitzer et al. 2009) in plants growing in a steady state. Very generally, the participation of a store in the substrate supply system of respiration then may become evident from the existence of a double(or multi)-exponential form of the tracer kinetics curve (Fig. 6.1). In that curve the store is reflected in the slower

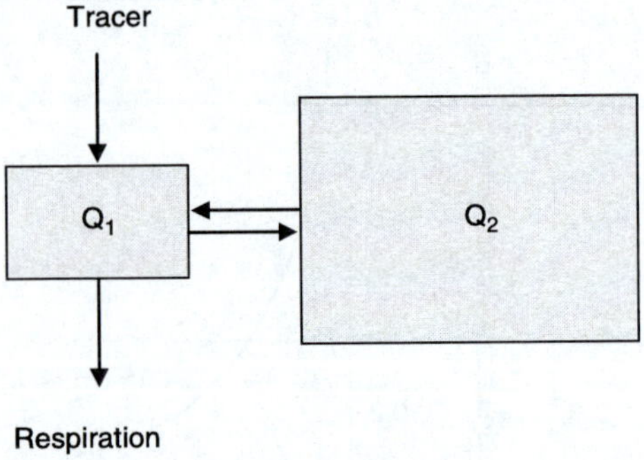

Fig. 6.2 Two-pool model of the substrate supply system of respiration (Lehmeier et al. 2010b). Symbols: Q_1, "metabolic and transport pool"; Q_2, "store". Tracer fixed in photosynthesis enters the metabolic and transport pool and then feeds respiration, or first exchanges with a store

exponential term(s) (see also Fig. 7.1 in Chap. 7). Such tracer data are commonly analysed by compartmental modelling, a mathematical tool for interpreting the tracer kinetics in terms of the number of pools, the topology/architecture/arrangement of the pools, and the size and half-life of the pools. For principles of this methodology see Atkins (1969), Moorby and Jarman (1975), Prosser and Farrar (1981), Rocher and Prioul (1987), Jacquez (1996), Bürkle et al. (1998), Lattanzi et al. (2005) and Chap. 7. The use of compartmental modelling for analysing tracer kinetics of respiratory CO_2 has been explored and discussed in detail by Lehmeier et al. (2008). An example of a compartmental model of the respiratory substrate supply system of a plant is given in Fig. 6.2. This two-pool model comprises a "metabolic and transport pool" and a "store". Tracer (i.e. labelled carbon assimilated in photosynthesis) enters a "metabolic and transport pool", which exchanges assimilate with the "store", or feeds it directly to respiration (for details of the model, see Sect. 6.2.2 below).

Although the tracer kinetics of respiratory CO_2 can demonstrate the *involvement* of a store, it does not prove its *chemical identity*. Unequivocal identification of the chemical identity of a respiratory substrate store is really difficult, particularly in intact plants. Still, comparison of the residence time of carbon in the respiratory substrate store with that of putative biochemical storage compounds can provide a clue, if the respiratory substrate store and a certain biochemical storage compound share the same half-life, pointing at this biochemical pool as the respiratory substrate store (Nogués et al. 2004; Mortazavi et al. 2009). Secondly, the size of the storage biochemical compound must be at least equivalent to that of the respiratory substrate store. If this is not the case, then some additional storage biochemical compound must be present. However, in general, we expect that stores of carbohydrate and protein also (or even predominantly) serve growth needs besides respiration. In this case the size of the respiratory store is only a proportion of the total amount of the stored material.

The half-life of carbon in the respiratory substrate store reveals *functional properties* of the store. Diurnal sucrose- or starch-storage pools have a half-life of less than 1 day (Borland and Farrar 1988; Gibon et al. 2009). Longer half-lives are expected for fructan or starch in longer-term stores (Winzeler et al. 1990; Schnyder

1993; Teixeira et al. 2007) or protein (Penning de Vries 1975; Dungey and Davies 1982; Irving and Robinson 2006).

6.1.4 Aim

In the following we review recent advances in the knowledge of functional properties of stores in the substrate supply system of respiration. In particular, we analyse the effects of day/night cycles (that is regular interruptions of the photosynthetic activity) and nutrient limitation on this system (Lehmeier et al. 2008, 2010a, b). These studies were performed with perennial ryegrass (*Lolium perenne* L.), an important species of temperate grasslands. The work specifically addresses the following questions:

1. How do nitrogen stress and diurnal interruptions of photosynthesis affect respiration and non-structural carbon pools?
2. Do shoot and roots have different respiratory substrate supply systems?
3. What are the sizes and kinetic properties of the storage pools of this system? Are they located in the shoot or in the roots?
4. How do nitrogen stress and diurnal interruptions of photosynthesis affect the size and kinetic properties of stores? How do these factors influence the contributions of stores to respiration?
5. What proportion of the non-structural biomass is used for respiration?

6.2 Experimental

6.2.1 Scenarios

We tested three scenarios: plants growing in continuous light (PPFD 275 μmol m^{-2} s^{-1}) with high nitrogen fertiliser supply (high N) or low nitrogen fertiliser supply (low N) and plants growing in diurnal cycles (day/night) with alternating 16-h light (PPFD 425 μmol m^{-2} s^{-1}) and 8-h dark periods. In all cases, nutrients and water were provided in the form of a modified Hoagland solution with either 1 mM NO_3^- (low N) or 7.5 mM NO_3^- (high N and day/night).

Except for these differences, the experimental conditions were the same in all scenarios. Thus, daily irradiance was 24 mol m^{-2} photosynthetic photon flux density; air temperature was kept at 20 ºC, and CO_2 concentration at 360 μL L^{-1}. Stands were grown from seed of *Lolium perenne* L. (cv. Acento), and kept in growth chambers. These formed part of a custom made $^{13}CO_2/^{12}CO_2$ gas exchange and labelling facility (Schnyder et al. 2003). The potted plants were held in large containers which were briefly flushed with nutrient solution every 3 h. Periodically, all stands were flushed with demineralised water to prevent salt accumulation.

In each scenario, half of the plants were grown in a chamber with ^{13}C-enriched CO_2, and the other half in ^{13}C-depleted CO_2. When closed stands were established, the $\delta^{13}C$ of CO_2 was changed for intervals ranging from 1 h to 29 days (^{13}C-enriched → ^{13}C-depleted atmosphere or vice versa). At the end of each "steady-state labelling" interval, the rates of shoot and root respiration and the $\delta^{13}C$ of shoot- and root-respired CO_2 of individual plants were measured in the dark using a system described by Lötscher et al. (2004) and Klumpp et al. (2005). After 5–6 h (high and low N) or 6–7 h (day/night) of respiration measurements, the plants were harvested. Carbon and nitrogen elemental contents and water-soluble carbohydrate (WSC) fractions were determined in shoot and root biomass. Respiration of non-labelled control plants, which grew continuously in the presence of ^{13}C-enriched or -depleted CO_2, was measured likewise. This determined the end members of a two-member mixing model which was used to calculate the fractions of unlabelled carbon ($f_{\text{unlabelled}}$) in CO_2 respired by the labelled plants.

6.2.2 Compartmental Modelling

One of the conditions for compartmental modelling is that the system is in a steady state (Lattanzi et al. 2005). Both continuous light scenarios met this condition for all time scales of the labelling. However, the steady-state condition was not satisfied by the day/night scenario for processes at sub-daily time-scales. For instance, the discontinuity of photosynthesis must have had consequences for metabolite pool sizes and fluxes, including rates of deposition and mobilisation of carbohydrate stores (Farrar and Farrar 1985; Borland and Farrar 1985). Yet, the steady-state condition was met to a close approximation for the day-by-day time scale. For instance, daily shoot and root specific respiration rates and nitrogen concentration in shoot and root biomass did not change over the experimental period (Lehmeier et al. 2010b). Therefore, we restricted the compartmental analysis for all three scenarios to the day-by-day timescale, similar as in Lattanzi et al. (2005). To this end, we first calculated the mean fraction of unlabelled carbon in CO_2 respired during a 1-day-long period (or multiples of that) as respiration-weighted averages of the fraction of unlabelled carbon of a given day or multiples of that, depending on the available time resolution of the data. For a detailed description of this procedure, see Supporting Information to Lehmeier et al. (2010b).

Then, the time course of $f_{\text{unlabelled}}$ was analysed with compartmental models according to principles discussed by Lattanzi et al. (2005) and Lehmeier et al. (2008). This assumed that the system was in steady state, pools were well mixed and all fluxes obeyed first-order kinetics. That is $F_{xy} = k_{xy} \times Q_x$, with F_{xy} the flux out of pool x into pool y; k_{xy}, the respective rate constant; and Q_x, the size of pool x. The term "pool" is defined as a set of compounds which possess the same proportion (fraction) of labelled carbon atoms. That is, a pool is an entity with uniform isotopic composition (Rescigno 2001). Thus, in principle, one pool can contain different biochemical compounds in different cellular and tissue locations, if the proportion of label is the same in all of them at all times.

One-, two- and three-pool models were tested. These models were defined as sets of differential equations describing the fluxes between pools and the environment.

Modelling was performed with the software ModelMaker (Cherwell Scientific, Oxford, UK). Differential equations were solved using the 4th order Runge–Kutta numerical method. Levenberg–Marquardt optimisation was used to find model parameters that minimised the sum of the squared differences between model predictions and observations. In addition, a custom-made computer programs (written in R; R Development Core Team 2007) was used to test a large number of parameter combinations, to ensure that global rather than local best fits were detected in the simulations (Lehmeier et al. 2008). The model selection process was guided by the principle of parsimony and the extra sum-of-squares F-test for comparing models as described by Motulsky and Christopoulos (2004). These procedures led to the two-pool model in Fig. 6.2 as the simplest biologically meaningful model that was able to describe the observed tracer time courses.

For this two-pool model, the fraction of tracer in the pools Q_1 and Q_2 with respect to time was given by (Supporting Information in Lehmeier et al. 2010b):

$$\mathrm{d}Q_1/\mathrm{d}t = \text{Tracer} + k_{21} \times Q_2 - k_{10} \times Q_1 - k_{12} \times Q_2 \tag{6.1a}$$

and

$$\mathrm{d}Q_2/\mathrm{d}t = k_{12} \times Q_1 - k_{21} \times Q_2, \tag{6.1b}$$

with k_{12} the rate constant governing the flux from pool Q_1 to Q_2, k_{21} that from Q_2 to Q_1 and k_{10} the respiratory flux out of Q_1. Tracer represents the flux of labelled carbon into the respiratory substrate system. In the steady-state, this influx equals the efflux by respiration. As the time course of $f_{\mathbf{unlabelled}}$ converged to an asymptotic value significantly larger than zero (see Fig. 6.1 and Table 6.3), Tracer is given by:

$$\text{Tracer} = \text{respiration rate} \times (1 - \text{asymptote}), \tag{6.2}$$

with respiration rate denoting the measured specific respiration rates of the plants and "asymptote" the fraction of respiration supplied by a substrate source that was not turned over by tracer within the labelling period.

In the steady state, that is, when $\mathrm{d}Q_1/\mathrm{d}t = \mathrm{d}Q_2/\mathrm{d}t = 0$, pool sizes were given by

$$Q_1 = \text{Tracer}/k_{10} \tag{6.3a}$$

and

$$Q_2 = \text{Tracer}/k_{10} \times k_{12}/k_{21}. \tag{6.3b}$$

The rate constants k_{xy} and the asymptote were the model-optimised parameters. The half-life $t_{1/2}(Q_x)$ of a pool Q_x was calculated from the rate constants as follows:

$$t_{1/2}(Q_x) = \ln(2)/k_x \tag{6.4}$$

with k_x the sum of all rate constants k_{xy} leaving the pool Q_x. The fractional contribution of current assimilates to respiration is defined as the probability that tracer leaves the system without cycling through Q_2. The fractional contribution of stores is given by the probability that tracer cycles through Q_2 at least once before it is respired. These fractional contributions are given by

$$\text{Contribution of current assimilates} = k_{10}/(k_{10} + k_{12}) \tag{6.5a}$$

and

$$\text{Contribution of temporary stores} = k_{12}/(k_{10} + k_{12}). \tag{6.5b}$$

The sum of the contribution of current assimilates, stores and the asymptote-value equals 1.

The mean residence time (τ, hours) of carbon in the respiratory substrate supply system was obtained as

$$\tau = (Q_1 + Q_2)/r_{\text{plant}}, \tag{6.6}$$

with r_{plant} the specific respiration rate of the plant.

6.3 Effects of Nitrogen Stress and Day Length on Growth, Respiration and Non-structural Carbon Pools

The different scenarios produced very distinctive effects on plant growth, respiration and carbohydrate stores (Tables 6.1 and 6.2):

Nitrogen stress greatly reduced the specific rates of growth, respiration and nitrogen uptake, the shoot/root ratio, and the efficiency of carbon use in structural biomass synthesis. On the other hand, nitrogen stress caused a strong increase of WSC concentration which was near-fully attributable to fructan storage. These effects agreed with typical nitrogen stress responses (e.g. Rufty et al. 1988; Amthor 1989; Makino and Osmond 1991; Morvan-Bertrand et al. 1999; Poorter and Nagel 2000).

Conversely, the *diurnal interruption of photosynthesis* by an 8-h-long dark period had no effect on the specific rates of growth and nitrogen uptake. Also, the shoot/root ratio and the efficiency of carbon use in structural biomass synthesis were not affected. However, WSC concentration was reduced, and—again—this effect was near-fully attributable to fructan.

6.4 Root and Shoot Respiration Are Served by the Same Pools

The kinetics of label release through root and shoot respiration were similar in all scenarios, except for an approx. 1-h delay for first label appearance in root respiration relative to that in the shoot (data not shown, but see Lehmeier et al. 2008, 2010a).

Table 6.1 Parameters of plant growth, nitrogen status and resource use efficiency as affected by day length and nitrogen fertiliser supply

	Day/night	Continuous light	
	High N	High N	Low N
Spec. growth rate (mg C g^{-1} plant-C h^{-1})	3.19^{A}	3.23^{A}	1.58^{B}
Spec. respiration rate (mg C g^{-1} plant-C h^{-1})	1.67^{A}	1.50^{B}	0.99^{C}
Spec. nitrogen uptake rate (mg N g^{-1} plant-C h^{-1})	0.116^{A}	0.124^{A}	0.035^{B}
Shoot:root ratio	4.15^{A}	3.84^{A}	2.96^{B}
Carbon-use efficiency in synthesis of structural biomass[a]	0.59^{A}	0.59^{A}	0.49^{B}

Plants of *Lolium perenne* were grown in a 16/8 h day/night regime (with a PPFD of 425 μmol m^{-2} s^{-1} during the light period) or in continuous light (with a PPFD of 275 μmol m^{-2} s^{-1}) with high (high N) or low nitrate supply (low N) in the nutrient solution. Except for these differences, the growth conditions, including the daily dose of photosynthetically active radiation, were the same in all scenarios (see Experimental and Lehmeier et al. 2010a, b). Different uppercase letters in a row signify differences between scenarios

[a]Estimated as specific growth rate × proportion of total structural carbon/(specific respiration rate + specific growth rate × proportion of total structural carbon)

Table 6.2 Mass of non-structural biomass components and respiratory substrate supply system (Lehmeier et al. 2010a, b)

	Day/night	Continuous light	
	High N (mg C g^{-1} plant-C)	High N (mg C g^{-1} plant-C)	Low N (mg C g^{-1} plant-C)
Fructan	106^{A}	242^{B}	332^{C}
Sucrose	32^{A}	23^{B}	20^{B}
Glucose	22^{A}	12^{B}	7^{C}
Fructose	23^{A}	14^{B}	10^{C}
Total WSC	183^{A}	276^{B}	369^{C}
Protein and amino acids[a]	71^{A}	48^{B}	30^{C}
Total non-structural biomass[b]	254^{A}	338^{B}	400^{B}
Respiratory substrate system[c]	73^{A}	163^{B}	246^{C}

For the definition of scenarios see Table 6.1 and Experimental. Values are means of six (continuous light) or eight (day/night) replicate plants. Pool estimates of respiratory substrate system were obtained with a fit of a two-pool model to daily mean $f_{unlabelled}$ data (Figs. 6.1 and 6.2). Different uppercase letters in a row signify differences between scenarios

[a]Estimated as 3.1 × N content (thus assuming a 3.1 *C*:*N* ratio in total amino-compounds and no nitrate in plants)

[b]Sum of total WSC and protein + amino acids

[c]Sum of the size of the "metabolic and transport pool" and the "store" (see Table 6.3)

The 1-h delay between shoot and root was probably due to phloem transport from the shoot to the roots (Windt et al. 2006). Compartmental modelling showed that respiratory substrate systems of similar structure (number, topology and relative size) and kinetic properties (half-life and relative contribution to respiration of the different pools) applied to both roots and shoot (Lehmeier et al. 2008, 2010a).

Table 6.3 Parameters of 2-pool compartmental models of the respiratory supply system of *Lolium perenne* plants (see Lehmeier et al. 2010a, b)

	Day/night	Continuous light	
	High N	High N	Low N
Size (mg C g^{-1} plant-C)			
Metabolic and transport	28^A	31^A	44^B
Store	45^A	132^B	202^C
Half-life (h)			
Metabolic and transport	5^A	6^A	20^B
Store	13^A	48^B	288^C
Mean residence time (h)	49^A	109^B	249^C
Fractional contribution (%)			
Current assimilation	36^A	44^A	67^B
Store	56^A	53^A	33^B
Asymptote	8^A	3^B	0^C

For the definition of scenarios see Table 6.1. Different uppercase letters in a row signify differences between scenarios

This suggested that the same substrate supply system served root and shoot respiration in all scenarios.

6.5 Respiratory Substrates Are Mainly Stored in the Shoot in the Form of Fructan

The two-pool compartmental model shown in Fig. 6.2 adequately distinguished a current assimilate pool and a store (compare Table 6.3 and Lehmeier et al. 2010b). This was particularly clear for the well-fertilised scenarios. However, at low N, the "metabolic and transport pool" exhibited a much slower half-life (20 h) than in the other scenarios. This half-life was similar to that of diurnal stores (Borland and Farrar 1985, 1988; Farrar and Farrar 1986; Lehmeier et al. 2010a). This raises the possibility that a short-term store was also involved in supplying substrate to respiration. However, in the absence of direct evidence for such a short-term store, we cannot exclude the possible existence of a truly slow metabolic and transport pool.

In all scenarios the size of the respiratory substrate store greatly exceeded that of the metabolic and transport pool, and a very large proportion of the respiratory substrate first cycled through a store before being respired (Table 6.3). Overall the respiratory substrate system represented a very significant fraction of the total plant carbon pool: 7% of total biomass carbon in the day/night cycle, 16% at high N and 25% at low N (Table 6.3).

In all likelihood, the fructan pool was the main component of the respiratory substrate store: it was the only biochemical fraction that could have accommodated the entire respiratory substrate store in all scenarios (Tables 6.2 and 6.3). Also, fructan storage occurred almost exclusively in the shoot (Lehmeier et al. 2010b), consistent with the model prediction that the respiratory substrate store must have resided in the shoot (Lehmeier et al. 2008).

6.6 Turnover of the Respiratory Substrate Store Is Slowed by Nitrogen Stress and Accelerated by Day/Night Cycles

The most significant result of this comparison of scenarios was the large variability of the stores' half-lives: 13 h in day/night, 48 h in high N and 288 h in low N (Table 6.3). Clearly, environmental conditions had a drastic effect on storage pool turnover. These relationships were also reflected in the mean residence time of carbon in the total respiratory substrate system: 49 h in day/night, 109 h in high N and 249 h in low N (Table 6.3). By contrast, the scenarios had little (if any) effect on the fractional contribution of stores to respiration. We are not aware of any previous demonstration of such enormous stress-dependent variability in turnover of respiratory substrate storage pools. This reduces opportunities for the discussion of the stress dependence of this turnover.

The variability in the mean residence time of carbon in the total respiratory substrate system (or turnover of the store) corresponded with large differences in fructan contents (Table 6.2). Mean residence time was long where fructan content was high. Also, we have found large and consistent effects of nitrogen stress on half-lives of fructan stores (Wild, Lehmeier, Lattanzi and Schnyder, unpublished data). This finding also concurs with the idea that fructan was the main contributor to the respiratory substrate store and that fructan turnover responded to the same environmental controls as the respiratory substrate store.

The very low half-life of the store in the day/night scenario was associated with a significantly elevated sucrose concentration. Much of this was probably stored in a diurnal vacuolar sucrose pool, which served to balance assimilate needs during the 8-h-long dark periods. The suggestion of a vacuolar sucrose pool serving as a respiratory substrate store is also supported by the similarity of the half-life of vacuolar sucrose in leaves of grasses (Farrar 1989). Obviously, such a requirement for a diurnal store did not exist in the continuous light scenarios. In fact, one may wonder if vacuolar sucrose storage did actually take place in the continuous light scenarios.

In principle, protein turnover could also contribute carbon skeletons for respiration. According to Dungey and Davies (1982) and Simpson et al. (1981) the half-life of leaf proteins in barley and maize is in the order of 6–9 days. Simpson et al. (1981) found no effect of continuous light (relative to day/night cycles) on turnover of total leaf protein or Rubisco. These data suggest that protein turnover is much less variable than the turnover of the respiratory substrate store in our investigations. Also, the strong continuous light effect on turnover of the respiratory substrate store (Table 6.3) is apparently not related to protein turnover. These comparisons suggest that, opposite to results from experiments with extended darkness (Brouquisse et al. 1998), proteins were not a (main) source of stored substrate for respiration in our studies. Nevertheless, we cannot exclude the possibility that the chemical identity of the respiratory substrate store could shift between the different stress scenarios. Importantly, the negative evidence for proteins provides further support for storage carbohydrates as the main source of stored substrate for respiration.

6.7 Conclusions

The most striking result of this work is the fivefold variation of observed mean residence time of carbon in the respiratory substrate system. This variation resulted from well-defined stresses: nitrogen stress caused a 2.3-fold increase, and continuous light *versus* day/night cycles caused a 2.2-fold increase. Other environmental factors (such as air temperature, relative humidity, CO_2 concentration and the daily sum of photosynthetically active radiation) did not interact/interfere with these effects, since they were the same in all scenarios.

There were fair correlations between mean residence time, the half-life of the store and fructan content. This result is consistent with the view that fructan turnover (perhaps in conjunction with that of vacuolar sucrose stores, which were possibly present in day/night cycles) was a major control of the storage-derived substrate supply for respiration.

The mean residence time correlated negatively with a number of parameters: specific growth rate, respiration rate, nitrogen uptake rate and nitrogen content of biomass. These negative relationships were particularly evident in the comparison of high N and low N in continuous light, and suggested that mean residence time increased with nitrogen limitation of growth (sink-limitation) and associated decreased metabolic activity. Obviously, mean residence time increased with increasing carbon saturation of plants or decreased carbon demand for synthesis of structural biomass.

Remarkably, the proportional contribution of stores to respiration was high in all scenarios, and independent of the size of the respiratory substrate store. For instance, comparing day/night *versus* continuous light we saw a threefold increase in the size of respiratory substrate store, but the contribution to respiration remained the same at approx. 55%. Evidently, the turnover of the respiratory substrate was controlled in such a way, that changes in the size of the store had no (or little) effect on the contribution of the store.

In conclusion, the regulation of substrate fluxes to and mobilisation fluxes from stores is important to meet the demands of growth (Lattanzi et al. 2005) and respiration (this chapter) and it is reasonable to assume that they also play a decisive role in the provision of substrate for the synthesis of carbon-based secondary defence compounds (Chaps. 2, 4 and 17).

References

Amthor JS (1989) Respiration and crop productivity. Springer, New York

Amthor JS (2000) The McCree – de Wit – Penning de Vries – Thornley respiration paradigms: 30 years later. Ann Bot 86:1–20

ap Rees T (1980) Assessment of the contribution of metabolic pathways to plant respiration. In: Davies DD (ed) The biochemistry of plants: a comprehensive treatise, vol 2. Academic, San Diego, pp 1–29
Atkins GL (1969) Multicompartment models in biological systems. Methuen, London
Avice JC, Ourry A, Lemaire G, Boucoud J (1996) Nitrogen and carbon flows estimated by ^{15}N and ^{13}C pulse-chase labeling during regrowth of alfalfa. Plant Physiol 112:281–290
Borland AM, Farrar JF (1985) Diel patterns of carbohydrate metabolism in leaf blades and leaf sheaths of *Poa annua* L. and *Poa jemtlandica* (Almq.) Richt. New Phytol 100:519–531
Borland AM, Farrar JF (1988) Compartmentation and fluxes of carbon in leaf blades and leaf sheaths of *Poa annua* L. and *Poa x jemtlandica* (Almq.) Richt. Plant Cell Environ 11:535–543
Brouquisse R, Gaudillere JP, Raymond P (1998) Induction of a carbon-starvation-related proteolysis in whole maize plants submitted to light/dark cycles and to extended darkness. Plant Physiol 117:1281–1291
Bürkle L, Hibberd JM, Quick WP, Kühn C, Hirner B, Frommer WB (1998) The H^+-sucrose cotransporter NtSUT1 is essential for sugar export from tobacco leaves. Plant Physiol 118:59–68
Davidson JL, Milthorpe FL (1966a) Leaf growth in *Dactylis glomerata* following defoliation. Ann Bot 30:173–184
Davidson JL, Milthorpe FL (1966b) The effect of defoliation on the carbon balance in *Dactylis glomerata*. Ann Bot 30:185–198
Dilkes NB, Jones DL, Farrar J (2004) Temporal dynamics of carbon partitioning and rhizodeposition in wheat. Plant Physiol 134:706–715
Dungey NO, Davies DD (1982) Protein turnover in the attached leaves of non-stressed and stressed barley seedlings. Planta 154:435–440
Farrar JF (1989) Fluxes and turnover of sucrose and fructans in healthy and diseased plants. J Plant Physiol 134:137–140
Farrar SC, Farrar JF (1985) Carbon fluxes in leaf blades of barley. New Phytol 100:271–283
Farrar SC, Farrar JF (1986) Compartmentation and fluxes of sucrose in intact leaf blades of barley. New Phytol 103:645–657
Gamnitzer U, Schäufele R, Schnyder H (2009) Observing ^{13}C labelling kinetics in CO_2 respired by a temperate grassland ecosystem. New Phytol 184:376–386
Gebbing T, Schnyder H, Kühbauch W (1999) The utilization of pre-anthesis reserves in grain filling of wheat. Assessment by steady-state $^{13}CO_2/^{12}CO_2$ labelling. Plant Cell Environ 22:851–858
Geiger DR, Saunders MA, Cataldo DA (1969) Translocation and accumulation of translocate in the sugar beet petiole. Plant Physiol 44:1657–1665
Gerhardt R, Stitt M, Heldt HW (1987) Subcellular metabolite levels in Spinach leaves – regulation of sucrose synthesis during diurnal alterations in photosynthetic partitioning. Plant Physiol 83:399–407
Gibon Y, Pyl ET, Sulpice R, Lunn JE, Höhne M, Günther M, Stitt M (2009) Adjustment of growth, diurnal starch turnover, protein content and central metabolism to a decrease of the carbon supply when Arabidopsis is grown in very short photoperiods. Plant Cell Environ 32:859–874
Gifford RM (2003) Plant respiration in productivity models: conceptualization, representation and issues for global terrestrial carbon-cycle research. Funct Plant Biol 30:171–186
Graber LF (1931) Food reserves in relation to other factors limiting the growth of grasses. Plant Physiol 6:43–71
Graf A, Smith AM (2011) Starch and the clock: the dark side of plant productivity. Trends Plant Sci 16:169–175
Hoch G (2007) Cell wall hemicelluloses as mobile carbon stores in non-reproductive plant tissues. Funct Ecol 21:823–834
Irving LJ, Robinson D (2006) A dynamic model of Rubisco turnover in cereal leaves. New Phytol 169:493–504
Jacquez JA (1996) Compartmental analysis in biology and medicine, 3rd edn. Biomedware, Ann Arbor

Klumpp K, Schäufele R, Lötscher M, Lattanzi FA, Feneis W, Schnyder H (2005) C-isotope composition of CO_2 respired by shoots and roots: fractionation during dark respiration? Plant Cell Environ 28:241–250

Kouchi H, Nakaji K, Yoneyama T, Ishizuka J (1985) Dynamics of carbon photosynthetically assimilated in nodulated soya bean plants under steady-state conditions. 3. Time-course study on ^{13}C incorporation into soluble metabolites and respiratory evolution of $^{13}CO_2$ from roots and nodules. Ann Bot 56:333–346

Kouchi H, Akao S, Yoneyama T (1986) Respiratory utilization of ^{13}C-labelled photosynthate in nodulated root systems of soybean plants. J Exp Bot 37:985–993

Kozlowski TT (1992) Carbohydrate sources and sinks in woody plants. Bot Rev 58:107–222

Lattanzi FA, Schnyder H, Thornton B (2005) The sources of carbon and nitrogen supplying leaf growth: assessment of the role of stores with compartmental models. Plant Physiol 137:383–395

Lea PJ, Ireland RJ (1999) Nitrogen metabolism in higher plants. In: Singh BK (ed) Plant amino acids. Marcel Dekker, New York, pp 1–47

Lehmeier CA, Lattanzi FA, Schäufele R, Wild M, Schnyder H (2008) Root and shoot respiration of perennial ryegrass are supplied by the same substrate pools: assessment by dynamic ^{13}C labeling and compartmental analysis of tracer kinetics. Plant Physiol 148:1148–1158

Lehmeier CA, Lattanzi FA, Schäufele R, Schnyder H (2010a) Nitrogen deficiency increases the residence time of respiratory carbon in the respiratory substrate supply system of perennial ryegrass. Plant Cell Environ 33:76–87

Lehmeier CA, Lattanzi FA, Gamnitzer U, Schäufele R, Schnyder H (2010b) Day-length effects on carbon stores for respiration of perennial ryegrass. New Phytol 188:719–725

Lötscher M, Gayler S (2005) Contribution of current photosynthates to root respiration of non-nodulated *Medicago sativa*: effects of light and nitrogen supply. Plant Biol 7:601–610

Lötscher M, Klumpp K, Schnyder H (2004) Growth and maintenance respiration for individual plants in hierarchically structured canopies of *Medicago sativa* and *Helianthus annuus*: the contribution of current and old assimilates. New Phytol 164:305–316

Makino A, Osmond B (1991) Effects of nitrogen nutrition on nitrogen partitioning between chloroplasts and mitochondria in pea and wheat. Plant Physiol 96:355–362

Moorby J, Jarman PD (1975) The use of compartmental analysis in the study of the movement of carbon through leaves. Planta 122:155–168

Mortazavi B, Conte MH, Chanton JP, Smith MC, Crumsey J, Ghashghaie J (2009) Does the ^{13}C foliage-respired CO_2 and biochemical pools reflect the ^{13}C of recently assimilated carbon? Plant Cell Environ 32:1310–1323

Morvan-Bertrand A, Boucaud J, Prud'homme MP (1999) Influence of initial levels of carbohydrates, fructans, nitrogen, and soluble proteins on regrowth of *Lolium perenne* L. cv Bravo following defoliation. J Exp Bot 50:1817–1826

Motulsky H, Christopoulos A (2004) Fitting models to biological data using linear and nonlinear regression. Oxford University Press, New York

Muntz K (1998) Deposition of storage proteins. Plant Mol Biol 38:77–99

Nogués S, Tcherkez G, Cornic G, Ghashghaie J (2004) Respiratory carbon metabolism following illumination in intact French bean leaves using $^{13}C/^{12}C$ isotope labeling. Plant Physiol 136:3245–3254

Penning de Vries FWT (1975) The cost of maintenance processes in plant cells. Ann Bot 39:77–92

Plaxton WC, Podestá FE (2006) The functional organization and control of plant respiration. Crit Rev Plant Sci 25:159–198

Pollock CJ, Cairns AJ (1991) Fructan metabolism in grasses and cereals. Annu Rev Plant Physiol Plant Mol Biol 42:77–101

Poorter H, Nagel O (2000) The role of biomass allocation in the growth response of plants to different levels of light, CO_2, nutrients and water: a quantitative review. Aust J Plant Physiol 27:595–607

Prosser J, Farrar JF (1981) A compartmental model of carbon allocation in the vegetative barley plant. Plant Cell Environ 4:303–307

R Development Core Team (2007) R: A language and environment for statistical computing. R Foundation for Statistical Computing, Vienna. URL: http://www.R-project.org

Rescigno A (2001) The rise and fall of compartmental analysis. Pharmacol Res 44:337–342

Rocher JP, Prioul JL (1987) Compartmental analysis of assimilate export in a mature maize leaf. Plant Physiol Biochem 25:531–540

Rufty TW, Huber SC, Volk RJ (1988) Alterations in leaf carbohydrate-metabolism in response to nitrogen stress. Plant Physiol 88:725–730

Ryle GJA, Cobby JM, Powell CE (1976) Synthetic and maintenance respiratory losses of $^{14}CO_2$ in uniculm barley and maize. Ann Bot 40:571–586

Schnyder H (1992) Long-term steady-state labelling of wheat plants by use of natural $^{13}CO_2/^{12}CO_2$ mixtures in an open, rapidly turned-over system. Planta 187:128–135

Schnyder H (1993) The role of carbohydrate storage and redistribution in the source-sink relationships of wheat and barley during grain filling – a review. New Phytol 123:233–245

Schnyder H, Schäufele R, Lötscher M, Gebbing T (2003) Disentangling CO_2 fluxes: direct measurements of mesocosm-scale natural abundance $^{13}CO_2/^{12}CO_2$ gas exchange, ^{13}C discrimination, and labelling of CO_2 exchange flux components in controlled environments. Plant Cell Environ 26:1863–1874

Sicher RC, Kremer DF, Harris WG (1984) Diurnal carbohydrate metabolism of barley primary leaves. Plant Physiol 76:165–169

Simpson E, Cooke RJ, Davies DD (1981) Measurement of protein degradation in leaves of Zea mays using [^{3}H]acetic anhydride and tritiated water. Plant Physiol 67:1214–1219

Smith AM, Stitt M (2007) Coordination of carbon supply and plant growth. Plant Cell Environ 30:1126–1149

Stitt M, Krapp A (1999) The interaction between elevated carbon dioxide and nitrogen nutrition: the physiological and molecular background. Plant Cell Environ 22:583–621

Sullivan JT, Sprague VG (1943) Composition of the roots and stubble of perennial ryegrass following partial defoliation. Plant Physiol 18:656–670

Tcherkez G, Nogués S, Bleton J, Cornic G, Badeck F, Ghashghaie J (2003) Metabolic origin of carbon isotope composition of leaf dark-respired CO_2 in French bean. Plant Physiol 131:237–244

Teixeira EI, Moot DJ, Mickelbart MV (2007) Seasonal patterns of root C and N reserves of lucerne crops (*Medicago sativa* L.) grown in a temperate climate were affected by defoliation regime. Eur J Agron 26:10–20

Van Iersel MW (2003) Carbon use efficiency depends on growth respiration, maintenance respiration and relative growth rate: a case study with lettuce. Plant Cell Environ 29:1441–1449

Volenec JJ, Ourry A, Joern BC (1996) A role for nitrogen reserves in forage regrowth and stress tolerance. Physiol Plant 97:185–193

Wagner W, Keller F, Wiemken A (1983) Fructan metabolism in cereals: induction in leaves and compartmentation in protoplasts and vacuoles. Zeitschrift für Pflanzenphysiologie 112:359–372

Windt CW, Vergeldt FJ, de Jager PA, Van As H (2006) MRI of long-distance water transport: a comparison of the phloem and xylem flow characteristics and dynamics in poplar, castor bean, tomato and tobacco. Plant Cell Environ 29:1715–1729

Winzeler M, Dubois D, Nösberger J (1990) Absence of fructan degradation during fructan accumulation in wheat stems. J Plant Physiol 136:324–329

Zeeman SC, Smith SM, Smith AM (2007) The diurnal metabolism of leaf starch. Biochem J 401:13–28

Chapter 7
Tracing Carbon Fluxes: Resolving Complexity Using Isotopes

H. Schnyder, U. Ostler, C. Lehmeier, M. Wild, A. Morvan-Bertrand, R. Schäufele, and F.A. Lattanzi

7.1 Introduction

All trophic systems, from a single cell to the global biosphere, depend on photosynthesis and metabolism of reduced carbon substrates. Cells and ecosystems are, in fact, interconnected and interdependent metabolic networks, which are operated by carbon substrate fluxes. Biotic and abiotic stresses can perturb these fluxes at different scales of biological organisation, from cell to organism (Chap. 1). Such perturbations can affect substrate partitioning between biochemical pathways and allocation between parts of an organism, potentially generating/involving tradeoffs between growth and defence activities (Chaps. 5, 10–12). Knowledge of these responses to stress can enhance our understanding of the controls and mechanisms of carbon fluxes in plants, plant–microbe associations and ecosystems. Moreover, it provides a mechanistic foundation for physiologically based models of plant growth and functioning (Chaps. 15, 17 and 18). At the organism level, the mechanisms underlying carbon cycling include metabolic pathways, transport processes, deposition and mobilisation of stores, synthesis of structural compounds, and environmental and genetic effects on these mechanisms. At the larger scale, such mechanisms may concern the structure and operation of trophic networks or the stability of ecosystems.

Isotope methodologies are useful tools for tracing carbon substrate fluxes, in autotrophic and heterotrophic organisms and trophic networks in ecosystems. A large diversity of (artificial and natural) tracer approaches is available for such investigations. These include feeding of position-labelled ^{13}C substrates; pulse- or

H. Schnyder (✉) • U. Ostler • C. Lehmeier • M. Wild • R. Schäufele • F.A. Lattanzi
Lehrstuhl für Grünlandlehre, Technische Universität München, Alte Akademie 12, 85350 Freising, Germany
e-mail: schnyder@wzw.tum.de

A. Morvan-Bertrand
UMR INRA-UCBN 950 EVA Ecophysiologie Végétale, Agronomie and Nutritions NCS, Université de Caen Basse-Normandie, Esplanade de la Paix, 14032 Caen Cedex, France

R. Matyssek et al. (eds.), *Growth and Defence in Plants*, Ecological Studies 220, DOI 10.1007/978-3-642-30645-7_7, © Springer-Verlag Berlin Heidelberg 2012 dynamic labelling with the radio-active short-lived ^{11}C and long-lived ^{14}C or the stable ^{13}C in CO_2; and tracing of the natural alteration of isotopic signals, in photosynthetic reactions and (post-photosynthetic) metabolic pathways. As we show below, there are specific isotope approaches to address questions on very different temporal and spatial scales. For instance the metabolic fluxes ("fluxome") in heterotrophic cells are usually analysed with position-labelled substrates (with high artificial enrichment of the rare isotope), whereas regional- and global-scale carbon fluxes are mostly traced on the basis of natural ^{13}C or ^{14}C signals.

Approach-specific mathematical tools are used to analyse the mechanisms underlying tracer time courses in organisms and ecosystems. In this chapter, we discuss general principles of different carbon isotope tracer methodologies and the specifics of their use in studies of processes at various time frames and scales of biological complexity. Then, we illustrate how the analytical tool "compartmental modelling" can help to analyse tracer time courses. In particular, we demonstrate how compartmental modelling can be used to (1) assess the relative merits of pulse- and dynamic (continuous) labelling for the quantification of carbon pools and fluxes, (2) constrain hypotheses of the topology (architecture, structure) of metabolic systems and (3) elucidate the effect of fructan turnover on the half-lives of fructose, glucose and sucrose in grass leaves. We point out constraints associated with scales of application of different approaches. Finally, we advocate the joint use of different isotope methodologies in future work.

7.2 Principles of Isotopic Tracer Methodologies

"Tracing carbon fluxes" means tracking carbon atoms in chemical reactions or during displacement. Isotopes are ideally suited for this purpose. The word "isotope" derives from the Greek words *isos* and *topos*, which refer to occupation of the "same place" in the periodic table of elements. The isotopes of an element differ in mass, because of a different number of neutrons, but they undergo the same chemical reactions and physical processes. Since they behave the same, alteration of the isotopic composition of a substrate does not (or only minimally) disturb the metabolic and transport pathways. Rather it provides an identifiable tag, or label, with which the course of a biological process can be traced or tracked without disturbance. Detection methods include mass spectrometry, spectroscopy, nuclear magnetic resonance, or radioactive decay measurements (De Groot 2004, 2008). Isotope techniques were proved useful to partition photosynthesis and respiration and to quantify carbon allocation to different compartments and partitioning into different biochemical compounds at various scales, from the cell to the globe (e.g. Bassham et al. 1950; Ludwig and Canvin 1971; Ryle et al. 1976; Geiger and Fondy 1979; Kouchi and Yoneyama 1984; Thorpe and Minchin 1991; Ciais et al. 1995; Yakir and Wang 1996; Schimel 1995; Fung et al. 1997; Gebbing et al. 1998; Gebbing and Schnyder 1999; Randerson et al. 1999; Hanson et al. 2000; Schnyder et al. 2003; Schnyder and Lattanzi 2005; Heinemeyer et al. 2006; Grimoldi et al. 2006; Tcherkez et al. 2009; Gamnitzer et al. 2009; Grams et al. 2011).

There are two principal ways by which isotopes can produce traceable signals in study objects. Either the signal is created artificially, by exposure to isotopically altered substrate, or it arises naturally in metabolism or transport processes. Artificial tracer approaches have made use of the radioactive short-lived ^{11}C (half-life 20.5 min) and long-lived ^{14}C (5,760 years) as well as the stable ^{13}C. Methods of label provision include exposure to isotopically altered CO_2 (Ludwig and Canvin 1971; Geiger 1980; Leavitt et al. 1994; Loreto et al. 1999; Haupt-Herting et al. 2001; Deléens et al. 1983; Gamnitzer et al. 2009) or feeding with uniformly or position-labelled organic substrates, such as sugars and amino acids (Libourel and Shachar-Hill 2008; Schwender 2009; Kruger and Ratcliffe 2009). In the latter, the intra-molecular labelling pattern, at metabolic and isotopic steady state, reflects the label redistribution in metabolic networks and, hence, the metabolic fluxes in the system.

Natural isotope signals are due to different reaction speeds of distinct isotopes in various biochemical and physical processes. These cause isotope fractionation (discrimination) in biochemical and physical processes in photosynthesis and metabolism (Deines 1980; Farquhar et al. 1989; Ehleringer et al. 2000; Ghashghaie et al. 2003; Hobbie and Werner 2004; Tcherkez and Farquhar 2005; Tcherkez and Hodges 2008). Thus, primary CO_2-fixation mechanisms (C3, C4 and marine systems) generate distinct isotopic signals (Bender 1971; O'Leary 1981; Farquhar et al. 1989). Furthermore, the isotope effect on pyruvate-dehydrogenase causes a depletion of ^{13}C in the metabolites of acetyl-CoA and lipids (DeNiro and Epstein 1977; Melzer and Schmidt 1987). Also, the fructose-producing aldolase reaction of the chloroplast prefers ^{13}C, which causes a ^{13}C-enrichment of leaf starch stored during photosynthesis (Gleixner and Schmidt 1997). Thus, sucrose produced from the remaining triose phosphates ("day sucrose") is ^{13}C-depleted, whereas that synthesised at night from depolymerised starch is ^{13}C-enriched, since it inherits the ^{13}C signal from starch (Cernusak et al. 2009). Such isotopic signals are useful tracers of metabolism (e.g. Tcherkez et al. 2003). A difficulty in the utilisation of natural isotope signals in primary photosynthate is their non-steadiness. For instance, the ^{13}C signal of phloem sap contents can vary significantly in diurnal cycles (Kodama et al. 2008). Such factors can complicate a quantitative evaluation and analysis of tracer data.

There are two popular methods of applying labelled CO_2 (or other substrates) and monitoring the propagation of the tracer: pulse(-chase)-labelling and dynamic (long-term) labelling. The latter method has also been referred to as "continuous" (Gamnitzer et al. 2009) or "steady-state" labelling (Geiger 1980; Schnyder 1992). However, in "fluxomics" studies the term "steady-state labelling" is used to denote a labelling principle in which the labelled precursor (usually a specific isotopomer of a substance) is supplied continuously at constant enrichment, and intra-molecular labelling patterns are measured when the system is in isotopic and metabolic steady state (Ratcliffe and Shachar-Hill 2006).

In dynamic labelling, the labelled substrate (e.g. CO_2, see Fig. 7.1) is supplied continuously during the time course of the studied process, at constant isotopic composition. The amount of tracer in the substance of interest increases continuously during label application until—eventually—all sources/pathways supplying

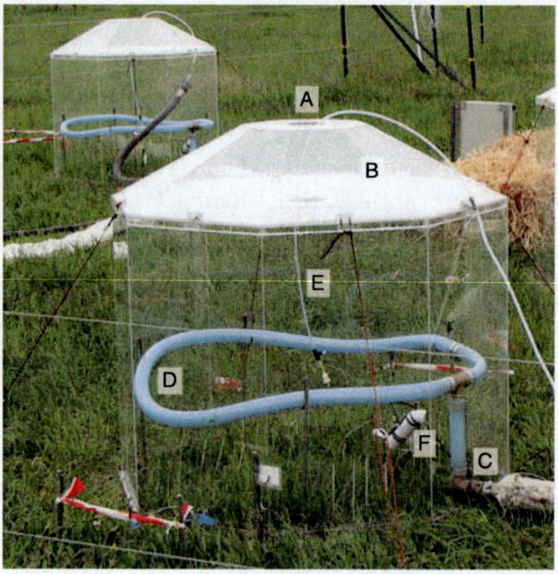

Fig. 7.1 Chamber system for $^{13}CO_2/^{12}CO_2$ labelling of a grassland ecosystem under field conditions (Gamnitzer et al. 2009). An ecosystem section is enclosed in a chamber with an opening in the top (A), therefore named "open-top chamber". A buffer volume (B) prevents ambient air incursion into the chamber headspace. Air with labelled CO_2 (CO_2 with altered $^{13}CO_2$ content) is provided to the chamber via the air supply tube (C) and distributed within the chamber headspace with the perforated tube (D). A sampling tube (E) allows sampling of chamber headspace air for monitoring of labelling conditions inside the chamber and for respiration measurements

the metabolite have reached label saturation (isotopic equilibrium with the labelled substrate). Observations of label content are performed simultaneously with labelling (Geiger 1980; Lattanzi et al. 2005; Lehmeier et al. 2008; Gamnitzer et al. 2009). The change in isotopic composition with time reflects the functional properties of the pool system supplying the synthesis of the metabolite (number and arrangement of pools and the size, turnover rate and contribution of each pool to the synthesis of the metabolite). Compartmental analysis (Atkins 1969; Jacquez 1996) is a useful tool to extract these functional characteristics from labelling kinetics. If the metabolite is not completely labelled, then the metabolite may not have turned over completely, or some unlabelled (or incompletely labelled) source is still contributing to its synthesis. Examples of the latter are metabolites which are synthesised from slowly turning over pools, such as stores or decomposing structural biomass (Lattanzi et al. 2005; Lehmeier et al. 2008).

In pulse-labelling the labelled substrate is provided for a period of time (pulse), which is very short in relation to the time course of the studied process, generally at high isotopic enrichment. Then follows an extended period in which again the unlabelled form of the substrate is provided, as prior to the pulse (chase period) (Austin et al. 1976; Jones et al. 1983; Gregory and Atwell 1991). In such an experiment, the evolution of tracer content in the substance of interest exhibits two phases: first, it increases as the labelled form of the substrate isotope becomes incorporated, then it decreases as the labelled substance is diluted by incorporation of the unlabelled form of the substrate ("washout" of the tracer). In general, the incorporation of the tracer is not monitored during the pulse-period. Typically, the first observation is made at, or shortly after, the end of the pulse. Further observations occur over the time scale of the process of interest.

Data analysis and interpretation are conducted with approach-specific mathematical tools/simulation models. They include compartmental analysis (Atkins 1969; Jacquez 1996; and see below) and modelling theory and computational methods of metabolic flux analysis (Ratcliffe and Shachar-Hill 2006; Sauer 2006; Schuetz et al. 2007; Libourel and Shachar-Hill 2008; Schwender 2009; Allen et al. 2009) for systems in metabolic steady state. Behaviour of non-steady systems—in natural conditions and usually much larger scales—is analysed with statistical methods such as wavelet coherence analysis (Vargas et al. 2010) or wiggle-matching procedures (Kilian et al. 2000).

Although carbon and its isotopes are the subject of this chapter, we recognise that isotopes of other bio-elements can be useful for carbon metabolism studies. For instance, dual labelling with ^{32}P and ^{14}C ascertained the nature and role of ribulose-1.5-bisphosphate in the reductive pentose phosphate cycle (Benson 1951). As another example, dual labelling experiments with nitrogen and carbon isotopes can help to partition amino-C and carbohydrate-C fluxes (Schnyder and de Visser 1999).

7.3 Processes, Time Frames and Scales of Biological Complexity

The use of carbon isotopes has advanced our understanding of carbon metabolism, allocation and cycling in a great variety of processes: metabolic pathways studies (including uptake/assimilation of CO_2 and biosynthesis of primary, secondary and structural compounds), synthesis and mobilisation of storage compounds, transport across membranes and through vascular conduits, autotrophic and heterotrophic respiration, carbon partitioning in ecosystems, and the roles of different photosynthetic types in the biogeochemistry and biogeography of the earth. These processes have characteristic and distinct time frames. The exchange of carbon in metabolic pathways, such as the Calvin cycle, occurs in minutes (Bassham et al. 1954). Transport of assimilate from leaves to roots in the phloem takes from several minutes to several days, with transport time correlating with plant size. Stores turn over at the scale of a day to many months, and the residence time of carbon in structural biomass varies from months to centuries. Accordingly, the kinetics of label propagation in a system is strongly affected by the types of participating processes.

Moreover, studies of the different processes are performed at different levels of biological organisation and complexity. Metabolic pathway studies (e.g. MFA) are commonly performed in components of cells (chloroplasts, mitochondria, vacuoles or components thereof), cells or excised tissues. Transport studies require more complex systems, such as whole organs or entire plants. Analyses of sink/source relationships, storage/mobilisation and tissue life span are performed with intact plants or plant stands/communities. On the other hand, carbon residence time was studied on various levels of biological integration from single plant (Lehmeier et al. 2008) to global scale (Bird et al. 1996).

Along with differences in time frames and biological complexity go different challenges in administration/exposure of the label and tracing its fate; metabolic pathway analyses are performed in highly controlled and reproducible conditions and terminated within minutes, whereas the studies of the residence time of carbon in ecosystems generally occur in non-reproducible conditions and require techniques with a resolution of days to centuries.

Also, the experimental methods for tracing label differ between different types of process studies and associated spatial scales (moreover and obviously, there have been transitions in the approaches used over the last 60 years). For instance, the pioneering works of photosynthetic metabolism have used dynamic labelling (*sensu* Ratcliffe and Shachar-Hill 2006) with $^{14}CO_2$ (Bassham et al. 1954), whereas metabolic flux analyses at organelle-, cell- or unicellular organism scales are mainly using steady-state labelling with ^{13}C-position-labelled organic substrates (Libourel and Shachar-Hill 2008) at time scales of seconds to days. Both used high isotopic enrichments. On the other hand, controlled-environment mesocosm studies have employed dynamic labelling with $^{13}CO_2$ at near-natural abundance levels for weeks to months (Deléens et al. 1983; Schnyder 1992). Studies of phloem transport have mainly used CO_2 pulse-labelling with the radioactive short-lived isotope ^{11}C (Minchin and Thorpe 2003) or long-lived ^{14}C (Geiger and Fondy 1979; Geiger 1980).

Studies focusing on long-term processes at ecosystem-level have often used pulse-labelling with $^{13}CO_2$ and $^{14}CO_2$ (Kuzyakov 2006) to investigate the residence time of carbon or the labelling kinetics of respiratory CO_2 (Ostle et al. 2000; Johnson et al. 2002; Carbone and Trumbore 2007; Carbone et al. 2007; Högberg et al. 2008; Bahn et al. 2009). Dynamic labelling experiments in ambient (free air) conditions are methodically challenging, particularly at the ecosystem level (Gamnitzer et al. 2009, 2011). Figure 7.1 shows an example of a field labelling system for weeks-long exposure of a grassland ecosystem to an atmosphere with altered $^{13}CO_2$ content. Yet, free air carbon dioxide enrichment (FACE) experiments have also employed continuous labelling with naturally ^{13}C-depleted CO_2 for CO_2 enrichment (Leavitt et al. 1994). This provides a measurable isotopic label which can be traced in the ecosystem (Glaser et al. 2006; Keel et al. 2006; Bock et al. 2007; Grams et al. 2011; Kuptz et al. 2011). However, the precision and accuracy of labelling (e.g. signal to noise ratio) of FACE systems is generally inferior to that of chamber-based systems (e.g. Gamnitzer et al. 2009).

At much larger scales, such as that of catchments, regions or the globe, the artificial alteration of isotopic content of CO_2 or carbon pools/substrates is generally not feasible. At field scale, C4 crops may be used to trace the fate of carbon in C3 soils (Buchmann and Ehleringer 1998; Bol et al. 2009). At regional and global scale, one must resort to natural isotopic signals, such as the different isotopic composition of terrestrial and oceanic CO_2 sink, which allow partitioning of land/ ocean contributions to the missing global carbon sink (Ciais et al. 1995; Fung et al. 1997; Randerson et al. 1999). However, natural (and bomb) ^{14}C signals provide a powerful tool for studies of soil carbon turnover at decadal to millennial time scales (Trumbore 2006, 2009). A remarkable exceptionality is given by the "anthropogenic" ^{14}C bomb spike that has been used as a tracer (Stenhouse and Baxter 1977) to address research questions from the scale of single organisms to that of the globe (e.g. Broecker et al. 1985; Bird et al. 1996; Richter et al. 1999; Spalding et al. 2008).

7.4 Pulse- *Versus* Dynamic Labelling

The relative merits of pulse- and dynamic labelling were discussed previously (Geiger 1980; Meharg 1994; Kuzyakov 2006; Paterson et al. 2009), but the two approaches were not compared directly and with quantitative methods. Here we quantitatively compare the tracer time courses (the so-called "tracer kinetics") of pulse and dynamic labelling. Based on data from a dynamic labelling experiment, the pool characteristics were determined by compartmental analysis (Fig. 7.2a). The dynamic labelling data suggested that the sink was supplied by a two-pool system, as shown in the inset of Fig. 7.2b. Compartmental analysis revealed that this system was composed of a "metabolic and transport pool" (P_1) with a half-life of 0.1 day, and a "store" with a half-life of 6 days (P_2).

These pool characteristics were then used to derive the tracer kinetics for pulse-chase labelling, based on a 0.8-day-long pulse (Fig. 7.2b). The kinetics of label uptake during the pulse is identical to the initial kinetics of dynamic labelling. The pulse caused a strong labelling of the rapidly turned-over P_1 (63 % label saturation just after the pulse), but a weak labelling of the slowly turned-over P_2 (5 % labelling just after the pulse). In the subsequent washout period, P_1 lost much more label than P_2, as the latter was much less labelled during the pulse. In consequence, detecting the contribution of P_2 in supplying the sink would require adjusting (*i.e.* increasing) the ^{13}C-enrichment of the labelling pulse. Otherwise, pulse-labelling experiments may be "biased" as the contribution of pools with slow turnover goes undetected. For instance, failure to recognise this restriction/disadvantage of pulse-labelling can lead to overestimation of the contribution of current assimilation in supplying a function (sink), and underrating of the role of slowly turned-over stores. Dynamic labelling avoids this problem, provided that the measurement frequency directly after the onset of labelling is high enough to resolve the fast pool(s) and the labelling is continued until (or close to) isotopic saturation of the slow pool(s).

However, there are also advantages for pulse-labelling. For instance, translocation velocity in plants can be assessed simply by providing a short pulse of $^{11}CO_2$ or $^{14}CO_2$ to photosynthesising leaves and monitoring the transit time of the labelled assimilate-pulse using radiation detectors placed at different positions along the translocation path (Geiger and Swanson 1965; Jahnke et al. 1981). Also, fluctuations in allocation patterns in non-steady systems can be detected by

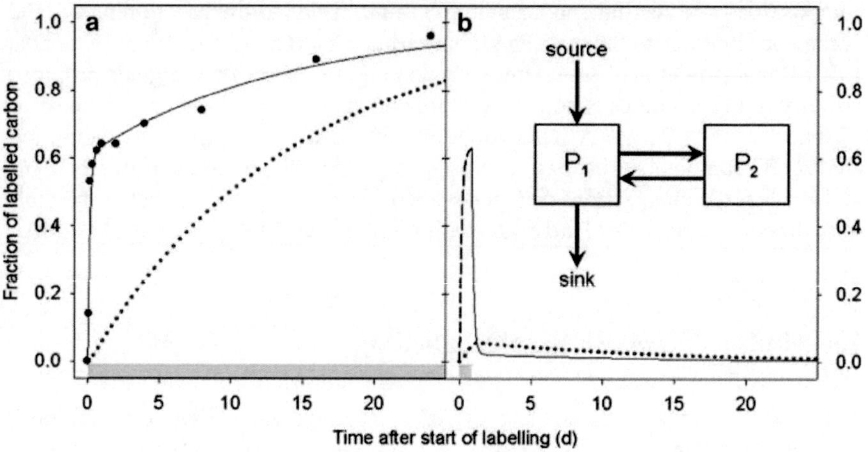

Fig. 7.2 Tracer time course (tracer kinetics) in a "dynamic labelling" (**a**) and a "pulse-chase labelling" experiment (**b**). Tracer kinetics of identical biological systems are compared in (**a**) and (**b**). This system conforms to a two-pool model, shown as an *inset* in (**b**). The system includes a "metabolic and transport pool" (P_1) and a "store" (P_2). Tracer taken up from the source must pass through P_1, before arriving in the sink. But, some of the tracer first cycles through P_2, before being passed on to the sink. In both panels, the labelling duration is indicated by a *grey shaded bar*. The measured data (*filled circles* in (**a**) give the fraction of labelled carbon in the amino-C flux supplying the leaf growth zone of a perennial ryegrass leaf (sink). Plants were grown in a (near-) steady-state in continuous light. The data were obtained from Wild et al. (unpublished) and analysed using procedures as described by Lattanzi et al. (2005). The *continuous line* in (**a**) represents the fit of the two-pool model to the data as obtained with compartmental analysis. In (**b**) the *dashed line* reflects the label increase during the pulse; the *solid line* gives the subsequent decay (washout) kinetics calculated using the same compartmental model as in (**a**), with identical pool characteristics. The *dotted line* in (**a**) and (**b**) give the labelling kinetics of P_2

sequential pulse-labelling of replicates of the system. Furthermore, pulse-labelling studies are experimentally less demanding than dynamic labelling, in particular under field conditions.

7.5 Using Compartmental Modelling to Assess Network Architecture/Topology and Metabolite Compartmentation

One of the present challenges to analysing and understanding metabolic fluxes in plants, as compared with unicellular organisms, is the much greater complexity of plant metabolic networks. To a significant extent, the greater complexity is related to compartmentation, which causes separation of networks, and to the existence and involvement of stores/storage compartments, which are a source of slowly labelled substrate (Kruger et al. 2007; Allen et al. 2009; see above). Failure to consider compartmentation can lead to misinterpretations of labelling patterns (Sweetlove et al. 2008).

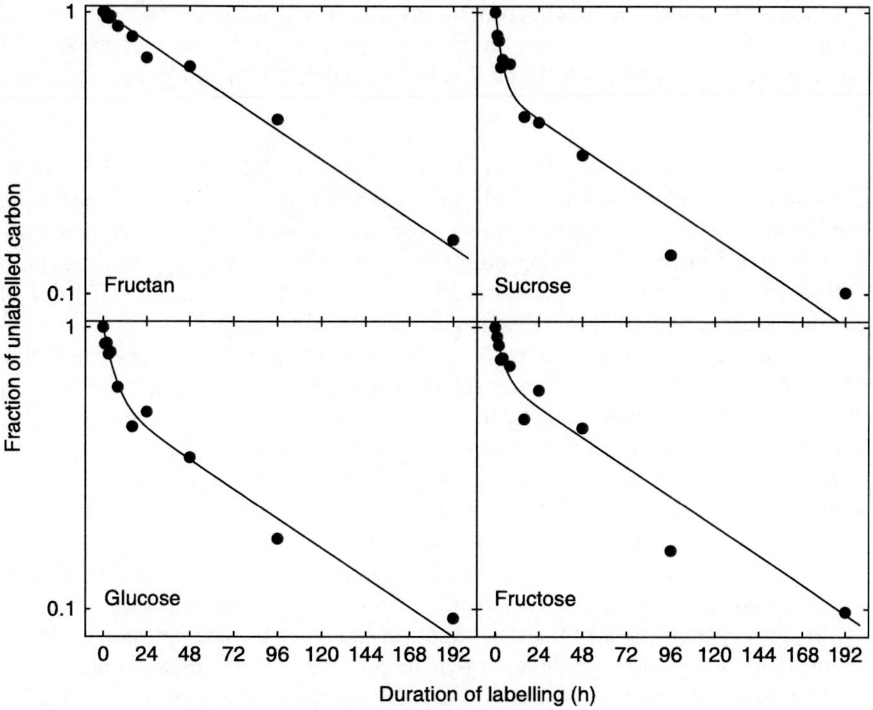

Fig. 7.3 Semi-logarithmic plot of the fraction of unlabelled carbon in fructan, sucrose, glucose and fructose during labelling. The data were obtained from a steady-state $^{13}CO_2/^{12}CO_2$ labelling experiment with *Lolium perenne* grown in continuous light with a high nitrogen supply (unpublished data). Carbohydrates were extracted from the youngest fully expanded leaf of mature tillers. Data modified from Lattanzi et al. (2012)

Compartmental modelling is one of the tools which may assist in resolving some of these problems. It can distinguish distinct pools of a metabolite, if the pools differ in the kinetics of labelling (i.e. slow- versus fast-labelled pools). Such differences are expected for metabolites originating from current assimilation and stores. Here we demonstrate the usefulness of compartmental modelling for this purpose using data from dynamic labelling of the water-soluble carbohydrates (glucose, fructose, sucrose and fructan) in the leaf blades of perennial ryegrass (Fig. 7.3).

The tracer kinetics of fructan, a vacuolar storage carbohydrate, fitted a one-pool (first-order kinetics) model with a half-life of 69 h ($r^2 = 0.98$). Conversely, the tracer kinetics of sucrose, glucose and fructose reflected two-pool systems. Their tracer kinetics fitted double exponential functions of the form $y = a{\cdot}e^{-b{\cdot}t} + (1 - a){\cdot}e^{-c{\cdot}t}$. The interpretation of the fit parameters depends on the system structure, which is discussed in detail below. However, to illustrate the power of the compartmental analysis tool, we discuss an example. This is represented by a system in which both pools incorporate and release tracer and in which no exchange occurs between the two pools. In such a system, the parameters *a* and

$1 - a$ represent the fractional contributions of pools 1 and 2 to the total concentration of the respective carbohydrates. The parameters b and c represent the turnover rates (h^{-1}) of pools 1 and 2, which are directly linked to the respective half-lives.

The fast pool (pool 1) of fructose, glucose and sucrose had very similar half-lives: 2.4 h for sucrose ($r^2 = 0.95$), 4.3 h for glucose ($r^2 = 0.98$) and 2.0 h for fructose ($r^2 = 0.96$), consistent with the expectation that they were formed from primary photosynthetic products. A more comprehensive analysis of central carbohydrate metabolism (considering both fructan metabolism and invertase activity) with a four-pool compartmental model demonstrated even faster half-lives of sucrose, glucose and fructose (Lattanzi et al. 2012).

The half-life of the carbon in pool 2 of these carbohydrates was the same as that of the fructan pool. This is consistent with the view that the residence time of carbon in pool 2 of these carbohydrates was controlled by the (vacuolar) fructan pool; the carbon in pools 2 of fructose, glucose and sucrose originated from the turnover of fructan. Fructan degradation yields (mainly) fructose. Part of this is used to form glucose via isomerisation, and both sugars are used for (re-)synthesis of sucrose (Pollock and Cairns 1991). The close similarity of the half-lives of pool 2 of fructose, glucose and sucrose indicates that the metabolic steps leading to sucrose re-synthesis occurred very rapidly. This interpretation was also supported by the low concentrations of pool 2 of fructose and glucose (data not shown). These results demonstrate the usefulness of dynamic labelling and compartmental analysis to unravel differences in the sub-cellular origin of metabolites in complex metabolic networks.

Compartmental modelling can also help to constrain predictions on the topology of networks. This is exemplified by different two-pool models fitted to the data of Fig. 7.2a. Table 7.1 shows the ten variants of two-pool models which differ in structure. One- and three-pool models were also fitted to the data shown in Fig. 7.2a. The one-pool model exhibited a significant lack of fit, whereas three-pool models were not supported by the data due to over-parameterisation (not shown). Among the two-pool models (Table 7.1), model 5 represented a system consisting of a storage compartment which exchanges with a metabolic and transport pool. Biological evidence supported the realism of this model (Wild 2010). The results of the model fits were consistent with the empirical expectation: model 5 fitted the data equally well or better than the other models.

Models 1 and 2 represented a serial arrangement of the two pools, with one pool receiving tracer from the source and the other pool releasing the tracer to the sink end of the system. These models fitted the data very poorly, compared to the other models, suggesting that a serial arrangement of the pools was unlikely. Models 7–10 fitted the data well, but the estimates of pool size and half-life were associated with large errors. These errors were a consequence of over-parameterisation of the models. This means that the models were more complex than was necessary to explain the tracer data. These models were therefore rejected, following the rule of parsimony. Simpler models (models 3–6) fitted the data equally well, but exhibited much less error than models 7–10. Therefore, these simpler models provided the

Table 7.1 Two-pool models fitted to the tracer time course shown in Fig. 7.1A

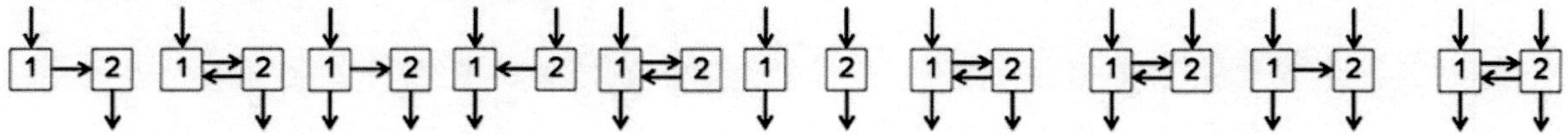

Parameter	Model 1	Model 2	Model 3	Model 4	Model 5	Model 6	Model 7	Model 8	Model 9	Model 10
Independent parameters	2	3	3	3	3	3	4	4	4	5
r^2	0.49	0.49	0.96	0.96	0.96	0.96	0.96	0.96	0.96	0.96
F-Value	8.78	3.84	97.84	98.13	97.83	98.12	57.07	57.25	57.25	36.80
$T_{1/2}$ Q_1 (h)	0.0 ± 1.8	11.4 ± 4.7	2.4 ± 0.4	2.4 ± 0.6	2.4 ± 0.4	2.4 ± 0.6	2.4 ± 5.0	2.4 ± 50.2	2.4 ± 2.1	2.4 ± 62.0
$T_{1/2}$ Q_2 (h)	10.1 ± 3.6	0.1 ± 0.0	234 ± 76	234. ± 77	144 ± 53	234 ± 77	198 ± 96,873	219 ± 3,508	234 ± 106	233 ± 1,015,800
Size Q_1	0.1 ± 2.6	16.5 ± 6.8	3.5 ± 0.6	3.5 ± 0.8	5.7 ± 1.2	2.1 ± 0.5	4 ± 505	4 ± 59	2.1 ± 2.2	3 ± 2,975
Size Q_2	14.6 ± 5.2	0.1 ± 0.1	131 ± 52	131 ± 45	129 ± 68	132 ± 43	130 ± 55,499	130 ± 6,307	132 ± 184	131 ± 572,421
p in Q_1	–	–	–	0.61	–	0.61	–	0.66	0.62	0.62
p in Q_2	–	–	–	0.39	–	0.39	–	0.34	0.38	0.38
p out Q_1	–	–	0.61	–	–	0.39	0.72	–	0.61	0.80
p out Q_2	–	–	0.39	–	–	0.61	0.28	–	0.39	0.20

The number of independent parameters is given by the number of adjustable fluxes (which is the total number of fluxes minus 1, as the influx is defined as the unit flux 1 h^{-1}) plus the number of adjustable pool sizes (which is 2) minus the number of constraints due to the steady-state assumption (1 for each pool). The r^2 and F values are the usual statistical information for the best fit of each model to the data. Pool sizes and half-lives ($T_{1/2}$) are given as means +/− SE. Pool sizes are given in relative units, as the import rate (the flux through the system) was assumed to be the unit flux. The parameter p denotes the fraction of the total influx/outflux into/out of pools 1 and 2, respectively. Data from Wild et al. (unpublished) best reflection of the topology of the system represented in Fig. 7.2a. Among these, models 3 and 6 represented systems serving two sinks, whereas models 4 and 5 served only one. As the experimental system considered here (Fig. 7.2) had only one physical sink, the leaf growth zone, models 3 and 6 seemed unapt. However, it is still possible that this single sink was fed by two distinct metabolic pathways, utilising two (groups of) metabolites with different origins in the supply system. Chemical analysis might reveal this possibility. This example demonstrates that modelling can guide experimentation, by pointing to system features which merit further analysis. Such work could lead to advances in hypothesis development.

Some system properties are sensitive to differences in topology, but others are not. For instance, the half-life estimate of pool 1 was the same for models 3–6. Also, predictions of pool contributions to the total sink flux (or shares of the total source flux) agreed perfectly. So these features were independent of differences in topology, meaning that uncertainties of topology were non-critical for the estimation of these parameters. Conversely, estimates of the half-life of pool 2 and of the size of pools 1 and 2 were dependent on topology, showing that knowledge of model topology is critical for accurate assessment of other system features.

7.6 Conclusions

Today, a wide range of isotope methodologies are available for tracing carbon fluxes at widely differing scales, from cellular metabolic pathways to global biogeochemical cycles. The development of the various methodologies has historic roots in different disciplines of bioscience, and the methods have been used to great advantage in their original disciplines. However, many methods have potential for application outside their traditional discipline. Furthermore, we can expect much benefit from applying different isotope methodologies to the same research questions. Investigations of natural intra-molecular isotope distributions can be combined with ^{13}C-labelling based metabolic flux analysis in microorganisms, plants or plant–microbe associations using paired experimental units for the two approaches. Such joint methodologies could be combined with dynamic labelling and compartmental modelling to shed light on the role of stores/recycling pools in metabolic networks. Inter alia, such work should be performed with plants and plant–microbe associations in non-stressed environments and conditions of abiotic and biotic stress, to further our mechanistic understanding of the real-world controls of tradeoffs in carbon substrate allocation and partitioning in these systems (Chaps. 1 and 20).

References

Allen DK, Libourel IGL, Shachar-Hill Y (2009) Metabolic flux analysis in plants: coping with complexity. Plant Cell Environ 32:1241–1257

Atkins GL (1969) Multicompartment models in biological systems. Methuen, London

Austin RB, Ford MA, Edrich JA (1976) Some effects of leaf posture on photosynthesis any yield in wheat. Ann Appl Biol 83:425–446

Bahn M, Schmitt M, Siegwolf R, Richter A, Brüggemann N (2009) Does photosynthesis affect grassland soil-respired CO_2 and its carbon isotope composition on a diurnal timescale? New Phytol 182:451–460

Bassham JA, Benson AA, Calvin M (1950) The path of carbon in photosynthesis. 8. The role of malic acid. J Biol Chem 185:781–787

Bassham JA, Benson AA, Kay LD, Harris AZ, Wilson AT, Calvin M (1954) The path of carbon in photosynthesis, 21. The cyclic regeneration of carbon dioxide acceptor. J Am Chem Soc 76:1760–1770

Bender MM (1971) Variations in $^{13}C/^{12}C$ ratios of plants in relation to pathway of photosynthetic carbon dioxide fixation. Phytochemistry 10:1239–1244

Benson AA (1951) The identification of Ribulose in $^{14}CO_2$ photosynthesis products. J Am Chem Soc 73:2971–2972

Bird MI, Chivas AR, Head J (1996) A latitudinal gradient in carbon turnover times in forest soils. Nature 381:143–146

Bock M, Glaser B, Millar N (2007) Sequestration and turnover of plant- and microbially derived sugars in a temperate grassland soil during 7 years exposed to elevated atmospheric pCO_2. Glob Change Biol 13:478–490

Bol R, Poirier N, Balesdent J et al (2009) Molecular turnover time of soil organic matter in particle-size fractions of an arable soil. Rapid Commun Mass Spectrom 23:2551–2558

Broecker WS, Peng TH, Ostlund G, Stuiver M (1985) The distribution of bomb radiocarbon in the ocean. J Geophys Res Oceans 90(C4):6953–6970

Buchmann NN, Ehleringer JR (1998) CO_2 concentration profiles, and carbon and oxygen isotopes in C3 and C4 crop canopies. Agric Forest Meteorol 89:45–58

Carbone MS, Trumbore SE (2007) Contribution of new photosynthetic assimilates to respiration by perennial grasses and shrubs: residence times and allocation patterns. New Phytol 176:124–135

Carbone MS, Czimczik CI, McDuffee KE, Trumbore SE (2007) Allocation and residence time of photosynthetic products in a boreal forest using a low-level ^{14}C pulse-chase labeling technique. Glob Change Biol 13:466–477

Cernusak LA, Tcherkez G, Keitel C, Cornwell WK, Santiago LS, Knohl A, Barbour MM, Williams DG, Reich PB, Ellsworth DS, Dawson TE, Griffiths HG, Farquhar GD, Wright IJ (2009) Viewpoint: why are non-photosynthetic tissues generally ^{13}C enriched compared with leaves in C_3 plants? Review and synthesis of current hypotheses. Funct Plant Biol 36: 199–213

Ciais P, Tans PP, White JWC, Trolier M, Francey RJ, Berry JA, Randall DR, Sellers PJ, Collatz JG, Schimel DS (1995) Partitioning of ocean and land uptake of CO_2 as inferred by $\delta^{13}C$ measurements from the NOAA climate monitoring and diagnostics laboratory global air sampling network. J Geophys Res Atmos 100:5051–5070

De Groot PA (ed) (2004) Handbook of stable isotope analytical techniques, vol I. Elsevier, Amsterdam

De Groot PA (ed) (2008) Handbook of stable isotope analytical techniques, vol II. Elsevier, Amsterdam

Deines P (1980) The isotopic composition of reduced organic carbon. In: Fritz P, Fontes JC (eds) Handbook of environmental isotope geochemistry, vol 1. The terrestrial environment A. Elsevier, Amsterdam, pp 329–406

Deléens E, Pavlides D, Queiroz O (1983) Natural ^{13}C abundance as a tracer for the determination of leaf matter turn-over in C3 plants. Physiologie Végétale 21:723–729

DeNiro MJ, Epstein S (1977) Mechanisms of carbon isotope fractionation associated with lipid synthesis. Science 197:261–263

Ehleringer JR, Buchmann N, Flanagan LB (2000) Carbon isotope ratios in belowground carbon cycle processes. Ecol Appl 10:412–422

Farquhar GD, Ehleringer JR, Hubick KT (1989) Carbon isotope discrimination and photosynthesis. Annu Rev Plant Physiol Plant Mol Biol 40:503–537

Fung I, Field CB, Berry JA, Thompson MV, Randerson JT, Malmstrom CM, Vitousek PM, Collatz GJ, Sellers PJ, Randall DA, Denning AS, Badeck F, John J (1997) Carbon 13 exchanges between the atmosphere and biosphere. Glob Biogeochem Cycles 11:507–533

Gamnitzer U, Schäufele R, Schnyder H (2009) Observing ^{13}C labelling kinetics in CO_2 respired by a temperate grassland ecosystem. New Phytol 184:376–386

Gamnitzer U, Moyes AB, Bowling DR, Schnyder H (2011) Measuring and modelling the isotopic composition of soil respiration: insights from a grassland tracer experiment. Biogeosciences 8:1333–1350

Gebbing T, Schnyder H (1999) Pre-anthesis reserve utilization for protein and carbohydrate synthesis in grains of wheat. Plant Physiol 121:871–878

Gebbing T, Schnyder H, Kühbauch W (1998) Carbon mobilization in shoot parts and roots of wheat during grain filling: assessment by $^{13}C/^{12}C$ steady-state labelling, growth analysis and balance sheets of reserves. Plant Cell Environ 21:301–313

Geiger DR (1980) Measurement of translocation. Methods Enzymol 69:561–571

Geiger DR, Fondy BR (1979) Methods for continuous measurement of export from a leaf. Plant Physiol 64:361–365

Geiger DR, Swanson CA (1965) Sucrose translocation in sugar beet. Plant Physiol 40:685–690

Ghashghaie J, Badeck FW, Lanigan G, Nogues S, Tcherkez G, Deléens E, Cornic G, Griffiths H (2003) Carbon isotope fractionation during dark respiration and photorespiration in C_3 plants. Phytochem Rev 2:145–161

Glaser B, Millar N, Blum H (2006) Sequestration and turnover of bacterial- and fungal-derived carbon in a temperate grassland soil under long-term elevated atmospheric pCO_2. Glob Change Biol 12:1521–1531

Gleixner G, Schmidt H-L (1997) Carbon isotope effects on the fructose-1,6-bisphosphate aldolase reaction, origin for non-statistical 13C distributions in carbohydrates. J Biol Chem 272:5382–5387

Grams TEE, Werner H, Kuptz D, Ritter W, Fleischmann F, Andersen CP, Matyssek R (2011) A free-air system for long-term stable carbon isotope labeling of adult forest trees. Trees 25:187–198

Gregory PJ, Atwell BJ (1991) The fate of carbon in pulse-labelled crops of barley and wheat. Plant Soil 136:205–213

Grimoldi AA, Kavanova M, Lattanzi FA, Schäufele R, Schnyder H (2006) Arbuscular mycorrhizal colonization on carbon economy in perennial ryegrass: quantification by $^{13}CO_2/^{12}CO_2$ steady-state labelling and gas exchange. New Phytol 172:544–553

Hanson PJ, Edwards NT, Garten CT, Andrews JA (2000) Separating root and soil microbial contributions to soil respiration: a review of methods and observations. Biogeochemistry 48:115–146

Haupt-Herting S, Klug K, Fock HP (2001) A new approach to measure gross CO_2 fluxes in leaves. Gross CO_2 assimilation, photorespiration, and mitochondrial respiration in the light in tomato under drought stress. Plant Physiol 126:388–396

Heinemeyer A, Ineson P, Ostle N, Fitter AH (2006) Respiration of the external mycelium in the arbuscular mycorrhizal symbiosis shows strong dependence on recent photosynthates and acclimation to temperature. New Phytol 171:159–170

Hobbie EA, Werner RA (2004) Intramolecular, compound-specific, and bulk carbon isotope patterns in C3 and C4 plants: a review and synthesis. New Phytol 161:371–385

Högberg P, Högberg MN, Göttlicher SG, Betson NR, Keel SG, Metcalfe DB, Campbell C, Schindlbacher A, Hurry V, Lundmark T, Linder S, Nasholm T (2008) High temporal resolution tracing of photosynthate carbon from the tree canopy to forest soil microorganisms. New Phytol 177:220–228

Jacquez J (1996) Compartmental analysis in biology and medicine. Elsevier, Amsterdam

Jahnke S, Stöcklin G, Willenbrink J (1981) Translocation profiles of ^{11}C-assimilates in the petiole of *Marsilea quadrifolia* L. Planta 153:56–63

Johnson D, Leake JR, Ostle N, Ineson P, Read DJ (2002) In situ $^{13}CO_2$ pulse-labelling of upland grassland demonstrates a rapid pathway of carbon flux from arbuscular mycorrhizal mycelia to the soil. New Phytol 153:327–334

Jones DA, Smith AM, Woolhouse HW (1983) An apparatus for pulse and pulse-chase experiments with $^{14}CO_2$ on attached leaves with known, steady-state rates of photosynthesis. Plant Cell Environ 6:161–166

Keel SG, Siegwolf RTW, Körner C (2006) Canopy CO_2 enrichment permits tracing the fate of recently assimilated carbon in a mature deciduous forest. New Phytol 172:319–329

Kilian MR, van Geel B, van der Plicht J (2000) ^{14}C AMS wiggle matching of raised bog deposits and models of peat accumulation. Quat Sci Rev 19:1011–1033

Kodama N, Barnard RL, Salmon Y, Weston C, Ferrio JP, Holst J, Werner RA, Saurer M, Rennenberg H, Buchmann N, Gessler A (2008) Temporal dynamics of the carbon isotope composition in a *Pinus sylvestris* stand: from newly assimilated organic carbon to respired carbon dioxide. Oecologia 156:737–750

Kouchi H, Yoneyama T (1984) Dynamics of carbon photosynthetically assimilated in nodulated soya bean-plants under steady-state conditions. 1. Development and application of $^{13}CO_2$ assimilation system at a constant ^{13}C abundance. Ann Bot 53:875–882

Kruger NJ, Ratcliffe RG (2009) Insights into plant metabolic networks from steady-state metabolic flux analysis. Biochimie 91:697–702

Kruger NJ, Le Lay P, Ratcliffe RG (2007) Vacuolar compartmentation complicates the steady-state analysis of glucose metabolism and forces reappraisal of sucrose cycling in plants. Phytochemistry 68:2189–2196

Kuptz D, Fleischmann F, Matyssek R, Grams TEE (2011) Seasonal patterns of carbon allocation to respiratory pools in 60-yr-old deciduous (*Fagus sylvatica*) and evergreen (*Picea abies*) trees assessed via whole-tree stable carbon isotope labeling. New Phytol. doi:10.1111/j.1469-8137.2011.03676.x

Kuzyakov Y (2006) Sources of CO_2 efflux from soil and review of partitioning methods. Soil Biol Biochem 38:425–448

Lattanzi FA, Schnyder H, Thornton B (2005) The sources of carbon and nitrogen supplying leaf growth. Assessment of the role of stores with compartmental models. Plant Physiol 137:383–395

Lattanzi FA, Ostler U, Wild M, Morvan-Bertrand A, Decau M-L, Lehmeier CA, Meuriot F, Prud'homme M-P, Schäufele R, Schnyder H (2012) Fluxes in central carbohydrate metabolism of source leaves in a fructan-storing C3 grass – rapid turnover and futile cycling of sucrose in continuous light under contrasted nitrogen nutrition status. J Exp Bot. doi:10.1093/jxb/ers020

Leavitt SW, Paul EA, Kimball BA, Hendrey GR, Mauney JR, Rauschkolb R, Rogers H, Lewin KF, Nagy J, Pinter PJ Jr, Johnson HB (1994) Carbon isotope dynamics of free-air CO_2-enriched cotton and soils. Agric Forest Meteorol 70:87–101

Lehmeier CA, Lattanzi FA, Schäufele R, Wild M, Schnyder H (2008) Root and shoot respiration of perennial ryegrass are supplied by the same substrate pools: assessment by dynamic ^{13}C labeling and compartmental analysis of tracer kinetics. Plant Physiol 148:1148–1158

Libourel IGL, Shachar-Hill Y (2008) Metabolic flux analysis in plants: from intelligent design to rational engineering. Annu Rev Plant Biol 59:625–650

Loreto F, Delfine S, Di Marco G (1999) Estimation of photorespiratory carbon dioxide recycling during photosynthesis. Aust J Plant Physiol 26:733–736

Ludwig LJ, Canvin DT (1971) An open gas-exchange system for simultaneous measurement of CO_2 and $^{14}CO_2$ fluxes from leaves. Can J Bot 49:1299–1313

Meharg AA (1994) A critical review of labeling techniques used to quantify rhizosphere carbon-flow. Plant Soil 166:55–62

Melzer E, Schmidt HL (1987) Carbon isotope effects on the pyruvate-dehydrogenase reaction and their importance for relative carbon-13 depletion in lipids. J Biol Chem 262:8159–8164

Minchin PEH, Thorpe MR (2003) Using the short-lived isotope ^{11}C in mechanistic studies of photosynthate transport. Funct Plant Biol 30:831–841
O'Leary MH (1981) Carbon isotope fractionation in plants. Phytochemistry 20:553–567
Ostle N, Ineson P, Benham D, Sleep D (2000) Carbon assimilation and turnover in grassland vegetation using an in situ $^{13}CO_2$ pulse labelling system. Rapid Commun Mass Spectrom 14:1345–1350
Paterson E, Midwood AJ, Millard P (2009) Through the eye of the needle: a review of isotope approaches to quantify microbial processes mediating soil carbon balance. New Phytol 184:19–33
Pollock CJ, Cairns AJ (1991) Fructan metabolism in grasses and cereals. Annu Rev Plant Physiol Plant Mol Biol 42:77–101
Randerson JT, Thompson MV, Field CB (1999) Linking ^{13}C-based estimates of land and ocean sinks with predictions of carbon storage from CO_2 fertilization of plant growth. Tellus B Chem Phys Meteorol 51:668–678
Ratcliffe RG, Shachar-Hill Y (2006) Measuring multiple fluxes through plant metabolic networks. Plant J 45:490–511
Richter DD, Markewitz D, Trumbore SE, Wells CG (1999) Rapid accumulation and turnover of soil carbon in a re-establishing forest. Nature 400:56–58
Ryle GJA, Cobby JM, Powell CE (1976) Synthetic and maintenance respiratory losses of $^{14}CO_2$ in uniculm barley and maize. Ann Bot 40:571–586
Sauer U (2006) Metabolic networks in motion: ^{13}C-based flux analysis. Mol Syst Biol 2:62
Schimel DS (1995) Terrestrial ecosystems and the carbon-cycle. Glob Change Biol 1:77–91
Schnyder H (1992) Long-term steady-state labelling of wheat plants by use of natural $^{13}CO_2/^{12}CO_2$ mixtures in an open, rapidly turned-over system. Planta 187:128–135
Schnyder H, de Visser R (1999) Fluxes of reserve-derived and currently assimilated carbon and nitrogen in perennial ryegrass recovering from defoliation. The regrowing tiller and its component functionally distinct zones. Plant Physiol 119:1423–1435
Schnyder H, Lattanzi FA (2005) Partitioning respiration of C3-C4 mixed communities using the natural abundance ^{13}C approach – testing assumptions in a controlled environment. Plant Biol 7:592–600
Schnyder H, Schäufele R, Lötscher M, Gebbing T (2003) Disentangling CO_2 fluxes: direct measurements of mesocosm-scale natural abundance $^{13}CO_2/^{12}CO_2$ gas exchange, ^{13}C-discrimination, and labelling of CO_2 exchange flux components in controlled environments. Plant Cell Environ 26:1863–1874
Schuetz R, Kuepfer L, Sauer U (2007) Systematic evaluation of objective functions for predicting intracellular fluxes in Escherichia coli. Mol Syst Biol 3:119
Schwender J (ed) (2009) Plant metabolic networks. Springer, Dordrecht, 331 p
Spalding KL, Arner E, Westermark PO et al (2008) Dynamics of fat cell turnover in humans. Nature 453:783–787
Stenhouse MJ, Baxter MS (1977) Bomb ^{14}C as a biological tracer. Nature 267:828–832
Sweetlove LJ, Fell D, Fernie AR (2008) Getting to grips with the plant metabolic network. Biochem J 409:27–41
Tcherkez G, Farquhar GD (2005) Carbon isotope effect predictions for enzymes involved in the primary carbon metabolism of plant leaves. Funct Plant Biol 32:277–291
Tcherkez G, Hodges M (2008) How stable isotopes may help to elucidate primary nitrogen metabolism and its interaction with (photo)respiration in C_3 leaves. J Exp Bot 59:1685–1693
Tcherkez G, Nogues S, Bleton J, Cornic G, Badeck F, Ghashghaie J (2003) Metabolic origin of carbon isotope composition of leaf dark-respired CO_2 in French Bean. Plant Physiol 131:237–244
Tcherkez G, Mahe A, Gauthier P, Mauve C, Gout E, Bligny R, Cornic G, Hodges M (2009) In folio respiratory fluxomics revealed by 13C isotopic labeling and H/D isotope effects highlight the noncyclic nature of the tricarboxylic acid "cycle" in illuminated leaves. Plant Physiol 151:620–630

Thorpe MR, Minchin PEH (1991) Continuous monitoring of fluxes of photoassimilate in leaves and whole plants. J Exp Bot 42:461–468

Trumbore SE (2006) Carbon respired by terrestrial ecosystems – recent progress and challenges. Glob Change Biol 12:141–153

Trumbore SE (2009) Radiocarbon and soil carbon dynamics. Annu Rev Earth Planet Sci 37:47–66

Vargas R, Detto M, Baldocchi DD, Allen MF (2010) Multiscale analysis of temporal variability of soil CO_2 production as influenced by weather and vegetation. Glob Change Biol 16:1589–1605

Wild M (2010) The carbon and nitrogen supply systems of leaf growth in perennial ryegrass – characterization by dynamic ^{15}N and ^{13}C labelling and compartmental analysis of tracer influx into the leaf growth zone. Doctoral thesis, Technische Universität München

Yakir D, Wang XF (1996) Fluxes of CO_2 and water between terrestrial vegetation and the atmosphere estimated from isotope measurements. Nature 380:515–517

Chapter 8
Solar Radiation as a Driver for Growth and Competition in Forest Stands

M. Leuchner, C. Hertel, T. Rötzer, T. Seifert, R. Weigt, H. Werner, and A. Menzel

8.1 Introduction

Solar radiation plays a key role for the growth and competition in plant ecosystems. It interacts with biomass by supplying phytoelements with the necessary energetic input for photosynthesis. Growth processes are triggered by the spectral composition of incoming radiation, referred to as light quality. This information is used to regulate growth by means of pigmentary absorption in certain wavebands, in particular in the UV, blue, and red fraction of the solar spectrum (Ammer 2000; Smith 1994). The allocation and partitioning of nutrients (see Chap. 9) as well as carbon (see Chap. 12) are directly dependent on the spatial distribution of radiation quality and quantity. Space occupation and exploitation is an important issue regarding competition for resources, below-ground as described in Chaps. 10 and 11 and above-ground (c.f. Chaps. 11–14) driven by radiation. Radiation is one of the main drivers for mechanistic models as applied in Chaps. 15, 17 and 18. This chapter intends to analyze properties of the radiation regime of a mature forest stand

M. Leuchner (✉) • C. Hertel • H. Werner • A. Menzel
Ecoclimatology, Technische Universität München, Hans-Carl-von-Carlowitz-Platz 2, 85354 Freising, Germany
e-mail: leuchner@wzw.tum.de

T. Rötzer
Chair of Forest Growth and Yield Science, Technische Universität München, Hans-Carl-von-Carlowitz-Platz 2, 85354 Freising, Germany

T. Seifert
Forest and Wood Science, Stellenbosch University, Private Bag X1, 7602 Matieland, South Africa

R. Weigt
Organismic Biology: Mycology, Ludwig-Maximilians-Universität München, Menzinger Str. 67, 80638 München, Germany

R. Matyssek et al. (eds.), *Growth and Defence in Plants*, Ecological Studies 220,
DOI 10.1007/978-3-642-30645-7_8, © Springer-Verlag Berlin Heidelberg 2012

achieved in an experimental study focusing on radiation–biomass interactions and taking into account that light is both consequence and determinant of competition within those stands.

Wavelength-selective reflection, transmission, and absorption of photons by heterogeneously distributed phytoelements determine the radiation field within a forest canopy and result in a high spatial and temporal variability. Thus, plant canopies operate as light filters concerning the spectral and spatial distribution of solar radiation at ground level (e.g., Grant 1997) and constitute important boundary conditions for regeneration. Several parameters are responsible for the properties of the light climate at a certain location within the canopy or at the forest floor. Differences in the incoming direct and diffuse radiation field, canopy architecture, composition and density, as well as phenological development account for most of the variability (Capers and Chazdon 2004; Ross et al. 1986; Baldocchi et al. 1984). Most of the absorption and spectral changes take place in the uppermost part of the sun crown depending on solar elevation and sky conditions that determine incoming radiation and the incident angle of direct irradiance. Those factors influencing the spatial distribution of radiation result in the occurrence of sunflecks and different levels of shading such as umbra and penumbra as well as shade flecks (i.e., short-term shading of otherwise exposed areas) with varying properties in regard to photon quantity and spectral quality.

The frequency and composition of the different solar radiation levels are crucial factors for ecological processes such as biomass allocation (Hendrich 2000), establishment, physiological processes of productivity, and natural and artificial regeneration. They provide photosynthetic energy and trigger morphogenetic processes (Smith and Whitelam 1997). Under favorable site fertility and without additional stress, radiation is the limiting factor for growth. Thus, dominant trees profit disproportionally more with increased growth rates as they make use of their enhanced access to radiation and by suppressing and reducing growth rates of smaller neighbors (see Sects. 14.2.2 and 14.4). The ecological relevance and importance of sunflecks and variability of different light and shading levels due to strong temporary enhancement of radiation has been emphasized before (e.g., Ammer 2003; Smith 2000; Grant 1997; Chazdon 1988). Due to high variability, snapshots at single moments and locations can only show the instantaneous conditions of the radiation climate in a stand. In order to estimate the principal patterns of the light distribution and parameterize the factor "light" for models, prolonged time intervals with respect to changing meteorological conditions and developmental stages with high spatiotemporal resolution must be used.

To address some of the mentioned aspects, an experiment at the research site "Kranzberger Forst"—about 35 km northeast of Munich, southern Germany—was conducted as part of a Collaborative Research Center (SFB607). A unique setup in terms of number, spatial and spectral resolution—with 130 spherical radiation sensors mounted in 25 vertical profiles consisting of six levels above and within an approximately 26-m tall mature European beech (*Fagus sylvatica* L.)–Norway spruce (*Picea abies* [L.] Karst) stand was chosen. For the main research area, as many as 194 trees and a single story structure are given with a total area of

30 × 30 m characterized by a projected leaf area index (LAI) of around 6 m^2 m^{-2} for both beech and spruce at full foliation (Pretzsch et al. 1998). The maximum leaf area density for beech is 6.0 situated in the upper third and 2.6 for spruce in the lower half of the canopy (Häberle et al. 2003). The basal area is 46.4 m^2 ha^{-1} the stand density 764 stems ha^{-1} (Wipfler et al. 2005). The resulting radiation data were available in a high spectral resolution of 0.8 nm within the waveband of 360–1,020 nm, covering a small fraction of UV-A (<380 nm), the entire visible range (380–780 nm) including PAR (400–700 nm), and parts of the near-infrared (>780 nm).

8.2 Interaction of Solar Radiation with Biomass

8.2.1 Spatial and Temporal Variability of Solar Radiation in a Mature Forest Stand

The spatiotemporal variability of solar radiation when penetrating a mature forest is extremely variable and depends on many influencing factors that determine the amount of transmitted photons and their spectral composition. In the following some of the variability is explained under the influence of meteorological and phenological conditions.

The vertical distribution of the transmitted PAR fraction ($PPFR_{rel}$: fraction of the transmitted photosynthetic photon fluence rate at the respective height) in the investigated stand is plotted in Fig. 8.1. The data are separated into subsets for beech and spruce as well as clear sky (CS) and overcast (OVC) conditions. Due to differences in the phenological properties of deciduous beech in comparison to coniferous spruce, the former is divided into a foliated (August and September) and defoliated (November and December) state, while for analysis of spruce the entire dataset from August to December is used (Leuchner et al. 2011).

During OVC conditions the fractions of PAR penetrating the canopy are significantly higher at all levels, which is in accordance with previous results (e.g., Méthy et al. 1987; Morgan et al. 1985). This is caused by the exclusively diffuse character of the incoming radiation under overcast skies. The scattered, diffuse radiation is able to enter the canopy gaps from all angles of the upper hemisphere which also occurs under CS conditions for the then much smaller diffuse part. The predominant direct, unscattered fraction that reaches the top of the canopy from only a small celestial angle is mostly extinct by the first interaction with phytoelements. This effect can be observed distinctively for beech with its more laminar upper canopy surface. The conical habit of spruce allows more direct radiation to penetrate into deeper layers of the canopy due to larger gaps. The extinction profile for both species follows quite well Beer's law as shown in previous studies for different kinds of vegetation covers (e.g., De Castro 2000; Roujean 1999; Baldocchi et al. 1984), although the transmittance levels for both European beech and Norway

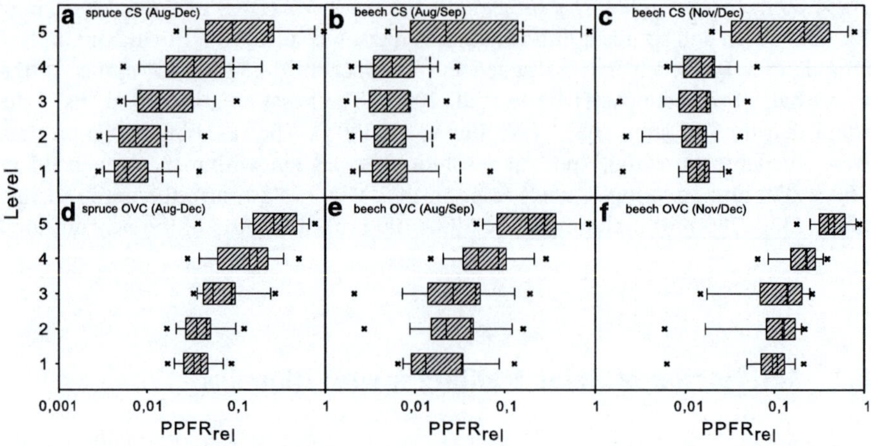

Fig. 8.1 Boxplots of non-attenuated PAR ($PPFR_{rel}$ on a log scale) at all canopy (2–5: 14, 17, 20, 23 m) and ground (1: 3 m) levels under clear sky (CS; **a–c**) and overcast (OVC; **d–f**) conditions for spruce (**a** and **d**), foliated (**b** and **e**), and defoliated (**c** and **f**) beech. The mean is represented by a *dashed vertical line*; the left hand side of the box is the lower quartile and the right hand side the upper quartile, with the median represented by the *vertical solid line*. Whiskers extend to 10th and 90th percentiles and crosses represent 5th and 95th percentiles (Leuchner et al. 2011)

spruce in our study are consistently lower. That can be explained by the high LAI of six in this dense stand. Under cloudless sky, the differences in mean and median values are quite large, with means considerably higher, indicating the influence of sunflecks that are characterized by short episodes (seconds to minutes) of high radiation up to 100% of above-canopy values.

The median light levels in such a dense, mature stand are in most cases quite low in all layers. The lowest values were encountered in the beech shade crown (levels 2–4) and at the ground (level 1) with transmitted PAR less than 1% of above-canopy radiation. Due to the mentioned sunflecks, the mean transmittance is higher with values starting above 1% and 5% for beech in respect to CS and OVC conditions and 2% and 5%, respectively, for spruce. Because of the different properties of the spherical sensors used in this work, the results are not directly comparable to other studies. Nevertheless, the transmittance levels are in good agreement with values found in the literature (e.g., Brantley and Young 2009; Capers and Chazdon 2004; De Castro 2000; Vierling and Wessman 2000; Gratani 1997; Chazdon 1988). Even in fall, where beech is defoliated, transmitted values are quite low due to extinction from branches and stems in combination with low solar elevation.

Due to the rapid enhancement of radiation in combination with an altered spectral composition, the sunfleck distribution and frequency of occurrence are of special ecological importance. The frequency of sunflecks correlates well with increased photosynthetic rates and stomatal conductance of leaves, a gain in biomass, and production of more propagules amongst others (Chazdon 1988).

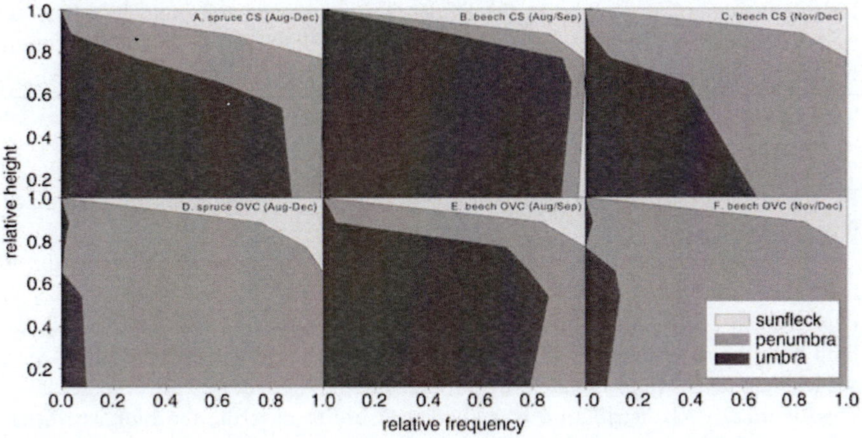

Fig. 8.2 Vertical distribution of sunflecks, penumbra, and umbra under clear sky (CS; **a–c**) and overcast (OVC; **d–f**) conditions for spruce (**a** and **d**), foliated (**b** and **e**), and defoliated (**c** and **f**) beech (Leuchner et al. 2011)

In the following, sunflecks are defined as more than 50% of photons of above-canopy occurrence, umbra as less than 2% of above-canopy photons, and penumbra as the region between 2 and 50% (Leuchner et al. 2011). Figure 8.2 shows the vertical distribution of sunfleck, penumbra, and umbra frequencies in a spruce (Fig. 8.2a, d), a foliated beech (Fig. 8.2b, e), and defoliated beech (Fig. 8.2c, f) canopy. Large differences between both species can be seen in the vertical distribution of penumbra and umbra, while sunflecks are distributed nearly similarly with maxima in the upper canopy and little influence in the shade crown and at the forest floor. Especially under overcast conditions, a much higher fraction of umbra is present throughout the crown in beech than in spruce. For defoliated beech under clear sky conditions, a higher reflectance of bark (Grant et al. 2003) and a lower solar elevation can lead to a much larger penumbral fraction compared to the foliated stage.

8.2.2 *Spectral Ratios as Indicators of Light Quality and Their Importance on Growth and Morphogenesis*

Leaves show high absorption of light in the blue and red range and high reflectance in the far red and near-infrared region of the spectrum (Combes et al. 2000). The ratio of red (R; 655–665 nm) and far red (FR; 725–735 nm) is a good approximation for the status of the phytochrome photoequilibrium (Smith 1994). R/FR is a good index for the detection of competition, structure and architecture, and internode expansion (Grant 1997; Ballaré et al. 1990). The behavior of R/FR under different tree species depends on the leaf coverage and leaf architecture.

Below deciduous tree canopies the R/FR declines greater than under coniferous trees, triggered by a higher selective transmission of light and a higher R absorption by the leaves (Federer and Tanner 1966). Blue light is primarily sensed by cryptochromes as photoreceptors and influences growth, development of higher plants, and promotes stomatal opening more than other spectral wavelengths (Matsuda et al. 2007). For the ratio of B_w/R_w (B_w: broadband blue 400–500 nm; R_w: broadband red 600–700 nm), only a few interaction studies were conducted (e.g., Navrátil et al. 2007).

An important aspect for the understanding of radiation–biomass interaction is the linkage of spectral radiation data with measured and modeled biomass data. As described in Sect. 9.3.4 biomass parameters such as specific leaf area (SLA) can help with the identification of sun and shade crowns. Sun leaves have lower SLA values than leaves in the shade crown (Sect. 11.5.3, Kitao et al. 2009; Nunn 2004). Also vertical LAI distribution is a good measure to describe the biomass distribution. LAI_z is the average leaf area within a vertical layer of 1 m in the respective height. For the following analysis both parameters (SLA, LAI_z) were calculated with the physiological, single-tree-based growth model BALANCE (see Chap. 18, Rötzer et al. 2010; Grote and Pretzsch 2002), based on a parameterization obtained at the research site (Grote 2002). As input variables for the biomass modeling tree height, crown width, crown base height, and stem diameter at breast height (1.3 m) for each individual tree were used. Based on the obtained tree individual SLA and LAI_z distributions a voxel space with 0.5-m sized voxels was created and linked to measured radiation.

Figure 8.3 shows the course of the B_w/R_w (b and e) and R/FR (c and f) for beech and spruce during OVC and CS conditions. Mean values of incoming R/FR during OVC are with 1.14 slightly higher than under CS conditions (1.12). The vertical profiles for spruce under OVC (Fig. 8.3f) show a stepwise declining course of R/FR towards the ground. Beech (Fig. 8.3c) also shows a declining ratio in the upper layers but an increase in R/FR in the lower layers. The increase of R/FR in beech below the shade crown (height of c. 14 m) down to the understory is more distinctive for beech. This is linked to the stand architecture and biomass distribution (Fig. 8.3a). The higher R/FR, especially during OVC conditions, was due to the fact that more diffuse unattenuated radiation can penetrate omnidirectionally from the upper hemisphere into the forest stand and foliage gaps. The results show that deciduous canopies are denser at a whole canopy level than the clumped coniferous stands even under diffuse conditions where the impact of scattered light from all directions is more homogenous. The lower R/FR values under broadleaved beech are an effect of higher absorption of the red fraction by the biomass. Even under spruce it is shown that the ratio in the middle and thus the denser part of the canopy is lower. The higher values in the upper layers of spruce during OVC and CS conditions are the effect of the species' special cone-shaped habit with larger gaps between trees.

The penetration of blue and red radiation into the canopy is displayed in Fig. 8.3b, e. The B_w/R_w ratios for beech and spruce varied under OVC conditions

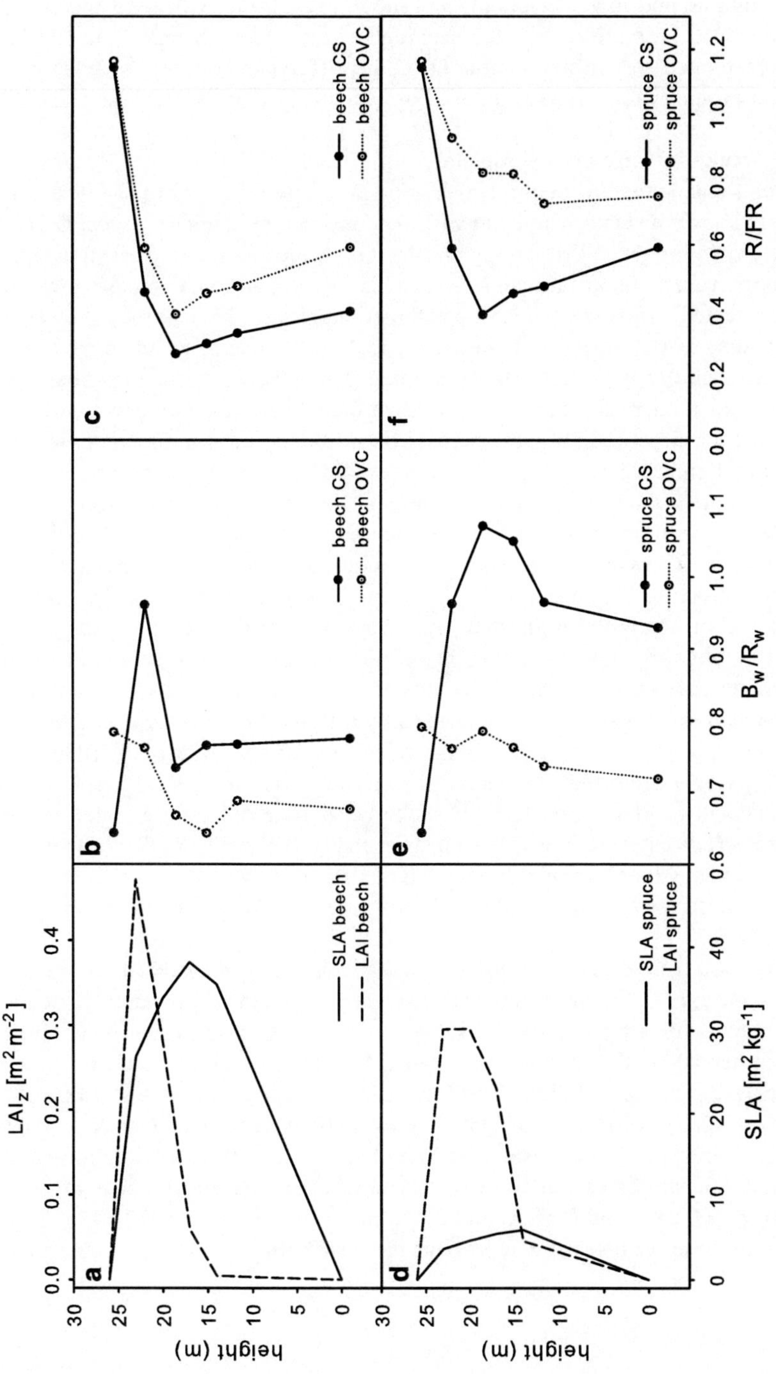

Fig. 8.3 Vertical distribution of the modeled leaf area index of layer z (LAI_z), the specific leaf area (SLA) (**a** and **d**) and observed ratios of B_w/R_w (**b** and **e**) and R/FR (**c** and **f**) for beech and spruce under different sky conditions (CS, OVC) (Hertel et al. in press). For abbreviations and ranges see text

within the middle and lower layers of the canopy (14–20 m). Values of incoming B_w/R_w under OVC conditions with 0.79 are higher than under CS conditions (0.64). For the investigated spruce stand under OVC the difference between mean above-canopy and ground values is very small. For beech (Fig 8.3b) the course shows a low distortion between 14 and 20 m in the area of the shade crown. The results under CS conditions display a completely different distribution of B_w/R_w among layers. One evident feature of the beech and spruce profiles was the increase of B_w/R_w from above the canopy to 23 m and 20 m, respectively, under spruce. B_w/R_w was enhanced within the entire canopy profile of both species in comparison to their above-canopy values. Especially under beech below 23 m a rapid decrease down to 20 m was detected with a small increase at the ground level (3 m). Spruce showed a different course of B_w/R_w throughout the vertical profile due to its stand architecture. The only similarity with beech was the increase of B_w/R_w in the upper canopy layers. All values under beech were lower than the comparable spruce values. On days with higher direct radiation an increase of the B_w/R_w ratio in the upper layers was evident (Hertel et al. 2011). The lower values for beech at 20 m are the result of larger shaded areas caused by higher biomass in the very top layer as seen in the vertical LAI_z distribution in Fig. 8.3a. The maximum LAI_z is higher (0.46 $m^2\ m^{-2}$) at the height of 23 m in beech, so more shaded areas are given linked to the broadleaved structure. For beech the sun crown (23 m) and for spruce the middle layers (17, 20 m) show the highest B_w/R_w values. According to this result LAI_z for spruce exhibits maxima at the heights of 17, 20, and 23 m. It seems that sun leaves and needles reflect more B in comparison to R; shade leaves show the opposite behavior. This is very likely caused by different reflectance properties of both species in terms of bark reflection and leaf wax (Grant et al. 2003; Gordon et al. 1998). It is evident that these shaded areas are characterized by higher SLA values. Especially under beech SLA values between 14 and 20 m are high and the according B_w/R_w values decrease. The most crucial factor for the B_w/R_w ratio is the availability and the distribution of biomass as described above. On the other hand, it is possible to describe the biomass distribution for beech and spruce with the help of the B_w/R_w ratio.

Additional statistical analyses utilizing linear regression between the modeled cumulative mean LAI and the spectral ratios were performed (Hertel et al. 2011). A dataset including all sky conditions was used. In the period of full foliation (August and September 2005) the cumulative mean LAI for beech and the mean B_w/R_w result in an R^2 of 0.58 ($p = 0.036$). The correlation of B_w/R_w with SLA shows an R^2 of 0.74 ($p = 0.005$). R/FR also correlates significantly with SLA with an R^2 of 0.63 ($p = 0.019$). For spruce the results deviate due to the different morphological crown habit. Linear regression analyses of cumulative LAI and B_w/R_w show an R^2 of 0.40 ($p = 0.025$) and for LAI and R/FR an R^2 of 0.69 ($p = 0.003$) (Fig. 8.4). According to these results, R/FR represents the distribution of biomass better than the B_w/R_w ratio for spruce.

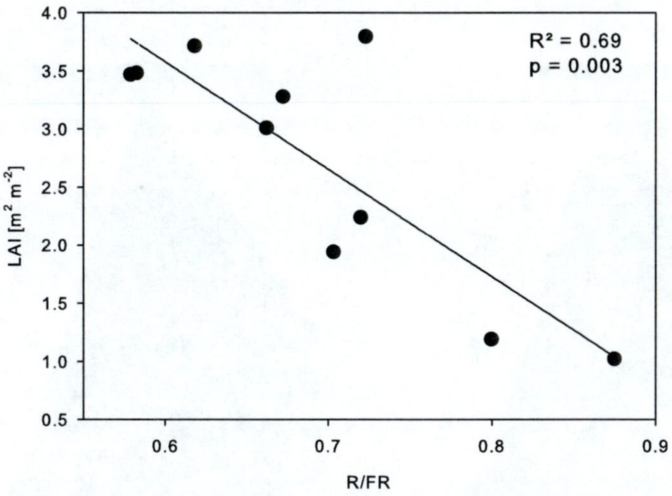

Fig. 8.4 Linear regression between cumulative LAI and R/FR for spruce (Hertel et al. in press)

8.2.3 Impacts of Thinning and Clearings to the Light Climate in a Mature Forest Stand: A Case Study

The spectral distribution and spatiotemporal variability of solar radiation are the drivers for natural and artificial regeneration. To get a better grip on the spatial heterogeneity and to evaluate impacts of thinning and clearings (see Chap. 13 and Sect. 14.5) on the solar radiation field, a three-dimensional model was developed (Knyazikhin et al. 1997), adapted, and validated for the Kranzberg Forest research site (Leuchner 2006), and different scenario calculations performed.

Figure 8.5a shows the horizontal variability of the transmitted PAR directly below the canopy space in 14-m height on a clear sky day in summer. The ellipses mark a beech, a spruce group, and a clearing, where up to 80% of PAR can still reach this level. The canopies of both species—even more pronounced for spruce—extinguish a high fraction of incoming PAR. The small-scale variability of solar radiation can be observed in particular in the area of beeches plotted as numerous peaks representing sunflecks and penumbral effects caused by rather small canopy gaps. A similar but more smoothed distribution can also be seen under overcast conditions (not shown), due to the decreased importance of sunflecks.

The applied model is based on a spectral approach and thus capable to calculate photomorphogenetically important spectral ratios such as the R/FR. Figure 8.5b shows exemplarily the horizontal distribution of R/FR in the same height and for the same conditions as for PAR. The variability in general is higher than for the broader PAR waveband. Again, ratios in the clearing match the observed values above the canopy. Under clear sky conditions the R/FR is much more reduced

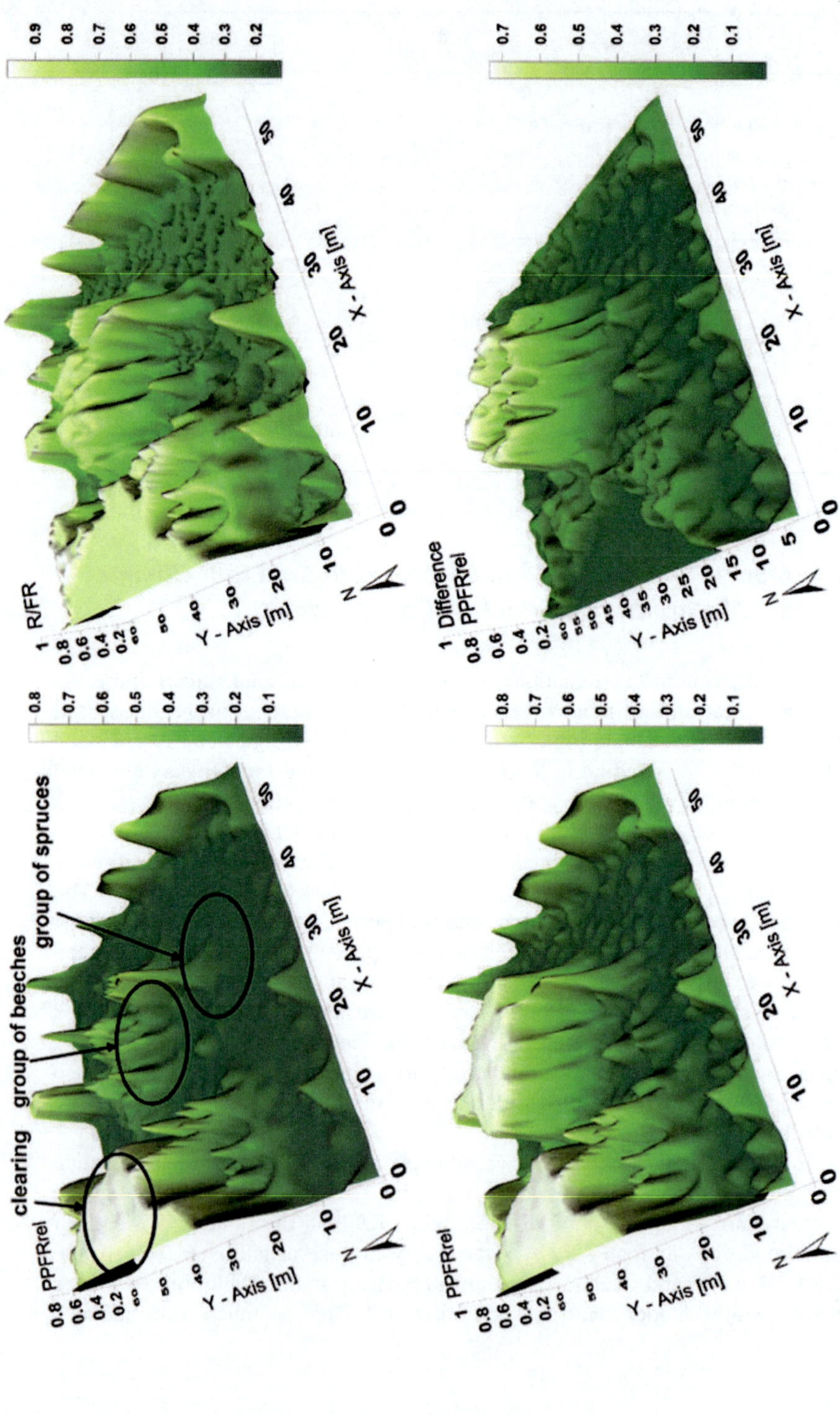

Fig. 8.5 Modeled horizontal distribution of transmitted PAR directly below the canopy (14 m height) on a clear sky day (**a**); modeled horizontal distribution of R/FR directly below the canopy (14 m height) on a clear sky day (**b**); modeled horizontal distribution of the transmitted PAR fraction directly below the canopy in 14 m height on a clear sky day with the scenario: all beeches removed from the stand (**c**); differences of the transmitted PAR fraction in **c** to the situation in **a** (**d**); method of interpolation: linear kriging (Leuchner 2006)

within and below the canopy than under overcast sky and much more enhanced in sunflecks with ratios closer to those in unattenuated radiation above the canopy. Validation of these data has shown that the rather high values of R/FR for beech are due to overestimation by the model (Leuchner 2006).

The PAR distribution below the shade crown from a model run during clear sky conditions is shown in Fig. 8.5c in a scenario where all beeches have been removed from the stand to simulate for thinning, creating a clearing of approximately 15 m in diameter. In comparison to the situation with all beeches present (Fig. 8.5a), where only small fractions of incoming PAR reach the bottom of the canopy, such a radical scenario allows most of the radiation to penetrate to the stem space of the new clearing. A significant increase of radiation can also be observed at the fringe of the adjacent spruces that fades after only a few meters. At the displayed height of 14 m, just at the bottom of the lower shade crown, a fraction of 80% of incoming PAR can still be seen in the clearing accompanied by a spectral composition quite similar to above-canopy values. The resulting differences of the modeled thinning scenario on the spatial PAR distribution are shown in Fig. 8.5d. In addition to the clearing itself, other adjoining and not directly neighboring areas within the canopy are to a certain degree affected by the supplemental radiation. During overcast sky, where only diffuse radiation is present, about 60% of incoming PAR penetrates the canopy within the clearing. Due to the diffuse nature of radiation, more photons can enter the canopy laterally. The transition to the adjacent spruce group is thus smoothed out and deeper compared to the clear sky run, where a high proportion of direct radiation and thus a strong distinction of zones with umbra and sunflecks occurs. These results of modeling have been partly confirmed by measurements at the experimental site.

Additional runs have been conducted with alternative thinning scenarios for both species, where similar effects were observed. The extent of radiation enhancement is directly dependent on the size of the modeled clearing. Also runs for R/FR, representing the quality of radiation, were performed showing strong alteration towards enhancement of the ratio up to values close to the open field.

The increased energy supply, as shown in Fig. 8.5 and by experimental observations, as well as the simultaneously altered spectral composition, plays the vital role for growth within the canopy of adjacent individuals, in particular for the PAR limited areas of the shade crown, and for regeneration at the ground.

8.2.4 *Estimation of Phenological Phases by Radiation Properties*

Traditionally, the determination of phenological phases is mainly done by ocular observations of predescribed events such as beginning of leaf unfolding. Lately, remote sensing, as well as ground-based techniques, was established for automated observations especially at remote locations (e.g., Ahrends et al. 2009; Richardson et al. 2009). Another method, utilizing the interaction of solar radiation with biomass, is shown here.

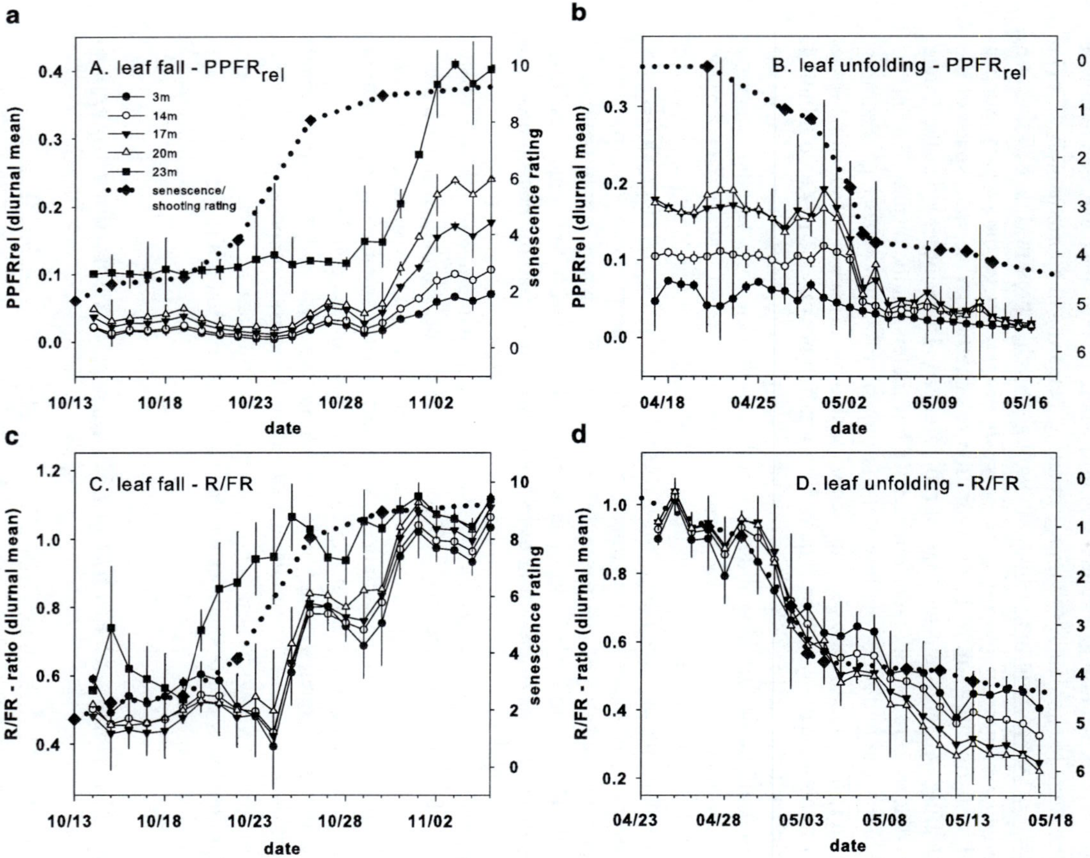

Fig. 8.6 Daily mean values (9–15 h CET) of $PPFR_{rel}$ (**a** and **b**) and R/FR (**c** and **d**) during leaf fall (**a** and **c**) and leaf unfolding (**b** and **d**) of beech in comparison with independently determined rating data of senescence (0–10; 10 = 100 % of leaves fallen or with changed color) and shooting (0–6; 6 = 100 % of leaves fully unfolded and young shoot erected) (Leuchner 2006)

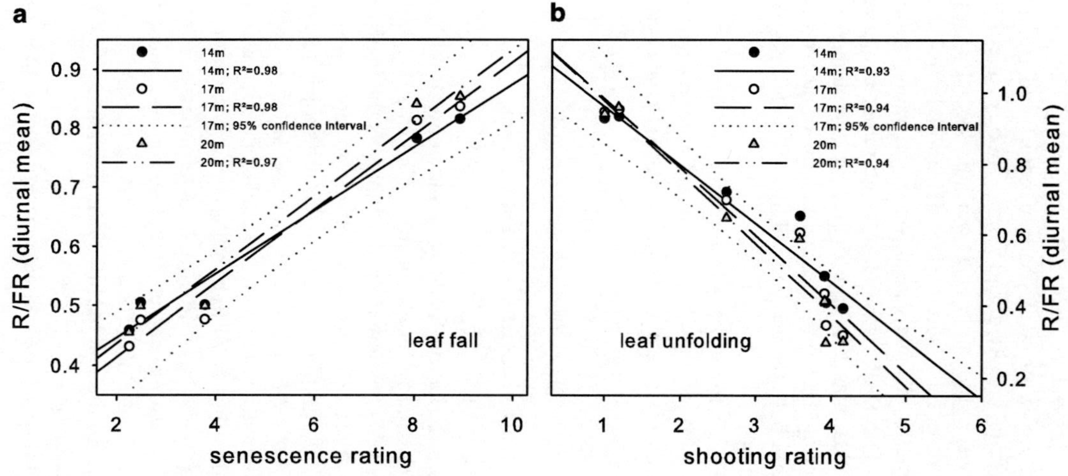

Fig. 8.7 Correlation of R/FR during leaf fall (**a**) and leaf unfolding (**b**) of beech with independently determined rating data of senescence and shooting (Leuchner 2006)

In Fig. 8.6 the relationship between the automated measurements of transmitted PAR as well as the spectral ratio R/FR and ocular observations of senescence and shooting rating, respectively, is shown during the course of leaf fall and leaf unfolding of beech. Daily mean values of transmitted PAR and R/FR are displayed for four levels within the canopy and one at the forest floor (3 m). In addition to radiation data, the respective value of the senescence rating (scale 0–10; 10 = 100% of leaves fallen or with changed color) and shooting rating (0–6; 6 = 100% of leaves fully unfolded and young shoot erected) are exhibited. A better agreement between the automated and the ocular observations can be found for the R/FR in comparison to the transmitted PAR especially during senescence. This is due to the fact that this spectral ratio can already sense the change in leaf color while PAR reacts slower and rather concurrent to the actual leaf fall. Figure 8.7 illustrates the high degree of accordance between the two methods when considering the R/FR. The coefficients of determination are slightly higher during leaf fall, but overall very high ($R^2 > 0.93$) independent of the location within the canopy.

To summarize, R/FR seems to be a better indicator for plant shade than PAR alone, due to the capability of already sensing leaf coloring before the actual leaf fall. Thus, it can be used to estimate phenological processes such as leaf fall and leaf unfolding to a certain degree.

8.3 Conclusions and Research Perspectives

The crucial role of solar radiation for growth and competition and the connected physiological processes and resource allocation (Chaps. 9, 11, 12, 17 and 18) were emphasized before. Besides the pure energetic input via the PAR waveband, certain smaller bands are of particular importance for information transfer such as red, far red, and blue. These photomorphogenetic triggers are subject to extreme spatial and temporal variability, since plant canopies operate as wavelength-selective light filters influenced by their structure creating small-scale gaps with rapidly changing light conditions. The distribution of light at the forest floor, in particular sunflecks, penumbra, and umbra, is the most important determinate for regeneration. For a better understanding of these processes in mature forests, an experiment has been conducted at the research site “Kranzberger Forst” pronouncing the importance of wavelength specific variability and the influence of boundary conditions such as sky condition and solar elevation.

One of the main findings of the unique experimental setup is a detailed description of the light climate in a mature forest stand allowing an assessment of the spatial and temporal variability of PAR and the photomorphogenetically important wavebands blue, red, and far red. A comparison between beech and spruce shows fundamental differences mainly due to the different morphological crown habit of the species. While the laminar shape of the dense upper beech canopy causes most of the incoming radiation to be absorbed in the very top layer, more radiation can penetrate into the spruce canopy due to larger gaps between trees, especially when a high fraction of direct radiation is present. In addition to the habit, parameters such as bark reflectance and leaf wax properties seem to play important roles for the spectral composition, although only few studies were performed outside the lab. Only a small number of sunflecks can be observed in the shade crown of both species. This indicates, despite their low occurrence, an important input in both terms, energy and signaling by rapidly increased photon quantity and altered spectral composition. Scattered, diffuse light is of special interest, because it can penetrate the canopy from the entire upper hemisphere more efficiently than the directionally defined unscattered fraction. It thus exhibits higher amounts of radiation throughout the whole canopy but in particular in the shade crown where almost no direct radiation is found. In terms of competition between both species the higher amount of radiation under spruce means that it can be rather beneficial for European beech to forage for radiation within or besides a Norway spruce crown than vice versa as mentioned in Sect. 13.3. Within a stand the distribution of light determines and is determined by the distribution of biomass. As shown in Sect. 14.4, this distribution is crucial for competition between dominant trees and their neighbors in particular under light limitation.

The attenuation of the energetic input of radiation in canopies has been investigated profoundly in the past decades. On the other hand, studies focusing on the spectral behavior within natural stands have been scarce and almost exclusively restricted to R/FR despite the known importance of their impact on growth and development. Especially the role of radiation quality for various ecophysiological processes and plant morphology with focus on different tree species will be matters for further investigation. The contribution of spectral properties for the development of the wood quality of young trees constitutes a very interesting field in silviculture, but only little is known. Another important topic for future research is the implementation of spectral components into radiation models for plant canopies. So far, only single models take into account different behavior of different wavebands. If spectral properties are considered, then only in a very simplistic form, not representative of the complex nature and its implications for photosynthesis and other processes.

References

Ahrends HE, Etzold S, Kutsch WL, Stoeckli R, Bruegger R, Jeanneret F, Wanner H, Buchmann N, Eugster W (2009) Tree phenology and carbon dioxide fluxes: use of digital photography for process-based interpretation at the ecosystem scale. Clim Res 39:261–274

Ammer C (2000) Untersuchungen zum Einfluss von Fichtenaltbeständen auf die Entwicklung junger Buchen. Berichte aus der Holz- und Forstwirtschaft. Shaker Verlag, Aachen

Ammer C (2003) Growth and biomass partitioning of *Fagus sylvatica* L. and *Quercus robur* L. seedlings in response to shading and small changes in the R/FR-ratio of radiation. Ann Forest Sci 60:163–171

Baldocchi DD, Hutchinson BA, Matt DR, McMillen RT (1984) Seasonal variations in the radiation regime within an oak-hickory forest. Agric Forest Meteorol 33:177–191

Ballaré CL, Scopel AL, Sánchéz RA (1990) Far-red radiation reflected from adjacent leaves – an early signal of competition in plant canopies. Science 247:329–332

Brantley ST, Young DR (2009) Contribution of sunflecks is minimal in expanding shrub thickets compared to temperate forest. Ecology 90:1021–1029

Capers RS, Chazdon RL (2004) Rapid assessment of understory light availability in a wet tropical forest. Agric Forest Meteorol 123:177–185

Chazdon RL (1988) Sunflecks and their importance to forest understorey plants. Adv Ecol Res 18:1–63

Combes D, Sinoquet H, Varlet-Grancher C (2000) Preliminary measurements and simulation of the spatial distribution of the Morphogenetically Active Radiation (MAR) within isolated tree canopy. Ann Forest Sci 57:497–511

De Castro F (2000) Light spectral composition in a tropical forest: measurements and model. Tree Physiol 20:49–56

Federer CA, Tanner CB (1966) Spectral distribution of light in the forest. Ecology 47:555–560

Gordon DC, Percy KE, Riding RT (1998) Effects of UV-B radiation on epicuticular wax production and chemical composition of four Picea species. New Phytol 138:441–449

Grant RH (1997) Partitioning of biologically active radiation in plant canopies. Int J Biometeorol 40:26–40

Grant RH, Heisler GM, Gao W, Jenks M (2003) Ultraviolet leaf reflectance of common urban trees and the prediction of reflectance from leaf surface characteristics. Agric Forest Meteorol 129:127–139

Gratani L (1997) Canopy structure, vertical radiation profile and photosynthetic function in a *Quercus ilex* evergreen forest. Photosynthetica 33:139–149

Grote R (2002) Foliage and branch biomass estimation of coniferous and deciduous tree species. Silva Fennica 36(4):779–788

Grote R, Pretzsch H (2002) A model for individual tree development based on physiological processes. Plant Biol 4:167–180

Häberle KH, Reiter IM, Nunn AJ, Gruppe A, Simon U, Gossner M, Werner H, Leuchner M, Heerdt C, Fabian P, Matyssek R (2003) KROCO, Freising, Germany: canopy research in a temperate mixed forest of Southern Germany. In: Basset Y, Horlyck V, Wright SJ (eds) Studying forest canopies from above: The International Canopy Crane Network. Smithsonian Tropical Research Institute and UNEP, Panama, pp 71–78

Hendrich C (2000) Ein kybernetisches Licht-Biomasse-Modell für Fichten-Buchen-Mischbestände. Dissertation, Munich

Hertel C, Leuchner M, Rötzer T, Menzel A (in press) Assessing stand structure of beech and spruce from measured spectral radiation properties and modeled leaf biomass parameters. Agric Forest Meteorol. doi: 10.1016/j.agrformet.2012.06.008

Hertel C, Leuchner M, Menzel A (2011) Vertical variability of spectral ratios in a mature mixed forest stand. Agric Forest Meteorol 151:1096–1105

Kitao M, Löw M, Heerdt C, Grams TEE, Häberle KH, Matyssek R (2009) Effects of chronic elevated ozone exposure on gas exchange responses of adult beech trees (*Fagus sylvatica*) as related to the within-canopy light gradient. Environ Poll 157:537–544

Knyazikhin Y, Mießen G, Panfyorov O, Gravenhorst G (1997) Small-scale study of three-dimensional distribution of photosynthetically active radiation in a forest. Agric Forest Meteorol 88:215–239

Leuchner M (2006) Messung und Modellierung der spektralen Strahlungsverteilung in einem Fichten-Buchen-Mischbestand. Dissertation, Munich

Leuchner M, Hertel C, Menzel A (2011) Spatial variability of photosynthetically active radiation in European beech and Norway spruce. Agric Forest Meteorol 151(9):1226–1232. doi:10.1016/j.agrformet.2011.04.014

Matsuda R, Ohashi-Kaneko K, Fukiwara K, Goto E, Kurata K (2007) Analysis of the relationship between blue-light photon flux density and the photosynthetic properties of spinach (*Spinacia oleracea* L.) leaves with regard to the acclimation of photosynthesis to growth irradiance. Soil Sci Plant Nutr 53:459–465

Méthy M, Fabreguettes J, Jardon F, Roy J (1987) Design of a simple instrument for the measurement of red/far-red ratios. Acta Oecologica/Oecol Plant 8:281–290

Morgan DC, Warrington IJ, Rook DA (1985) Some observations on the spectral distribution characteristics of short-wave radiation within *Pinus radiata* D. Don canopies. Plant Cell Environ 8:201–206

Navrátil M, Spunda V, Marková I, Janous D (2007) Spectral composition of photosynthetically active radiation penetrating into a Norway spruce canopy: the opposite dynamics of the blue/red spectral ration during clear and overcast days. Trees 21:311–320

Nunn A (2004) Risiko-Einschätzung der chronisch erhöhten Ozonbelastung mittels "Free-Air"-Begasung von Buchen (*Fagus sylvatica*) und Fichten (*Picea abies*) eines forstlich begründeten Mischbestandes. Dissertation, Munich

Pretzsch H, Kahn M, Grote R (1998) Die Fichten-Buchen-Mischbestände des Sonderforschungsbereiches "Wachstum oder Parasitenabwehr?" im Kranzberger Forst. Forstw Cbl 117:241–257

Richardson AD, Braswell BH, Hollinger DY, Jenkins JP, Ollinger SV (2009) Near-surface remote sensing of spatial and temporal variation in canopy phenology. Ecol Appl 19:1417–1428

Ross MS, Flanagan LB, La Roi GH (1986) Seasonal and successional changes in light quality and quantity in the understory of boreal forest ecosystems. Can J Bot 64:2792–2799

Rötzer T, Leuchner M, Nunn AJ (2010) Simulating stand climate, phenology, and photosynthesis of a forest stand with a process based growth model. Int J Biometeorol 54(4):449–464

Roujean JL (1999) Measurements of PAR transmittance within boreal forest stands during BOREAS. Agric Forest Meteorol 93:1–6

Smith H (1994) Sensing the light environment: the functions of the phytochrome family. In: Kendrick RE, Kronenberg GHM (eds) Photomorphogenesis in plants, 2nd edn. Kluwer Academic Publishers, Dordrecht Boston London, pp 377–416

Smith H (2000) Phytochromes and light signal perception by plants - an emerging synthesis. Nature 407:585–591

Smith H, Whitelam GC (1997) The shade avoidance syndrome: multiple responses mediated by multiple phytochromes. Plant Cell Environ 20:840–844

Vierling LA, Wessman CA (2000) Photosynthetically active radiation heterogeneity within a monodominant Congolese rain forest canopy. Agric Forest Meteorol 103:265–278

Wipfler P, Seifert T, Heerdt C, Werner H, Pretzsch H (2005) Growth of adult Norway Spruce (*Picea abies* [L.] Karst.) and European Beech (*Fagus sylvatica* [L.]) under free-air ozone fumigation. Plant Biol 7:611–618

Chapter 9
Site Conditions and Tree-Internal Nutrient Partitioning in Mature European Beech and Norway Spruce at the Kranzberger Forst

A. Göttlein, M. Baumgarten, and J. Dieler

9.1 Introduction

Information about site properties and nutritional status of trees are important for a reasonable interpretation of experimental results and also give hints, to which extent results may be extrapolated to other forest stands. Furthermore, another important stand parameter is the distribution of above-ground biomass and nutrients in different tree organs. Thus, one aim of this chapter is to provide additional information about the experimental site Kranzberger Forst, extending the data given by Pretzsch et al. (1998) who characterised the stand especially with a focus on growth characteristics.

Information about nutrient concentrations in different compartments of tree biomass are crucial for tree and stand characterisation, for estimations of nutrient availability, as indicator for forest health and stand productivity, and ultimately for understanding strategies of functioning of species and forest ecosystems and to determine sustainability criteria (Hagen-Thorn et al. 2004). Usually foliar nutrient concentrations from defined crown regions are determined for evaluating the nutritional status of the stand and for site characterisation. Although more detailed studies of nutrient concentrations in different tree organs and in foliage are

A. Göttlein (✉)
Forest Nutrition and Water Resources, Technische Universität München,
Hans-Carl-von-Carlowitz-Platz 2, 85354 Freising, Germany
e-mail: goettlein@forst.tu-muenchen.de

M. Baumgarten
Chair of Ecophysiology of Plants, Technische Universität München,
Hans-Carl-von-Carlowitz-Platz 2, 85354 Freising, Germany

J. Dieler
Chair of Forest Growth and Yield Science, Technische Universität München,
Hans-Carl-von-Carlowitz-Platz 2, 85354 Freising, Germany

R. Matyssek et al. (eds.), *Growth and Defence in Plants*, Ecological Studies 220,
DOI 10.1007/978-3-642-30645-7_9, © Springer-Verlag Berlin Heidelberg 2012

available (e.g. Jacobsen et al. 2003; Rademacher 2005; Wyttenbach et al. 1995), detailed data for the comparison of beech and spruce with respect to nutrient allocation on the same site are rather scarce. For a better understanding of the differences between these two species, which may also reflect different ecological strategies, a comparative study of mature beech and spruce trees growing at the Kranzberger Forst, a site with nearly optimal nutrient supply was carried out.

The distribution of the macronutrients N, P, K, Ca and Mg was analysed in different tree organs, such as wood, bark, branches, twigs and foliage at a whole tree harvest in 2004. Moreover, nutrient contents in the foliage were determined along a vertical gradient along the crown to finally evaluate nutrient allocation in dependence to crown position, and in relation to different leaf parameters.

Following the growth–differentiation-balance theory (GDB), the understanding of the tree-internal nutrient distribution in different plant organs is essential for a comprehensible explanation of characteristic trade-offs of nutrient resources between growth and defence investments within beech and spruce. The knowledge of spatial nutrient allocation in tree species provides fundamental information about overall strategies of a species. Hence, the findings are a prerequisite for the species-dependent evaluation of resource (biomass and nutrients) allocation efficiency and to elucidate strategies under stress conditions.

9.2 Material and Methods

At the experimental site "Kranzberger Forst" (Pretzsch et al. 1998) in 2004, three representative trees each of beech and spruce were chosen for total harvest and detailed nutrient analysis of the above-ground biomass (Table 9.1, Fig. 9.1). Tree ages were determined as 53 $\pm$ 2 for spruce and 53 $\pm$ 2 for beech at the time of harvest (Pretzsch et al. 2010). Sample trees were taken within the radius of the crane installed on this site, however, outside of the area of intensive measurement of the ozone fumigation experiment (Werner and Fabian 2002; Nunn et al. 2002; Karnosky et al. 2007). According to common practices (BMELF 1994), beech was harvested at the beginning of August (fully developed leaves before senescence) and spruce in November (vegetation dormancy).

For the three harvested trees of beech, three main branches each were chosen and samples from the following compartments were taken:

Leaves: representative samples at distances of 10 cm, 50 cm, 100 cm, 200–300 cm, >300 cm from the top of branches

Twigs: representative samples of diameter classes <0.5 cm, 0.5–1 cm, 1–1.5 cm, 1.5–2 cm, 2–2.5 cm; for samples <1 cm diameter wood and bark were not separated for analyses, for samples >1 cm diameter the respective separation was made

Branches: six samples at equal distances from 2.5 cm diameter down to the stem; samples were grouped by diameter classes <5.0 cm, 5–20 cm, >20 cm and analysed separately for bark and wood; for diameters >5 cm wood was additionally divided into heart wood and sap wood

Table 9.1 Characteristics of the trees harvested for nutrient analysis

Species	Tree number	DBH (cm)	Tree height (m)	Crown length (m)
Beech	101	30.7	23.3	11.6
	189	26.0	24.8	7.4
	205	28.7	25.2	16.7
Spruce	56	32.7	27.2	9.8
	87	23.5	26.8	7.7
	264	28.0	25.9	6.1

DBH diameter at breast height; crown length as determined from the first living branch to the total tree height

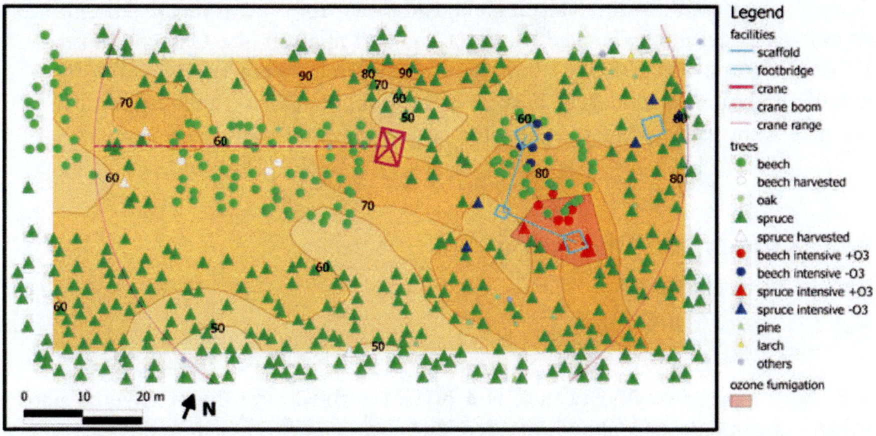

Fig. 9.1 Overview of the experimental site Kranzberger Forst with tree distribution of the year 2004 (stem positions), installations and interpolated thickness of the loess layer (cm). The ozone fumigation area indicates, where the respective facilities were installed, not the dimensions of the ozone cloud in the canopy region; intensive $+O_3$: intensively investigated trees at elevated ozone concentrations; intensive $-O_3$: intensively investigated trees at ambient ozone concentrations

Stem: discs were cut in different heights of the stem and separately analysed for bark, heart wood and sap wood; for discs with a diameter >20 cm additionally a sample of wood between heart and sap wood was taken

For the three harvested trees of spruce, one representative branch each of the whorls 1, 4, 7, 10, 15 and 20 was chosen for analysis and samples from the following compartments were taken:

Needles: needles of each age class were analysed separately; in this study the first needle year (1.NY) is presented to document the current situation of nutrient distribution, and the third needle year (3.NY) is representing the long-term development of nutrient distribution within our sample trees

Twigs: twigs <1 cm were analysed separately for each age class; wood and bark were not separated for analysis

Branches: samples were grouped by diameter classes <5.0 cm, 5–20 cm, >20 cm and analysed separately for bark and wood; for diameters >5 cm wood was additionally divided in heart wood and sap wood

Stem: discs were cut at different heights of the stem and separately analysed for bark, heart wood and sap wood; for discs with a diameter >20 cm additionally a sample of wood between heart and sap wood was taken

In the very rare case, that within a whorl no adequate branch could be chosen, a suitable branch from the whorl above or below was sampled.

All samples were dried, ground to powder and analysed for nutrient elements by acid digestion with HNO_3 and subsequent ICP-OES-spectroscopy (for metals, P and S) or by combustion in an elemental analyser (for N). For detailed description of the analytical methods used for plant and soil analysis see Gutachterausschuss Forstliche Analytik (2005). Biomasses of the trees were calculated using the detailed measurements during harvest and the model Silva (Pretzsch et al. 2002). All data were specified resp. calculated on a dry weight basis.

9.3 Results and Discussion

9.3.1 Site Conditions

The soil of the experimental site is a luvisol derived from loess, layered above tertiary sediments and showing a tendency to stagnant moisture in the deeper mineral soil (according to the German soil classification: "im Unterboden schwach haftnässe-pseudovergleyte Parabraunerde aus Löß über tertiären Molassesedimenten"). Some essential characteristics of a soil profile from the edge of the experimental site are given in Table 9.2. Soil acidity is strong to medium (according to AG Standortskartierung 1996) and classified to the Fe/Al buffer range (according to Ulrich 1981) which is characterised by the dissolution of secondary clay minerals and Fe and Al hydroxides. The cation exchange capacity (CEC) is in a medium range (according to AG Standortskartierung 1996); the proportion of alkaline and earth alkaline metals ("base saturation" BS) is low in the upper soil profile (down to the BvBt horizon) and medium to high in the deeper horizons (classified according BMELF 1996b). This depth profile of BS is typical for luvisols of this region (Kölling 2010).

The thickness of the loess layer was mapped in a 10 × 10 m grid on the whole experimental plot, interpolated by a geographic information system (Quantum GIS Version 1.6.0) and ranges from 45 to 100 cm (Fig. 9.1). Underneath the sampled trees, the loess layer amounts to about 50–70 cm. In general on the western site of the experimental plot the loess layer is shallower than on the eastern site, where the experimental towers are set up. Also in the eastern part the heterogeneity of thickness in tendency is higher. The water holding capacity for plant available water (WHC_P) of the loess layer is in the range of 22.0–27.5 vol % (Table 9.2).

Table 9.2 Characteristics of mineral soil at the experimental site Kranzberger Forst

Depth (cm)	Symbol[a]	Texture[a]	Density ($g\ cm^{-3}$)	WHC_P (vol %)[a]	pH_{H2O}	pH_{KCl}	CEC ($\mu eq\ g^{-1}$)	BS (%)
0–9	Ah	Ut2	0.55	27.5	4.10	3.17	141.20	9.4
9–28	AlBv	Ut3	0.96	25.5	4.41	3.71	59.25	7.5
28–44	AlBv2	Ut3	0.96	25.5	4.28	3.77	54.09	7.3
44–56	BvBt	Ut3	1.06	25.5	4.32	3.78	53.28	6.7
56–80	Bt	Ut4	1.17	22.0	4.61	3.69	98.53	24.1
80–102	SgBt	Tu4	1.09	17.0	4.99	3.67	141.08	57.4
102–120	BtCv	Lu	1.23	19.5	4.76	3.76	122.77	63.7
>120	II Cv	Ls3	1.31	19.5	5.12	3.65	108.52	71.0

WHC_P water holding capacity for plant available water; pH_{H2O} soil pH determined with distilled water; pH_{KCl} soil pH determined with 1 M KCl-solution; *CEC* cation exchange capacity determined by NH_4Cl-extraction; *BS* percentage of alkaline and earth alkaline metals at CEC; the main texture classes are *S* sand, *U* silt, *L* loam, *T* clay

[a]According to German classification (AG Standortskartierung 1996)

Taking the value of texture class Ut3 (silt with a medium clay content) as an average, within the loess layer 25.5 mm (equals litre per m) of plant available water can be stored per 0.1 m soil depth. Because the main rooting zone in Kranzberger Forst is within the loess layer, the minimal storage capacity of plant available water can be estimated from the thickness of the loess layer, thus ranging from 115 to 255 mm. This high storage capacity in combination with sufficient precipitation (annual precipitation 750–850 mm; mean annual temperature 7.0–8.0 °C; data from the nearby forest observation plot (Waldklimastation) Freising operated by the Bavarian Forest Service ("Bayerische Landesanstalt für Wald und Forstwirtschaft") warrants sufficient water supply of the forest stand throughout the whole year. The only exception may be very extreme and long-lasting dry periods in spring or summer months, like in the year 2003. Although extreme drought events are unlikely to occur in this region, it is predicted that they will occur more frequently under future climate conditions (e.g. Ciais et al. 2005; AK KLIWA 2006; Löw et al. 2006). Conversely, after events of high precipitation or at the end of the winter period the low proportion of coarse-sized soil pores in the loess layer can cause some oxygen deficiency in deeper mineral soil, leading to the observed formation of slightly stagnic properties in the SgBt horizon.

Because it is well known, that top soil properties are different under beech and spruce (Rehfuess 1990) soil samples were taken under pure spruce ($n = 30$) or pure beech ($n = 23$), respectively (Table 9.3, Schuhbäck 2004). The humus under spruce is characterised as moder and significantly thicker than under beech, where mulltype moder is found. Due to the higher content of base cations in beech leaves (Table 9.4), base saturation in the humus layer is significantly higher under beech than under spruce (Table 9.3). The positive effect of beech on pH and base saturation is significant for the whole top soil down to 30 cm depth, however, with differences getting smaller with increasing soil depth. Under spruce highest fine root density is found at the transition zone from humus to mineral soil (Rastin and Ulrich 1990), leading to a high turnover of root biomass coupled with an input of organic matter in the upper mineral soil. In combination with the higher activity

Table 9.3 Influence of tree species on top soil properties

	Thickness (cm)		pH_{H2O}		CEC (μeq g^{-1})		BS (%)		C (%)		N (%)	
	S	B	S	B	S	B	S	B	S	B	S	B
L/Of	**5.16**	2.89	4.12	**4.93**	247	279	62.0	**79.5**	**35.0**	28.7	**1.36**	1.17
Oh	**1.87**	1.07	3.77	**4.36**	**189**	131	26.7	**40.2**	**18.0**	9.7	**0.82**	0.54
M1	10	10	3.88	**4.22**	**120**	89	5.6	**9.9**	**3.7**	2.7	0.22	0.19
M2	10	10	3.91	**4.09**	**62**	56	5.1	**7.8**	**1.4**	1.0	0.05	0.04
M3	10	10	4.07	**4.22**	58	52	6.6	**9.1**	0.9	0.6	0.04	0.03

L/Of, Oh horizons of the humus layer; *M1–M3* mineral soil in depth class 0–10 cm, 10–20 cm, 20–30 cm; pH_{H2O} soil pH determined with distilled water; *CEC* cation exchange capacity determined by NH_4Cl-extraction; *BS* percentage of alkaline and earth alkaline metals at CEC; significant differences ($p < 0.05$) between spruce (S) and beech (B) within a single horizon are indicated by bold printing of the higher value

Table 9.4 Nutrient concentrations (per dry weight) and nutritional status of beech (leaves of the sun crown) and spruce (current year needles whorl 7) harvested in the year 2004; optimal range according to BMELF (1996a)

	N (%)	P (mg g^{-1})	K (mg g^{-1})	Ca (mg g^{-1})	Mg (mg g^{-1})
Beech	2.31 ± 0.03	1.13 ± 0.02	5.95 ± 0.18	5.73 ± 0.19	1.91 ± 0.04
Optimal range	***1.8–2.5***	***1.0–1.7***	***5.0–7.5***	***4.0–8.5***	***0.7–1.4***
Spruce	1.48 ± 0.08	1.28 ± 0.05	4.98 ± 0.04	3.96 ± 0.25	0.97 ± 0.10
Optimal range	***1.3–1.7***	***1.2–2.0***	***3.5–7.0***	***1.0–5.0***	***0.75–1.5***

of earth worms under beech (Schlenker and Denno 1971), incorporating mineral particles into the humus layer, the higher C and N content as well as the higher CEC in the top soil under spruce may be explained.

9.3.2 *Nutritional Status of Beech and Spruce*

Although the upper mineral soil is acidic and has a low base saturation the nutrition of beech and spruce with K, Ca, Mg is in an optimal range (Table 9.4), because tree roots can easily reach the horizons with higher base saturation. Also for N and P the harvested trees showed optimal nutritional values, however, for P close to the lower threshold. The combination of low pH in the topsoil and high base saturation in the deeper soil has a positive impact on tree nutrition, because for all nutrients within the rooting zone regions for their favourable uptake are accessible. So also for the micronutrients (Fe, Mn, Cu, Zn, B) the nutritional status of beech and spruce was optimal according to BMELF (1996a) (data not shown). Also the intensively observed trees of the ozone fumigation experiment (treatment and control) were, with some single exceptions over the years, in the range of optimal nutrition (Göttlein et al. 2009). So when interpreting the treatments of the Kranzberg experiment in normal years, it is very unlikely that nutritional effects (deficiency or surplus) are of importance. However, extreme drought events may also have an impact on the nutritional status of the trees, even in the subsequent year (Göttlein et al. 2009), and thus have to be considered separately.

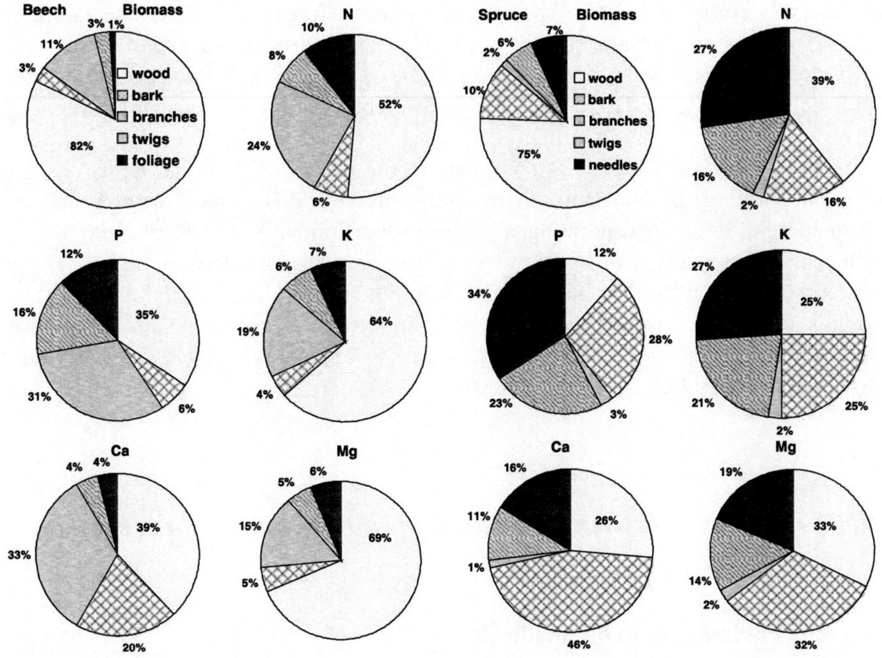

Fig. 9.2 Biomass and nutrient partitioning in different tree organs in beech and spruce (data from authors)

9.3.3 Nutrient Distribution in Beech and Spruce

For the sampled beech and spruce trees about three-quarter of the biomass is in stem wood (Fig. 9.2). The rest of the biomass is distributed between branches, twigs, bark and foliage, with differences emerging between species. Beech shows high proportion of biomass in the branches, whereas in spruce the proportion is high in needles and bark. Although the proportion of stem wood is similar for beech and spruce there are large differences in the proportion of nutrients stored in this compartment. Beech puts much more nutrients into the wood than spruce, with the biggest differences found for K and Mg. Adding the proportions of stem wood and stem bark, which corresponds to the conventional harvesting, the differences between beech and spruce for biomass, N and P vanish. For K and Mg beech still puts 18–19 % more into these compartments, whereas for Ca spruce has a 13 % higher proportion incorporated to wood and bark. Due to its evergreen character spruce incorporates a higher proportion of nutrients to the green biomass, whereas beech holds a considerable amount of nutrients in branches and twigs. It is known since long time (Wolff 1880; Fiedler et al. 1973) that usually nutrient concentrations in plant tissues of beech are higher than in spruce. However, studies on nutrient partitioning within mature trees of beech and spruce are scarce when the focus is to compare the two species on the same forest site. Weis and Göttlein (2002) compared nutrient storage in beech and spruce at the experimental site "Höglwald", however, in the time of vegetation dormancy so that the green biomass of beech was not included in their study. For beech a study on nutrient partitioning was done in Brandenburg/Germany (Krauß and Heinsdorf 2008), resulting in a very similar biomass distribution as compared to the Kranzberger Forst. However, for the partitioning of nutrients there were marked differences with beech at Kranzberger Forst incorporating a lower proportion of P and K to stem wood, whereas the proportion of Ca in this compartment was considerably higher. These differences are on the one hand due to different site conditions, on the other hand genetic differences (provenance) may be of importance. Thus the results obtained at Kranzberger Forst may only be extrapolated to comparable site conditions, cautioning generalisation for beech or spruce.

9.3.4 Foliage Nutrient Distribution in Beech and Spruce Along a Vertical Crown Profile

9.3.4.1 Defining Sun and Shade Crown

Prior to the evaluation of nutrient distribution in foliage it is essential to characterise crown architecture, leaf biomass allocation and light conditions within the crown profile. This will allow a comprehensible definition of sun and shade crown of beech and spruce.

Due to high stand density at the Kranzberger Forst beech leaves were mainly abounded within the first 3 m from the top of the crown. In lower parts of the foliated part of the crown (>3 m in distance from crown top) only few leaves were found. So, for beech it has to be distinguished between foliated crown and total crown, the latter being defined as the region from crown base (base of the first living branch) to the top of the tree. In spruce living needles of up to 8 years age could be found usually down to the 20th whorl (depending on tree) with a main needle biomass accumulation within 27–17 m tree height (see also Chap. 8, Reiter 2004; Reiter et al. 2005).

For beech, according to the investigations of Reiter et al. (2005), in the Kranzberger Forst light is increasing exponentially from the non-foliated up to the foliated crown parts, with a more pronounced and almost linear increase from the foliated crown base along the foliated crown area upwards to the fully light exposed crown top. The crown area just below the top has the highest leaf density, the highest carbon gain and, thus, is occupying the main canopy space (more than 35 %) within a small region of the maximum tree height (75–90 %). In contrast, for spruce an almost linear increasing light exposure upwards along the foliated crown was described, as well as high leaf density and high carbon gain nearly along the whole area below the upper sun crown from about 20 to 80 % maximum tree height

(Reiter et al. 2005; Reiter 2004). Rötzer et al. (2010) determined leaf biomass distribution with measured and modelled Leaf Area Index (LAI) and Specific Leaf Area (SLA) for beech and spruce trees in the Kranzberger Forst and demonstrated a similar crown compartmentation. The vertical profile of the photomorphologicaly important wave length ratio Bw/Rw (Bw, broadband blue 400–500 nm; Rw, broadband red 600–700 nm) supported these differences in biomass distribution and crown architecture of beech and spruce (Chap. 8). A distinct rise of Bw/Rw was found in the upper crown area (23 m height in beech and 20–17 m height in spruce) and an abrupt decline in beech below 20 m due to high shading below the main leaf biomass, while Bw/Rw is decreasing slowly downwards according to the LAI for spruce (see Fig. 8.3 in chapter 8).

The inverse of SLA, i.e. leaf mass per area (LMA), is reflecting growth strategies, as leaf development under full sunlight exposure and high photosynthetic efficiency is generally associated with high leaf masses per area and high area-based leaf nitrogen contents as compared to shade leaves (Oren et al. 1986; Reich et al. 1991; Ellsworth and Reich 1992; Ogaya and Penuelas 2007). This could be also demonstrated for nitrogen, phosphorus and potassium contents per leaf in our study (data not shown). Thus, LMA can be used for stand level analyses, e.g. to determine leaf differentiation in sun and shade leaves, and hence, to estimate vertical light gradients within a forest canopy, with its heterogeneous tree/canopy architecture depending on light availability. Kitao et al. (2009) showed a positive correlation of LMA with mean integrated photosynthetically active photon flux density in beech trees of the Kranzberg experiment. A similar relationship was demonstrated by Ogaya and Penuelas (2007) for mature *Quercus ilex* under a wide range of climatic conditions. Also, for coniferous tree species a causal relationship between light gradient and LMA was described (Oren et al. 1986; Bond et al. 1999; Temesgen and Weiskittel 2006). Sun leaves performed a lower SLA than shaded leaves as described for Kranzberg Forest previously by Nunn et al. (2002).

A compilation of data about the photosynthetic capacity of beech leaves from previous investigations at Kranzberger Forst (Kitao et al. 2009; Nunn 2005; Koch 2005; Löw 2007) is shown in Fig. 9.3 with the light-saturated rate of electron transport ($J_{\max}$) as expression for photosynthetic efficiency. LMA is reflecting the varying light conditions within the crown profile.

In our study, LMA for beech was highest in leaves from the crown top and linearly decreasing downwards to about 1 m distance from the crown top (Fig. 9.4). LMA is further decreasing down to the lowest part of the foliated crown below 2.5 m distance from the crown top, however, with a less steep slope. In spruce LMA in needles (first and third year) decreased more or less linearly from the top crown downwards to about 18 m tree height, below LMA was rather constant (Fig. 9.5).

In summary, we define at Kranzberger Forst

- For beech:
  The crown part <1 m from the crown top (24.4–23.4 m) as the light exposed sun crown, 1–2.5 m (23.3–21.9 m) as the upper shade crown, and below 2.5 m (21.9 m) as the lower shade crown

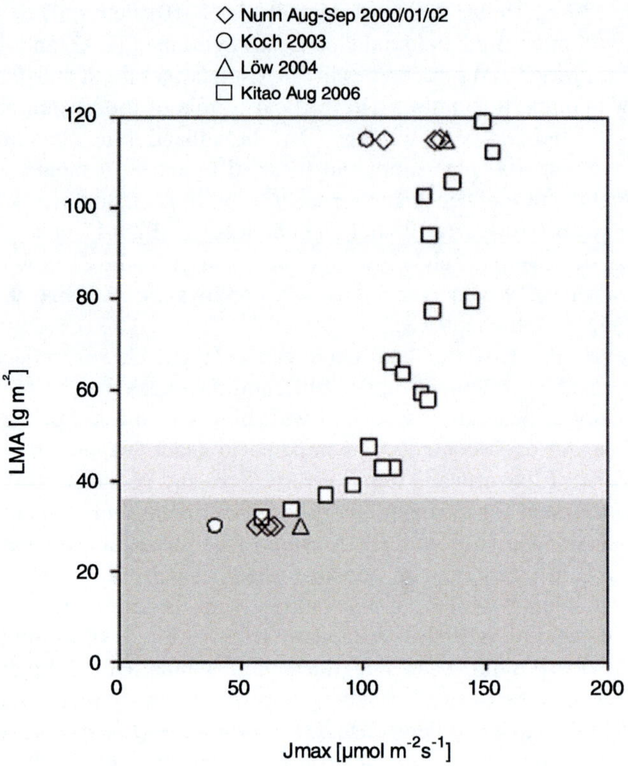

Fig. 9.3 Light-saturated rate of electron transport (J_{max}) as an indicator for photosynthetic capacity in beech leaves with different leaf mass area (LMA) from trees within the Kranzberger Forst experiment, measured in different sample years, background shading: *white*—sun crown, *light grey*—upper shade crown, *grey*—lower shade crown (data from: Löw 2007; Koch 2005; Nunn 2005; Kitao et al. 2009)

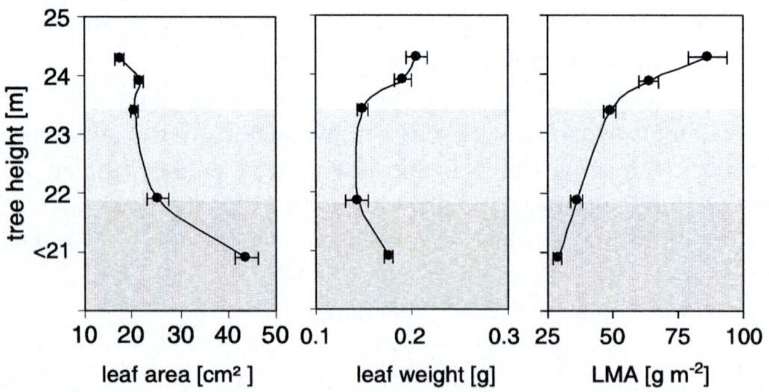

Fig. 9.4 Leaf characteristics of beech in different compartments along the vertical crown gradient (*LMA* leaf mass area); values given as mean ± standard error, background shading: *white*—sun crown, *light grey*—upper shade crown, *grey*—lower shade crown (data from authors)

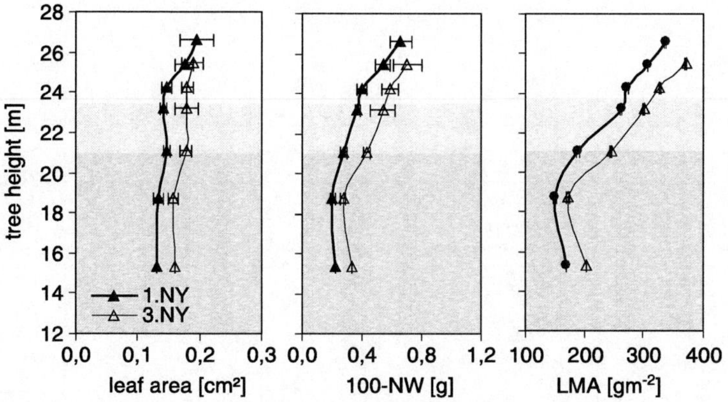

Fig. 9.5 Leaf characteristics of spruce needles in different compartments along the vertical crown gradient (*LMA* leaf mass area); values given as mean ± standard error, background shading: *white*—sun crown, *light grey*—upper shade crown, *grey*—lower shade crown (data from authors)

- For spruce:
  The first nine whorls (26.6–23.5 m) as the sun crown, the 10th–15th whorl (23–21 m) as the upper shade and 16th–24th whorl (20–15.3 m) as the lower shade crown

9.3.4.2 Vertical Gradients in Beech

Nutrient distribution of beech foliage was analysed vertically along the crown profile (Fig. 9.6). Concentrations of N, P and K followed more or less the same tendency with slightly lower concentrations in the top crown as compared to the subsequently slightly increasing concentrations in the upper shade crown. N, P and K concentrations were about 16% lower in the sun as compared to the upper shade crown, Ca and Mg only about 7%. High leaf area and proportionally low leaf weight of leaves (Fig. 9.4) in the lower shade crown led to higher nutrient concentrations of P and K compared to the leaves in the upper crown. Because Ca is strongly related to cell walls and membranes (Marschner 1995), the decreasing concentrations of Ca may be due to a lower need of structural functions in these leaves showing lowest LMA. Furthermore plant-internal transport of Ca is strongly correlated to the water transport in the xylem and thus to transpiration (Lyr et al. 1992). So, consequently, concentrations of Ca in the lower shade crown are comparably low. Although mobility of Mg in the phloem and thus the chance for redistribution is higher than for Ca (Steucek and Koontz 1970), the crown profile of Mg concentrations was very similar to the one of Ca. Because Mg is also highly correlated to photosynthesis its low concentrations in the lower shade crown correspond well with the low values of J_{max} (Fig. 9.3) in this crown region.

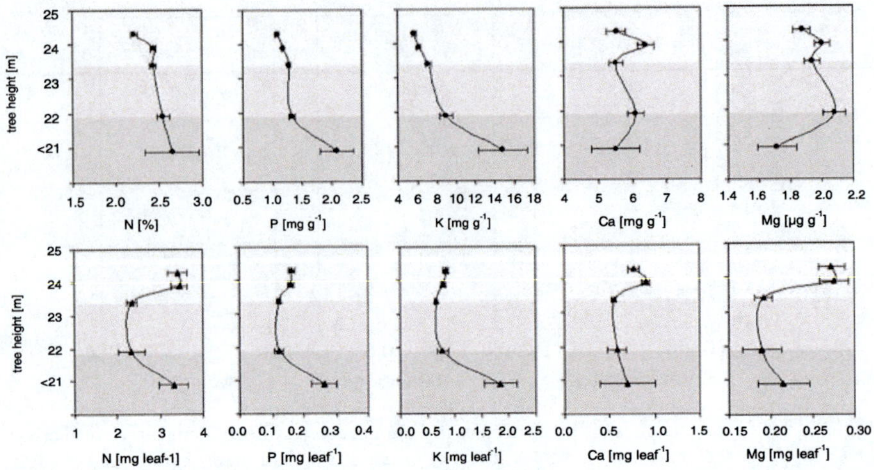

Fig. 9.6 Nutrient concentrations (per dry weight) in beech leaves (above) and nutrient contents per leaf (below) along the vertical crown gradient; values given as mean ± standard error, background shading: *white*—sun crown, *light grey*—upper shade crown, *grey*—lower shade crown (data from authors)

In Fig. 9.6, nutrient concentrations were also converted to nutrient contents per leaf. Lowest nutrient contents per leaf were found in the region of the upper shade crown, with increasing values in the lower shade crown for N, P and K contents. In the sun crown N, Ca and Mg contents were markedly higher, compared to the shade crown. Enhanced electron transport rates J_{max} in sun leaves (Fig. 9.3, leaves with high LMA) are corresponding well with the increased nutrient contents per accordingly differentiated leaf, necessary for high photosynthetic performance and transport capacities in this crown domain (see Fig. 9.3). The high nutrient contents per leaf in the lower shade crown correspond to the large size of these leaves (Fig. 9.4)

9.3.4.3 Vertical Gradients in Spruce

Nutrient distribution in spruce needles (first and third year needles) was analysed vertically along a crown profile from the youngest (first) to the 24th whorl, i.e. downwards from 26.6 to 15.3 m mean tree height (Fig. 9.7).

LMA of the older needles was higher due to further incorporation of structural components (Fig. 9.5), leading to lower concentrations of N, P, K and Mg in older needles. This effect, however, may also be due to year-to-year variations of needle weight, size and nutrient incorporation. The nearly immobile cation Ca is accumulating with needle age (Lyr et al. 1992). The gradient along the profile was quite similar for N, P and K. Within the sun crown region high concentrations were found not in the top crown area but in the needles from adjacent whorls. This crown domain is characterised by the highest photosynthetic primary production and

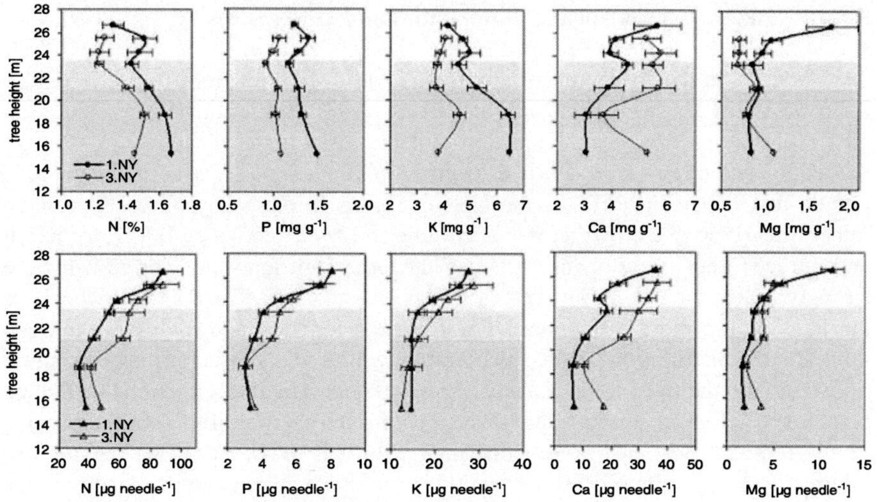

Fig. 9.7 Nutrient concentrations in spruce needles (above) and nutrient contents per needle (below) of the first and third needle years along the vertical crown gradient; values given as mean ± standard error, background shading: *white*—sun crown, *light grey*—upper shade crown, *grey*—lower shade crown (data from authors)

canopy occupation (Reiter 2004, Chap. 10) and the highest LAI (Chap. 8). An abrupt decline of nutrients was found beneath this "high productivity" crown domain, followed by a continuous increase downwards in the lower shade crown (Fig. 9.7). For Ca and Mg, the concentration gradient within the crown was oriented in the opposite way, with a very steep decrease from the top of the sun crown to the beginning of the shade crown. Subsequently, there was only a moderate decrease down to the lower end of the crown. Here again, the crown parts with highest radiation input and thus highest transpiration incorporate higher amounts of Ca and Mg.

Looking at the contents per needle, similar performance for all nutrients with high contents in the sun crown and declining contents downwards along the vertical gradient could be observed. This is in accordance with a decreasing leaf mass and thus decreasing LMA from the crown top to the shaded crown region (Fig. 9.5). Similar to beech, in spruce in the deepest shade crown there is a tendency for increasing nutrient contents per assimilation organ. In the literature, the decreasing nutrient concentrations (N, P, K, Mg) in older needles often is explained by retranslocation to the current-year needles (Fiedler et al. 1973). As shown by the calculation of the nutrient content per needle in our study this cannot be the reason, because for these elements the content of the older needles is not lower than the content of the current year needles. Thus the observed differences in concentrations can only be attributed to the increasing weight of the older needles. Also BMELF (1996a) stated that for Mg a notable retranslocation to the current-year needles can only be observed for trees with suboptimal Mg supply. Only for Ca the expected enrichment with increasing needle age (BMELF 1996a) can be stated.

9.3.4.4 Beech Versus Spruce: Similarities and Differences

As described in the literature for coniferous and deciduous trees (e.g. Oren et al. 1986; Reich et al. 1991; Ellsworth and Reich 1992; Stith et al. 1993; Bond et al. 1999) LMA, light intensity, nutrition and photosynthetic capacity are strongly related to each other across a broad spectrum of plant species, with photosynthetic photon flux density (PPFD) as the driving factor. In our study N was correlated rather linearly with a crown profile according to LMA, and hence PPFD, for beech and spruce (data not shown), whereas the other nutrients are correlated more strongly to leaf area or leaf weight (see Figs. 9.4 and 9.5). For P and K, which are nutrients with a more pronounced physiological function and high plant-internal mobility, the distribution of the concentration values was similar to the distribution of leaf area in the tree crown. Ca and Mg have more structural functions within the plant (especially Ca) and show a restricted plant-internal mobility. Thus their crown profile (content per leaf) rather correlated with leaf mass. N can be regarded as intermediate, with similarities to P and K in the shade crown and similarities to Ca and Mg in the sun crown.

Apparently, beech and spruce follow different strategies concerning biomass distribution (Fig. 9.8). Beech at the Kranzberger Forst has to realise its photosynthetic production within a vertical range of not much more than 3 m of the foliated crown (see Fig. 9.3), whereas spruce can distribute its photosynthetic activity over a vertical range of about 12 m of foliated crown (see Sect. 9.3.4.1). These findings are consistent with investigations by Reiter (2004) at the same site (see Sect. 9.3.4.1).

For both tree species the elements N, P and K show increasing concentrations under decreasing light availability to the assimilation organs, Ca and Mg in both species show similar response in terms of increasing concentrations and contents in the sun crown (Figs. 9.6 and 9.7). Remarkably for both species, beech and spruce, the partitioning of nutrients is very closely related to the partitioning of the green biomass (Table 9.5). There are only two exceptions: For beech the proportion of K is in tendency lower in the sun crown, probably due to the higher leaching of this element out of leaves with higher photosynthetic stress (Lyr et al. 1992). For spruce the exception is Mg, which is enriched in the lower shade crown. This may be due to the need of keeping enough chlorophyll also in the older needles of the lower shade crown, because in this crown domain there are up to eight needle years on one branch.

Looking at the amount of nutrients per unit area of assimilation organs (Table 9.6), spruce allocates two to six times more nutrients. The ratio of nutrients per area foliage in the sun crown is two to four times higher in spruce than in beech foliage, in the upper shade crown even four to six times, and in the lower shade crown again two to four times higher. It is assumed that the high values for Ca are due to a subsequent accumulation in the evergreen needles with rising needle age. In summary, the leaf area-based evaluation of nutrient distribution in

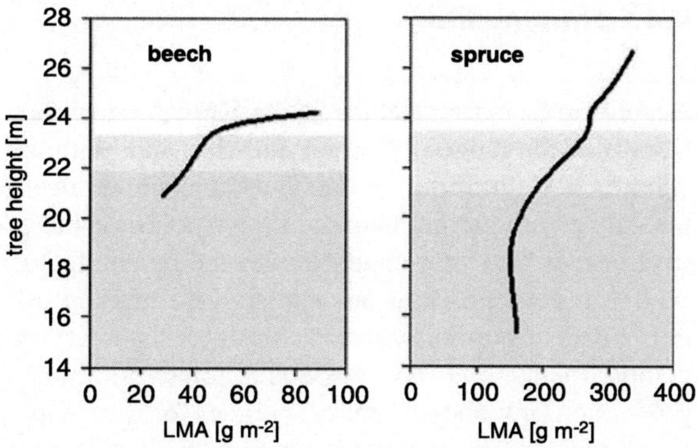

Fig. 9.8 Comparison of leaf mass area (LMA) of beech and spruce foliage in different compartments of the vertical crown area, background shading: *white*—sun crown, *light grey*—upper shade crown, *grey*—lower shade crown (data from authors)

Table 9.5 Distribution of green biomass (leaves/needles) and incorporated nutrients to the different crown compartments (values in %)

	Biomass	N	P	K	Ca	Mg
Beech						
Sun crown	35	34	32	29	36	35
Upper shade crown	55	56	56	57	54	56
Lower shade crown	10	10	12	14	10	9
Spruce						
Sun crown	27	26	27	26	29	25
Upper shade crown	48	48	47	47	50	43
Lower shade crown	25	26	26	27	21	32

Table 9.6 Nutrient content per cm^2 assimilation organ in different crown compartments of beech and spruce; values are given as mean ± standard error

	N	P	K	Ca	Mg
Beech ($\mu g\ cm^{-2}$ *leaf*)					
Sun crown	178.0 ± 5.8	8.5 ± 0.3	42.6 ± 1.4	42.4 ± 1.5	14.5 ± 0.6
Upper shade crown	111.9 ± 4.5	5.8 ± 0.2	32.2 ± 0.9	25.3 ± 1.1	9.0 ± 0.4
Lower shade crown	75.2 ± 4.6	5.9 ± 0.6	41.6 ± 4.9	15.7 ± 1.9	4.9 ± 0.5
Spruce ($\mu g\ cm^{-2}$ *needle*)					
Sun crown	400.9 ± 7.7	32.6 ± 1.0	129.6 ± 2.8	179.3 ± 6.3	26.8 ± 1.7
Upper shade crown	354.8 ± 12.4	26.8 ± 0.8	100.0 ± 3.4	152.2 ± 8.8	22.0 ± 0.8
Lower shade crown	275.7 ± 11.3	21.7 ± 0.8	98.6 ± 5.0	72.5 ± 4.9	15.3 ± 0.6
Ratio spruce/beech					
Sun crown	2.3	3.8	3.0	4.2	1.9
Upper shade crown	3.2	4.6	3.1	6.0	2.5
Lower shade crown	3.7	3.7	2.4	4.6	3.1

beech and spruce foliage exhibits a distinct difference between the two species in nutrient (resource) utilisation. This again underlines the high efficiency of space utilisation of the foliated crown of beech for assimilation as compared to spruce at our study site.

9.4 Conclusions

According to current tables at the Kranzberger Forst the nutritional status of the trees is in the range of normal nutrition and, with the exception of extreme years, water is not a limiting factor. Thus, this site is ideal for factorial experiments like the ozone fumigation, because the response of the ecosystem to the experimental treatment is only marginally influenced by site factors.

Beech and spruce at our site showed marked differences in the plant-internal nutrient distribution, with beech allocating more nutrients to the non-productive compartments, like stem wood and branches. So the effectivity of nutrient use for production of woody biomass is higher for spruce as compared to beech. However, beech has to gain its assimilates in a smaller vertical region of the canopy with a markedly lower amount of green biomass. This, together with the lower amount of nutrients per unit leaf area demonstrates the high effectivity of nutrient use with respect to photosynthesis of beech as compared to spruce.

Probably this growth behaviour contributes to the fact that beech is the naturally dominant tree species in Central Europe with a high physiological tolerance and competiveness (Ellenberg 1996). The fast built-up and low construction and maintenance costs for the deciduous leaf biomass increase competitiveness with respect to carbon gain. On the other hand, under stress conditions this competitive advantage might turn to a disadvantage, because increased stress defence should much more influence growth. This might explain the reduced growth of beech under chronic O_3-stress (Pretzsch et al. 2010).

Unclear is the benefit or disadvantage of the relatively high proportion of nutrients incorporated in beech wood (stem and branches). The investment for acquiring relatively high nutrient contents in the woody compartments, however, could also be a competitive strategy to withhold nutrients from potential competitors, to keep nutrients in the habitat and to slowly reallocate them for the next beech generation during decomposition. Such a strategy of beech may be of advantage in natural forests, even on nutrient poor sites. In managed forests, however, high nutrient contents in the woody parts lead to higher nutrient export during harvest. Thus, forest management has to consider such differences in nutrient contents between beech and spruce in order to warrant nutrient sustainability for future tree generations.

Acknowledgements Data for drawing Fig. 9.1 were provided by Dr. J. Zander, Dr. K.-H. Häberle and Dr. R. Weigt.

References

AG Standortskartierung (1996) Forstliche Standortsaufnahme, 5th edn. IHW, Eching

AK KLIWA (2006) Regionale Klimaszenarien für Süddeutschland. Heft 9: Herausgeber: Landesanstalt für Umwelt, Messungen, und Naturschutz Baden-Württemberg, Bayerisches Landesamt für Umwelt, Deutscher Wetterdienst

BMELF (1994) Bundesweite Bodenzustandserhebung im Wald (BZE) – Arbeitsanleitung, 2nd edn. Bundesministerium f. Ernährung, Landwirtschaft u. Forsten, Bonn

BMELF (1996a) Deutscher Waldbodenbericht 1996, vol 1. Bundesministerium f. Ernährung, Landwirtschaft u. Forsten, Bonn

BMELF (1996b) Deutscher Waldbodenbericht 1996, vol 2. Bundesministerium f. Ernährung, Landwirtschaft u. Forsten, Bonn

Bond JB, Farnsworth BT, Coulombe RA, Winner WE (1999) Foliage physiology and biochemistry in response to light gradients in confers with varying shading tolerance. Oecologia 120:183–192

Ciais P, Reichstein M, Viovy N, Granier A, Ogée J, Allard V, Aubinet M, Buchmann N, Bernhofer C, Carrara A, Chevallier F, De Noblet N, Friend AD, Friedlingstein P, Grünwald T, Heinesch B, Keronen P, Knohl A, Krinner G, Loustau D, Manca G, Matteucci G, Miglietta F, Ourcival JM, Papale D, Pilegaard K, Rambal S, Seufert G, Soussana JF, Sanz MJ, Schulze ED, Vesala T, Valentini R (2005) Europe-wide reduction in primary productivity caused by the heat and drought in 2003. Nature 437:529–533

Ellenberg H (1996) Vegetation Mitteleuropas mit den Alpen, 4th edn. Verlag Eugen Ulmer, Stuttgart

Ellsworth DS, Reich PB (1992) Leaf mass per area, nitrogen content and photosynthetic carbon gain in Acer saccharum seedlings in contrasting forest light environments. Funct Ecol 6 (4):423–435

Fiedler HJ, Nebe W, Hoffmann F (1973) Forstliche Pflanzenernährung und Düngung. Gustav Fischer, Stuttgart

Göttlein A, Rodenkirchen H, Häberle KH, Matyssek R (2009) Nutritional effects triggered by the extreme summer 2003 in the free air ozone fumigation experiment at the Kranzberger Forst. Eur J Forest Res 128:129–134

Gutachterausschuss Forstliche Analytik (2005) Handbuch Forstliche Analytik. Bundesministerium für Verbraucherschutz, Ernährung und Landwirtschaft, Bonn

Hagen-Thorn A, Armolaitis K, Callesens I, Stjernquist I (2004) Macronutrients in tree stems and foliage: a comparative study of six temperate forest species planted at the same sites. Ann Forest Sci 61:489–498

Jacobsen C, Rademacher P, Meesenburg H, Meiwes K-J (2003) Gehalte und chemische Elemente in Baumkompartimenten – Literaturstudie und Datensammlung. Berichte des Forschungszentrums Waldökosysteme der Universität Göttingen, Reihe B, Band 69

Karnosky DF, Werner H, Holopainen T, Percy K, Oksanen T, Oksanen E, Heerdt C, Fabian P, Nagy J, Heilman W, Cox R, Nelson N, Matyssek R (2007) Free-air exposure systems to scale up ozone research to mature trees. Plant Biol 9(2):181–190

Kitao M, Löw M, Heerdt C, Grams TEE, Häberle KH, Matyssek R (2009) Effects of chronic elevated ozone exposure on gas exchange responses of adult beech trees (Fagus sylvatica) as related to the within-canopy light gradient. Environ Pollut 157:537–544

Koch N (2005) Skalierung innerhalb ontogenetischer Stadien und Wachstumsszenarien von Buche (*Fagus sylvatica*) und Fichte (*Picea abies*) unter Ozoneinfluss. Dissertation Technische Universität München, Fakultät Wissenschaftszentrum Weihenstephan, Lehrstuhl für Ökophysiologie der Pflanzen

Kölling C (2010) Macht sauer wirklich lustig? LWFaktuell 78:21–24

Krauß HH, Heinsdorf D (2008) Herleitung von Trockenmassen und Nährstoffspeicherungen in Buchenbeständen, vol 38, Eberswalder Forstliche Schriftenreihe. Landesforstanstalt Eberswalde, Eberswalde

Löw M (2007) Sensitivity of photosynthesis and stomatal conductance in mature beech (*Fagus sylvatica* L.) to ambient or experimentally enhanced ozone regimes. Dissertation Technische Universität München, Fakultät Wissenschaftszentrum Weihenstephan, Lehrstuhl für Ökophysiologie der Pflanzen

Löw M, Herbinger K, Nunn AJ, Häberle K-H, Leuchner M, Heerdt C, Werner H, Wipfler P, Pretzsch H, Tausz M, Matyssek R (2006) Extraordinary drought of 2003 overrules ozone impact on adult beech trees (*Fagus sylvatica*). Trees 20(5):539–548

Lyr H, Fiedler HJ, Tranquillini W (1992) Physiologie und Ökologie der Gehölze. Gustav Fischer, Jena
Marschner H (1995) Mineral nutrition of higher plants, 2nd edn. Academic, London
Nunn AJ (2005) Risikoeinschätzung der chronisch erhöhten ozonbelastung mittesl "Free-Air"-Begasung von Buchen (Fagus sylvatica) und Fichten (Picea abies) eines forstlich begründeten Mischbestands. Dissertation Technische Universität München, Fakultät Wissenschaftszentrum Weihenstephan, Lehrstuhl für Ökophysiologie der Pflanzen
Nunn AJ, Reiter IM, Häberle KH, Werber H, Langebartels C, Sandermann H, Heerdt C, Fabian P, Matyssek R (2002) Free-air ozone canopy fumigation in an old mixed forest: concept and observations in beech. Phyton 42:105–119
Ogaya R, Penuelas J (2007) Leaf mass per area ratio in Quercus ilex leaves under a wide range of climatic conditions. The importance of low temperatures. Acta Oecol 31:168–173
Oren R, Schulze ED, Matyssek R, Zimmermann R (1986) Estimating photosynthetic rates and annual carbon gain in conifers from specific leaf weight and leaf biomass. Oecologia (Berlin) 70:187–193
Pretzsch H, Kahn M, Grote R (1998) Die Fichten-Buchen-Mischbestände des Sonderforschungsbereiches "Wachstum und Parasitrenabwehr?" im Kranzberger Forst. Forstw Cbl 117:241–257
Pretzsch H, Biber P, Dursky J (2002) The single tree-based stand simulator SILVA: construction, application and evaluation. Forest Ecol Manage 162:3–21
Pretzsch H, Dieler J, Matyssek R, Wipfler P (2010) Tree and stand growth of mature Norway spruce and European beech under long-term ozone fumigation. Environ Pollut 158:1061–1070
Rademacher P (2005) Nährelementgehalte in den Kompartimenten wichtiger Wirtschaftsbaumarten und deren Bedeutung für die Reststoffverwertung. Holz als Roh- und Werkstoff 63:285–296
Rastin N, Ulrich B (1990) Durchwurzelungstiefe und Wurzelverteilung der Fichte auf drei unterschiedlichen Standorten. Forst u Holz 45(22):665–668
Rehfuess KE (1990) Waldböden, 2nd edn. Parey, Hamburg
Reich PB, Walters MB, Ellsworth DS (1991) Leaf age and season influence the relationships between leaf nitrogen, leaf mass per area and photosynthesis in maple and oak trees. Plant Cell Environ 14:251–259
Reiter IM (2004) Space-related resource investments and gains of adult beech (Fagus sylvatica) and spruce (Picea abies) as a quantification of aboveground competitiveness. Dissertation Technische Universität München, Fakultät Wissenschaftszentrum Weihenstephan, Lehrstuhl für Ökophysiologie der Pflanzen
Reiter IM, Häberle KH, Nunn AJ, Heerdt C, Reitmayer H, Grote R, Matyssek R (2005) Competitive strategies in adult beech and spruce: space-related foliar carbon investment versus carbon gain. Oecologia 146:337–349
Rötzer T, Leuchner M, Nunn AJ (2010) Stimulating stand climate, phenology, and photosynthesis of a forest stand with a process based growth model. Int J Biometeorol 54(4):449–464
Schlenker G, Denno D (1971) Die Auswirkung der Laubholz- bzw. Fichtenbestockung auf den Regenwurmbesatz in Parabraunerden und Pseudogleyen des Neckarlandes. Mitt Ver Forstl Standortskunde u Forstpflanzenzüchtung 21:60–66
Schuhbäck T (2004) Nährelementstatus und Bodenzustand an der Bestandesgrenze Buche-Fichte im Mischbestand Kranzberger Forst. Diplomarbeit, Fachgebiet Waldernährung u. Wasserhaushalt, TU München, Freising-Weihenstephan
Steucek GL, Koontz HV (1970) Phloem mobility of magnesium. Plant Physiol 46:50–52
Stith T, Reich PB, Son Y (1993) Canopy dynamics and aboveground production of five tree species with different leaf longevities. Tree Physiol 12:327–345
Temesgen H, Weiskittel AR (2006) Leaf mass per area relationships across light gradients in hybrid spruce crowns. Trees 20:522–530
Ulrich B (1981) Ökologische Gruppierung von Böden nach ihrem chemischen Bodenzustand. Z Pflanzenernähr Bodenk 144:289–305

Weis W, Göttlein A (2002) Vergleich von Biomasse, Elementgehalten und Elementvorräten von Fichte (Picea abies (L.) Karst) und Buche (Fagus sylvatica L.) am Standort Höglwald zu Zeiten der Vegetationsruhe. Forstliche Forschungsberichte München 186:163–167

Werner H, Fabian P (2002) Free-air fumigation of mature trees. Environ Sci Pollut Res 9 (2):117–121

Wolff E (1880) Aschen-Analysen von land- und forstwirthschaftlichen Producten, Fabrik-Abfällen und wildwachsenden Pflanzen. Zweiter Theil. Verlag von Wiegand, Hempel & Parey, Berlin

Wyttenbach A, Schleppi P, Tobler L, Bajo S, Bucher J (1995) Concentrations of nutritional and trace elements in needles of Norway spruce (Picea abies [L.] Karst.) as functions of the needle age class. Plant Soil 168–169:305–312

Chapter 10
Plants and Their Ectomycorrhizosphere: Cost and Benefit of Symbiotic Soil Organisms

R. Agerer, A. Hartmann, K. Pritsch, S. Raidl, M. Schloter, R. Verma, and R. Weigt

10.1 Introduction

"Most higher plants have no roots — they have mycorrhizae". This is one of the most challenging statements in ecology (Begon et al. 1986). This entry holds true till now, since many publications corroborate the crucial importance of this mostly mutualistic symbiosis of tracheophytes with fungi. Mycorrhizae are no homogeneous entity. They differ in their modes of interaction with the plant and of nutrition. These features are used to distinguish mycorrhizal classes (Agerer 1993; Smith and Read 2008). One of the best known and most frequently studied mycorrhizal classes are ectomycorrhizae (ECM), which we focus on in this chapter.

R. Agerer (✉) • S. Raidl • R. Verma
Department of Biology I and GeoBio-Center, Organismic Biology, Mycology, Ludwig-Maximilians-Universität München, Menzinger Str. 67, 80638 München, Germany
e-mail: reinhard.agerer@bio.lmu.de

A. Hartmann
Research Unit Microbe-Plant Interactions, Helmholtz Zentrum München, German Research Center for Environmental Health, Ingolstädter Landstrasse 1, 85764 Neuherberg, Germany

K. Pritsch
Institute of Soil Ecology, Helmholtz Zentrum München, German Research Center for Environmental Health, Ingolstädter Landstr. 1, 85764 Neuherberg, Germany

M. Schloter
Environmental Genomics, Helmholtz Zentrum München, German Research Center for Environmental Health, Ingolstädter Landstr. 1, 85764 Neuherberg, Germany

R. Weigt
Department of Biology I and GeoBio-Center, Organismic Biology, Mycology, Ludwig-Maximilians-Universität München, Menzinger Str. 67, 80638 München, Germany

Department of Ecology and Ecosystem Management, Ecophysiology of Plants, Technische Universität München, Hans-Carl-von-Carlowitz-Platz 2, 85354 Freising, Germany

R. Matyssek et al. (eds.), *Growth and Defence in Plants*, Ecological Studies 220,
DOI 10.1007/978-3-642-30645-7_10,

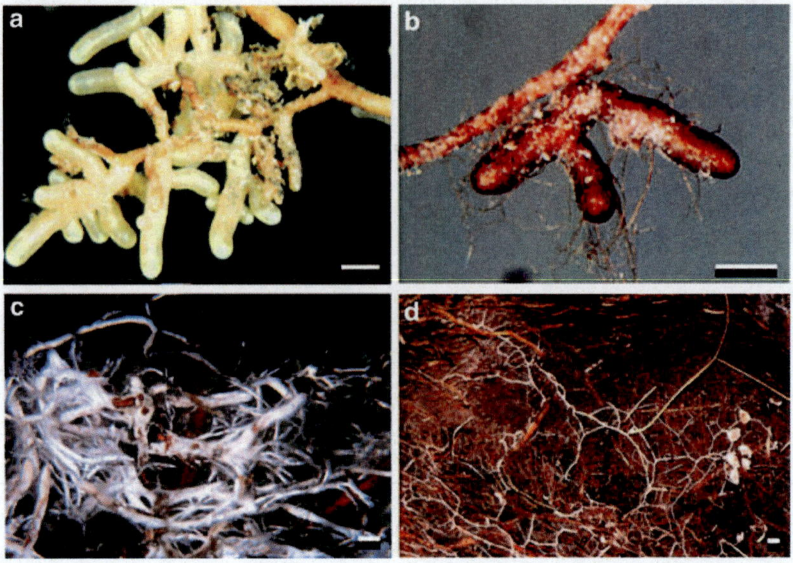

Fig. 10.1 Four different exploration types (ET). (**a**) Contact ET (smooth mantle surface, no rhizomorphs, hydrophilic, *Lactarius* cf. *uvidus*), (**b**) short-distance ET (hairy mantle surface due to long emanating hyphae, no rhizomorphs, hydrophilic, *Genea hispidula*), (**c**) medium-distance fringe ET (with many filamentous rhizomorphs with fringy margins, hydrophobic, *Cortinarius alboviolaceus*), (**d**) long-distance ET (tuberculate ECM systems with very far-reaching, ramifying, interconnected rhizomorphs, hydrophobic, *Suillus plorans*). *Bar* represents for **a–c**: 0.5 mm; **d**: 5 mm

ECM are characterised by intercellular growth of hyphae in the root cortex (Agerer 1991) and a hyphal mantle that envelopes the root. It cuts off the root from the surrounding soil in a physical but not physiological manner. The mantle and, if present, its extramatrical mycelium (EMM) explore and exploit the soil, and acquire, take up, transport and transfer water and nutrients from the soil to the tree roots. These fungal structures represent the almost exclusive aid of the tree to get access to the necessary nutrients from natural soil. The fungus in reverse earns carbohydrates for growth and reproduction (Smith and Read 2008). Economically spoken, the tree invests carbohydrates, obtained through photosynthesis, into its partner to receive essential nutrients. Costs for the tree depend upon the amount of carbohydrates used by the fungus for the production of mycelium. The benefits depend on the amount of nutrients and water the tree receives for its carbohydrate investment.

A generalising classification of ECM regarding putative ecologically and functionally important characters has been established by Agerer (2001), and refers to the extent, differentiation and amount of EMM that emanates from the hyphal mantle of the ECM (Fig. 10.1). The "contact exploration type" (C-ET) forms smooth mantles with only a very limited amount of mostly solitary, not very evident emanating hyphae. This type is predominantly hydrophilic, i.e. water can easily moisten the ECM. Organic as well as inorganic substrate can be directly contacted.

Many emanating hyphae and mostly hydrophily are characteristics of the "short-distance ET" (SD-ET). The emanating hyphae grow to a rather restricted distance, but frequently as a dense mycelium into the surrounding soil (Agerer and Raidl 2004). The "medium-distance ET" (MD-ET) forms additional rhizomorphs which can reach considerable distances in the soil. The most frequent subgroup of this type is the "fringe subtype" (MDf-ET) with many rhizomorphs that are often frequently interconnected by thinner rhizomorphs and by variably dense emanating hyphae; the fringe subtype is hydrophobic (Agerer and Raidl 2004). In contrast, the "smooth subtype" (MDs-ET) forms smooth rhizomorphs and is mostly hydrophilic. In the "mat subtype" (MDm-ET), ECM and their emanating hyphae and rhizomorphs are so densely aggregated that there is apparently no space for any other ECM species to colonise roots within these mats. This subtype is otherwise very similar to the MDf-ET. The "long-distance ET" (LD-ET) is characterised by very long, highly differentiated rhizomorphs with vessel-like hyphae (Agerer 1987–2008, 1991, 1995; Agerer and Rambold 2004–2011). These rhizomorphs can obtain a length of several decimetres (Schramm 1966), and are, including their ECM, generally hydrophobic. Which exploration type an ECM belongs to is species- and fungal relationship-dependent (Agerer 2006, 2007).

Rhizomorphs, particularly those of the LD-ET, are highly appropriate for transport of solutes (Duddridge et al. 1980). Only a limited number of studies regarding transport function and efficiency are available yet. Based on the rhizomorphs' anatomical differentiation (Agerer 1991, 1999, 2001, 2006), it might be concluded on their differential function (Kammerbauer et al. 1989). Apart from being suitable devices for uptake and transport, mycorrhizal hyphae, rhizomorphs and hyphal mantles can provide valuable support for bacterial growth. Bacteria apparently influence the formation of ECM or they may contribute to dissolve minerals or may be responsible for nitrogen/nutrient acquisition (Calvaruso et al. 2007; Frey-Klett et al. 2007; Korkama et al. 2006; Mogge et al. 2000; Poole et al. 2001; Schelkle et al. 1996; Timonen et al. 1998). Nutrient uptake by ECM fungi often occurs via enzymatic activities by degrading organic nutrient-rich material. Thereby, a range of enzymes specific to certain molecules or chemical bonds are available to ECM (Agerer et al. 2000; Pritsch et al. 2004). Differences in degradation specificities and capacities are proven for a diversity of ECM (Courty et al. 2006, 2007; Pritsch et al. 2004).

ECM species assemblages are often composed of different ET and can occupy different ecological niches (Agerer and Göttlein 2003; Agerer et al. 2002; Dickie et al. 2002; Genney et al. 2006). A change in species composition, often coinciding with changes in proportions of ET, can be caused by environmental changes (Brand et al. 1992; Godbold and Berntson 1997; Rey and Jarvis 1997; Weigt et al. 2012).

Based preferentially on the results obtained during 12 years of ECM research in the SFB 607 "Growth and Parasite Defence – Competition of Resources in Economic Plants from Forestry and Agronomy", this chapter focuses on (a) ectomycorrhizal space occupation and potential space exploitation, (b) niche occupation and potential nutrient mobilisation and (c) carbon costs of ECM under a changing environment due to elevated within-tree crown concentrations of carbon dioxide (CO_2) and ozone (O_3).

10.2 Space Occupation and Potential Space Exploitation by Ectomycorrhizal Fungi

Space occupation is an important issue regarding competition for resources, well known above-ground, e.g. for light and space within tree crowns (Chaps. 8, 11–13). In contrast, within the soil, space occupation is less apparent. As almost all finest roots (≤1 mm) of European woodland and forest tree species (e.g. *Fagus*, *Picea*, *Pinus*, *Quercus*) are converted to ECM (Smith and Read 2008), space occupation for nutrient acquisition in forest soils is prevailingly identical with space occupation by ECM.

The importance of extramatrical mycelia of ECM, particularly of rhizomorphs, has already been evidenced several times. Read (1992), who was the first to point out the ecological and functional importance of the mycelium, calculated 200 m g^{-1} dry forest soil for *Suillus bovinus* (Pers.) Roussel grown in root chambers. The pioneering synthesis experiments with *Suillus bovinus* ECM on pine seedlings provided first insights into space occupation capacity of ECM fungi. Similar experiments highlighted the extent of the extramatrical mycelia repeatedly. Although quantifications of mycelia in soils have been performed (Schubert et al. 2003; Wallander et al. 2001), there were no investigations available to compare ECM fungi for their capacity to occupy space in the soil, until Raidl (1997) characterised several species regarding differentiation and range of their extramatrical mycelia. This path-breaking study in combination with the high diversity of thoroughly documented ECM types (comp. Agerer 1987–2008), triggered the distinction of exploration types (Agerer 2001) and studies focussing on space occupation of extramatrical mycelia.

Agerer and Raidl (2004) showed for ECM of *Cortinarius obtusus* (Fr.) Fr. and *Tylospora asterophora* (Bonyord.) Donk, synthesised with Norway spruce (*Picea abies* (L.) Karst.) seedlings in flat rhizotrons, that mycelial range and density of *C. obtusus* (MDf-ET) and *T. asterophora* (SD-ET) differed considerably. Biased by the limited space of the rhizotrons used, the range of *T. asterophora* mycelium was approximately 12 mm that of *C. obtusus* 19 mm. Identical investigations on the MDf-ET *Piloderma croceum* J. Erikss. & Hjortstam and the LD-ET *Rhizopogon roseolus* (Corda) Th.Fr. supported on the one hand the values obtained for *C. obtusus* and resulted on the other hand in a mycelial range of at least 400 mm for the LD-ET. Compared to the C-ET with almost no EMM (Agerer 2001, 2007) as a negligible mycelial space occupation, the mycelium of the SD-ET covers 89 ± 18 (SE) mm cm^{-1} ECM, the MDf-ET 165 ± 39 mm cm^{-1} ECM and the LD-ET 1,336 ± 354 mm cm^{-1} ECM (Table 10.1 and Weigt et al. 2012). When calculating the potential space occupation per unit of ECM length, i.e. the complete area that is covered by the extramatrical mycelial systems, irrespective of the density of the hyphal layers within the mycelial systems (Weigt et al. 2012 and “Erratum”), the values between the SD- and the MDf-ET are not much different with 321 ± 104 mm cm^{-1} ECM and 271 ± 69 mm cm^{-1} ECM, respectively. The LD-ET, however, differs strongly with a 15- to 17-fold higher potential space occupation of 4,787 ± 1,322 mm cm^{-1} ECM. The different space ranges are compared as

Table 10.1 Specific values of ECM based on a length of 10 mm and exploration types

	C-ET	SD-ET	MDf-ET	LD-ET
Mantle biomass (MA-BM)	0.94 μg[a]	0.64 μg[b]	0.55 μg[c]	1.01 μg[d]
Extramatrical mycelial BM (EMM-BM)	0 μg	3.2 μg	6.0 μg	48.7 μg
Ectomycorrhiza BM (ECM-BM)	0.94 μg	3.84 μg	6.55 μg	49.71 μg
EMM length (EMM-L)	0 m[1]	3.7 m	6.9 m	55.9 m
Actual mycelial space occupation (aMSO)	0 mm^2[e]	89 ± 18 mm^2	165 ± 39 mm^2	1,336 ± 354 mm^2
Potential MSO (pMSO)	0 mm^2[e]	321 ± 104 mm^2[f]	271 ± 69 mm^2[f]	4,787 ± 1,322 mm^2[f]
pPO_4-Depletion	8.7 mm^2[g,a]	321 mm^2[f]	271 mm^2[f]	4,787 mm^2[f]
pN_{org}-Depletion	4.4 mm^2[g,a]	23.1 mm^2[f]	42.9 mm^2[f]	346.6 mm^2[f]
pNH_4-Depletion	62.2 mm^2[g,a]	321 mm^2[f]	271 mm^2[f]	4,787 mm^2[f]
pNO_3-Depletion	613 mm^2[g,a]	321 mm^2[f]	271 mm^2[f]	4,787 mm^2[f]

MA-BM biomass of mycorrhizal mantle (based on diameter of *Picea*-ECM and on their mantle thickness as obtained from Agerer and Rambold 2004–2009, and on the specific hyphal dry mass of 228.9 mg cm^{-3} according to Bakken and Olson (1983)); *EMM-BM* biomass of extramatrical mycelium (according to Weigt et al. 2010, 2012); *ECM-BM* biomass of ECM comprising mantle and extramatrical mycelium, without Hartig net; *EMM-L* length of extramatrical mycelium (according to Weigt et al. 2010, 2012); *aMSO* actual mycelial space occupation (according to Weigt et al. 2012); *pMSO* potential mycelial space occupation, i.e. the complete area covered by an extramatrical mycelial system, irrespective of the density of individual hyphae (according to Weigt et al. 2012 and "Erratum"); *pPO_4-depletion* potential depletion area for phosphate, assuming a specific depletion zone of 0.2 mm around an uptaking surface (Nye and Tinker 1977) and a mean hyphal density less than 0.05 mm (Weigt et al. 2012 and "Erratum"); *pN_{org}-depletion* potential depletion area of organic nitrogen, assuming 0.002 mm of wood decomposition around a saprotrophic hypha (concluded from Liese (1964, 1970) and Schmid and Liese (1964) indicating 1–2 μm), laccase activity of ECM (Courty et al. 2006), and an uptake of nitrogen bound to this plant cell wall material, and a mean hyphal density of less than 0.05 mm; *pNH_4-depletion* potential depletion area for ammonium, assuming a depletion zone of 2 mm around an uptake surface (according to Chapin et al. (2002): 1–2 mm) and a mean hyphal density less than 0.05 mm; *pNO_3-depletion* potential depletion area for nitrate, assuming a depletion zone of 10 mm around an uptake surface (according to Chapin et al. (2002): 6–10 mm) and a mean hyphal density less than 0.05 mm

[a]Mean of $n = 15$ species

[b]Mean of $n = 12$

[c]Mean of $n = 7$

[d]Mean of $n = 5$

[e]Possibly too low, as ECM of the C-ET often possess a few solitary emanating hyphae, too

[f]Due to the dense arrangement of the extramatrical hyphae with a mean distance of <0.05 mm (Weigt et al. 2012 and "Erratum"), the whole area occupied by the extramatrical mycelium (pMSO) and not the actual occupied area (aMSO) determines the space of acquisition

[g]Based on the mean diameter of C-ET ECM of 0.44 mm

shown in Fig. 10.2, where mycelia were squeezed between two perspex plates containing peat as growth substrate (Agerer and Raidl 2004; Weigt et al. 2010, 2011). In view of these growth conditions, transfer of the results to natural soils appears not unproblematic, but as the litter in the Of fraction of the organic soil is layered, too, due to periodic litter fall, and as ECM and extramatrical mycelia are preferentially squeezed horizontally between such natural layers (Agerer, pers. obs.), an application of results obtained in flat rhizotrons to natural soil appears acceptable. The obvious individual variability of space occupation mirrors ontogenetical differences, and, therefore, the mean value may rather well represent ET-dependent space occupation in nature.

Space occupation is crucial for resource exploitation and uptake (see also Chap. 11). Colpaert and van Tichelen (1996) pointed out that "probably one of the best ways of studying the effect of environmental stress factors on mycorrhizas is to focus on the growth of the external mycelium". A rhizotron filled with natural soil (organic rich mineral soil of the A horizon), and a root connected to a living mature spruce tree, formed many systems of *Xerocomus badius* (Fr.) Kühner ECM (LD-ET), Weigt et al. (unpubl.). Porter roots of the ECM possessed a projection area of 13 cm^2, and the ECM of 1.8 cm^2. If the depletion zones for phosphate (0.2 mm; Nye and Tinker 1977) are added, then the potential space to be exploited by the ECM increases to 9 cm^2. Assuming that the rhizomorphs and the hyphae between the rhizomorphs are less than 0.4 mm apart from each other, which seems reasonable (Weigt et al. 2010), the complete space covered by the mycelial complex can be considered as the potential exploitation surface with an area of 94 cm^2, i.e. approximately 82 % of the rhizotron used (data not shown). If the ECM had been formed by a C-ET, the potential exploitation area, as represented by the projected area of the smooth ECM, would be only 9 cm^2 with regard to phosphate, roughly 10 % of the space potentially occupied by the LD-ET. This highlights the importance of the EMM for tree nutrition and the possible influence of the ET within an ECM community. The potential space occupation as calculated for the LD-ET *X. badius* corresponds well with the data provided by Leake et al. (2004) who calculated the EMM of the LD-ET *Pisolithus tinctorius* (Agerer and Rambold 2004–2011) to representing 75 % of the absorptive area of the whole soil compartment occupied by roots.

The ET-specific values of space occupation (Table 10.1) can be directly applied to the stand to evaluate the mycelial extension and biomass when knowing the species community and ECM length (Weigt et al. 2012). In a first application, the ECM community under mature Norway spruce trees fumigated with twice-ambient O_3, as compared to an untreated control (Häberle et al. 1999), resulted in a shift from MD-ET and LD-ET to C-ET and SD-ET after 5 years, reducing the actual and potential mycelial space occupation area by 77 % (Weigt et al. 2012). Similar values were obtained in a more detailed investigation following 8 years of O_3 treatment. There, the reduction was even more pronounced and decreased to 19% of the control plot under ambient O_3 (unpubl. data). This decrease is somewhat diminished, when the space occupation of the ECM without EMM is added, i.e. only the mantle surface without EMM is considered, since the absolute abundance of ECM in the plot under twice-ambient O_3 was found to be higher than in the control plot (data not shown).

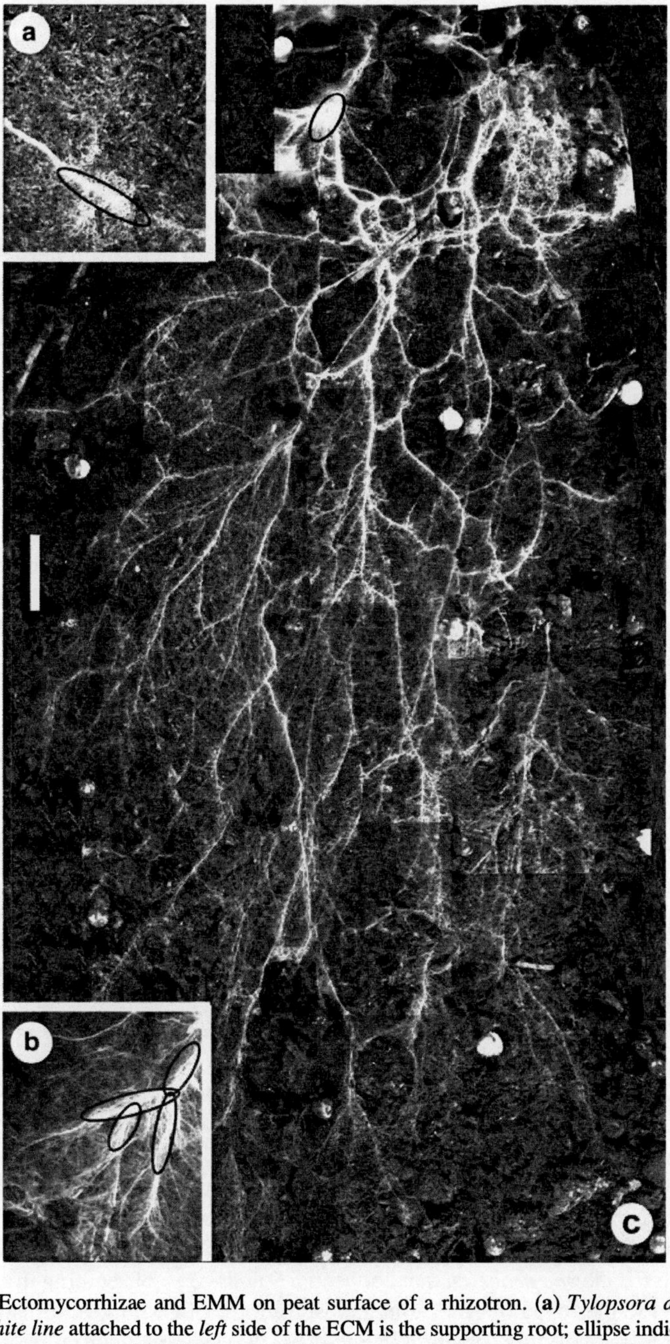

Fig. 10.2 Ectomycorrhizae and EMM on peat surface of a rhizotron. (**a**) *Tylopsora asterophora* (SD-ET; *white line* attached to the *left* side of the ECM is the supporting root; ellipse indicate ECM). (**b**) *Cortinarius obtusus* (MDf-ET; the *ellipses* indicate the position of the ECM within the extramatrical mycelium). (**c**) *Rhizopogon roseolus* (LD-ET; *ellipse* indicate ECM, filamentous structures represent hyphae and rhizomorphs; *small round* patches in **c** are pins used for orientation during photography). *Bar* represents for all figs. 10 mm

10.3 Carbon Costs of ECM Under Elevated Concentrations of Carbon Dioxide and Ozone

The calculation of carbon costs of ECM that have to be invested by the tree necessitates a quantification of the standing biomass, turnover and respiration rate of ECM including their EMM, and the biomass of fruitbodies. Very limited data are available for these issues and allow only rough estimations about the carbon allocation at ECM community level, ending up with 12.5–15 % of annual primary production (Smith and Read 2008). Other estimations ranged between 9 and 50% of the trees' net primary production (Markkola 1995; Rillig et al. 2002; Simard et al. 2002).

Such estimations suffer for various reasons. Standing biomass of ECM or their total lengths or projection areas under natural conditions are sporadically measured at best or only based exclusively upon very limited data sets (Dahlberg et al. 1997; Fogel and Hunt 1979; Wöllecke 2001). The proportion of mantle and Hartig net biomass on ECM total biomass remained almost unconsidered. Vogt et al. (1982) generalised the fungal proportion of ECM as being 40 %. With respect to total carbon investment, however, the quantification of fungal biomass via hyphal mantle alone is misleading as the mycelium in proportion to the mantle biomass may dramatically differ between species (Table 10.1). Since the establishment of the phospholipid fatty acid (PLFA) method by Wallander et al. (2001), quantifications of extramatrical mycelia are more frequent now, but remain in most cases non-compared to absolute abundance or biomass of ECM, and may still underestimate the total mycelial biomass, as external mycelia production was shown to be 300 % higher in natural soil than in acid-washed sand (Hendricks et al. 2006). This method can roughly discriminate between saprotrophic and mycorrhizal hyphae, but is not able to distinguish the contribution of mycelia of different ECM, which would be necessary to calculate species-specific contributions of ECM to soil mycelia (Agerer 2001; Weigt et al. 2012).

Turnover and respiration rates represent further restrictions in estimating the carbon transfer of trees to the fungal partners. The turnover rate of EMM is assumed as being 7 days (Smith and Read 2008). Longevity, and consequently turnover of ECM, is species- and site-dependent and varies between 51 and 139 days (Rygiewicz et al. 1997; Sittig 1999). Fine roots themselves have much longer lifetimes (Smith and Read 2008), and their biomass cannot be used as a basis of ECM biomass calculations. The measurement of respiration of ECM and separately of the extramatrical mycelia are suggested as amounting to 60 % of the carbon allocated to the fungus or 4.3 % of total carbon assimilated (Rygiewicz and Andersen 1994). Soil respiration appeared to be predominantly influenced by ECM and particularly by those species that produced rhizomorphs (Hasselquist et al. 2010). Weigt et al. (2010) estimated for 7-month-old Norway spruce seedlings ECM with *Piloderma croceum* that between 5.9 and 8.3 % of seedling dry mass was transferred to the fungal partner. But as respiration of ECM mycelium and ECM is known to be temperature sensitive (Koch et al. 2007), climate, weather conditions, seasonality and species-dependence make generalisations ambitious. When fruitbody dry matter was included in carbon budget estimations, it referred either mostly to epigeous or occasionally to hypogeous fungi (Agerer 1985; Dahlberg et al. 1997; Fogel and Hunt 1979; Markkola et al. 1995).

These limitations also apply to treatment-specific influences on ECM. Positive impacts of elevated CO_2 concentrations, applied to trees or seedlings on ECM biomass, have been shown several times (e.g. Godbold et al. 1997; Parrent et al. 2006; Weigt et al. 2010). It is generally concluded that elevated CO_2 causes increases in ECM colonisation and the amount of produced EMM (Alberton and Kuyper 2009; Parrent et al. 2006; Tingey et al. 2000). For example, Weigt et al. (2010) found a slight but not significant increase in total ECM length by 25 and 61 % of the MDf-ET *Piloderma croecum* and the MDs-ET *Tomentellopsis submollis*, respectively, in response to twice ambient CO_2 and moderate nitrogen amendment. Simultaneously, the EMM of *P. croceum* ECM increased slightly from approximately 50.8 to 75.5 $\mu g\ g^{-1}$ soil. This increment was caused by higher numbers of ECM, as the hyphal biomass and length per cm ECM remained rather stable at approximately 6 $\mu g\ cm^{-1}$ and 6.9 $m\ cm^{-1}$, respectively.

With respect to the influence of tree crown O_3 exposure on root systems, only the changes in general root parameters have often been investigated, such as longevity and turnover (e.g. King et al. 2001; Phillips et al. 2009). At the study site "Kranzberger Forst", altered C allocation to below-ground compartments of mature trees under elevated crown O_3 exposure was shown by increased soil respiration and fine root production, and a shift in vertical fine root distribution (Nikolova et al. 2010 and Chap. 11). Studies regarding the performance of ECM often concentrate on seedlings and only a very limited number on older trees. For mature trees, Haberer et al. (2007) found an increase in the density of vital ECM and a change in ECM community structure of European beech (*Fagus sylvatica* L.) at the "Kranzberger Forst", as well as a reduced specific N uptake under twice-ambient O_3 and concluded on a change in nitrogen nutrition of the trees by ECM under O_3 treatment. Species-specific changes in nitrogen acquisition could not be documented. Reduced N uptake under elevated O_3 by mature beech and, less pronounced, in spruce was also shown by a soil-^{15}N-labelling study (Weigt 2010, see Chap. 11). At the same site, Grebenč and Kraigher (2007) found a significantly increased number of vital ECM on beech under twice-ambient O_3 concentrations. Although not consistently from year to year, the number of ECM types and species richness increased in addition. Kasurinen et al. (2005) concluded from their studies that increasing tropospheric O_3 levels can represent an important stress factor in northern birch forests, as they might alter mycorrhizal morphotype assemblages, mycorrhizal colonisation rates and sporocarp production.

Changes in ECM diversity have never been reported with respect to their ET-related amount of EMM, although it can be regarded as an important functional unit (Agerer 2001; Leake et al. 2004). Our recent experiments apply the ET-classification to analyse the influence of O_3 treatments (Weigt et al. 2010, 2012) and focus on potential changes in ET-specific biomass of the EMM in natural Norway spruce stands.

Although it appears generally difficult to calculate absolute carbon allocation to ECM, treatment-dependent relative changes might be successfully compared. Weigt et al. (2010, 2012) calculated generally the EMM biomass of SD-ET, MDf-ET and LD-ET as 3.2, 6 and 48.7 μg cm^{-1} ECM (Table 10.1), respectively. Applying these ET-specific values to the Norway spruce stand at the "Kranzberger Forst" (Häberle et al. 1999, Chap. 11), the total mycelial biomass under twice-ambient O_3-fumigated trees decreased considerably as compared to an untreated plot after a 5 years (Weigt et al. 2012) and, similarly, after 8 years of O_3 treatment (unpubl. results, see above). As C-ET generally form thicker ECM with thicker mantles (Table 10.1), the above-mentioned biomass decrease is partially but inconsiderably diminished when mantle biomass is added, since absolute abundance of ECM increased under twice-ambient O_3 (unpubl. data).

10.4 Niche Occupation and Nutrient Mobilisation

ECM preferentially occur in the organic layer, especially in the Of-horizon (Meyer 1962). In addition, the quality of the organic substrate plays an important role for ECM assemblage and abundance (Tedersoo et al. 2003) as well as decomposition status of organic litter (Aneja et al. 2006). Fine-scale studies suggested that ECM species are not evenly distributed vertically and horizontally (Agerer et al. 2002; Baier et al. 2006; Dickie et al. 2002; Gebhardt 2005; Genney et al. 2006; Scattolin et al. 2008; Wöllecke 2001). Site occupation patterns on micro-scale appeared to be often dependent upon soil nutrient contents (Agerer and Göttlein 2003; Rosling and Rosenstock 2008).

10.4.1 Distribution in a Heterogeneous Soil Environment

Some recent approaches demonstrated species-specific niche occupation of ECM (Gebhardt 2005), but only a few ECM types occasionally showed a stand-dependent correlation with soil nutrients (e.g. *Cenococcum geophilum* Fr., *Piloderma croceum*, *Tomentella* sp.). The rather few positive results were possibly because of the large cores for ECM studies (5 cm diameter) and the even larger cores (10 cm diameter) surrounding the 5-cm-cores taken before, for ECM studies. They likely levelled off heterogeneities in nutrient contents and ECM distribution that could be expected at micro-scale level (comp. Agerer and Göttlein 2003). Genney et al. (2006) found an ECM morphotype-specific vertical distribution and differences in the corresponding extramatrical mycelia reminiscent of the EMM patterns of ECM exploration types. Although nutrient status of the soil horizons was not analysed, a correlation of some ECM types to deviating soil conditions was evident, at least concluding from the provided soil profile pictures. The method "micromapping of ECM" (Agerer et al. 2002) related mycorrhizal abundances to nutrient contents of micro-soil cores of 8 mm diameter (Agerer and Göttlein 2003). Some of the studied nutrients were heterogeneously distributed, showing considerable differences at distances of 2.5 cm, e.g. for Ca (193 vs. 565 mg kg^{-1}), K (93 vs. 260 mg kg^{-1}), Mg (29 vs. 93 mg kg^{-1}) and Fe + Mn (94 vs. 141 mg kg^{-1}), along with a different pH (4.8 vs. 5.2). Abundances of ECM were partially significantly correlated with these soluble nutrients (Agerer and Göttlein 2003): *Cortinarius obtusus* positively with NH_4^+ and Mg, *Lactarius theiogalus* positively with NH_4^+, K, Na, Mg, Fe + Mn, and negatively with pH. Wallander et al. (2003) found evidence for different colonisation patterns of apatite and mineral use by a diversity of ECM fungal mycelia.

Leaves of O_3-treated trees may reveal an altered quality as indicated by leaf litter colonising microbial communities (Chung et al. 2006), although mineralisation rates of beech and spruce litter were similarly independent of the O_3 treatment (Aneja et al. 2007). The amounts of lignin and cellulose in O_3-treated versus control spruce needles and beech leaves, however, did not differ after 8 weeks of exposition to microbial communities (Aneja et al. 2007). As litter from O_3-treated spruce and beech plots had higher contents in starch at unchanged *C/N* ratio, a preferential colonisation by saprotrophic fungi can be hypothesised, possibly outcompeting ECM fungi (Gadgil and Gadgil 1975; Lindahl et al. 1999). The results obtained by Aneja et al. (2007), showing a shift from basidiomycote dominated fungal communities to ascomycotous ones, indicate that the ECM fungi, predominantly basidiomycotes (Rinaldi et al. 2008), are possibly disadvantaged in comparison to saprotrophic Ascomycota. The shift from ECM species with high amount of extramatrical mycelia to those with less under O_3-treated Norway spruce, considerably less pronounced in beech (see Sect. 10.3), might, apart from a decreased sugar availability in roots of O_3-treated plants, be the result of hyphal competition between saprotrophic and mycorrhizal fungi for carbohydrates and nutrients in the litter. An apparent direct access to phosphate from a saprotroph by the LD-ET fungus *Paxillus involutus* (Batsch) Fr. was evidenced by Lindahl et al. (1999), an additional indication that the decrease of LD-ET ECM under elevated O_3 could favour saprotrophic fungi. Additionally, also changing nutrient contents under elevated O_3 treatment (Haberer et al. 2007) and their influence on leaf-inhabiting microbial communities are important. Since changes in bacterial assemblages have also influenced the results obtained by Aneja et al. (2007), interpretation of litter quality influences on ectomycorrhizal communities, particularly in a heterogeneous environment, is difficult.

10.4.2 Nutrient Acquisition and Transport

Carbon transfer from trees to their ECM partners for support of the extraradical and particularly of the extramatrical mycelia are with no doubt crucial investments for getting access to nutrients, especially to the macronutrients nitrogen (N) and phosphorus (P), and likely to potassium (K), magnesium (Mg) and possibly calcium (Ca) (Smith and Read 2008). Nitrogen limitations are typical for many ecosystems dominated by ECM plants in temperate and boreal forests and woodlands (Smith and Read 2008) and strongly influence the productivity of trees. Most of the nitrogen (N) is organically bound in the organic matter (Smith and Read 2008), and this organically bound N is preferentially the target of ECM fungi (Read 1992). Apart from differential enzymatic activities of the fungi (Pritsch et al. 2004), mineral weathering by lowering the pH, often caused by exudation of organic acids like oxalate (van Schöll et al. 2006) frequently deposited as crystals on the hyphae, and the extent of the extramatrical mycelia (Read 1992; Rousseau et al. 1994; Wallander et al. 2003) play important roles for nutrient acquisition. Transport capacities are particularly important for those ECM that form extended and long-reaching rhizomorphs (Duddridge et al. 1980; Kammerbauer et al. 1989; Rousseau et al. 1994).

10.4.2.1 Enzymatic Activities of Ectomycorrhizae

Potential enzyme activities of ECM are indicators of traits related to functioning of ECM in the ecosystem (Pritsch and Garbaye 2011). Extracellular enzymes are expressed by soil organisms to mobilise not only nutrients bound in nitrogen or phosphorus-containing polymers such as e.g. chitin, proteins, phospholipids, or DNA, but also mineral nutrients contained in dead plant material. Ectomycorrhizal species show a range of enzyme activities that can be displayed as their enzyme activity profiles obtained by using excised ECM tips in a series of enzyme assays in microplate assays under laboratory conditions (Courty et al. 2005; Pritsch et al. 2004, 2011). An enzyme activity profile indicates the potential activity of a given number of enzymes measured at individual mycorrhizal tips and can be used to compare profiles of different species (Courty et al. 2005) or of species and ECM communities under different treatments or environmental scenarios. The method has been successfully applied in many studies and revealed for example a strong influence of the plant host species (Pritsch et al. 2006), host genotype (Courty et al. 2011), soil conditions i.e. liming (Rineau and Garbaye 2009) or heavy metal contamination (Pritsch et al. 2006).

To study the influence of elevated ozone concentrations on below-ground functioning of ECM, enzyme activity profiles were determined for communities of ECM collected underneath O_3-treated mature Norway spruce trees and a corresponding group of non-treated trees at the "Kranzberger Forst" free-air ozone fumigation site. A set of eight enzyme activities was tested including leucine amino peptidase (EC 3.4.11.1), β-xylosidase (EC 3.2.1.37), β-glucuronidase (EC 3.2.1.31), cellobiohydrolase (EC 3.2.1.91), *N*-acetylglucosaminidase (EC 3.2.1.14), β-glucosidase (EC 3.2.1.3), acid phosphatase (EC 3.1.3.2) and as oxidative enzyme laccase (EC 1.10.3.2). Almost all hydrolytic enzyme activities except for glucuronidase were decreased under elevated O_3, while laccase activity, an oxidative enzyme involved in degradation of lignin and other phenolic substances, significantly increased (Jana Ernst, unpubl. results). Thus the above-mentioned differences in carbon allocation to the mycorrhizal communities with a drastic decrease in carbon costly exploration types in spruce also resulted in differences in the enzyme activity profiles of the ECM communities. Stimulated laccase activity may point towards the necessity to access more recalcitrant nutrients for example an increase in protein–phenol complexes (Theuerl and Buscot 2010). This may either be due to altered litter quality or a higher demand of nutrients for the plants under stress conditions thus requiring associated organisms with more efficient potential for mobilising nutrients.

Insights into the complexity of plant stress reactions and their influence on mycorrhizosphere functioning were also obtained from experiments under more controlled conditions in phytotron experiments of young spruce and beech trees grown in containers (see Chap. 12). In a phytotron study with juvenile spruce and beech grown in mixture, soil samples collected in the mycorrhizosphere of the plants were used to measure enzyme activities as functional parameter (Pritsch et al. 2005). The mycorrhizosphere is enriched in active microorganisms because of the priming effect of plant carbon input (Kuzyakov 2002) and therefore was assumed to reflect altered carbon allocation from stressed plants (Andersen 2003). Enzyme activities in mycorrhizosphere soil integrate activities of all present contributors such as roots, ECM fungi, associated microorganisms and free-living microorganisms in this compartment. In the phytotrone study of Luedemann et al. (2005, 2009), O_3 stress was combined with the root pathogen *Phytophthora citricola* as a second stressor. This study showed that the competitive behaviour of spruce and beech was modified by the combined stress (Luedemann et al. 2005, 2009) and that altered enzyme activity patterns indicated stress reactions in the mycorrhizosphere (Pritsch et al. 2005). For example, increased activities of phosphatase and chitinase in the mycorrhizosphere of spruce were interpreted as a consequence of the plant reaction towards this combined stress that weakened spruce more than beech. This study revealed the complexity of interactions with co-occurring factors such as competition between plant species and interaction with a pathogen and their influence on enzyme activities in the mycorrhizosphere. It also showed that reactions in the mycorrhizosphere of young plants with an increase of some enzyme activities can differ from that of older plants with a decrease of the same enzyme activities. Thus, the link between plant stress due to ozone and enzyme activities in the mycorrhizosphere can be masked by other factors such as plant age and tree species.

10.4.2.2 Nutrient Mobilisation and Transfer

Soil densely occupied by ECM and extramatrical mycelia contained significantly higher concentrations of labile N, PO_4, SO_4, H, Al, Fe, Cu, Mn and Zn compared to soil lacking ECM (Aguilera et al. 1993; Griffiths et al. 1991,1994). This indicates an evident influence of mycorrhizal hyphae on free soil chemistry by dissolving more nutrients than have been taken up by the mycelium (Smits et al. 2008). ECM mycelia have been shown to grow towards mineral grains and direct energy flow towards the minerals (Hedh et al. 2008; Paris et al. 1996; Rosling and Rosenstock

2008; Smits et al. 2008; Wallander 2000; Wallander et al. 2003). The amount of removal is depending upon the fungal species involved (Bending and Read 1995; Perez-Moreno and Read 2000). This is supported by studies that could correlate hyphal length occupying the substrate with phosphate nutrition and growth parameters of seedlings (Ekblad et al. 1995; Rousseau et al. 1994). ECM communities comprising species with higher amount of extramatrical mycelia and with a greater proportion of mycorrhizal total lengths, thus able to occupy larger soil areas, should have higher capabilities for transfer of nutrients to the trees. Although species- and strain-specific differences are known (Bending and Read 1995; Gorissen and Kuyper 2000; Wallander 2000), the dramatic decrease in potential soil occupation of the ECM communities under O_3-treated mature spruce, as seen in the "Kranzberger Forst" (see Sect. 10.2) (Weigt et al. 2012, and unpubl. data), should be mirrored by lower nutrient contents of the trees, not necessarily expressed simply by nutrient concentrations, because lower standing biomass of the trees may have accumulated lower total amounts of nutrients in spite of similar concentrations.

As shown by Weigt et al. (2010, and "Erratum") the mean distance of hyphae in a potential mycelial space occupation (pMSO) area, i.e. the complete area that is covered by the extramatrical mycelial systems, irrespective of the density of individual hyphae, is less than 0.05 mm and therefore, the complete area can be exploited in PO_4 (Table 10.1), as the PO_4-specific depletion zone is approximately 0.2 mm (Nye and Tinker 1977, see above). Since the potential depletion zones for NH_4 and NO_3 are even larger (Table 10.1), the occupied area can potentially be impoverished of nitrogen, too.

When ECM are mapped in their natural position and their potential mycelial space occupation area is depicted (Fig. 10.3), the range and intensity of mycelial influence on soil compartments become evident. Both *Tylospora fibrillosa* (SD-ET; Fig. 10.3c) and *Cortinarius obtusus* (MDf-ET; Fig. 10.3d) occupy rather large areas, being of a wider range and higher density at least close to the ECM in the latter species (Weigt et al. 2012). There is a zone of intermixture of both mycelia (Fig. 10.3e), but the distribution of the ECM indicates also separate regions. The MDs-ET *Piceirhiza internicrassihyphis* and the C-ET *Russula ochroleuca* (Agerer and Rambold 2004–2011) are just within both space occupation areas. This arrangement appears at a first glance as a severe competition problem between the species, but might not be that serious at least with *C. obtusus,* as the C- and MDs-ET ECM are hydrophilic whereas the mycelium of *C. obtusus* is hydrophobic (Fig. 10.3b, f). However, hydrophobic ECM are generally hydrophilic at the periphery of their EMM (Raidl 1997). As *T. fibrillosa* is hydrophilic, its hyphae possibly can function even in the close vicinity of the hydrophobic ones of *C. obtusus* without being crucially competitive, as hydrophily is a prerequisite for intimate contact with the substrate and for nutrient uptake. Such a first analysis of an ECM community indicates that hydrophilic and hydrophobic ECM can apparently occupy different ecological niches, even if they are sympatric. *Cortinarius obtusus* with its hydrophilic mycelial front covers the position of the hydrophilic ECM only in earlier ontogenetic stages, when the EMM is not yet such far reaching. The hydrophilic ECM will have the possibility to contact soil particles with their

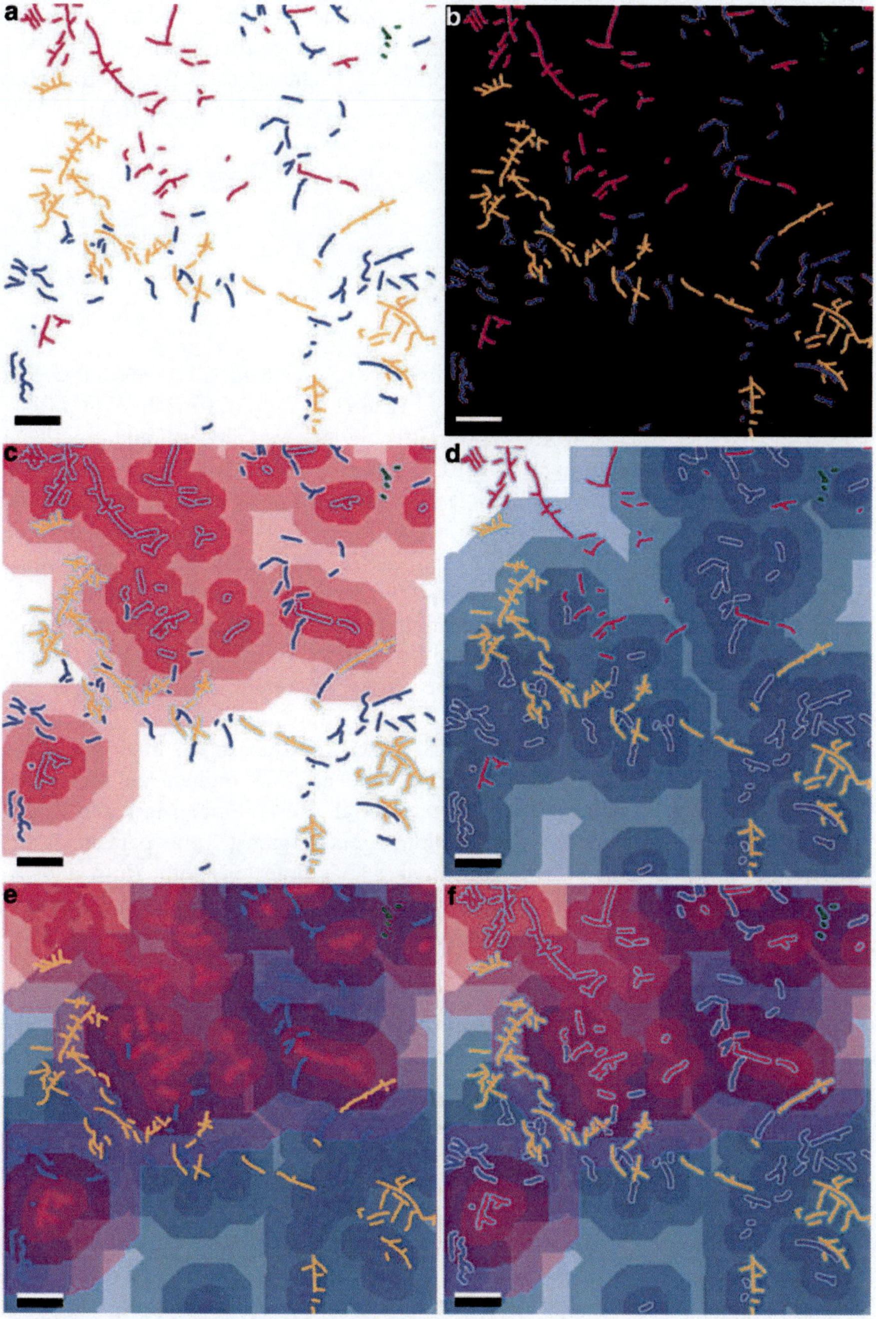

Fig. 10.3 In situ distribution and potential space occupation of ECM: *Cortinarius obtusus* (*blue*), *Tylospora fibrillosa* (*pink*), *Piceirhiza internicrassihyphis* (*ochre*), *Russula ochroleuca* (*green*).

hydrophilic surface, before or after the hydrophilic front of *C. obtusus* hyphae having passed and having become hydrophobic at this spot after ageing. They either can recur on nutrients not completely removed in advance by the hydrophobic competitor, or new nutrients become available, or they are specialised to a different set of nutrients, as shown, e.g. for the hydrophobic *C. obtusus* and the hydrophilic C-ET *Lactarius decipiens* and *L. theiogalus* (Agerer and Göttlein 2003).

10.4.3 Competition and Interaction

Ectomycorrhizal fungi are facing a great diversity of organisms as agents and targets of interactions. Since the investigations by Leake et al. (2002), Lindahl et al. (1999, 2001), and Werner and Zadworny (2003), it is obvious that physical and functional interactions occur between mycelia of ECM and of saprotrophs. Prokaryotes of the domain Bacteria isolated from fruitbodies, the so-called mycorrhiza helper bacteria, promote ECM formation (Duponnois and Garbaye 1991). Bacteria can also be included within tuberculate ECM (Li et al. 1992) and live on mantle and hyphal surface of ECM (Bomberg and Timonen 2009; Mogge et al. 2000; Schelkle et al. 1996; Timonen and Hurek 2006). Timonen and Hurek (2006) conclude that external mycelia can expand the habitat favourable for common rhizosphere bacteria into the soil far from the immediate rhizosphere. More important, it is suspected that prokaryotes play a role in nitrogen nutrition and mineral dissolution (Smith and Read 2008; Timonen and Hurek 2006). Even soil mesofauna and microfauna can influence fungi and ECM formation (Chakraborty et al. 1985; Ingham and Massicotte 1994; Mitchell and Parkinson 1976; Timonen et al. 2004). Since Marx and Davey (1969), ECM are regarded as protecting agents against root parasites, e.g. *Phytophthora cinnamomi*. *Castanea sativa* stands infected with *Phytophthora cambivora* revealed less severe reactions (Branzanti et al. 1994, 1999), and ECM with less extramatrical mycelia are formed as compared to a healthy stand, indicating

Fig. 10.3 (continued) (**a**) Micromap of ECM of a Norway spruce in Of-layer drawn in their natural position, shape and dimensions (taken from Agerer et al. 2002). (**b**) *Lines* around ECM indicate hydrophoby (*white*) and hydrophily (*blue*); only ECM of *C. obtusus* are hydrophobic. (**c**) ECM of *T. fibrillosa* (SD-ET) shown with their potential space occupation by the EMM; intensity of *pink colour* indicates density of mycelium as shown by Weigt et al. (2011); *boderlines* between intensity changes indicate distances from ECM surface in multiples of their diameter (0.3 mm), ×1, ×4, ×9, ×16 (the outmost). (**d**) ECM of *C. obtusus* (MDf-ET) shown with the potential space occupation of the EMM; intensity of *blue colour* indicates density of mycelium as shown by Weigt et al. (2011); *boderlines* between intensity changes indicate distances from ECM surface in multiples of their diameter (0.3 mm), ×1, ×4, ×9, ×16, ×25 (the outmost). (**e**) Combination of **c** and **d**, indicating the position of *P. internicrassihyphis* (MDs-ET) in *ochre* and *R. ochroleuca* (C-ET) in *green*; *R. ochroleuca* without space-occupying EMM; potential space occupation area of *P. internicrassihyphis* (MDs-ET) not known and therefore not shown. (**f**) The same as **e**, but hydrophoby/hydrophily of ECM indicated; only ECM of *C. obtusus* are hydrophobic. *Bar* = 5 mm a shift in ECM community (Blom et al. 2009). Although all such interactions contribute to the multivariate relations within the soil, to ecology and function of trees, most of the recent ECM studies relevant to this topic focus on the influences of bacteria on ectomycorrhizal fungi.

10.4.3.1 ECM-Associated Bacteria

The close association of bacteria with mycorrhizal structures and the localisation of bacteria on the surface of ECM mantle and hyphae have been demonstrated repeatedly with several techniques, including cultural techniques (Timonen et al. 1998), fluorescence in situ hybridisation (FISH) and confocal laser scanning microscopy (Mogge et al. 2000). In several cases, it could be demonstrated that bacteria may even enter mycorrhizal fungi, as in the case of *Laccaria bicolor*, when retrieved from forest soil (Bertaux et al. 2005). Figure 10.4 demonstrates the colonisation of the surface of mycorrhizal mantles of different ECM from forest soil using FISH analysis and epifluorescence microscopy in a quantitative manner. The bacterial community of the mature Norway spruce and European beech stand at the "Kranzberger Forst" was analysed using different phylogenetic probes directed against α-, β- and γ-proteobacteria (Gram-negative bacteria), which are usually very frequently found on the surface of ECM. The major environmental parameter differing between the samples as presented in Table 10.2 was the water content of the soil, which was extremely low for several months until the sampling in December 2003, and wet to even water saturated in April 2004. It is clearly visible that the total number of bacteria, as determined with the DNA-staining dye DAPI (diamidino-phenylindole), was dependent both on the ECM fungus and the environmental conditions. In the case of *Lactarius subdulcis* the number of mycorrhizae-associated bacteria was much higher under wet than under dry conditions. In contrast, the number of bacteria were reduced under wet conditions in *R. ochroleuca* and *X. chrysenteron*. Only in the case of *Fagirhiza pallida*, the number of bacteria was extremely high under both conditions (Table 10.2). Interestingly, the occurrence of Gram-positive HighGC- and LowGC-bacteria were more frequent under dry conditions in the case of *F. pallida* and (almost) not present at wet conditions particularly on the surface of *X. badius* (Fr.) Kühner, *R. ochroleuca* (Pers.) Fr. and *L. subdulcis*. The latter two ECM represent hydrophilic ECM, whereas *X. badius* is hydrophobic. Beta- and gamma-proteobacteria are usually most frequently found in biofilms located on the surface of ECM, while α-proteobacteria were lower in number (Table 10.2). Although a high diversity of different genera and species is contributing to these large families, differences in the colonisation of mycorrhizal surfaces by bacteria can be envisaged. More detailed cloning and sequencing studies revealed that another huge class of soil bacteria, the *Acidobacterium* lineage is very frequently present on the mycorrhizal surface — in same cases amounting to 40–50 % of all bacteria (C. Kellermann, unpubl.). Calvaruso et al. (2007) reported on a *Scleroderma citrinum* Pers. ectomycorrhizosphere that it significantly structures the culturable bacterial

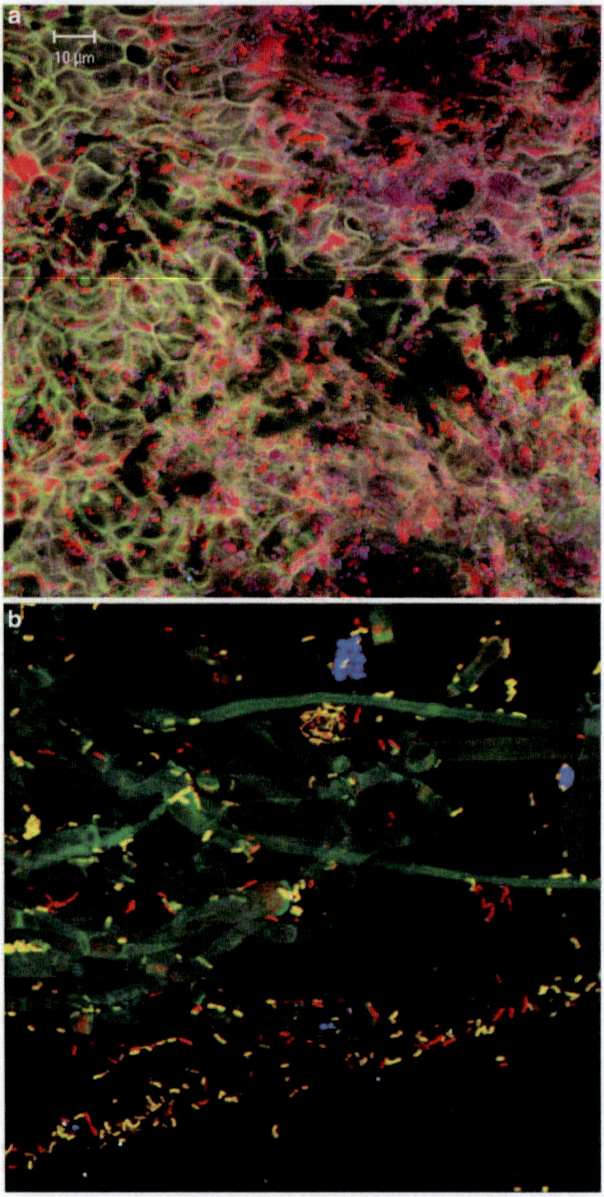

Fig. 10.4 Confocal laser scanning microscopic images of ectomycorrhiza–bacteria associations after specific visualisation of the associated bacteria using FISH analysis: (**a**) ectomycorrhizal mantle of *Lactarius subdulcis* in surface view; (**b**) emanating hyphae and cystidia of *Fagirhiza pallida*. FISH analysis was performed each with a combination of two oligonucleotide probes: EUB388-Cy3 and Alf1B-FLUOS in **a** and EUB388-Cy3 and BET43-FLUOS in **b**. Using these combinations of probes and fluorescent dyes, α-proteobacteria or β-proteobacteria appear *yellow-coloured*. All other cells are labelled *red*. *Bar* for both figs. 10 μm (Claudia Kellermann, unpubl.)

Table 10.2 In situ identified bacterial groups associated with ectomycorrhizae from forest soil (FISH analysis and epifluorescence microscopy)

Ectomycorrhizae	Bacterial numbers (cell counts per mm^2)	α-Proteobacteria	β-Proteobacteria	γ-Proteobacteria	High GC-bacteria	Low GC-bacteria
Xerocomus chrysenteron	5,000–12,000	–	++	+	(+)	–
	1,000–5,000	+	++	+	(+)	–
Fagirhiza pallida	>12,000	+	+++	++	(+)	(+)
	>12,000	++	+++	++	–	–
Xerocomus badius	1,000–5,000	–	++	+	(+)	–
	1,000–5,000	–	++	–	–	–
Russula ochroleuca	1,000–5,000	–	++	+	–	–
	Up to 1,000	+	+	+	–	–
Lactarius subdulcis	0–100	–	–	–	–	–
	5,000–12,000	+	+++	++	–	–

First sampling date (*upper row*; December 2003; very dry soil conditions!) and second plus third sampling date (*lower row*; April 2004; after rewetting during winter). – = lacking, (+) = found only occasionally (5–10 %), + = present in minor amounts (10–20 %), ++ = present in major amounts (20–40 %), and +++ = present in dominant amounts (around 60 %). 100 % = all bacteria stained with general DNA-stain DAPI; Claudia Kellermann, unpubl.

communities in the two soil horizons studied by selecting very efficient strains for phosphorus and iron mobilisation. Thus, bacteria are regularly associated with the surface of mycorrhizal hyphae. According to their diverse physiological potentials, bacteria certainly contribute substantially to the impact of mycorrhiza on soil–plant interactions and the performance of plants.

10.4.3.2 ECM Fungi as Competitors and as Potential Parasites

Environmental stress or changes in growth conditions, whether soil-derived or via above-ground plant components, may trigger impacts on the balance of organismic dependences within the soil, and could also influence mutualistic systems (Andersen 2003). Stress scenarios that change carbohydrate allocation into the root system and alter carbon availability for the fungal partner can influence competition, too (Alberton et al. 2007).

Some ECM species apparently associate with or exclude each other when tested in nature at micro-scale dimensions in squares of 2.5 × 2.5 mm (Agerer et al. 2002). Although statistically not significant, *Cortinarius obtusus* and *Piceirhiza internicrassihyphis* ECM preferentially occurred in such a close neighbourhood, whereas the latter species did not associate with *Russula ochroleuca*. *Russula ochroleuca*, in turn, preferred *Xerocomus badius* (Fr.) Kühner ECM (Agerer et al. 2002). Co-inoculation of seedlings under artificial or semi-natural conditions with two or several ECM species evidenced a preferential formation of a subset of ECM species (Kennedy and Bruns 2005; Parladé and Alvarez 1993), indicating that competition may play an important role during ECM formation. Mycelia of a non-identified ECM isolate (Wu et al. 1999) showed an aggressive replacement of the mycelium of *P. tinctorius* and a progressive formation of ECM in areas where *P. tinctorius* had resided earlier in the experiment. Direct evidence of hyphal interaction between two different ECM fungi was obtained by Agerer (2002).

Changes in carbon allocation to roots and ECM via altered photosynthetic capacities due to negative influence of O_3 or enhanced carbohydrate allocation due to increased CO_2 concentrations might influence ECM communities (Andersen 2003; Weigt et al. 2010). A striking example has been found in spruce under twice-ambient O_3 at the "Kranzberger Forst". ECM communities in the O_3-treated spruce experienced a dramatic increase in ECM of *Hygrophorus olivaceoalbus* (Fr.) Fr. (Agerer 2011). Apparently the roots react strongly against this fungus, as the cells produce extraordinarily high amounts of tannins. The hyphae form a Hartig net, but in older root regions, colonise cells and form microsclerotia within the root cells that apparently function as asexual propagules. Finally, the meristem decomposes and in old ontogenetic stages the ECM break open (Agerer 2011). As this phenomenon is not restricted to O_3-treated spruce but occurs similarly in higher altitudes of the National Park Bavarian Forest, different stressors have to be considered for this peculiar ECM features.

10.4.4 Microbial Activities in the Rhizosphere and Root Exudates

The volume of soil influenced by plant roots was first termed "rhizosphere" by Hiltner in 1904 and is specified as a zone of high microbial activity due to large quantities of carbon and other nutrients. Large amounts of carbon (C) are released from roots into the soil in the form of root exudates, containing 5–21% of photosynthetically fixed C (Marschner 1995). This release of C compounds and nutrients into the rhizosphere, the so-called "rhizodeposition", comprises the total C entering the soil in the form of water-soluble exudates, secretions, lysates, gases and mucilage (Grayston et al. 1996). Within rhizodeposits, water soluble exudates, mainly carbohydrates, carboxylic acids and amino acids (Lynch and Whipps 1990), are probably the most attractive components for microorganisms and therefore highly responsible for microbial growth (Lynch and Whipps 1990). As microbial communities are strongly influenced by root exudates (Brant et al. 2006), it has been hypothesised that plants may select beneficial microbial communities in their rhizosphere (Singh et al. 2007). The data published in the literature so far, which support this hypothesis, are mainly related to *Arabidopsis*, where specific mutants were used. For example, there is biochemical evidence that the tricarboxylic acid cycle intermediate L-malic acid secreted from roots of *Arabidopsis* selectively signals and recruits the beneficial rhizobacterium *Bacillus subtilis* FB17 in a dose-dependent manner (Rudrappa et al. 2008). When mycorrhizal fungi are associated with the roots the root vicinity is aptly referred to as "mycorrhizosphere" (Smith and Read 2008).

In a study by Esperschütz et al. (2009a), C fluxes between young European beech trees and soil were investigated following a chilling period of 3 weeks, using a continuous labelling set-up of the plants with ^{13}C labelled CO_2. It has been postulated that due to low concentrations of easily degradable carbon during tree dormancy, growth conditions for microbes in the rhizosphere are limited. Therefore, after chilling, when exudates are released into the mycorrhizosphere, a strong response of microbial communities utilising plant C is expected. Photosynthetically fixed carbon could be traced into plant tissue, dissolved organic carbon and total microbial biomass, where it was utilised by different microbial communities. Results from ^{13}C analyses of different plant parts reflected the allocation and distribution of recently fixed C within the young trees; 3.5 days after the start of labelling, the ^{13}C had been detected in leaves, where the assimilated carbon is used for leaf formation in the initial growth phase (Dyckmans et al. 2002) after dormancy. In leaves, assimilates are transformed into sugars and amino acids, which are transported via twigs and stems to other parts of the plant. Probably due to the length of the translocation pathway and anabolism of carbohydrates of woody plants (Kozlowski et al. 1991), the labelled C was detected in the belowground plant parts after 20.5 days. Coarse roots basically serve as a storage site of carbon, resulting in a lower amount of incorporated newly assimilated C compared to fine roots. Conversely, fine roots, known as a highly active plant tissue, require high amounts of carbon for growth, resulting in high extents of recently assimilated carbon. Significant incorporation of labelled ^{13}C into the microbial biomass was observed 10.5 days after the experiment has been started. Due to carbon allocation into the rhizosphere, nutrient stress decreased. It became evident that exudates were preferentially used by Gram-negative bacteria, resulting in an enhanced growth of these microbes (comp. Sect. 10.4.3.1). In accordance with other studies (Arao 1999; Butler et al. 2003), mycorrhizal fungi were also enriched in ^{13}C. Overall, the obtained results indicate a fast turnover of exudates and the development of initial food web structures. Additionally, a transport of assimilated carbon into bulk soil by mycorrhizal fungi was observed. Rhizodeposition may vary in response not only to environmental (water potential, light, soil compaction, temperature) and biological parameters (plant species, stage of development; Baudoin et al. 2003), but also to the presence of microorganisms (Grayston et al. 1996).

Esperschütz et al. (2009b) studied the consequences of increased O_3 exposure of trees on carbon fluxes and microbial community structure in the rhizosphere. Therefore, young beech trees exposed to ambient or twice-ambient O_3 concentrations over a 3 years' period were continuously labelled in the last vegetation period using ^{13}C-labelled CO_2. Harvesting of rhizosphere soil from individual trees was performed on a monthly basis, and analysis of the C fluxes was performed by following the ^{13}C label into the dissolved organic carbon fraction of the rhizosphere soil as well as into the phospholipid fraction of the rhizosphere microbes. It was postulated that an increase in O_3 concentration in the atmosphere would alter the microbial community structure and carbon fluxes in the rhizosphere soil. Results from this study demonstrate the high dynamic of microbial communities utilising plant-derived carbon in the rhizosphere of beech trees over a vegetation period and in response to plant stressors. It could be shown that microbial biomass in the mycorrhizosphere as well as individual microbial communities and their activity pattern in the mycorrhizosphere of young beech trees are mainly driven by seasonal variability. An increase in total microbial biomass as well as in individual microbial communities was detected during the vegetation period from June through September 2006. However, a clear O_3 effect was also visible mainly at the end of the vegetation period. PLFA data indicated earlier induced plant senescence as a response to the twice-ambient O_3 treatment. In particular, Gram-positive bacteria and fungi showed higher incorporation of plant C. Coherently, low incorporation of plant C into Gram-negative bacteria suggested the utilisation of other C sources rather than plant C at these time points. Furthermore, higher microbial biomass and abundance of plant C utilising microbes were observed under the O_3 treatment over the whole vegetation period. Surprisingly, the total amount of rhizodeposits did not change between O_3-treated and untreated plants, indicating that probably the quality of the exudates drives the growth of microbial communities rather than their total amount.

10.5 Conclusions

The ectomycorrhizosphere of roots plays a key role in functional ecology of trees and is a target and responder of environmental stress, irrespective of whether caused above- or below-ground. Space occupation of ECM is an important issue for function, e.g. resource capture, of trees and fungi and can be influenced by environmental stress. Future studies on ECM should therefore include length measurements of the standing crop of ECM with respect to their mycelial exploration types, calculations of potential space occupation, and of carbon cost investment in ECM including the EMM, as well as of the mantle and of the Hartig net biomass. The presented data for ET-specific EMM biomass and potential space occupation can be used for such estimations, but additional studies are necessary to support the presently applied values. Particular focus has to be laid on the "smooth" subtype of the "medium distance" ET, as for that type biomass, hyphal length and space occupation of its EMM are not known yet. For fully understanding the functional relationships of ECM fungi and plants, turnover rates of ECM and their EMM, as well as ECM respiration at least under laboratory conditions, should be included. In situ mapping of ECM can be combined with ET-specific range and densities of EMM in combination with their hydrophobic or hydrophilic properties to unravel competition phenomena and niche occupation. Physical and functional relations of ECM to saprotrophic and parasitic as well as between ectomycorrhizal fungi, and to bacteria should be intensely studied. Especially the function of the bacteria — hyphal surface bound ones as well as endobacteria — should be in focus regarding ectomycorrhizal nitrogen nutrition, as ectomycorrhizal hyphae can apparently select for bacterial populations efficient for particular nutrient mobilisation. As the soil comprises an assemblage of micro-niches, heterogeneous in their nutrient and ion composition, micro-soil cores as small as possible should be analysed directly where the ECM are growing. Enzymatic activities of ECM communities should be recorded and related to total ECM length in order to obtain an overall picture of the enzymatic capacity of these symbiotic organs in the soil. Diversity studies on ECM can be supported by DNA-sequencing and a subsequent taxon identification. Finally, in order to provide information about ecological relevance, ECM quantification and last but not least consideration of their putatively functionally important features, predominantly revealed by exploration-type specific mycelial features and the organisation of rhizomorphs, need to be focussed on.

References

Agerer R (1985) Zur Ökologie der Mykorrhizapilze, vol 97, Bibliotheca Mycologica. Cramer, Vaduz

Agerer R (1987–2008) Colour atlas of ectomycorrhizae. 1st–14th delivery. Einhorn, Schwäbisch Gmünd

Agerer R (1991) Characterization of ectomycorrhiza. In: Norris JR, Read DA, Varma AK (eds) Techniques for the study of mycorrhiza, vol 23, Methods in microbiology. Academic, San Diego, pp 25–73

Agerer R (1993) Mycorrhizae: ectomycorrhizae and ectendomycorrhizae. Prog Bot 54:505–529

Agerer R (1995) Anatomical characteristics of identified ectomycorrhizas: an attempt towards a natural classification. In: Varma K, Hock B (eds) Mycorrhiza: structure, function, molecular biology and biotechnology. Springer, Heidelberg, pp 685–734

Agerer R (1999) Never change a functionally successful principle: the evolution of Boletales s. l. (Hymenomycetes, Basidiomycota) as seen from below-ground features. Sendtnera 6:5–91

Agerer R (2001) Exploration types of ectomycorrhizae. A proposal to classify ectomycorrhizal mycelial systems according to their patterns of differentiation and putative ecological importance. Mycorrhiza 11:107–114

Agerer R (2002) The ectomycorrhiza *Piceirhiza internicrassihyphis*: a weak competitor of *Cortinarius obtusus*? Mycol Prog 1:291–299

Agerer R (2006) Fungal relationships and structural identity of their ectomycorrhizae. Mycol Prog 5:67–107

Agerer R (2007) Diversity of ectomycorrhizae as seen from below and above ground: the exploration types. Z Mykol 73:61–88

Agerer R (2011) Asexual reproduction of *Hygrophorus olivaceoalbus* by intracellular microsclerotia in root cells of *Picea abies* – a winner of ozone stress? Mycol Prog. 11:425–435 doi:10.1007/s11557-011-0757-y

Agerer R, Göttlein A (2003) Correlations between projection area of ectomycorrhizae and H_2O extractable nutrients in organic soil layers. Mycol Prog 2:45–52

Agerer R, Raidl S (2004) Distance-related semi-quantitative estimation of the extramatrical ectomycorrhizal mycelia of *Cortinarius obtusus* and *Tylospora asterophora*. Mycol Prog 3:57–64

Agerer R, Rambold G (2004–2011) DEEMY – an information system for characterization and determination of ectomycorrhizae. München, Germany.http://www.deemy.de [first posted on 2004-06-01; most recent update: 2011-01-10]

Agerer R, Schloter M, Hahn C (2000) Fungal enzymatic activity in fruitbodies. Nova Hedwigia 71:315–336

Agerer R, Grote R, Raidl S (2002) The new method 'micromapping', a means to study species-specific associations and exclusions of ectomycorrhizae. Mycol Prog 1:155–166

Aguilera LM, Griffiths RP, Caldwell BA (1993) Nitrogen in ectomcycorrhizal mat and non-mat soils of different-age Douglas-fir forests. Soil Biol Biochem 25:1015–1019

Alberton O, Kuyper TW (2009) Ectomycorrhizal fungi associated with *Pinus sylvestris* seedlings respond differently to increased carbon and nitrogen availability: implications for ecosystem responses to global change. Glob Change Biol 15:166–175

Alberton O, Kuyper TW, Corissen A (2007) Competition for nitrogen between *Pinus sylvestris* and ectomycorrhizal fungi generates potential for negative feedback under elevated CO_2. Plant Soil 296:159–172

Andersen CP (2003) Source-sink balance and carbon allocation belowground in plants exposed to ozone. New Phytol 157:213–228

Aneja MK, Sharma S, Fleischmann F, Stich S, Heller W, Bahnweg G, Munch JC, Schloter M (2006) Microbial colonization of beech and spruce litter - influence of decomposition site and plant litter species on the diversity of microbial community. Microb Ecol 52:127–135

Aneja MK, Sharma S, Fleischmann F, Stich S, Heller W, Bahnweg G, Munch JC, Schloter M (2007) Influence of ozone on litter quality and its subsequent effects on the initial structure of colonizing microbial communities. Microb Ecol 54:151–160

Arao T (1999) In situ detection of changes in soil bacterial and fungal activities by measuring ^{13}C incorporation into soil phospholipids fatty acids from ^{13}C acetate. Soil Biol Biochem 31:1015–1020

Bakken LR, Oisen RA (1983) Buoyant densities and dry-matter contents of microorganisms: conversion of a measured biovolume into biomass. Appl Environ Microbiol 45:1188–1195

Baier R, Ingenhaag J, Blaschke H, Göttlein A, Agerer R (2006) Vertical distribution of an ectomycorrhizal community in upper soil horizons of a young Norway spruce (*Picea abies* [L.] Karst.) stand of the Bavarian Limestone Alps. Mycorrhiza 16:197–206

Brand F, Taylor AFS, Agerer R (1992) Quantitative Erfessung bekannter Ektomykorrhizen in Fichtenversuchsfiächen nach Behandlung mit saurer Beregnung und Kalkung: Die Reakton der natürlichen Ektmykorrhiza-Population der Fichte auf saure Beregnung und Kalkung. Bericht BMFT-Projekt Nr. 0339175F

Baudoin E, Benizri E, Guckert A (2003) Impact of artificial root exudates on the bacterial community structure in bulk soil and maize rhizosphere. Soil Biol Biochem 35:1183–1192

Begon M, Harper JJ, Townsend CR (1986) Ecology. Blackwell, Oxford

Bending GD, Read DJ (1995) The structure and function of the vegetative mycelium of ectomycorrhizal plants. V. Foraging behaviour and translocation of nutrients from exploited litter. New Phytol 130:401–409

Bertaux J, Schmid M, Hutzler P, Hartmann A, Garbaye J, Frey-Klett P (2005) Occurrence and distribution of endobacteria in the plant-associated mycelium of the ectomycorrhizal fungus *Laccaria bicolor* S238N. Environ Microbiol 7:1786–1795

Blom JM, Vannini A, Vettraino AM, Hale MD, Godbold DL (2009) Ectomycorrhizal community structure in a healthy and a *Phytophthora*-infected chestnut (*Castanea sativa* Mill.) stand in central Italy. Mycorrhiza 20:25–38

Bomberg M, Timonen S (2009) Effect of tree species and mycorrhizal colonisation on the archaeal population of boreal forest rhizospheres. Appl Environ Microbiol 75:308–315

Brant JB, Myrold DD, Sulzman EW (2006) Root controls on soil microbial community structure in forest soils. Oecologia 148:650–659

Branzanti MB, Rocca E, Zambonelli A (1994) Influenza di funghi ectomicorrizici su *Phytophthora cambivora* e *P. cinnamomi* del castagno. Mycol Ital 1994:47–52

Branzanti MB, Rocca E, Pisi A (1999) Effect of ectomycorrhizal fungi on chestnut ink disease. Mycorrhiza 9:103–109

Butler JL, Williams MA, Bottomley PJ, Myrold DD (2003) Microbial community dynamics associated with rhizosphere carbon flow. Appl Environ Microbiol 69:6793–6800

Calvaruso C, Turpault M-P, Leclerc E, Frey-Klett P (2007) Impact of ectomycorrhizosphere on the functional diversity of soil bacterial and fungal communities from a forest stand in relation to nutrient mobilization processes. Microb Ecol 54:567–577

Chapin SF III, Matson PA, Mooney HA (2002) Principles of terrestrial ecosystem ecology. Springer, Berlin

Chakraborty S, Theodorou C, Bowen GD (1985) The reduction of root colonization by mycorrhizal fungi by mycophagous amoeba. Can J Microbiol 31:295–297

Chung H, Zak DR, Lilleskov EA (2006) Fungal community composition and metabolism under elevated CO_2 and O_3. Oecologia 147:143–154

Colpaert JV, van Tichelen KK (1996) Mycorrhizas and environmental stress. In: Frankland JC, Magan N, Gadd GM (eds) Fungi and environmental change. Symposium of the British Mycological Society. Cambridge University Press, Cambridge, pp 109-128

Courty PE, Pritsch K, Schloter M, Hartmann A, Garbaye J (2005) Activity profiling of ectomycorrhiza communities in two forest soils using multiple enzymatic tests. New Phytol 167:309–319

Courty P-E, Pouysegur R, Buée M, Garbaye J (2006) Laccase and phosphatase activities of the dominant ectomycorrhizal types in a lowland oak forest. Soil Biol Biochem 38:1219–1222

Courty P-E, Bréda N, Garbaye J (2007) Relation between oak tree phenology and the secretion of organic matter degrading enzymes by *Lactarius quietus* ectomycorrhizas before and during bud break. Soil Biol Biochem 39:1655–1663

Courty P-E, Labbé J, Kohler A, Marçais B, Bastien C, Churin JL, Garbaye J, Le Tacon F (2011) Effect of poplar genotypes on mycorrhizal infection and secreted enzyme activities in mycorrhizal and non-mycorrhizal roots. J Exp Bot 62:249–260

Dahlberg A, Jonsson L, Nylund J (1997) Species diversity and distribution of biomass above and below ground among ectomycorrhizal fungi in an old-growth Norway spruce forest in South Sweden. Can J Bot 75:1223–1335

Dickie IA, Xu B, Koide RT (2002) Vertical niche differentiation of ectomycorrhizal hyphae in soil as shown by T-RFLP analysis. New Phytol 156:527–535

Duddridge JA, Malibari A, Read DJ (1980) Structure and function of mycorrhizal rhizomorphs with special reference to their role in water transport. Nature (London) 287:834–836

Duponnois R, Garbaye J (1991) Mycorrhiza helper bacteria associated with Douglas fir - *Laccaria laccata* symbiosis: effects in aseptic and in glasshouse conditions. Ann Forest Sci 48:239–251

Dyckmans J, Flessa H, Brinkmann K, Mai C, Polle A (2002) Carbon and nitrogen dynamics in acid detergent fibre lignins of beech (*Fagus sylvatica* L.) during the growth phase. Plant Cell Environ 25:469–478

Ekblad A, Wallander H, Carlsson R, Huss-Danell K (1995) Fungal biomass in roots and extramatrical mycelium in relation to macronutrients and plant biomass of ectomycorrhizal *Pinus sylvestris* and *Alnus incana*. New Phytol 131:443–451

Esperschütz J, Pritsch K, Gattinger A, Buegger F, Winkler JB, Munch JC, Schloter M (2009a) Microbial response to exudates in the rhizosphere of young beech trees (*Fagus sylvatica* L.) after dormancy. Soil Biol Biochem 41:976–1985

Esperschütz J, Pritsch K, Gattinger A, Welzl G, Haesler J, Buegger F, Winkler JB, Munch JC, Schloter M (2009b) Influence of chronic ozone stress on carbon translocation pattern into rhizosphere microbial communities of beech trees (*Fagus sylvatica* L.) during a growing season. Plant Soil 323:85–95

Fogel R, Hunt G (1979) Fungal and arboreal biomass in a western Oregon Douglas-fir ecosystem: distribution patterns and turnover. Can J Forest Res 9:245–256

Frey-Klett P, Garbaye J, Tarkka M (2007) The mycorrhiza helper bacteria revisited. New Phytol 176:22–36

Gadgil RL, Gadgil PD (1975) Suppression of litter decomposition by mycorrhizal roots of *Pinus radiata*. N Z J Forest Sci 5:35–41

Gebhardt S (2005) Räumliche Struktur und zeitliche Dynamik von Ektomykorrhizagemeinschaften in Roteichenökosystemen der Niederlausitz, Cottbuser Schriften. vol 34, Brandenburgische Technische Universität Cottbus, Cottbus

Genney DR, Anderson IC, Alexander IJ (2006) Fine-scale distribution of pine ectomycorrhizas and their extramatrical mycelium. New Phytol 170:381–390

Godbold DL, Berntson GM, Bazzaz FA (1997) Growth and mycorrhizal colonization of three North American tree species under elevated CO_2. New Phytol 137:433–440

Godbold DL, Berntson GM (1997) Elevated atmospheric CO_2 concentration changes ectomycorrhizal morphotype assemblages in *Betula papyrifera*. Tree Physiol 17:347–350

Gorissen A, Kuyper TW (2000) Fungal species-specific responses of ectomycorrhizal Scots pine (*Pinus sylvestris*) to elevated CO_2. New Phytol 146:163–168

Grayston SJ, Vaughan D, Jones D (1996) Rhizosphere carbon flow in trees, in comparison with annual plants: the importance of root exudation and its impact on microbial activity and nutrient availability. Appl Soil Ecol 5:29–56

Grebenč T, Kraigher H (2007) Changes in the community of ectomycorrhizal fungi and increased fine root number under adult beech trees chronically fumigated with double ambient ozone. Plant Biol 9:279–287

Griffiths RP, Ingham ER, Caldwell BA, Castellano MA, Cromack K (1991) Microbial characteristics of ectomycorrhizal mat communities in Oregon and California (USA). Biol Fertil Soils 11:196–202

Griffiths RP, Baham JE, Caldwell BA (1994) Soil solution chemistry of ectomycorrhizal mats in forest soil. Soil Biol Biochem 26:331–337

Haberer K, Grebenc T, Alexou M, Gessler A, Kraigher H, Rennenberg H (2007) Effects of long-term free-air ozone fumigation on d15N and total N in *Fagus sylvatica* and associated mycorrhizal fungi. Plant Biol 9:242–252

Häberle K-H, Werner H, Fabian P, Pretzsch H, Reiter I, Matyssek R (1999) 'Free-air' ozone fumigation of mature forest trees: a concept for validating AOT40 under stand conditions. In: Fuhrer J, Achermann B (eds) Critical level for ozone – level II. Swiss Agency for the Environment, Forests and Landscape (SAEFL), Bern, pp 133–137

Hasselquist NJ, Varagas R, Allen MF (2010) Using soil sensing technology to examine interactions and controls between ectomycorrhizal growth and environmental factors on soil CO_2 dynamics. Plant Soil 331:17–29

Hedh J, Wallander H, Erland S (2008) Ectomycorrhizal mycelial species composition in apatite amended and non amended mesh bags buried in a phosphorus poor spruce forest. Mycol Res 112:681–688

Hendricks JJ, Mitchell RJ, Kuehn KA, Pecot SD, Sims SE (2006) Measuring external mycelia production of ectomycorrhizal fungi in the field: the soil matrix matters. New Phytol 171:179–186

Hiltner L (1904) Über neuere Erfahrungen und Probleme auf dem Gebiete der Bodenbakteriologie unter besonderer Berücksichtigung der Gründüngung und Brache. Arb DLG 98:59–78

Ingham ER, Massicotte HB (1994) Protozoan communities around conifer roots colonized by ectomycorrhizal fungi. Mycorrhiza 5:53–61

Kammerbauer H, Agerer R, Sandermann H (1989) Studies on ectomycorrhiza XXII. Mycorrhizal rhizomorphs of *Thelephora terrestris* and *Pisolithus tinctorius* in association with Norway spruce *(Picea abies)*: formation in vitro and translocation of phosphate. Trees 3:78–84

Kasurinen A, Keinänen MM, Kaipainen S, Nilsson L-O, Vapaavuori E, Kontro MH, Holopainen T (2005) Below-ground responses of silver birch trees exposed to elevated CO_2 and O_3 levels during three growing seasons. Glob Change Biol 11:1167–1169

Kennedy PG, Bruns TD (2005) Priority effects determine the outcome of ectomycorrhizal competition between two *Rhizopogon* species colonizing *Pinus radiata* seedlings. New Phytol 166:631–638

King JS, Pregitzer KS, Zak DR, Sober J, Isebrands JG, Dickson RE, Hendrey GR, Karnosky DF (2001) Fine-root biomass and fluxes of soil carbon in young stands of paper birch and trembling aspen as affected by elevated atmospheric CO_2 and tropospheric O_3. Oecologia 128:237–250

Koch N, Andersen CP, Raidl S, Agerer R, Matyssek R, Grams TEE (2007) Temperature-respiration relationships differ in mycorrhizal and non-mycorrhizal root systems of *Picea abies* (L.) Karst. Plant Biol 9:545–549

Korkama T, Fritze H, Pekkanen A, Pennanen T (2006) Interactions between extraradical ectomycorrhizal mycelia, microbes associated with the mycelia and growth rate of Norway spruce (*Picea abies*) clones. New Phytol 173:798–807

Kozlowski TT, Krammer PJ, Pallardy SG (1991) The physiological ecology of woody plants. Academic, San Diego

Kuzyakov Y (2002) Review: factors affecting rhizosphere priming effects. J Plant Nutr Soil Sci Z Pflanzenernahr Bodenkd 165:382–396

Leake JR, Donnelly DP, Boddy L (2002) Interactions between ectomycorrhizal and saprotrophic fungi. In: van der Heijden MGA, Sanders IR (eds) Mycorrhizal ecology. Ecological studies vol 157 Springer, Berlin, pp 346–372

Leake J, Johnson D, Donelly D, Muckle G, Boddy L, Read D (2004) Networks of power and influence: the role of mycorrhizal mycelium in controlling plant communities and agroecosystems. Can J Bot 82:1016–1045

Li CY, Massicotte HB, Moore LVH (1992) Nitrogen-fixing *Bacillus* sp. associated with Douglas-fir tuberculate ectomycorrhizae. Plant Soil 140:35–40

Liese W (1964) Über den Abbau verholzter Zellwände durch Moderfäulepilze. Holz Roh Werkstoff 22:289–295

Liese W (1970) Ultrastructural aspects of woody tissue disintegration. Ann Rev Phytopathol 8:231–258

Lindahl B, Stenlid J, Olsson S, Finlay R (1999) Translocation of 32p between interacting mycelia of a wood-decomposing fungus and ectomycorrhizal fungi in microcosm systems. New Phytol 144:183–193

Lindahl B, Stenlid J, Finlay R (2001) Effects of resource availability on mycelial interactions and ^{32}P transfer between a saprotrophic and an ectomycorrhizal fungus in soil microcosms. FEMS Microbiol Ecol 38:43–52

Luedemann G, Matyssek R, Fleischmann F, Grams TEE (2005) Acclimation to ozone affects host/ pathogen interaction and competitiveness for nitrogen in juvenile *Fagus sylvatica* and *Picea abies* trees infected with *Phytophthora citricola*. Plant Biol 7:640–649

Luedemann G, Matyssek R, Winkler J, Grams TEE (2009) Contrasting ozone × pathogen interaction as mediated through competition between juvenile European beech (*Fagus sylvatica*) and Norway spruce (*Picea abies*). Plant Soil 323:47–60

Lynch JM, Whipps JM (1990) Substrate flow in the rhizosphere. Plant Soil 129:1–10

Markkola AM, Othonen R, Tarvainen O, Ahonen-Jonnarth U (1995) Estimates of fungal biomass in Scots pine stands on an urban pollution gradient. New Phytol 131:139–147

Marschner H (1995) Mineral nutrition of higher plants. Academic, London

Marx DH, Davey CB (1969) The influence of ectotrophic mycorrhizal fungi on the resistance of pine roots to pathogenic infections. III. Resistance of aseptically formed mycorrhizae to infection by *Phytophthora cinnamomi*. Phytopathology 59:549–558

Meyer FH (1962) Die Mykorrhiza der Waldbäume. Mitt Deutsch Dendrol Ges 62:55–59

Mitchell MJ, Parkinson D (1976) Fungal feeding of orbatid mites (Acari: Cryptostigmata) in an aspen woodland soil. Ecology 57:02–312

Mogge B, Loferer C, Agerer R, Hutzler P, Hartmann A (2000) Bacterial community structure and colonization patterns of *Fagus sylvatica* L. ectomycorrhizospheres as determined by fluorescence in situ hybridization and confocal laser scanning microscopy. Mycorrhiza 9:271–278

Nikolova PS, Andersen CP, Blaschke H, Matyssek R, Häberle K-H (2010) Belowground effects of enhanced tropospheric ozone and drought in a beech/ spruce forest (*Fagus sylvatica* L./*Picea abies* [L.] Karst). Environ Pollut 158:1071–1078

Nye PH, Tinker PB (1977) Solute movement in the soil-root system, Sudies in ecology. vol 4, Blackwell, Oxford

Parrent JL, Morris WF, Vilgalys R (2006) CO_2-enrichment and nutrient availability alter ectomycorrhizal fungai communities. Ecology 87:2278–2287

Paris F, Botton B, Lapeyrie F (1996) In vitro weathering of phlogopite by etomycorrizal fungi. II. Effect of K^+ and Mg^{2+} deficiency and N sources on accumulation of oxalate and H^+. Plant Soil 179:141–150

Parladé J, Alvarez JF (1993) Coinoculation of aseptically grown Douglas fir with pairs of ectomycorrhizal fungi. Mycorrhiza 3:93–96

Perez-Moreno J, Read DJ (2000) Mobilization and transfer of nutrients from litter to tree seedlings via the vegetative mycelium of ectomycorrhizal plants. New Phytol 145:301–309

Phillips DL, Johnson MG, Tingey DT, Storm MJ (2009) Elevated CO_2 and O_3 effects on fine-root survivorship in ponderosa pine mesocosms. Oecologia 160:827–837

Poole EJ, Bending GD, Whipps JM, Read DJ (2001) Bacteria associated with *Pinus sylvestris-Lactarius rufus* etomycorrhizas and their effects on mycorrhiza formation in vitro. New Phytol 151:743–751

Pritsch K, Courty P, Churin J-L, Cloutier-Hurteau B, Ali M, Damon C, Duchemin M, Egli S, Ernst J, Fraissinet-Tachet L, Kuhar F, Legname E, Marmeisse R, Müller A, Nikolova P, Peter M, Plassard C, Richard F, Schloter M, Selosse M-A, Franc A, Garbaye J (2011) Optimized assay and storage conditions for enzyme activity profiling of ectomycorrhizae. Mycorrhiza 21:589–600

Pritsch K, Garbaye J (2011) Enzyme secretion by ECM-fungi and exploitation of mineral nutrients from soil organic matter. Ann Forest Sci 68:25–32

Pritsch K, Raidl S, Marksteiner E, Blaschke H, Agerer R, Schloter M, Hartmann A (2004) A rapid and highly sensitive method for measuring enzyme activities in single mycorrhizal tips using 4-methylumbelliferone-labelled fluorogenic substrates in a microplate system. J Microbiol Methods 58:233–241

Pritsch K, Luedemann G, Matyssek R, Hartmann A, Schloter M, Scherb H, Grams TEE (2005) Mycorrhizosphere responsiveness to atmospheric ozone and inoculation with *Phytophthora citricola* in a phytotron experiment with spruce/beech mixed cultures. Plant Biol 7:718–727

Pritsch K, Günthardt-Goerg MS, Munch JC, Schloter M (2006) Influence of heavy metals and acid rain on enzymatic activities in the mycorrhizosphere of model forest ecosystems. Water Snow Landscape Res 80:289–304

Pritsch K, Esperschütz J, Haesler F, Raidl S, Winkler B, Schloter M (2009) Structure and activities of ectomycorrhizal and microbial communities in the rhizosphere of *Fagus sylvatica* under ozone and pathogen stress in a lysimeter study. Plant Soil 323:97–109

Raidl S (1997) Studien zur Ontogenie an Rhizomorphen von Ektomykorrhizen, Bibliotheca Mycologica. vol 169, Cramer, Berlin, pp 1–184

Read DJ (1992) The mycorrhizal mycelium. In: Allen MF (ed) Mycorrhizal functioning. An integrative plant-fungal process. Chapman & Hall, New York, pp 102–133

Rey A, Jarvis PG (1997) Growth response of young birch trees (*Betula pendula* Roth.) after four and a half years of CO_2 exposure. Ann Bot 80:809–816

Rillig CM, Treseder KK, Allen MF (2002) Global change and mycorrhizal fungi. In: van der Heijden MGA, Sanders IR (eds) Mycorrhizal ecology, Ecological studies. vol 157, Springer, Berlin, pp 135–160

Rinaldi AC, Comandini O, Kuyper TW (2008) Ectomycorrhizal fungal diversity: separating the wheat from the chaff. Fungal Divers 33:1–45

Rineau F, Garbaye J (2009) Does forest liming impact the enzymatic profiles of ectomycorrhizal communities through specialized fungal symbionts? Mycorrhiza 19:493–500

Rosling A, Rosenstock N (2008) Ectomycorrhizal fungi in mineral soil. Miner Mag 72:127–130

Rousseau JV, Sylvia DM, Fox AJ (1994) Contribution of ectomycorrhiza to the potential nutrient-absorbing surface of pine. New Phytol 128:639–644

Rudrappa T, Czymmek KJ, Paré PW, Bais HP (2008) Root-secreted malic acid recruits beneficial soil bacteria. Plant Physiol 148:1547–1556

Rygiewicz PT, Andersen CP (1994) Mycorrhiza alter quality and quantity of carbon allocated below ground. Nature (London) 369:58–60

Rygiewicz PT, Johnson MG, Ganio LM, Tingey DT, Storm MJ (1997) Lifetime and temporal occurrence of ectomycorrhizae on ponderosa pine (*Pinus ponderosa* Laws.) seedlings grown under varied atmospheric CO_2 and nitrogen levels. Plant Soil 189:275–287

Scattolin L, Montecchio L, Mosca E, Agerer R (2008) Vertical distribution of the ectomycorrhizal community in the top soil of Norway spruce stands. Eur J Forest Res 127:347–357

Schelkle M, Ursic M, Farquhar M, Peterson RL (1996) The use of laser scanning confocal microscopy to characterize mycorrhizas of *Pinus strobus* L. and to localize associated bacteria. Mycorrhiza 6:431–440

Schmid R, Liese W (1964) Über die mikromorphologischen Veränderungen der Zellwand-strukturen von Buchen- und Fichtenholz beim Abbau durch *Polyporus versicolor* (L.) Fr. Archiv Mikrobiol 47:260–276

Schramm JR (1966) Plant colonization studies on black wastes from anthracite mining in Pennsylvania. Trans Am Philos Soc 56:5–189

Schubert R, Raidl S, Funk R, Bahnweg G, Müller-Starck G, Agerer R (2003) Quantitative detection of agar-cultivated and rhizotron-grown *Piloderma croceum* Erikss. & Hjortst. by ITS-based fluorescent PCR. Mycorrhiza 13:159–165

Simard SW, Durall DM, Jones MD (2002) Carbon and nutrient fluxes within and between mycorrhizal plants. In: van der Heijden MGA, Sanders IR (eds) Mycorrhzal ecology, Ecological studies. vol 157, Springer, Berlin, pp 33–74

Singh BK, Munro S, Potts JM, Millard P (2007) Influence of grass species and soil type on rhizosphere microbial community structure in grassland soils. Appl Soil Ecol 36:147–155

Sittig U (1999) Zur saisonalen Dynamik von Ektomykorrhizen der Buche (*Fagus sylvatica* L.). Ber Forsch Waldökosyst 162:1–119

Smith SE, Read DJ (2008) Mycorrhizal symbiosis, 3rd edn. Academic, San Diego

Smits MM, Bonneville S, Haward S, Leake JR (2008) Ectomycorrhizal weathering, a matter of scale? Miner Mag 72:131–134

Tedersoo L, Kõjalg U, Hallenberg N, Larsson K-H (2003) Fine scale distribution of ectomycorrhizal fungi and roots across substrate layers including coarse woody debris in a mixed forest. New Phytol 159:153–165

Theuerl S, Buscot F (2010) Laccases: toward disentangling their diversity and functions in relation to soil organic matter cycling. Biol Fertil Soils 46:215–225

Timonen S, Hurek T (2006) Characterization of culturable bacterial populations associating with *Pinus sylsvestris - Suillus bovinus* mycorrhizospheres. Can J Microbiol 52:769–778

Timonen S, Jørgensen K, Haahtela K, Sen R (1998) Bacterial community structure at defined locations of the *Pinus sylvestris-Suillus bovinus* and *Pinus sylvestris-Paxillus involutus* mycorrhizospheres in dry pine forest humus and nursery peat. Can J Microbiol 44:499–513

Timonen S, Christensen S, Ekelund F (2004) Distribution of protozoa in scots pine mycorrhizospheres. Soil Biol Biochem 36:1087–1093

Tingey DT, Phillips DL, Johnson MG (2000) Elevated CO_2 and conifer roots: effects on growth, life span and turnover. New Phytol 147:87–103

van Schöll L, Hoffland E, van Beeren N (2006) Organic anion exudation by ectomycorrhizal fungi and *Pinus sylvestris* in response to nutrient deficiencies. New Phytol 170:153–163

Vogt KA, Grier CC, Meier CE, Edmonds RL (1982) Mycorrhizal role in net primary production and nutrient cycling in *Abies amabilis* ecosystems in western Washington. Ecology 63:370–380

Wallander H (2000) Uptake of P from apatite by *Pinus sylvestris* seedlings colonised by different ectomycorrhizal fungi. Plant Soil 218:249–256

Wallander H, Nilsson LO, Hagerberg D, Bååth E (2001) Estimation of the biomass and seasonal growth of external mycelium of ectomycorrhizal fungi in the field. New Phytol 151:753–760

Wallander H, Mahmood S, Hagerberg D, Johansson L, Pallon J (2003) Elemental composition of ectomycorrhizal mycelia identified by PCR-RFLP analysis and grown in contact with apatite or wood ash in forest soil. FEMS Microbiol Ecol 44:57–65

Weigt R (2010) Effects of elevated ground-level ozone on nitrogen acquisition of mature European beech (*Fagus sylvatica*) and Norway spruce (*Picea abies*) trees. Doctoral thesis, Weihenstephan Center of Life Sciences, TU München

Weigt R, Raidl S, Verma R, Rodenkirchen H, Göttlein A, Agerer R (2010) Effects of twice-ambient carbon dioxide and nitrogen amendment on biomass, nutrient contents and carbon costs of Norway spruce seedlings as influenced by mycorrhization with *Piloderma croceum* and *Tomentellopsis submollis*. Mycorrhiza 21:375–391

Weigt R, Verma R, Raidl S, Agerer R (2012) Exploration type specific standard values of extramatrical mycelium - a step towards assessing ectomycorrhizal space occupation and biomass in natural soil. Mycol Prog 11:287–297, 'Erratum'

Werner A, Zadworny M (2003) In vitro evidence of mycoparasitism of the ectomycorrhizal fungus *Laccaria laccata* against *Mucor hiemalis* in the rhizosphere of *Pinus sylvestris*. Mycorrhiza 13:41–47

Wöllecke J (2001) Charakterisierung der Mykorrhizazönosen zweier Kiefernforste unterschiedlicher Trophie, Cottbuser Schriften zu Bodenschutz und Rekultivierung. vol 17, Brandenburgische Technische Universität Cottbus, Cottbus

Wu B, Nara K, Hogetsu T (1999) Competition between ectomycorrhizal fungi colonizing *Pinus densiflora*. Mycorrhiza 9:151–159

Chapter 11
Case Study "Kranzberger Forst": Growth and Defence in European Beech (*Fagus sylvatica* L.) and Norway Spruce (*Picea abies* (L.) Karst.)

K.-H. Häberle, R. Weigt, P.S. Nikolova, I.M. Reiter, J. Cermak, G. Wieser, H. Blaschke, T. Rötzer, H. Pretzsch, and R. Matyssek

11.1 Introduction

The hypothesis that a trade-off exists in the resource allocation of plants between growth and defence according to the growth–differentiation balance theory (GDB, Herms and Mattson 1992; see Chap. 1) was tested during 1998 through 2010 at the research site "Kranzberger Forst", exemplifying adult beech (*Fagus sylvatica*) and spruce (*Picea abies*) trees under mixed-stand conditions (Matyssek et al. 2005b, 2007a, b, 2010; Häberle et al. 2009). Choosing the comparative analysis of biomass partitioning of the two competing tree species with their contrasting crown architecture and foliage habit (i.e. deciduous vs. evergreen, Reiter et al. 2005) in this chapter as a starting point, we comprehend the outcome from an 8-year free-air canopy exposure experiment to an enhanced ozone (O_3) regime, employed for

K.-H. Häberle (✉) • R. Weigt • P.S. Nikolova • H. Blaschke • R. Matyssek
Chair of Ecophysiology of Plants, Technische Universität München, 85350 Freising, Germany
e-mail: haeberle@wzw.tum.de

I.M. Reiter
CEA/Cadarache, DSV, DEVM, Laboratoire d'Ecophysiologie Moléculaire des Plantes, UMR 6191 CNRS-CEA-Université de la Méditerranée, 13108 Saint-Paul-lez-Durance, Cedex, France

J. Cermak
Institute of Forest Botany, Mendel University of Agriculture and Forestry, Zemedelska 3, Brno 63800, Czech Republic

G. Wieser
Department of Alpine Timberline Ecophysiology, Federal Office and Research Centre for Forests, Rennweg 1, 6020 Innsbruck, Austria

T. Rötzer • H. Pretzsch
Chair of Forest Growth and Yield Science, Technische Universität München, 85350 Freising, Germany

R. Matyssek et al. (eds.), *Growth and Defence in Plants*, Ecological Studies 220,
DOI 10.1007/978-3-642-30645-7_11,

interfering with resource allocation *sensu* GDB (Matyssek et al. 2005b). This experiment represented a nucleus for many partner groups for testing and extending GDB as based on both empirical assessments and model development and validation (see Chaps. 2, 8–10, 12–18). Clarification of growth performance within intra-specific tree groups or between beech and spruce groups neighbouring each other in the forest is considered as a prerequisite for the analysis of tree response to abiotic and biotic stress. Therefore, after introducing into the research site and characterising above- and belowground biomass partitioning, both tree species will be compared for their competitiveness in view of space-related resource use (see Sect. 11.2.1). This kind of examination allows cost–benefit analyses of competition-associated resource turnover and, in addition, standardises comparisons between the contrasting foliage habits. Subsequently a conceptual approach will be presented on how defence costs may be estimated for each species. O_3 responses will be discussed in view of stress-signalling mechanisms and potential discrepancies in responsiveness between ontogenetic stages. Conclusions will be drawn on the different strategies of beech and spruce in coping with conflicting resource demands between growth and defence.

The trace gas ozone was chosen as experimental tool for two reasons: (1) as a stressor that has the capacity for re-adjusting the trees' carbon allocation, enabling the disclosure of regulatory mechanisms; (2) as an abiotic surrogate for stress by pathogens in terms of similarity in the biochemical response pathways in either case (Sandermann 1996; Sandermann et al. 1998; Matyssek et al. 2005b).

Pathogens and ozone are known to induce the production of reactive oxygen species (ROS) that trigger programmed cell death, metabolic signalling cascades and the production of ethylene, jasmonic acid and salicylic acid production, which altogether reflect measures of defence (Matyssek et al. 2005a; Nunn et al. 2005a). Although new insights have been elaborated on the plants' regulation of stress tolerance against biotic and abiotic impacts (see Chaps. 2 and 3), the overall conclusion is corroborated that plant-internal means are restricted to central ROS-scavenging pathways, irrespective of the particular elicitor. Progress has been made, however, in relating physiological activity to its control at the molecular level, aiming at understanding plant performance as an outcome from balanced gene regulation (Chap. 2).

Beyond the role as a study tool, ground-level O_3 as occurring at enhanced concentrations currently represents the air pollutant with the highest detrimental potential to vegetation (Matyssek et al. 2012). Concentrations are predicted to even increase further in many regions of the northern and southern hemisphere, due to anthropogenic emissions from industries, traffic and land use change (Sitch et al. 2007). By mitigating the carbon sink-strength of forests and modifying the metabolic response of trees to increasing CO_2 concentrations, ground-level O_3 has become a global factor of climate change (Sitch et al. 2007; Matyssek et al. 2010, 2012). Knowledge about the O_3 response of adult forest trees has been lacking, given that evidence is mostly available from juvenile trees experimentally exposed to controlled environments (Kolb and Matyssek 2001). As a consequence, needs were high for parameterising and validating modelling of carbon flux under current and predicted O_3 impact with ecologically meaningful databases, and hence, for providing reliable decision tools for environmental policy making (IPCC 2007).

Findings from "Kranzberger Forst" about adult beech and spruce, as summarised in this chapter, need to refer to assessments at the site reported by other chapters of this book for extending interpretations and warranting consistency examination. In particular, information is crucial on the edaphic conditions determining nutrient supply (Chap. 9), the light regimes across canopies (Chap. 8) and the competitive interactions (Chap. 12), including those within the mycorrhizosphere (Chap. 10). Tree-level allometry (Chap. 13) as the reference for stand-level structural development (Chap. 14) will support the evaluation of findings on beech and spruce reported here for their general validity. The outcome from the O_3 experiment contributed to the development of the models PLATHO (Chap. 17) and BALANCE (Chap. 18) and stimulated novel approaches in statistical analysis (Chap. 16), while profiting from modelling with respect to tree/stand-level process scaling (see Sects. 11.4 and 11.5).

11.2 Biomass Partitioning at the Research Site "Kranzberger Forst"

11.2.1 Site Description

The forest research site "Kranzberger Forst" of TU München is located in southern Bavaria, 6 km west of the university campus of Freising-Weihenstephan at an altitude of 485 m a.s.l (11°39′41″E, 48°25′08″N). Mean annual temperature is 7.8 °C and mean precipitation 786 mm $year^{-1}$ (Pretzsch et al. 1998). On average, transpiration and interception nearly exhaust the amount of rainfall during the growing season so that the soil capacitance for water storage is refilled in winter (Matyssek et al. 2009). The soil is a luvisol derived from loess over tertiary sediments with a humus layer varying in thickness according to species composition (see Chap. 9). Within the managed forest a fenced area of 0.5ha size was reserved for experiments. A formerly pure spruce plantation had been converted into a mixed forest by thinning parts of the mature spruce stand and introducing groups of beech trees. Later the old spruce canopy had been harvested step by step and thinning activities had been extended to the surrounding spruce-dominated area thus allowing natural spruce regeneration fostered by planting spruce seedlings. Consequently spruce trees were determined to be 60 years old in 2011 by dendrochronological analysis, and beech trees are 20 years older. Stand characteristics are summarised by Pretzsch et al. (2010). As stand thinning by silvicultural management had been suspended during 20 years prior to experimentation, tree competition for light was at maximum (see Chap. 8) and conducive to self-thinning (Pretzsch et al. 2010).

The site had been chosen because of the tree species combination which is common to managed forests in southern Germany and propagated as a silvicultural

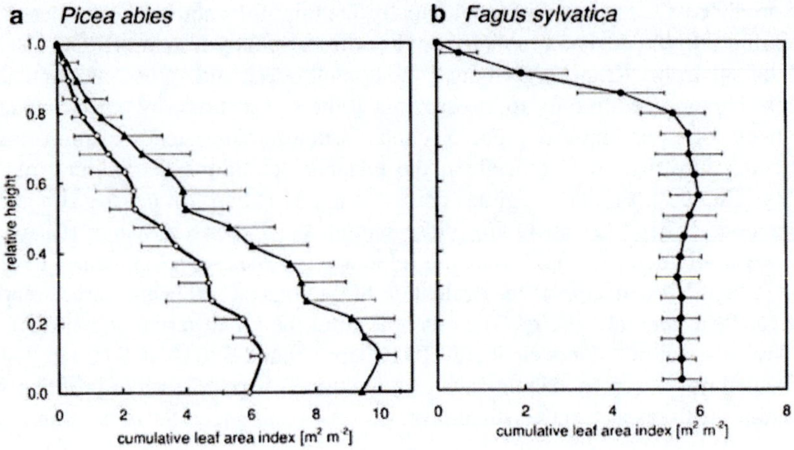

Fig. 11.1 Mean cumulative leaf area index in the normalised vertical profile of the foliated crown, *error bars* denote standard deviation. (**a**) Cumulative projected (*open diamonds*) and hemi surface (*solid triangles*) leaf area of 3 spruce profiles, which had been corrected according to Pokorny and Marek (2000) and Fassnacht et al. (1994). (**b**) Mean cumulative projected leaf area index of 7 profiles in beech

concept for the future. The encountered tree age at the site was crucial for experimentation, representing adultness while trees still displayed distinct annual growth increments as a prerequisite for resolving stress response.

11.2.2 Aboveground Biomass Partitioning

Planting beech in groups between spruce (Rebel 1922) leads to a mosaic pattern of inter- and intra-specific competition with varying effects on the developing crown structure of trees as depending on their individual light exposure. Structural architecture and biomass as assessed on representative beech and spruce reference trees in close vicinity to the research site allowed to model canopy biomass distribution of both tree species under the highly structured, mixed-stand conditions (Grote and Reiter 2004). Although forming a closed canopy, the tree crowns did not intermingle. Neighbouring beech trees jointly formed homogenous canopies contrasting with adjacent spruce, where heterogeneity was caused by gaps between branches and stems that overtopped beech (although being 20 years older than spruce; see above) by up to 5 m in height in 2007. Beech concentrated its foliage in a compact canopy layer at around 4 m below the tree tops. The cumulative leaf area index stagnated already below the upper 20 % of the foliated vertical crown length (Fig. 11.1b, see also Chap. 8) underlining the negligible contribution of leaves and branches in the deep shade. These may abruptly become important if disturbances occur that result in gap formation (Reiter et al. 2005). Conversely, the cumulative leaf area index of spruce increased steadily across the upper 80 % of the foliated crown. Crown length of spruce exceeded that of beech by a factor of about three (Fig. 11.1a). Given the

Fig. 11.2 Coarse roots visualised by ground-penetrating radar (GPR) at a square section of the research site "Kranzberger Forst" (10 × 10 m). Edges are marked by *yellow columns*. *Dotted lines* connect stems of living trees (picture above) with their root stocks (*graph* below, roots reconstructed from GPR data by Fa. GEOLOG, Augsburg, Germany)

contrasting crown shapes of the two species, both displayed maximum canopy closure (expressed in proportion of occupied canopy space) at a similar stem height of 23 m aboveground (year 2000), indicating the zone of most intense competition for light and space (Reiter et al. 2005).

11.2.3 Belowground Biomass Partitioning

Belowground root extension was assessed through several methodologies. Scanning by ground penetrating radar (Hruska et al. 1999) traced coarse roots of >20 mm in diameter down to the rooting depth of 1 m at the site. The analysis showed roots to pervade the soil rather homogeneously, and that belowground parts of root stocks upon tree felling resisted decomposition for more than 10 years (Fig. 11.2). Coarse root biomass of both species resembled each other in allometric relations to the breast height diameter of stems (as derived from entire tree-level coarse root excavations, Fig. 11.3).

Across the soil profile (see Chap. 9), roots of both species were traced down to 90-cm depth, although the vertical profile varied distinctly between beech and

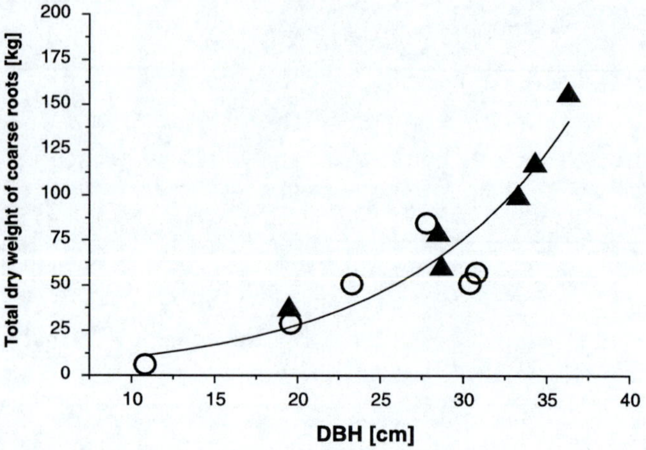

Fig. 11.3 Biomass of coarse roots (>5 mm) from six beech and spruce trees each excavated at an adjacent plot outside the research site but in the same stand. *DBH* diameter at breast height. *Circles* indicate European beech, *triangles* Norway spruce

spruce: Spruce roots of all diameter classes occupied mainly upper soil horizons (*O* and *A*), whereas beech roots displayed presence mainly in the *A* and *B* horizon down to about 60 cm in depth (see below: Fig. 11.8).

Vertical root distribution in sample trees was assessed on the basis of radial sap flow patterns in tree stems (Čermák et al. 2008), making use of mathematical treatment as originally developed for peak separation in gas-chromatography (Novák 1975). Sap flow densities along radii were splained through two Gaussian curves towards best fit assessment. Verification was based on severing individual roots at different soil depths, local watering as well as comparisons with fine root distribution as provided by scanning technology (Čermák et al. 2008). Two trees of similar size (diameter at breast height around 30 cm) were compared (Fig. 11.4). The approach yields sap flow rates as provided by roots at different soil depths along with proportions of water supplied to transpiration from shallow and deep soil layers. On such grounds, beech turned out to rely on sinker rather than on superficial roots in water acquisition, distinctly contrasting with spruce in such respects.

By removing the upper 10 cm of the soil by using compressed air under high pressure ("air spade"), it became evident that some kind of "shyness" as known to exist between tree crowns (as also observed at "Kranzberger Forst") was not detectable between root systems irrespective of the species of neighbouring trees (Fig. 11.5).

Comparing above- and belowground biomass partitioning of beech and spruce showed root–shoot ratios to be quite similar (0.11 in beech vs. 0.14 in spruce, Fig. 11.6). The different foliage types (deciduous vs. evergreen) are reflected in the ratios of above- and belowground absorbing tissues (fine roots/foliage). In beech, fine root biomass is nearly similar to foliage biomass (ratio 0.81), whereas in spruce carbon investments into the foliage exceeds that into fine roots by a factor of 4 (ratio 0.24) (Fig. 11.6).

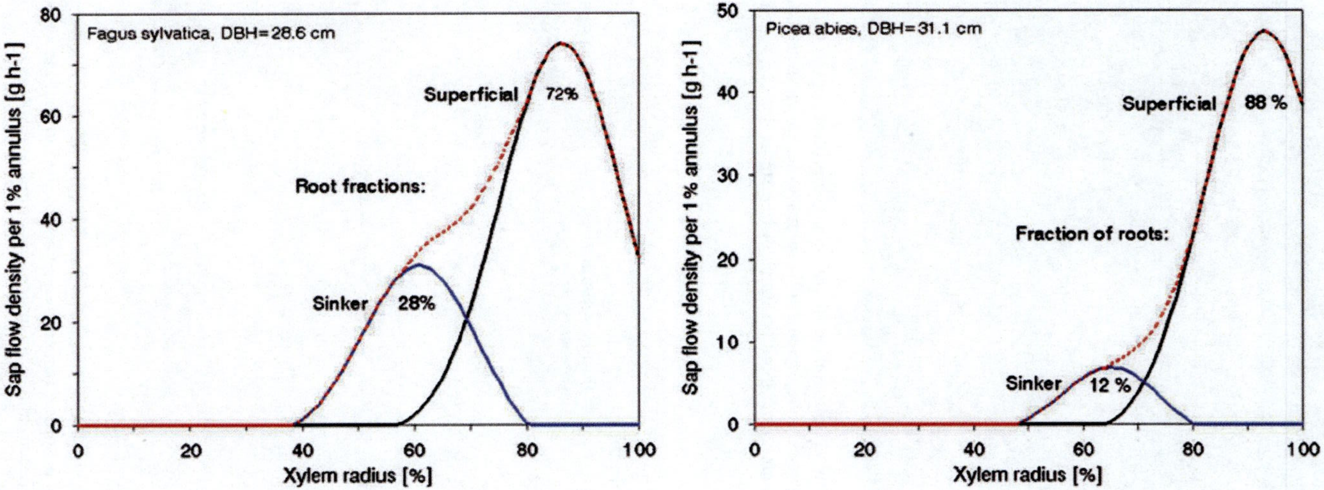

Fig. 11.4 (**a, b**) Radial pattern of sap flow per 1 % annulus in *Fagus* and *Picea* sample trees (*dotted mantle curve*). The *mantle curve* is split into two fractions characterising flow coming from roots growing in different depths; only shallow and deep soil layers with superficial and sinker roots are distinguished (in % of total flow)

Fig. 11.5 Overlapping root systems of beech and spruce are foraging the soil homogenously as shown by the "air spade" technique, when soil particles were blown away under high pressure exposing the fine roots. *White ovals* mark positions of tree stems

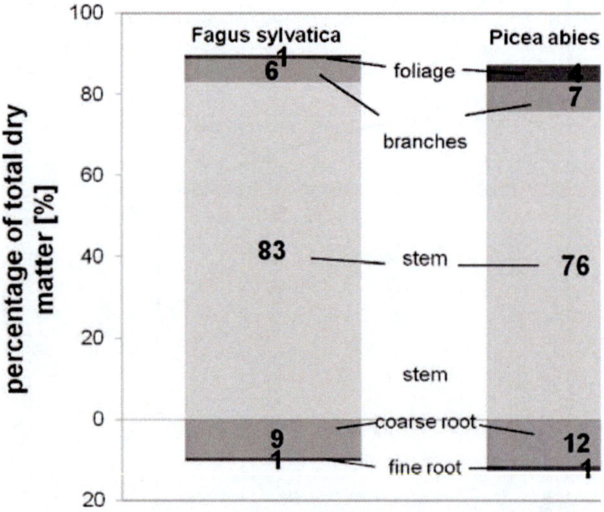

Fig. 11.6 Fractions of above- and belowground biomass (dry weight). Root–shoot ratios account for 0.11 in beech and 0.14 in spruce, respectively, whereas the ratios of the absorbing tissues (fine roots/foliage) were 0.81 (beech) and 0.24 (spruce). Fractions of the trees ($n = 5$) were calculated using the model BALANCE (see Chap. 18)

11.3 Comparative Analysis of Competitiveness Between Beech and Spruce

Competition for resources is inevitably competition for space. A space-related approach was used to compare the competitiveness of beech and spruce quantitatively (Grams et al. 2002, see Chap. 12). Accordingly, above- and belowground costs and benefits of competition-related resource turnover were assessed in view of the contrasting foliage habit and crown and root architecture of beech and spruce—as (1) canopy and soil space (i.e. volume) occupied per unit of carbon investment in leaves and roots, respectively, (2) resource gain (carbon, water, nutrients) per unit of space occupied and (3) running costs (respiration, transpiration) for sustaining the space occupied of relevance for competition (i.e. crown or root volume per unit of consumed resources) (Table 11.1). Aboveground the most distinct difference between the species was associated with annual space-related C investments into foliage biomass. The advantage of beech over spruce to fill the space with low-mass, i.e. low-cost leaves (Reiter et al. 2005), is reversed when considering the higher C investments of beech into branch and twig axes in the upper crown for proficient leaf positioning into light (Reiter 2004). The higher efficiency of space occupation by spruce even increases further if the average needle life span of about 5 years is accounted for (Matyssek et al. 1995; Nunn et al. 2006). Carbon gain and running costs per volume were quite similar in both species. Nevertheless, what turns the balance at the end in favour of beech is its higher annual branch volume increment (rather than total investments in foliage and woody axes). This means faster access to new resources (light), enabling swift intrusion into gaps upon disturbance (Reiter et al. 2005). Hence, time (i.e. response dynamics) associates with space as determinants of plant competitiveness.

Belowground, both tree species exploited the available rooting depth, which is restricted at the site to 1 m due to a dense clay layer. Fine root distribution peaks beneath beech in deeper soil horizons than beneath spruce (Fig. 11.8). Comparing cost–benefit ratios between beech and spruce, relations are changing with soil depth. Due to lower C investments in upper horizons beech is more efficient in space occupation and "running" respiratory C costs, whereas spruce performs more efficiently in deeper horizons (Nikolova 2007). However, space exploitation appears to be crucial in competitiveness belowground because investments are only profitable in relation to acquired returns (water, nutrients). Water is captured by both species most efficiently at the location of their respective bulk of fine roots, i.e. where the efficiency of space occupation is low: in spruce directly below the soil surface (black A_h horizon), in beech somewhat deeper (brown A_l/B_v horizon, Nikolova 2007).

Comparing the above- and belowground cost–benefit ratios it is striking, to which extent C investments resemble each other per unit of occupied volume regarding resource-absorbing organs (light energy and water or nutrients, respectively). Summarised over the species, ranges are investments of 5–90 m^3 (kmol C)$^{-1}$ into foliage and 10–70 m^3 (kmol C)$^{-1}$ regarding fine roots (Reiter et al. 2005; Nikolova 2007). Only foliar C investments can be species specifically differentiated

Table 11.1 Ranges of space-related carbon and water investments (into foliage and fine roots) and gains calculated as ratios ("efficiencies") as described in the text

	Aboveground (foliage)		Belowground (fine roots)	
	Beech	Spruce	Beech	Spruce
Efficiency of space occupation (m^3 $kmol^{-1}$ C)	5–75	5–35	15–50 ($O + A_h$) 12–19 (A_l/B_v)	10–20 ($O + A_h$) 18–32 (A_l/B_v)
Efficiency of space exploitation [$^{(1)}$ mol C m^{-3} a^{-1} or $^{(2)}$ l H_2O m^{-3} day^{-1}]	10–1,300 (1)	10–1,400 (1)	7–8 ($O + A_h$) 15–16 (A_l/B_v) (2)	12–15 ($O + A_h$) 4–5 (A_l/B_v) (2)
Space-related running costs (C) (m^3 $kmol^{-1}$ C)	1–18	3–23	12–40 ($O + A_h$) 12–17 (A_l/B_v)	10–30 ($O + A_h$) 20–44 (A_l/B_v)
Space-related running costs (water) (m^3 $kmol^{-1}$ H_2O)	10–60	10–120		

According to definitions in physics, efficiency is conceived as a ratio of gain versus concurrent resource use (Reiter et al. 2005). Depending on the status of the resources (investment or return) the volume shows up as denominator or numerator, respectively. $O + A_h$ = humus layer + upper mineral soil, A_l/B_v = lower mineral soil (down to a depth of maximal 0.2 m)

with beech occupying twice the volume per kmol C than spruce. Regarding respiration ("running" costs) more rooted volume fell upon one kmol C in the soil (10–44 m^3 (kmol C)$^{-1}$) than foliated crown volume upon one kmol C in the air (1–23 m^3 (kmol C)$^{-1}$) in both species (Table 11.1).

When relating the rates of water uptake from the soil and water loss by transpiration in the crowns to the respective volume occupied by root or leaf biomass it turned out that on average the volume-related water uptake exceeded the volume-related water loss by a factor of ten in both tree species (span water uptake 750–2,250 l m^{-3} a^{-1}, span water loss 22–2,70 l m^{-3} a^{-1}). These levels are plausible in view of a tenfold larger volume occupied by the foliage compared to the roots.

11.4 Costs of Defence

A central feature of GDB is the presumed trade-off between growth- and defence-related metabolism in the plant-internal competition in resource allocation for carbon compounds (Herms and Mattson 1992). Such kind of trade-offs have repeatedly been concluded from qualitative comparisons between species (Fine et al. 2006; Villar et al. 2006) but have hardly been quantified (Mutikainen et al. 2002; Gayler et al. 2004, see also Chaps. 1 and 17). An approximation of the amount of carbon which is potentially transferable from defence to growth accounted in spruce foliage for 2.2–4.9 %, and in beech foliage for 3.9–4.4 % of the annual gross primary production depending on light exposure (Häberle et al. 2009). Such ranges of carbon trade-off are consistent with conclusions from meta-analyses (Koricheva et al. 1998). The approximation was based on the analysis of nine defence-related groups of compounds which are commonly classified as "secondary metabolites" (*sensu latu*, cf. Schulze et al. 2002), including allelochemicals (Clancy et al. 1995). Contrasting with Poorter and Villar (1997), who had studied the chemical composition of the biomass of different tree organs, we focused on the foliage and its defence-related carbon pool. The metabolite groups were categorised as three classes with different turnover intervals (Häberle et al. 2009). The minimum leaf dry mass-related level of each compound encountered during the 4-year study period was regarded to reflect the indispensable carbon demand in the defence metabolism of sun and shade leaves in beech and spruce. Hence, the difference between minimum and maximum levels of each compound, i.e. the range of variation, was conceived as the amount of carbon which is potentially transferable between growth and defence-related processes (cf. Schulze et al. 2002; Matyssek et al. 2005b). Such a view does not depend on knowledge about the molecular control nor the distinction between constitutive and induced defence (Stamp 2003).

The C amount invested in defence is reflected only partly by the defence-related metabolites. Respiration, providing the energy for metabolite synthesis and turnover, needs to be considered as well in relation to turnover and gross primary production (GPP) as the ultimate reference for estimating C transferability between growth and defence (Häberle et al. 2009, see Chap. 6). Partitioning GPP into foliage mass formation, foliar respiration and C translocation to the other tree organs was taken from annual C balances at the branch level (Reiter et al. 2005).

In summary, the potential for trading off carbon between opposing needs for growth and defence turned out to be much lower at least in the case of beech and spruce as predicted in the conceptual approach of Herms and Mattson (1992).

11.5 The Long-Term Free-Air Ozone Fumigation Experiment at "Kranzberger Forst"

The majority of O_3 effects on forest trees reported *prior to* the "Kranzberger Forst" study highlighted here has been derived from chamber studies conducted with juvenile trees (Reich 1987; Sandermann et al. 1997; Kolb and Matyssek 2001). Young trees, however, are uncertain surrogates for adult, field-grown trees (Samuelson and Edwards 1993; Grulke and Miller 1994; Samuelson et al. 1996; Fredericksen et al. 1996; Kolb et al. 1997; Kolb and Matyssek 2001; Grulke and Retzlaff 2001; Wieser et al. 2002a; Matyssek and Sandermann 2003) due to morphological and physiological traits changing during ontogeny (Taylor and Hanson 1992; Fredericksen et al. 1995; Ferdinand et al. 2000; Wellburn et al. 1996; Wieser et al. 2002a, b, 2003). As another shortcoming, microclimatic conditions in chambers differ significantly from those in the field (Sandermann et al. 1997), the more so as young trees typically develop in semi-shaded environments on the forest floor, whereas canopies of adult trees may become sun-exposed. Hence, it is challenging to distinguish between ontogenetic and microclimatic effects in tree response.

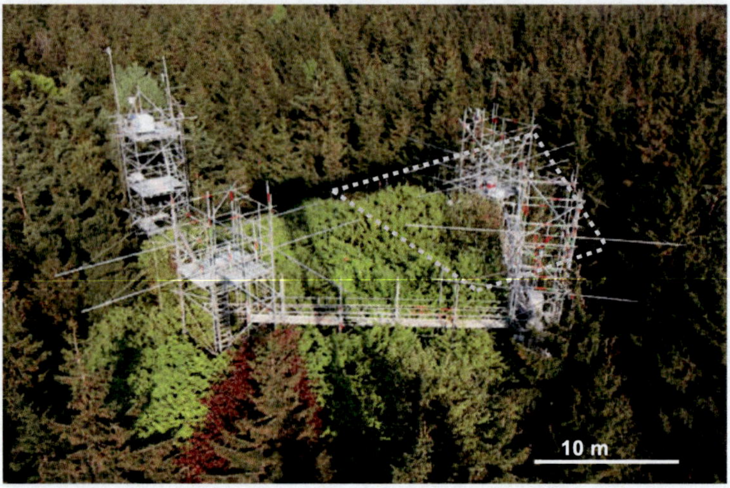

Fig. 11.7 Free-air ozone fumigation experiment at "Kranzberger Forst", scaffolding consisting of four towers with walkways, May 2006. Ozone was released by a dense gridnet of over 100 teflon tubes hanging from horizontal poles and ropes in the area marked by *dotted lines*

Therefore, the microclimatic bias was accounted for at "Kranzberger Forst" through the examination of (1) differently sized trees, (2) exposure to an experimentally enhanced O_3 regime within the canopy of adult spruce and beech trees and (3) placement of container-grown saplings of spruce (Wieser et al. 2002a, b) and beech (Wieser et al. 2003) into the sun-exposed and shaded stand canopy.

11.5.1 Experimental Set-up

At "Kranzberger Forst" adult beech and spruce trees were exposed to ambient air (control; $1 \times O_3$), and an enhanced twice-ambient O_3 regime ($2 \times O_3$) generated continuously by means of a free-air O_3 canopy fumigation system throughout the growing seasons of 2000 through 2007 (Fig. 11.7, Nunn et al. 2002; Werner and Fabian 2002; Karnosky et al. 2007). To prevent acute O_3 injury under $2 \times O_3$ maximum levels were restricted to 150nl O_3 l^{-1}. Scaffolding and a canopy crane provided access to the sun and shade crowns of five adult trees per O_3 treatment and species (for discussion on experimental restrictions and limitations see Chap. 16). Consistent microclimatic conditions for young trees were ensured by placing the containers (nine per O_3 regime with six plants each) for 2 years within the uppermost sun crowns (25 m above ground) of adult trees and on the forest floor (with the microclimate including light being comparable to shaded stand canopy conditions). The containers were regularly watered to prevent soil drought.

The local $1 \times O_3$ regime, expressed as cumulative ozone exposure (SUM0), slightly varied around 140 μl l^{-1} h each year during the study period, except for 2003 with SUM0 = 194 μl l^{-1} h. SUM0 as an external index was almost doubled

Table 11.2 Weather conditions and ozone exposure exemplified during the growing seasons of 2003 and 2004, and as long-term means (1970–2000) at "Kranzberger Forst"

Year	Mean air temperature (°C)	Mean precipitation (mm)	SUM0 ($\mu l\ l^{-1}$ h)	AOT40 ($\mu l\ l^{-1}$ h)
2003	18.0	279	193.4	33.0
2004	15.1	337	142.7	17.3
1970–2000	14.8	442		

SUM0 = complete cumulative O_3 exposure, AOT40 = cumulative O_3 exposure above the threshold of 40 $\mu l\ l^{-1}$

under 2 × O_3, whereas the difference in cumulative ozone uptake (COU, as the internal O_3 dose) was proportionally smaller between the two O_3 regimes due to stomatal regulation (Matyssek et al. 2010). This latter phenomenon was conspicuous, in particular, in 2003 because of stomatal closure during this exceptionally dry year (Ciais et al. 2005).

The container plants were studied during the growing seasons of 2003, i.e. the dry year, and 2004, a year with average precipitation (cf. Löw et al. 2006; Matyssek et al. 2007a; Table 11.2). The O_3 exposure under 1 × O_3 (and hence, also 2 × O_3) was high in 2003 relative to that during the humid season of 2004. Repeatedly and at the end of both growing seasons, leaf gas exchange and biochemical characteristics were assessed in juvenile and adult trees (cf. Herbinger et al. 2005, 2007; Matyssek et al. 2005b), along with root parameters and ectomycorrhizal associations (cf. Grebenc and Kraigher 2007; Zeleznik et al. 2007).

11.5.2 Ozone Response of Adult Forest Trees

At the end of the long-term O_3 experiment, the total stem volume increment of beech reflected a 44 % reduction on average under 2 × O_3 (Pretzsch et al. 2010). Such an effect was absent in spruce. In absolute terms, stem productivity in beech was reduced by 10 $m^3\ ha^{-1}\ a^{-1}$ and slightly, but insignificantly increased in spruce by 0.5 $m^3\ ha^{-1}\ a^{-1}$ under 2 × O_3. Findings were based on radial wood cores taken at four different heights along the stem, as the analysis of annual stem diameter increments at breast height did not indicate any O_3 effect in both species (Wipfler et al. 2005). Instead, stem height growth was promoted, leading to changed stem shapes (neiloidal in beech, cone-shaped in spruce), and by this, resulting in the different stem volume increments (Pretzsch et al. 2010).

Which were the underlying physiological processes affected by ozone, preceding the integrative response of stem productivity (Matyssek et al. 2007a)? Such were reduced stomatal conductance (Kitao et al. 2009; Nunn et al. 2005a) and starch and sucrose levels (Blumenröther et al. 2007), but increased dark respiration rates of leaves (Kitao et al. 2009). In addition, reduced length of growing seasons occurred, as resulting from delayed leaf flush and accelerated senescence (Nunn et al. 2005a), although such effects varied between years, tree species and leaf types (sun vs. shade foliage). Pools of defence-related metabolites differentiated into nine classes of compounds were not significantly affected by ozone in statistical terms at any sampling date and accordingly the potential trade-off between growth and defence remained unchanged (Häberle et al. 2009, see Sect. 11.3).

Elevated ozone appeared to harden beech leaves against fungal invasion as shown by reduced *Apiognomonia errabunda* colonisations in young leaves in early summer (Bahnweg et al. 2005; Olbrich et al. 2010b). This protective effect disappeared in the course of the vegetation period, in particular in sun-exposed leaves. *Apiognomonia* is the most prominent and widespread beech endophyte normally not causing disease symptoms. Its infection and penetration strategies, however, resemble those of certain biotrophic fungal pathogens and occasionally, indeed, severe leaf injury by necrotic lesions occurs ("beech blight disease") leading to premature leaf loss and die-back of young shoots of the tree (Butin 1995). Neither in beech nor in spruce an ozone-dependent difference in lost leaf area due to herbivoral insects was found (Nunn et al. 2005b).

In parallel, physiologically mediated O_3 effects also occurred belowground, in the root system and in the soil. Ozone, which does not physically penetrate soil, acted belowground through changed sink–source relationships and shoot–root signalling at the whole-tree level. Although belowground C allocation was often observed to become limited upon aboveground O_3 impact, the C sink strength of roots may increase, if leaf-level N relations are disturbed under O_3 stress (Andersen 2003). Consistently, reduced N uptake by fine roots was found with different ^{15}N labelling approaches, an effect stronger in beech than in spruce (Haberer et al. 2007; Weigt 2010). At "Kranzberger Forst" belowground C allocation appeared in total to be enhanced under $2 \times O_3$ in both species, although to different extents. While the standing root biomass across the whole rooting zone and the three root diameter classes stayed unaffected (Weigt 2010; Nikolova et al. 2010), growth and maintenance appeared to be increased under $2 \times O_3$ as indicated by enhanced soil respiration underneath both species (Nikolova et al. 2010) as well as increases in fine root production and number of mycorrhizal root tips in beech (Grebenc and Kraigher 2007; Haberer et al. 2007; Nikolova et al. 2010). In spruce, also coarse root respiration was slightly enhanced (Ritter W, Freising, pers. comm.), and a considerable reduction in mycorrhizal mycelium was found as response to $2 \times O_3$ (see Chap. 10).

In addition, a shift in the vertical fine root distribution was indicated in both species under $2 \times O_3$, towards increased root density in the *A* horizon, particularly in beech, and reduced density in the *B* horizon at a depth of 20–30 cm (Fig. 11.8). Within the organic layers, fine root density of spruce under $2 \times O_3$ was decreased in the upper organic layer (Oi), but increased in the lower organic layer (Oa), indicating restriction of spruce fine roots under $2 \times O_3$ to more favourable growth conditions with high nutrient and water availability. As late summer and autumn prior to root sampling were marked by low precipitation, the reduced root biomass of fine roots (<2 mm and $\geq 2 < 5$ mm) in the upper organic layer perhaps also reflected a higher susceptibility of $2 \times O_3$ spruce trees to drought conditions under $2 \times O_3$ rather than $1 \times O_3$. Root density, as assessed by total root harvest (within 1 m^2 plots down to 0.5 m depth), stayed similar independent of species or O_3 regime

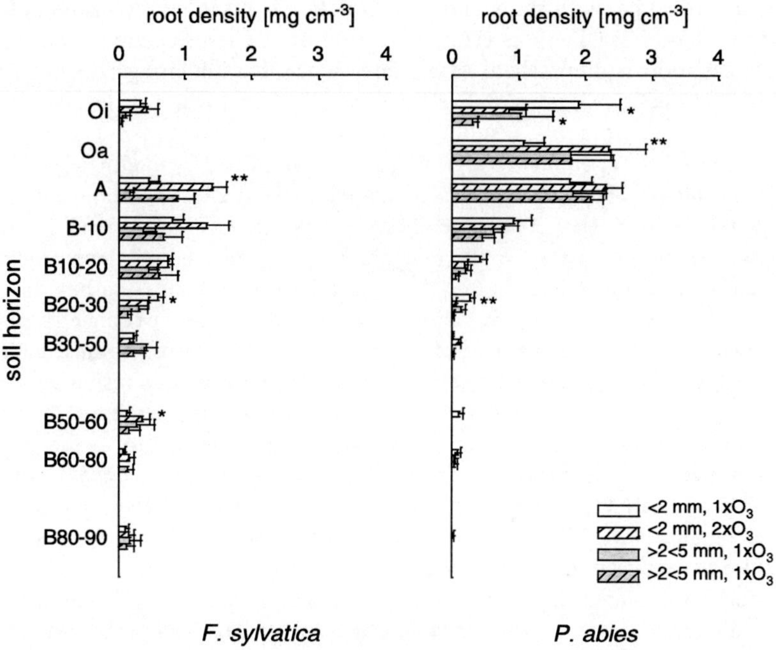

Fig. 11.8 Root density of the root diameter classes < 2 mm and ≥ 2 < 5 mm of beech (*left*) and spruce (*right*) across the soil profile (see Chap. 9), separated by ozone treatments (1 × O_3, 2 × O_3), given as means ± SE, with $n = 5$ plots. *Asterisks* indicate significant differences between ozone treatments at $p < 0.05$ (*), 0.01 (**) or 0.001 (***)

(fine roots < 2 mm: 193 ± 25 (*SE*) g m^{-2}, medium-sized roots ≥ 2 < 5 mm: 294 ± 36 g m^{-2}, coarse roots > 5 mm: 2,307 ± 488 g m^{-2}).

Drought modified the ozone response. As found in 2003, O_3 uptake was decoupled from O_3 exposure, i.e. SUM0 increased in 2003 by 41% whereas seasonal COU was slightly less than during humid years. Stomatal closure due to drought superimposed the effect of ozone on stomatal conductance (Löw et al. 2006), levelling out some ozone impacts as e.g. stimulation of soil respiration (Nikolova et al. 2009, 2010).

The link between the various responses to ozone is to be considered in the plant-internal sink–source relationships and signalling pathways, with the underlying mechanisms challenging research (Matyssek et al. 2008). Although a full-cost analysis which covers the entirety of resource investment into synthesis and turnover of cell metabolites, including stress defence, is illusionary (Lerdau and Gershenzon 1997), some findings can ascertain parts, at least, of the analysis. For example, O_3-induced stomatal closure (Kitao et al. 2009) was consistent with an increased expression of the gene NCED1, which controls the biosynthesis of abscisic acid (ABA) (Jehnes et al. 2007). Winwood et al. (2007) proposed a mechanism by which stimulation of root growth is triggered by circular flow of cytokinins (CK) between shoot and root. Due to CK depletion by ozone in leaves leading to drain from the roots via xylem transport, CK translocation in the phloem back to the roots is diminished. As a consequence, the inhibiting effect of CK on fine-root production decreases (Riefler et al. 2006), as low CK levels typically signal N limitation to roots, resulting in a growth stimulus (Winwood et al. 2007). The O_3-caused effect on fine-root growth may be amplified by modified glutathione levels in roots (Haberer et al. 2008) which are known to affect dividing cells in root meristems (Zellnig et al. 2000). Mehlhorn and Wellburn (1987) found that ozone response was triggered via the ethylene pathway. O_3-induced ethylene production originating from the precursors ACC (1-amino-cyclopropane-1-carboxylic acid) and methyl-ACC was found to cause programmed cell death (Langebartels and Kangasjärvi 2004), a mechanism of defence primarily evolved against pathogen attack, which like O_3 induces oxidative stress. Hence, the plant's response is rather unspecific in view of the agent that actually imposes the oxidative stress. At "Kranzberger Forst", O_3 in some years also reduced the length of the growing season of beech leaves, mainly as a result of premature leaf senescence (Nunn et al. 2005b). The latter O_3 effect is linked with the metabolic ethylene pathway, too (Miller et al. 1999). Consistently, elevated ACC concentrations were found in beech leaves exposed to $2 \times O_3$ (Nunn et al. 2005a).

Perhaps, stimulated shoot elongation growth (Pretzsch et al. 2010) in beech and spruce under enhanced O_3 impact is an effect of increased ethylene formation as well. Ethylene is known to promote shoot elongation in submerged or flooded plants of a range of genera (i.e. *Arabidopsis*, *Lotus*, *Poplar*, *Oryza*, *Potamogeton* and *Rumex*) suggesting that response mechanisms exist that induce growth via ethylene-triggered alterations in gene expression (Bailey-Serres and Voesenek 2010). Hence, consistency appears to exist with the finding of stimulated stem height growth in the adult beech and spruce trees under $2 \times O_3$ (Pretzsch et al. 2010). Nevertheless reports contrast in that several terrestrial plants displayed ethylene-mediated suppression of stem elongation growth (e.g. in *Acer negundo*, Yamamoto and Kozlowski 1987; *Pisum sativum*, Ross and Reid 1986; mutants of *Solanum lycopersicum*; Sharp et al. 2000).

For reconciling the ambiguity, growth-related plant differentiation processes were suggested to be regulated through extensive crosstalk among multiple phytohormonal signalling pathways (Gazzarrini and McCourt 2003). Varying associations of signalling pathways may act as hormonal integrators in "tuning" growth response. Conclusions based only on the concentration of individual phytohormones appear, therefore, to be too restricted for interpreting O_3-related growth response. Rather, evidence increases that several phytohormones act in parallel through targeting cascades of genes that belong to different gene clusters, eventually arriving at such sets of genes which altogether then become responsible in shaping the growth response. In such terms, phytohormones, although acting independently, would appear to become effective through one common integrative signal (Nemhauser et al. 2006; Ross et al. 2011). The resulting signal appears to be translated by decomposing protein inhibitors first of the "DELLA" type, enabling the initiation of shoot extension growth (Achard et al. 2003).

It is still to be clarified to what extent findings on hormonal regulation from *Arabidopsis* and *Pisum* can be generalised for woody plants like *Picea abies* and *Fagus sylvatica*, underlining the need for comparative analyses across broad ranges of species and genotypes (Ross et al. 2011). Such analyses need to functionally link the molecular with the physiological process level in understanding plant growth on a mechanistic basis (Matyssek et al. 2005a; Nemhauser et al. 2006).

In conclusion it becomes apparent that ethylene can act both as an inhibitor and promoter of elongation growth. Regarding ozone stress, ethylene interaction with other phytohormones may be crucial for explaining the stimulatory effect on stem height growth in mechanistic terms. It appears that ozone activates evolutionarily old response mechanisms against oxidative stress (cf. Ross and Reid 2010), involving ethylene, ABA and cytokinins without distinguishing between specific agents of this kind of stress. Such mechanisms may be as old as the beginnings of the oxidative atmosphere, which since then has contained ground-level ozone although at low, pre-industrial levels.

11.5.3 Role of Ontogeny in Ozone Responses

Without resource limitation trees inevitably increase in size during their ontogenetic development until reaching their species-specific growth limit. Physiological changes including altered ozone response during the lifetime of a tree can be attributed mainly to the requirements of adjusting internal resource allocation to the size of the tree organs and the adjustment between them (Pretzsch 2010); however, physical age may play an additional minor role (Genet et al. 2011).

Leaf traits were found to differ significantly between young and adult trees. Under comparable micro-climatic conditions, including light exposure, leaves of beech saplings possessed higher specific leaf area (SLA) than leaves of adult trees (Wieser et al. 2003). Irrespective of tree species, SLA of young and adult trees increased under progressive shading within the canopy (Wieser et al. 2002b, 2003). Similar size- and light-related differences in SLA were reported also from spruce at the same study site (Wieser et al. 2002a, b).

Tree size significantly affected foliar gas exchange and biochemical parameters (Herbinger et al. 2005, 2007). Under sun exposure, net photosynthesis (P_n) of beech saplings was consistently lower as compared to that of adult trees, although it was mostly higher in saplings in the shade under $1 \times O_3$ (but not so under $2 \times O_3$, Fig. 11.9). Also the CO_2 concentration in the leaf intercellular air space (c_i) and stomatal conductance for water vapour (g_{H2O}; Fig. 11.9) differed under sun exposure, to annually variable extents, between juvenile and adult trees, with the former displaying higher parameter levels in 2003, but lower in 2004. Shading, however, generally led to higher levels in the juvenile trees.

Saplings had typically higher levels of total glutathione, but lower levels of ascorbate than adult trees. Ascorbate tended to be mostly oxidised in the saplings, whereas glutathione was typically reduced under $1 \times O_3$ and sun exposure in 2003 (Herbinger et al. 2005, 2007). Sun exposure also favoured, in the saplings, high

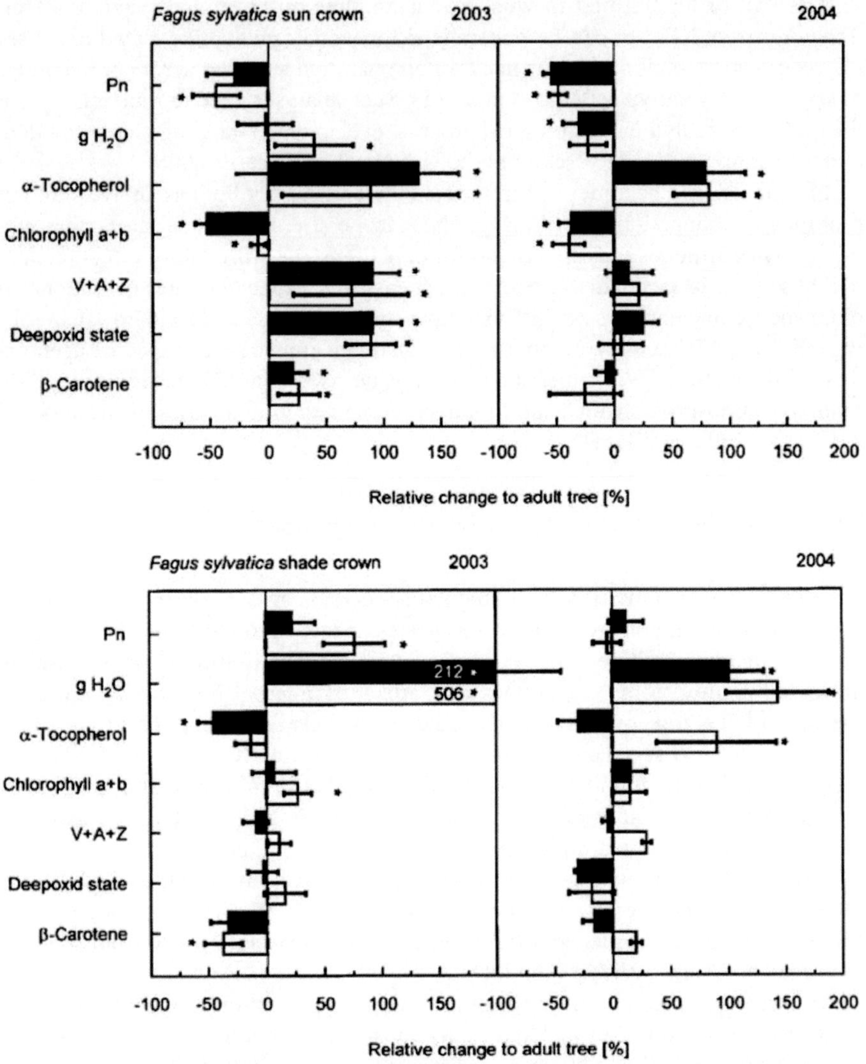

Fig. 11.9 Net photosynthesis (P_n), stomatal conductance for water vapour (g_{H2O}), contents of α-tocopherol, chlorophyll *a* + *b*, xanthophyll cycle carotenoids (violaxanthin [V], antheraxanthin [A] + zeaxanthin [Z]), de-epoxidation state and β-carotene in the sun (*top*) and shade crown (*bottom*) of *Fagus sylvatica* leaves of young seedlings and adult trees after exposure to 1 × O_3 (*solid bars*) and 2 × O_3 (*open bars*) throughout the growing seasons of 2003 and 2004. Data are means of 3–6 trees ± SD and expressed as percent difference to adult trees (= 100 %). *Asterisks* denote significant differences between young and adult trees at $p < 0.05$ (compiled after data from Herbinger et al. 2005, 2007)

levels of α-tocopherol, but reductions in chlorophyll *a* + *b* and xanthophyll cycle carotenoids (violaxanthin [V], antheraxanthin [A] + zeaxanthin [Z]) which were in a more de-epoxidised state and more or less equal β-carotene contents as compared to adult trees (Fig. 11.9, effects of size, i.e. relative differences of young trees from adult trees). These differences in antioxidative compounds between saplings and adult trees faded away in the shade (except for chlorophyll *a* + *b*, Fig. 11.9). Regarding gas exchange and biochemical parameters, tree size-dependent differences were primarily found in the shade-crown in 2003 (Herbinger et al. 2005, 2007), as P_n, g_{H2O}, xanthophyll cycle carotenoids, their de-epoxidised state and β-carotene levels were higher in saplings than adult trees (Fig. 11.9) with a modulating role of elevated O_3 (Fig. 11.10, effects of ozone, i.e. relative differences of $2 \times O_3$ from $1 \times O_3$).

These size-dependent O_3 effects may be ascribed to the fact that saplings had a better water supply in 2003 than adult trees (Herbinger et al. 2007; Matyssek et al. 2007a, g_{H2O} in Fig. 11.9). Therefore, the higher O_3 susceptibility of seedlings compared to mature trees can partly be attributed to higher g_{H2O}, and thus higher O_3 influx into the foliage (Kolb et al. 1997; Wieser et al. 2003; Herbinger et al. 2005). Furthermore, leaf area-based antioxidant concentrations tend to increase during tree ontogeny (Wieser et al. 2002a, b, 2003; Tegischer et al. 2002). Hence, the ratio of potentially available antioxidants on a leaf area basis to O_3 influx increases with tree size. This fact possibly not only explains differences in O_3 susceptibility with respect to tree age and size (Wieser et al. 2002a) but also such between sun and shade-adapted leaves and plants. Beech saplings consistently displayed low antioxidant capacity per unit of O_3 uptake, in particular in the shade (Wieser et al. 2003; Herbinger et al. 2005, 2007). The latter appears to be plausible, as detoxification requires energy and substrate through photosynthesis (Wieser et al. 2003; Herbinger et al. 2005, 2007; Löw et al. 2006) to maintain antioxidants in a reduced state (cf. Matyssek et al. 2007b; and further references therein).

Sun-exposed saplings showed significantly higher leaf, shoot and root biomass (Wieser G, Then C, Innsbruck, unpublished; Zeleznik et al. 2007) as compared to shaded saplings. Independent of light exposure, $2 \times O_3$ caused a decline in foliage dry mass (81 %), leaf number (66 %), shoot biomass (91 %), height growth (90 %), root dry mass (87 %) and the leaf to root biomass ratio (71 %) after 2 years, whereas the root vs. shoot biomass ratio was not affected by elevated O_3. In addition to the observed reduction in biomass increment, mycorrhization diversity also was lowered under $2 \times O_3$ in both, sun- and shade-exposed saplings (Zeleznik et al. 2007).

No differences between small and large trees were detected by various labelling experiments using stable isotopes. A comparison between containerised beech seedlings in the phytotron and adult beech trees labelled at "Kranzberger Forst" with a pulse of CO_2 (depleted in ^{13}C) by means of the newly developed "isoFACE" system (Grams et al. 2011) revealed a significantly smaller percentage of carbon to be allocated to stem respiration under elevated ozone—a reaction independent of tree size (Ritter et al. 2011). Also consistently between young and old trees N-uptake was reduced in beech as detected by ^{15}N labelling in a phytotron study (Luedemann et al. 2005, 2009) and at "Kranzberger Forst" (Weigt 2010).

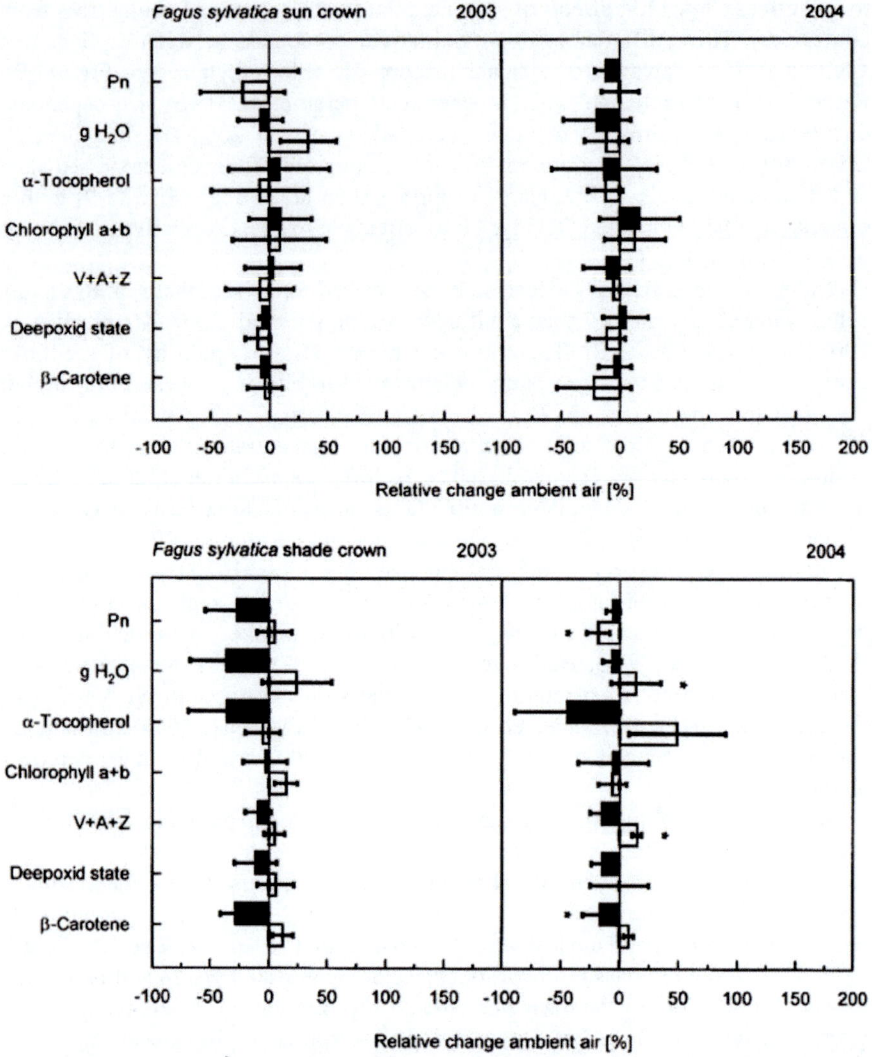

Fig. 11.10 Net photosynthesis (P_n), stomatal conductance for water vapour (g_{H2O}), contents of α-tocopherol, chlorophyll *a* + *b*, xanthophyll cycle carotenoids (violaxanthin [V], antheraxanthin [A] + zeaxanthin [Z]), de-epoxidation state and β-carotene in the sun (*top*) and shade crown (*bottom*) of *Fagus sylvatica* leaves of young seedlings (*open bars*) and adult trees (*solid bars*) after exposure to 1 × O_3 and 2 × O_3 throughout the growing seasons of 2003 and 2004. Data are means of 3–6 trees ± SD and expressed as percent difference to ambient air (1 × O_3 = 100 %). *Asterisks* denote significant differences between 1 × O_3 and 2 × O_3 at $p < 0.05$ (compiled after data from Herbinger et al. 2005, 2007)

11.5.4 Ozone Susceptibility of Beech and Spruce

After 8 years of exposure to the twice-ambient ozone regime (2 × O_3), no sweeping effects were found at cell, tissue and organ levels, neither in beech nor in spruce (Nunn et al. 2005a, b; Matyssek et al. 2010). Both tree species responded to ozone significantly, but in an inconsistent manner with substantial variation between parameters, crown parts (sun *vs.* shade), seasons and annual climatic conditions (Löw et al. 2006; Kitao et al. 2009; Olbrich et al. 2010a, Chap. 2). Nevertheless, a remarkable loss in wood production of beech occurred by 44 % at the whole-tree level under 2 × O_3, whereas spruce remained almost unaffected in this respect (Pretzsch et al. 2010). It is concluded therefore that adult trees of beech and spruce are at least as susceptible to ozone as juvenile ones, which the risk assessment on the European level has been based on (Karlsson et al. 2004), and although response mechanisms may differ (Matyssek et al. 2012).

The higher O_3 susceptibility of beech compared to spruce is in agreement with results from controlled experiments with beech and spruce saplings (Kozovits et al. 2005; Luedemann et al. 2009). The strong reduction of stem increment appears to correspond to the carbon allocation to the roots and associated carbon drain to the soil (Nikolova et al. 2010; Ritter et al. 2011), whereas leaf-level carbon gain was only marginally affected (Kitao et al. 2009; Matyssek et al. 2010, 2012). Thus, not the reduced carbon fixation in the mesophyll was responsible for the dramatically reduced carbon storage capacity of beech at "Kranzberger Forst" due to ozone as often reported from experiments under controlled conditions, but the changed carbon allocation and eventually associated root exudates, driven by O_3-caused phytohormonal perturbance (Winwood et al. 2007; Matyssek et al. 2012). To provoke plant response, ozone has to enter the leaves via stomata (Wittmann et al. 2007). When the detoxification capacity of the leaves is exceeded, conservative mechanisms are induced, developed to mitigate oxidative stress independent of the stress factor (e.g. pathogens, light, O_3; Matyssek et al. 2008) and leading to programmed cell death associated with a cascade of hormonal signals (see Sect. 11.5.2). Apart from phytohormonal signalling under O_3 stress, carbon allocation depends on C sink–source relationships and is triggered by sugar sensing mechanisms (Rolland et al. 2006, see also Chaps. 7 and 17).

Ozone uptake (COU) allows cause–effect based risk assessments as opposed to the cumulative external O_3 exposure (Löw et al. 2006; Matyssek et al. 2007b), still in use officially for that purpose. Future efforts must concentrate on calculating the ecologically meaningful "effective" doses, i.e. integrating the O_3 load taken up into the leaves and their antioxidative detoxification capacity as the actually functional basis of O_3 sensitivity (Matyssek et al. 2008). A challenge still is the quantification of volatile organic substances (VOCs) emitted by the trees, possessing the capacity for both promoting O_3 production and O_3 depletion within the leaf boundary layer, as depending on physico-chemical circumstantialities of the air (Stockwell et al. 1997; Matyssek et al. 2012). Spruce is known to produce high quantities of isoprene and monoterpenes, beech only monoterpenes and to a minor extent (Kesselmeier and Staudt 1999). Beech at "Kranzberger Forst" was found to increase aldehyde release under 2 × O_3 (Cojocariu et al. 2005). It remains to be examined, however, of whether VOC emissions are the reason for the lower O_3 susceptibility of spruce compared to beech.

Although the existence of beech and spruce does not seem to be endangered by enhanced O_3 loads on a short-term scale, as they have ample capacities of acclimation and adaptation, the functioning of the forest ecosystem as a whole—including the belowground compartment—appears to be prone to O_3-mediated perturbance as a long-term component of environmental change.

11.6 Conclusion: Contrasting Strategies of Beech and Spruce

In spite of the contrasting foliage type, crown structure and root system, European beech and Norway spruce resembled each other in many respects, identifying both as successfully competing late-successional species. Aboveground competitiveness in terms of carbon gain per unit of occupied space around foliated branches and space-related "running costs" of transpiration and respiration of the aboveground structures were similar in beech and spruce. The advantage of beech in filling the crown space with less C investment into the foliage is counterbalanced by the longer leaf longevity and lower C investments into the branch structure of spruce. The physical interaction of swaying tree crowns in both tree species as well as break-down of suppressed trees upon self-thinning resulted in the formation of crown gaps. Since the relative annual increment of crown volume was larger in sun branches of beech than spruce, the formation of crown gaps was considered to be advantageous to beech regarding further space acquisition, as trees of this species are able to more rapidly (re-)conquer abandoned canopy space relative to spruce. Therefore, the crucial trait favouring beech appears to be the fast conquest of new crown space and the more flexible, i.e. plastic response to disturbance (e.g. gap formation, drought). Findings from belowground competitiveness conformed to the foliage-level results. However belowground, competitive species performance depended on the soil layer and edaphic conditions. The vertical fine root distribution peaked in beech along the soil profile below the respective position of spruce, which is even accentuated when both species share the same soil volume (Nikolova 2007). In terms of vertical fine-root distribution, beech again is favoured when disturbances (like drought) prevail (Nikolova et al. 2010). Eventually, the competitive success of beech also belowground appears to be associated with the capacity to exploit available resources swiftly from soil (i.e. through constructing fine-roots with high specific fine-root length, Nikolova 2007). This finding supports the theory of Grime (1977) that "the competitive success is a reflection of the individual capacity to exploit resources rapidly".

Given enhanced regimes of tropospheric ozone as a stress factor, the traits mentioned above are not advantageous to beech. The species turned out to be more susceptible to ozone uptake than spruce regarding adult trees as well as seedlings, irrespective of ontogeny leading to decreased wood production. Affected by chronic ozone exposure primarily is not the carbon fixation in the mesophyll but the changed carbon allocation. The latter is triggered by hormonal signalling in response to oxidative stress, which is incited both by ozone and pathogen attack.

The amount of carbon which is available to be traded-off from investments into defence towards growth turned out to be rather limited both in beech and spruce (about 2–5 %) and, hence, is less pronounced than suggested by GDB. The proportion was not even increased under the impact of enhanced, chronic ozone exposure (Häberle et al. 2009).

We conclude that the costs of ozone tolerance are not only determined by direct C investments and turnover of defence metabolites and/or repair of ozone-induced injury but to some major extent also by redirected C allocation as part of complex sink–source relationships. The question arises as to whether stress resistance is intrinsically coupled with C allocation? May be stress avoidance mechanisms are only marginally influenced by adapted C allocation as compared to mechanisms of stress tolerance. Examples of both means of response do not give a consistent picture. Tolerance against heat (through heat shock proteins) and frost stress (by restructuring of membranes) appears to be rather inexpensive in terms of C allocation whereas tolerance against nutrient or drought stress can require high investments into the mycorrhizal root system. Costs are low, when existing structures or features can be used, e.g. switching between C3- and CAM photosynthesis pathways in the genus *Clusia* in between hours or days under drought stress (Lüttge et al. 2010).

Unravelling the plant's expenses for various stress responses remains a challenging task especially regarding the involvement of hormonal and sugar signalling processes which have been shown in this chapter to be decisive in understanding the complex tree response to ozone between the tasks of growth and defence.

Acknowledgements We gratefully acknowledge the technical assistance by Thomas Feuerbach, Peter Kuba, Josef Heckmair and Ilse Süß. We thank Prof. Dr. J. Fromm, Prof. Dr. C. Körner and Prof. Dr. U. Lüttge for helpful suggestions. The work had been funded within the interdisciplinary research programme "Growth and Parasite Defence" (SFB 607) by "Deutsche Forschungsgemeinschaft" and within the project CASIROZ (FP 5) by the European Commission (EVK2-2002-00165). We thank "Bayerische Staatsforsten" for providing the research site.

References

Achard P, Vriezen WH, Van Der Straeten D, Harberd NP (2003) Ethylene regulates Arabidopsis development via the modulation of DELLA protein growth repressor function. Plant Cell 15:2816–2825

Andersen CP (2003) Source–sink balance and carbon allocation below ground in plants exposed to ozone. New Phytol 157:213–228

Bahnweg G, Heller W, Stich S, Knappe C, Betz G, Heerdt C, Kehr RD, Ernst D, Langebartels C, Nunn AJ, Rothenburger J, Schubert R, Muller-Starck G, Werner H, Matyssek R, Sandermann H Jr (2005) Beech leaf colonization by the endophyte *Apiognomonia errabunda* dramatically depends on light exposure and climatic conditions. Plant Biol 7:659–669

Bailey-Serres J, Voesenek LACJ (2010) Life in the balance: a signaling network controlling survival of flooding. Curr Opin Plant Biol 13:489–494

Blumenröther MC, Löw M, Matyssek R, Oβwald W (2007) Flux-based response of sucrose and starch in leaves of adult beech trees (*Fagus sylvatica* L.) under chronic free-air O_3 fumigation. Plant Biol 9:207–214

Butin HH (1995) Tree diseases and disorders: causes, biology, and control in forest and amenity trees. Oxford University Press, Oxford

Čermák J, Nadezhdina N, Meiresonne L, Ceulemans R (2008) Scots pine root distribution derived from radial sap flow patterns in stems of large leaning trees. Plant Soil 305:61–75

Ciais P, Reichstein M, Viovy N, Granier A, Ogée J, Allard V, Aubinet M, Buchmann N, Bernhofer C, Carrara A, Chevallier F, De Noblet N, Friend AD, Friedlingstein F, Grünwald T, Heinesch B, Keronen P, Knohl A, Krinner G, Loustau D, Manca G, Matteucci G, Miglietta F, Ourcival JM, Papale D, Pilegaard K, Rambal S, Seufert G, Soussana JF, Sanz MJ, Schulze ED, Vesala T, Valentini R (2005) Europe-wide reduction in primary productivity caused by the heat and drought in 2003. Nature 437:529–533

Clancy KM, Wagner MR, Reich PB (1995) Ecophysiology and insect herbivory. In: Smith WK, Hinckley TM (eds) Ecophysiology of coniferous forests. Academic Press, San Diego, pp 125–180

Cojocariu C, Escher P, Häberle KH, Matyssek R, Rennenberg H, Kreuzwieser J (2005) The effect of ozone on the emission of carbonyls from leaves of adult *Fagus sylvatica*. Plant Cell Environ 28:603–611

Fassnacht KS, Gower ST, Norman JM, McMurtie RE (1994) A comparison of optical and direct methods for estimating foliage surface area in forests. Agric Forest Meteorol 71:183–207

Ferdinand JA, Fredericksen TS, Kouterick KB, Skelly JM (2000) Leaf morphology and ozone sensitivity of two open pollinated genotypes of black cherry (*Prunus serotina*) trees. Environ Pollut 108:297–302

Fine PVA, Miller ZJ, Mesones I, Irazuzta S, Alppel HM, Stevens MHH, Sääksjärvi I, Schultz JC, Coley PD (2006) The growth-defense trade-off and habitat specialization by plants in Amazonian forests. Ecology 87(Suppl 7):150–162

Fredericksen TS, Joyce BJ, Skelly JM, Steiner KC, Kolb TE, Kouterick KB, Savage JE, Snyder KR (1995) Physiology, morphology, and ozone uptake of leaves of black cherry seedlings, saplings, and canopy trees. Environ Pollut 89:273–283

Fredericksen TS, Skelly JM, Steiner KC, Kolb TE, Kouterick KB (1996) Size-mediated foliar response to ozone in black cherry trees. Environ Pollut 91:53–63

Gayler S, Leser C, Priesack E, Treutter D (2004) Modelling the effect of environmental factors on the "trade-off" between growth and defensive compounds in young apple trees. Trees 18:363–371

Gazzarrini S, McCourt P (2003) Cross-talk in plant hormone signalling: what Arabidopsis mutants are telling us. Ann Bot (Lond) 91:605–612

Genet A, Wernsdörfer H, Jonard M, Pretzsch H, Rauch M, Ponette Q, Nys C, Legout A, Ranger J, Vallet P, Saint-André L (2011) Ontogeny partly explains the apparent heterogeneity of published biomass equations for *Fagus sylvatica* in central Europe. Forest Ecol Manage 261:1188–1202

Grams TEE, Kozovits AR, Reiter IM, Winkler JB, Sommerkorn M, Blaschke H, Häberle K-H, Matyssek R (2002) Quantifying competitiveness in woody plants. Plant Biol 4:153–158

Grams TEE, Werner H, Kuptz D, Ritter W, Fleischmann F, Andersen C, Matyssek R (2011) A free-air system for long-term stable carbon isotope labeling of adult forest trees. Trees 25:187–198

Grebenc T, Kraigher H (2007) Changes in the community of ectomycorrhizal fungi and increased fine root number under adult beech trees chronically fumigated with double ambient ozone concentration. Plant Biol 9:279–287

Grime JP (1977) Evidence for the existence of three primary strategies in plants and its relevance to ecological and evolutionary theory. Am Nat 111:1169–1194

Grote R, Reiter IM (2004) Competition-dependent modelling of foliage biomass in forest stands. Trees 18:596–607

Grulke NE, Miller PR (1994) Changes in gas exchange characteristics during the life span of giant sequoia – implications for response to current and future concentrations of atmospheric ozone. Tree Physiol 14:659–668

Grulke NE, Retzlaff WA (2001) Changes in physiological attributes of ponderosa pine from seedling to mature tree. Tree Physiol 21:275–286

Haberer K, Grebenc T, Alexou M, Gessler A, Kraigher H, Rennenberg H (2007) Effects of long-term free-air ozone fumigation on δ15N and total N in *Fagus sylvatica* and associated mycorrhizal fungi. Plant Biol 9:242–252

Haberer K, Herbinger K, Alexou M, Rennenberg H, Tausz M (2008) Effects of drought and canopy ozone exposure on antioxidants in fine roots of mature European beech (*Fagus sylvatica*). Tree Physiol 28:713–719

Häberle K-H, Nunn AJ, Reiter IM, Werner H, Heller W, Bahnweg G, Gayler S, Lütz C, Matyssek R (2009) Variation of defence-related metabolites in the foliage of adult beech and spruce: a conceptual approach to approximating traded-off carbon. Eur J Forest Res 128:99–108

Herbinger K, Then C, Löw M, Haberer K, Alexous M, Koch N, Remele K, Heerdt C, Grill D, Rennenberg H, Häberle K-H, Matyssek R, Tausz M, Wieser G (2005) Tree age dependence and within-canopy variation of leaf gas exchange and antioxidative defence in *Fagus sylvatica* under experimental free-air ozone exposure. Environ Pollut 137:476–482

Herbinger K, Then C, Haberer K, Alexou M, Löw M, Remele K, Rennenberg H, Matyssek R, Grill D, Wieser G, Tausz M (2007) Gas exchange and antioxidative compounds in young beech trees under free-air ozone exposure and comparisons to adult trees. Plant Biol 9:288–297

Herms DA, Mattson WJ (1992) The dilemma of plants: to grow or defend. Q Rev Biol 67:283–335

Hruska J, Cermak J, Sustek S (1999) Mapping tree root systems with ground-penetrating radar. Tree Physiol 19:125–130

IPCC (2007) Climate change 2007: impacts, adaptations and vulnerability. In: Parry ML, Canziani OF, Palutikof JP, Van der Linden PJ, Hanson CE (eds) Contribution of the Working Group II for the fourth assessment report of the Intergovernmental Panel on Climate Change. Cambridge University Press, Cambridge

Jehnes S, Betz G, Bahnweg G, Haberer K, Sandermann H, Rennenberg H (2007) Tree internal signalling and defence reactions under ozone exposure in sun and shade leaves of European beech (*Fagus sylvatica* L.) trees. Plant Biol 9:253–264

Karlsson PE, Uddling J, Braun S, Broadmeadow M, Elvira S, Gimeno B, Le Thiec D, Oksanen E, Vandermeiren K, Wilkinson M, Emberson LD (2004) Dose-response relationships for ozone impact on the biomass accumulation of young trees of different European species based on AOT40 and cumulative leaf uptake of ozone. Atmos Environ 38:2283–2295

Karnosky DF, Werner H, Holopainen T, Percy K, Oksanen T, Oksanen E, Heerdt C, Fabian P, Nagy J, Heilman W, Cox R, Nelson N, Matyssek R (2007) Free-air exposure systems to scale up ozone research to mature trees. Plant Biol 9:181–190

Kesselmeier J, Staudt M (1999) Biogenic volatile organic compounds (VOC): an overview on emission, physiology and ecology. J Atmos Chem 33:23–88

Kitao M, Löw M, Heerdt C, Grams TEE, Häberle K-H, Matyssek R (2009) Effects of chronic elevated ozone exposure on gas exchange responses of adult beech trees (*Fagus sylvatica*) as related to the within-canopy light gradient. Environ Pollut 157:537–544

Kolb TE, Matyssek R (2001) Limitations and perspectives about scaling ozone impact in trees. Environ Pollut 115:373–393

Kolb TE, Fredericksen TS, Steiner KC, Skelly JM (1997) Issues in scaling tree size and age responses to ozone: a review. Environ Pollut 98:195–208

Koricheva J, Larsson S, Haukioja E, Keinänen M (1998) Regulation of woody plant secondary metabolism by resource availability: hypothesis testing by means of meta-analysis. Oikos 83:212–226

Kozovits AR, Matysssek R, Blaschke H, Göttlein A, Grams TEE (2005) Competition increasingly dominates the responsiveness of juvenile beech and spruce to elevated CO_2 and/or O_3 concentrations throughout two subsequent growing seasons. Glob Change Biol 11:1387–1401

Langebartels C, Kangasjärvi J (2004) Ethylene and jasmonate as regulators of cell death in disease resistance. In: Sandermann H (ed) Molecular ecotoxicology of plants, vol 170. Springer, Heidelberg, pp 75–109

Lerdau M, Gershenzon J (1997) Allocation theory and chemical defense. In: Bazzaz FA, Grace J (eds) Plant resource allocation. Academic Press, San Diego, pp 265–277

Löw M, Herbinger K, Nunn AJ, Häberle K-H, Leuchner M, Heerdt C, Werner H, Wipfler P, Pretzsch H, Tausz M, Matyssek R (2006) Extraordinary drought of 2003 overrules ozone impact on adult beech trees (*Fagus sylvatica*). Trees 20:539–548

Luedemann G, Matyssek R, Fleischmann F, Grams TEE (2005) Acclimation to ozone affects host-pathogen interaction and competitiveness for nitrogen in juvenile *Fagus sylvatica* and *Picea abies* trees infected with *Phytophthora citricola*. Plant Biol 7:640–649

Luedemann G, Matyssek R, Winkler JB, Grams TEE (2009) Contrasting ozone × pathogen interaction as mediated through competition between juvenile European beech (*Fagus sylvatica*) and Norway spruce (*Picea abies*). Plant Soil 323:47–60

Lüttge U, Kluge M, Thiel G (2010) Botanik. Wiley-VCH, Weinheim

Matyssek R, Sandermann H (2003) Impact of ozone on trees: an ecophysiological perspective, vol 64, Progress in botany. Springer, Heidelberg, pp 349–404

Matyssek R, Reich PB, Oren R, Winner WE (1995) Response mechanisms of conifers to air pollutants. In: Smith WK, Hinckley TH (eds) Physiological ecology of coniferous forests, Physiological ecology series. Academic, New York, pp 255–308

Matyssek R, Agerer R, Ernst D, Munch JC, Osswald W, Pretzsch H, Priesack E, Schnyder H, Treutter D (2005a) The plant's capacity in regulating resource demand. Plant Biol 7:560–580

Matyssek R, Wieser G, Nunn A, Löw M, Then C, Herbinger K, Blumenröther M, Alexou M, Jehnes S, Heerdt C, Koch N, Häberle K-H, Haberer K, Werner H, Tausz M, Fabian P, Rennenberg H, Grill D, Oßwald W (2005b) How sensitive are forest trees to ozone? – new research on an old issue. In: Omasa K, Nouchi I, de Kok LJ (eds) Plant response to air pollution and global change. Springer, Tokyo, pp 21–28

Matyssek R, Bahnweg G, Ceulemans R, Fabian P, Grill D, Hanke DE, Kraigher H, Osswald W, Rennenberg H, Sandermann H, Tausz M, Wieser G (2007a) Synopsis of the CASIROZ case study: carbon sink strength of *Fagus sylvatica* L. in a changing environment – experimental risk assessment of mitigation by chronic ozone impact. Plant Biol 9:163–180

Matyssek R, Bytnerowic A, Karlsson P-E, Paoletti E, Sanz M, Schaub M, Wieser G (2007b) Promoting the O_3 flux concept for forest trees. Environ Pollut 146:587–607

Matyssek R, Sandermann H, Wieser G, Booker F, Cieslik S, Musselman R, Ernst D (2008) The challenge of making ozone risk assessment for forest trees more mechanistic. Environ Pollut 156:567–582

Matyssek R, Wieser G, Patzner K, Blaschke H, Häberle K-H (2009) Transpiration of forest trees and stands at different altitude: consistencies rather than contrasts. Eur J Forest Res 128:579–596

Matyssek R, Wieser G, Ceulemans R, Rennenberg H, Pretzsch H, Haberer K, Löw M, Nunn AJ, Werner H, Wipfler P, Oßwald W, Nikolova P, Hanke DE, Kraigher H, Tausz M, Bahnweg G, Kitao M, Dieler J, Sandermann H, Herbinger K, Grebenc T, Blumenröther M, Deckmyn G, Grams TEE, Heerdt C, Leuchner M, Fabian P, Häberle K-H (2010) Enhanced ozone strongly reduces carbon sink strength of adult beech (*Fagus sylvatica*) Resume from the free-air fumigation study at Kranzberg Forest. Environ Pollut 158:2527–2532

Matyssek R, Kosovits AR, Schnitzler J-P, Pretzsch H, Dieler J, Wieser G (2012) Forest trees under air pollution as a factor of climate change. In: Tausz M, Grulke N (eds) Trees in a changing environment: ecophysiology, adaptation and future survival, Plant ecophysiology series. Springer, Berlin (in press)

Mehlhorn H, Wellburn AR (1987) Stress ethylene formation determines plant sensitivity to ozone. Nature 327:417–418

Miller JD, Richard N, Arteca RN, Pell EJ (1999) Senescence-associated gene expression during ozone-induced leaf senescence in Arabidopsis. Plant Physiol 120(4):1015–1024

Mutikainen P, Walls M, Ovasaka J, Keinänen M, Julkunen-Tiitto R, Vapaavuori E (2002) Costs of herbivore resistance in clonal saplings of *Betula pendula*. Oecologia 133:364–371

Nemhauser JL, Fangxin Hong F, Chory J (2006) Different plant hormones regulate similar processes through largely nonoverlapping transcriptional responses. Cell 126:467–475

Nikolova PS (2007) Below-ground competitiveness of adult beech and spruce trees: resource investments versus returns. Doctoral thesis, Weihenstephan Center of Life Sciences, TU München

Nikolova PS, Raspe S, Andersen CP, Mainiero R, Helmut Blaschke H, Matyssek R, Häberle K-H (2009) Effects of the extreme drought in 2003 on soil respiration in a mixed forest. Eur J Forest Res 128:87–98

Nikolova PS, Andersen CP, Blaschke H, Matyssek R, Häberle K-H (2010) Belowground effects of enhanced tropospheric ozone and drought in a beech/spruce forest (*Fagus sylvatica* L./*Picea abies* [L.] Karst). Environ Pollut 158:1071–1078

Novák J (1975) Quantitative analysis by gas-chromatography. Marcel Dekker, New York

Nunn AJ, Reiter IM, Häberle K-H, Werner H, Langebartels C, Sandermann H, Heerdt C, Fabian P, Matyssek R (2002) "Free-air" ozone canopy fumigation in an old-growth mixed forest: concept and observations in beech. Phyton 42:105–119

Nunn AJ, Anegg S, Betz G, Simons S, Kalisch G, Seidlitz HK, Grams TEE, Häberle KH, Matyssek R, Bahnweg G, Sandermann H, Langebartels C (2005a) Role of ethylene in the regulation of cell death and leaf loss in ozone-exposed European beech. Plant Cell Environ 28:886–897

Nunn AJ, Reiter IM, Häberle K-H, Langebartels C, Bahnweg G, Pretzsch H, Sandermann H, Matyssek R (2005b) Response pattern in adult forest trees to chronic ozone stress: identification of variations and consistencies. Environ Pollut 136:365–369

Nunn AJ, Wieser G, Reiter IM, Häberle K-H, Grote R, Havranek WM, Matyssek R (2006) Testing the unifying theory of ozone sensitivity with mature trees of *Fagus sylvatica* and *Picea abies*. Tree Physiol 26:1391–1403

Olbrich M, Gerstner E, Bahnweg G, Häberle K-H, Matyssek R, Welzl G, Ernst D (2010a) Transcriptional signatures in leaves of adult European beech trees (*Fagus sylvatica* L.) in an experimentally enhanced free air ozone setting. Environ Pollut 158:977–982

Olbrich M, Knappe C, Wenig M, Gerstner E, Häberle K-H, Kitao M, Matyssek R, Stich S, Leuchner M, Werner H, Schlink K, Muller-Starck G, Welzl G, Scherb H, Ernst D, Heller W, Bahnweg G (2010b) Ozone fumigation (twice ambient) reduces leaf infestation following natural and artificial inoculation by the endophytic fungus *Apiognomonia errabunda* of adult European beech trees. Environ Pollut 158:1043–1050

Pokorny R, Marek MV (2000) Test of accuracy of LAI estimation by LAI-2000 under artificially changed leaf to wood area proportions. Biol Plant 43:537–544

Poorter H, Villar R (1997) The fate of aquired carbon in plants: chemical composition and construction costs. In: Bazzaz FA, Grace J (eds) Plant resource allocation. Academic Press, San Diego, pp 39–71

Pretzsch H (2010) Re-evaluation of allometry: state-of-the-art and perspective regarding individuals and stands of woody plants, vol 71, Progress in botany. Springer, Heidelberg, pp 339–369

Pretzsch H, Kahn M, Grote R (1998) Die Fichten-Buchen-Mischbestände des Sonderforschungsbereiches "Wachstum oder Parasitenabwehr?" im Kranzberger Forst. Forstwiss Cblt 117:241–257

Pretzsch H, Dieler J, Matyssek R, Wipfler P (2010) Tree and stand growth of mature Norway spruce and European beech under long-term ozone fumigation. Environ Pollut 158:1061–1070

Rebel K (1922) Wiederaufforstung der 1920er Windwurfflächen auf der schwäbischen-bayrischen Hochebene. Faksimile Ausgabe. Roland, Bremen

Reich PB (1987) Quantifying plant response to ozone: a unifying theory. Tree Physiol 3:63–91

Reiter IM (2004) Space-related resource investments and gains of adult beech and spruce as quantification of aboveground competitiveness. Doctoral thesis, Weihenstephan Center of Life Sciences, TU München

Reiter IM, Häberle K-H, Nunn AJ, Heerdt C, Reitmayer H, Grote R, Matyssek R (2005) Competitive strategies in adult beech and spruce: space-related foliar carbon investment versus carbon gain. Oecologia 146:337–349

Riefler M, Novak O, Strnad M, Schmülling T (2006) Arabidopsis cytokinin receptor mutants reveal functions in shoot growth, leaf senescence, seed size, germination, root development, and cytokinin metabolism. Plant Cell 18:40–54

Ritter W, Andersen CP, Matyssek R, Grams TEE (2011) Carbon flux to woody tissues in a beech/spruce forest during summer and in response to chronic elevated O_3 exposure. Biogeosci Discuss 8:4131–4161

Rolland F, Baena-Gonzalez E, Sheen J (2006) Sugar sensing and signaling in plants: conserved and novel mechanisms. Ann Rev Plant Biol 57:675–709

Ross JJ, Reid JB (1986) Intemode length in *Pisum.* The involvement of ethylene with the gibberellin-insensitive erectoides phenotype. Physiol Plant 67:673–679

Ross JJ, Reid JB (2010) Evolution of growth-promoting plant hormones. Funct Plant Biol 37:795–805

Ross JJ, Weston DE, Davidson SE, Reid JB (2011) Plant hormone interactions: how complex are they? Physiol Plant 141:299–309

Samuelson LJ, Edwards GS (1993) A comparison of sensitivity to ozone in seedlings and trees of Quercus rubra L. New Phytol 125:373–379

Samuelson LJ, Kelly JM, Mays PA, Edwards GS (1996) Growth and nutrition of Quercus rubra seedlings and mature trees after three seasons of ozone exposure. Environ Pollut 91:317–320

Sandermann H (1996) Ozone and plant health. Ann Rev Phytopathol 34:347–366

Sandermann H, Wellburn AR, Heath RL (1997) Forest decline and ozone: a comparison of controlled chamber and field experiments, vol 127, Ecological studies. Springer, Berlin

Sandermann H, Ernst D, Heller W, Langebartels C (1998) Ozone: an abiotic elicitor of plant defence reactions. Trends Plant Sci 3:47–50

Schulze ED, Beck E, Müller-Hohenstein K (2002) Pflanzenökologie. Spektrum Akademischer Verlag, Heidelberg

Sharp RE, LeNoble ME, Else MA, Thorne ET, Gherardi F (2000) Endogenous ABA maintains shoot growth in tomato independently of effects on plant water balance: evidence for an interaction with ethylene. J Exp Bot 51:1575–1584

Sitch S, Cox PM, Collins WJ, Huntingford C (2007) Indirect radiative forcing of climate change through ozone effects on the land-carbon sink. Nature 448:791–794

Stamp N (2003) Out of the quagmire of plant defense hypotheses. Q Rev Biol 78:23–55

Stockwell WR, Kirchner F, Kuhn M, Seefeld S (1997) A new mechanism for regional atmospheric chemistry modeling. J Geophys Res 102:25847–25879

Taylor GE Jr, Hanson PJ (1992) Forest trees and tropospheric ozone. Role of canopy deposition and leaf uptake in developing exposure-response relationships. Agric Ecosyst Environ 42:255–273

Tegischer K, Tausz M, Grill D, Wieser G (2002) Tree-age and needle-age dependent variations of antioxidants and photoprotective pigments in spruce needles at the alpine timberline. Tree Physiol 22:591–596

Villar R, Robleto JR, de Jong Y, Poorter H (2006) Differences in construction costs and chemical composition between deciduous and evergreen woody species are small as compared to differences among families. Plant Cell Environ 29:1629–1643

Weigt R (2010) Effects of elevated ground-level ozone on nitrogen acquisition of mature European beech (*Fagus sylvatica*) and Norway spruce (*Picea abies*) trees. Doctoral thesis, Weihenstephan Center of Life Sciences, TU München

Wellburn FAM, Lau K-K, Milling MK, Wellburn AR (1996) Drought and air pollution affect nitrogen cycling and free-radical scavenging in *Pinus halepensis* Mill. J Exp Bot 47:1361–1367

Werner H, Fabian P (2002) Free-air fumigation of mature trees. Environ Sci Pollut Res 9:117–121

Wieser G, Tegischer K, Tausz M, Häberle K-H, Grams TEE, Matyssek R (2002a) Age effects on Norway spruce (*Picea abies*) susceptibility to ozone uptake: a novel approach relating stress avoidance to defense. Tree Physiol 22:583–590

Wieser G, Hecke K, Tausz M, Häberle K-H, Grams TEE, Matyssek R (2002b) The role of antioxidative defense in determining ozone sensitivity of Norway spruce (*Picea abies* [L.] Karst.) across tree age: implications for the sun and shade crown. Phyton 42:245–253

Wieser G, Hecke K, Tausz M, Häberle K-H, Grams TEE, Matyssek R (2003) The influence of microclimate and tree age on the defense capacity of European beech (*Fagus sylvatica* L.) against oxidative stress. Ann Forest Sci 60:131–135

Winwood J, Pate AE, Price AJ, Hanke DE (2007) Effects of long-term, free-air ozone fumigation on the cytokinin content of mature beech trees. Plant Biol 9:265–278

Wipfler P, Seifert T, Heerdt C, Werner H, Pretzsch H (2005) Growth of adult Norway spruce (*Picea abies* [L.] Karst.) and European beech (*Fagus sylvatica* L.) under free-air ozone fumigation. Plant Biol 7:611–618

Wittmann C, Matyssek R, Pfanz H, Humar M (2007) Effects of ozone impact on the gas exchange and chlorophyll fluorescence of juvenile birch stems (*Betula pendula* Roth.). Environ Pollut 150:258–266

Yamamoto F, Kozlowski TT (1987) Regulation by auxin and ethylene of responses of *Acer negundo* seedlings to flooding of soil. Environ Exp Bot 27:329–340

Zeleznik P, Hrenko M, Then C, Koch N, Grebenc T, Levanic T, Kraigher H (2007) CASIROZ: root parameters and types of ectomycorrhiza of young beech plants exposed to different ozone and light regimes. Plant Biol 9:298–308

Zellnig G, Tausz M, Pesec B, Grill D, Müller M (2000) Effects of glutathione on thiol redox systems, chromosomal aberrations, and the ultrastructure of meristematic root cells of *Picea abies* (L.) Karst. Protoplasma 212:227–235

Chapter 12
Growth and Space Use in Competitive Interactions Between Juvenile Trees

T.E.E. Grams, M.J. Daigo, J.B. Winkler, S. Gayler, and R. Matyssek

12.1 Introduction

Plants are exposed to a plethora of environmental influences that are affecting their life cycle by either abiotic, e.g., physicochemical or biotic factors such as associations with microorganisms or interactions with animals. Those influences create a dilemma in resource allocation (see Chaps. 1 and 14) as they rarely act as single impacts but are typically multifactorial. For instance, under conditions of climate change, increase of atmospheric carbon dioxide (CO_2) concentration is paralleled by rising temperature and enhanced risk of drought events. These multifactorial influences do not affect plants as isolated individuals but as parts of complex interactions with their neighboring plants, comprising positive and negative, i.e., facilitative and competitive interactions, respectively.

This chapter discusses the effects of elevated concentrations of CO_2 and ozone (O_3, as an intrinsic component of climate change; Fowler 2008; Sitch et al. 2007) on growth and resource allocation of two ecologically and economically important tree species in Central Europe, European beech (*Fagus sylvatica* L.), and Norway spruce (*Picea abies* Karst.). Focus is on the mechanisms of intra- and inter-specific competition between the two tree species and on how competitive interactions may

T.E.E. Grams (✉) • M.J. Daigo • R. Matyssek
Chair of Ecophysiology of Plants, Technische Universität München, Hans-Carl-von-Carlowitz-Platz 2, 85354 Freising, Germany
e-mail: grams@wzw.tum.de

J.B. Winkler
Research Unit Environmental Simulation, Institute of Biochemical Plant Pathology, German Research Center for Environmental Health, Helmholtz Zentrum München, Ingolstädter Landstraße 1, 85764 Neuherberg, Germany

S. Gayler
Water & Earth System Science (WESS) Competence Cluster, c/o Universität Tübingen, Hölderlinstr. 12, 72074 Tübingen, Germany

R. Matyssek et al. (eds.), *Growth and Defence in Plants*, Ecological Studies 220,
DOI 10.1007/978-3-642-30645-7_12,

modulate impacts of elevated O_3 and CO_2. To this end, we review a series of growth chamber experiments where beech and spruce were grown in isolation or under intra- and inter-specific competition.

Competition of plants for resources is conceived as the integral of spatio-temporal resource use (Küppers 1989; Matyssek and Schulze 1987; Schwinning 1996). Along this line, an approach to interpret competition as space-related resource investments and gains was introduced for woody plants to quantify competitiveness by space-related efficiencies of (1) resource investment into standing biomass (space occupation) and (2) resource gain (space exploitation) (Grams et al. 2002). Subsequently, this approach has been employed and promoted (Grams and Matyssek 2010; Kozovits et al. 2005b; Luedemann et al. 2009; Rodenkirchen et al. 2009). Understanding competitiveness of plants as an optimization process in space use has been challenging the conventional consideration of competitive success as a function of maximization of resource gain. Or in other terms, as expressed by Fakhri A. Bazzaz during the first international meeting of SFB 607 in 2001 "You (i.e. the plant) don't have to optimize, you have to maximize." In this chapter we will (1) valuate this question, i.e., whether the competitive success of a plant is related to an efficient space use, i.e., optimization strategy, or conversely to the maximization of resource gain as such, i.e., maximization strategy. It will (2) be examined further of whether effects of elevated O_3 and CO_2 concentrations and the combination of both are modified by the different competitive settings, i.e., growth in isolation, mono- or mixed culture.

12.2 Material and Methods

12.2.1 Plants, Climate Conditions, and O_3/CO_2 Treatments

We report on a total of four consecutive phytotron experiments which were performed under similar environmental conditions (regarding climate and O_3/CO_2 regimes) in four walk-in phytotrons (size ca. 2.8 m × 3.4 m, Fig. 12.1a; Kozovits et al. 2005a; Payer et al. 1993; Thiel et al. 1996). The experiments were run during the years 1995 through 2005 (see Table 12.1) and are described in detail elsewhere (Grams et al. 1999; Kozovits et al. 2005a; Luedemann et al. 2005; Ritter et al. submitted). Trees were planted in forest soil (dystric cambisol, pH of about 4.5). In the first experiment beech trees were grown individually in 10 L pots, whereas in the subsequent three experiments 20 plants were grown together in one container (soil volume of 62 L, with an area of 0.56 m × 0.37 m, soil depth of 0.30 m). The container experiments (Exp. 2–4, see Table 12.1) used either monocultures or "one-by-one" beech/spruce mixtures. In each case, the 20 trees were arranged in rows of 4 × 5 individuals (Fig. 12.1b).

After planting, trees were kept for one growing season in a climate-controlled greenhouse under outside conditions at either ambient or elevated

Fig. 12.1 (**a**) One of the four walk-in phytotrons of the "GSF—National Research Center for Environment and Health" (present name: "Helmholtz Zentrum München—German Research Center for Environmental Health") in Munich/Germany. (**b**) *Top view* on juvenile spruce and beech trees (10 each) grown in mixed culture in Experiment 3. (**c–f**) Photographs of the six central trees of beech and spruce (three each) grown together in mixed culture in Experiment 3. The 14 trees grown at the edge of the container have been cut away to give better view on the central six tree individuals

(i.e., ambient + 300 μL L^{-1}) CO_2 concentrations. During the following spring, plants were transferred to the four phytotrons. We reproduced the climate conditions and O_3 regimes throughout entire growing seasons as previously recorded at forest sites with fluctuating ambient or elevated (i.e., twice-ambient but restricted to 150 nL L^{-1}) O_3 levels. In the first and second experiment

Table 12.1 Overview on experiments summarized in Chap. 12

	Duration[a] (years)	Planting pattern	Gaseous treatments: Ambient air (control)	$+O_3$	$+CO_2$	$+O_3/+CO_2$	Planting density[b]: Plants/m^2	DM/m^2 ($g\,m^{-2}$)	DM/soil volume ($g\,L^{-1}$)
Exp. 1	1 + 1 1995–1996	Individual trees of beech	✓	✓	✓	✓	25	1,640 ± 106	6.61 ± 0.48
Exp. 2	1 + 2 1998–2000	Mono- and mixed cultures of beech and spruce	✓	✓	✓	✓	96	3,947 ± 196	12.89 ± 1.55
Exp. 3	1 + 2 2001–2003	Mixed culture of beech and spruce	✓	✓	–	–	96	2,448 ± 213	8.18 ± 1.10
Exp. 4	1 + 1 2004–2005	Mono- and mixed culture of beech and spruce	✓	✓	–	–	96	644 ± 23	2.16 ± 0.22

[a]Pretreatment adaptation years in a climate-controlled glasshouse plus experimental years in the four phytotrons of the "GSF—National Research Center for Environment and Health" (present name: "Helmholtz Zentrum München—German Research Center for Environmental Health")

[b]Dry mass at the end of the experiment. Data originate from Jungermann (1998), Kozovits et al. (2005a, b), Luedemann et al. (2005, 2009) and Ritter et al. (submitted)

(Table 12.1, Grams et al. 1999; Kozovits et al. 2005a) four gaseous treatments were established, denoted as follows: ambient CO_2/ambient O_3 = "gaseous control", ambient CO_2/elevated O_3 = "+O_3", elevated CO_2/ambient O_3 = "+CO_2" and elevated CO_2/elevated O_3 = "+CO_2/+O_3". In the third and fourth experiment (Table 12.1, Luedemann et al. 2005, 2009; Ritter et al. submitted), focus was on effects of ambient and elevated O_3 under ambient CO_2 concentrations ("gaseous control" and "+O_3", respectively). During winter, plants were kept outdoors in open-top chamber, where corresponding CO_2 concentrations were maintained (Exp. 1 and 2) or under a pergola as shelter against hard frost events (Exp. 3 and 4).

12.2.2 Whole-Tree Relative Growth Rate

At the end of each experimental growing season, biomass of the single potted trees (Exp. 1) or of the six central trees of each container (Exp. 2–4) was determined through destructive harvests or allometric relations (Kozovits et al. 2005a). Together with the starting biomass of trees, the annual whole-tree RGR was calculated as (Hunt et al. 2002):

$$\mathrm{RGR} = \frac{\ln(\mathrm{Biomass}_{t_1}) - \ln(\mathrm{Biomass}_{t_0})}{t_1 - t_0}, \tag{12.1}$$

where $\mathrm{Biomass}_{t_0}$ and $\mathrm{Biomass}_{t_1}$ represent the biomass at the end of two subsequent years, i.e., the years t_0 and t_1, respectively.

12.2.3 Analysis of Competitiveness

Above- and belowground competitiveness was quantified by two space-related efficiencies of resource use. The efficiency of space occupation aboveground was calculated as the space occupied by the crown per unit of biomass investments into stem, branches, and foliage (for details see Grams et al. 2002; Kozovits et al. 2005b). Belowground the occupied space was assumed as the soil volume around the roots and calculated from the total fine root length and the radius of the depletion zone of water around roots (Daigo et al. submitted). A radius of 20 mm was considered as an adequate approximation (Garrigues et al. 2006; Syring and Claassen 1995; see also Chap. 10). The second parameter of competitiveness, the efficiency of space exploitation, was calculated as the resources acquired from the occupied space, i.e., per unit volume. Aboveground this parameter was calculated as the annual C gain retrieved per unit of occupied crown space. Annual C gain was quantified through a photosynthesis model parameterized for the study trees (Falge et al. 1996; Kozovits et al. 2005b). Belowground the uptake of water from the soil, assessed through the photosynthesis model as annual transpiration, was expressed per unit of the occupied root space.

12.2.4 Statistical Analysis

We rely on the statistical analysis of biomass development performed in each phytotron study (Grams et al. 1999; Grams and Matyssek 1999; Kozovits et al. 2005a, b; Luedemann et al. 2005, 2009; Ritter et al. submitted). In this present synthesis, coefficients of correlations between RGR and efficiencies of space-related competitiveness were calculated using regressions in SPSS (SPSS Inc., Chicago, USA).

12.3 Results

The intended synthesis requires the comparison of the key findings on the biomass development (i.e., whole-tree relative growth rate, RGR) and competitive efficiencies (i.e., space use). Therefore, corresponding datasets from the above-mentioned experiments are presented in the following.

The whole-tree relative growth rate (RGR) of juvenile beech trees grown isolated (Exp. 1) was significantly reduced from 1.92 under gaseous control conditions to 1.68 under $+O_3$ (Table 12.2). Elevated CO_2 did not enhance RGR but diminished the adverse O_3 impact (RGR of 1.85 under $+O_3/+CO_2$). The overall range of RGR was similar for juvenile beech trees grown in mono- or mixed culture (20 trees per container) during the first experimental year of Exp. 2 when biomass density was still low. In the second year with a final biomass of about 4,000 g m^{-2} soil surface, all three gaseous treatments resulted in reduction of RGR by 20–40% compared to the control in monoculture (Fig. 12.2). These adverse effects were intensified in mixture with spruce as the reductions in RGR relative to the control were about 60% under $+CO_2$ and $+O_3/+CO_2$ and about 95% under $+O_3$. The enhancement of adverse O_3 effects on beech grown in competition with spruce was confirmed by Exp. 3 and 4 where RGR at the end of the experiment under $+O_3$ was reduced by about 70% and 50%, respectively (Fig. 12.2). Hence, adverse O_3 effects on beech were stronger when growing in mixed culture with spruce than in beech monoculture.

The range of RGR of beech observed under the various gaseous treatments was distinctly larger than of spruce under the same conditions (see RGR of spruce plotted versus RGR of beech in Fig. 12.3). Juvenile beech trees displayed a range from almost 0.0 to about 2.5 with the highest RGR in monoculture and lowest when grown in mixture with spruce. In contrast, RGR of juvenile spruce trees (see *y*-axis in Fig. 12.3) was restricted to a much narrower range from about 0.4 to 1.3 (with the exception of one high RGR of 1.7). This illustrates the larger phenotypic plasticity of beech compared to spruce. In general, data on RGR of the two species under the different treatments in mono- and mixed culture (Exp. 2–4) do not follow the 1-to-1 line in Fig. 12.3, indicating different RGR of the two species under identical conditions, i.e., same treatments and experiments. At lower growth rates of

Table 12.2 Whole-tree relative growth rate (RGR) of juvenile beech trees under the four gaseous treatments (means ± SE) in isolation (i.e., one tree per pot) in Exp. 1 or in mono- and mixed cultures in Exp. 2–4

			RGR							
	Year	Culture	Control		$+O_3$		$+CO_2$		$+O_3/+CO_2$	
Exp. 1	1996	Isolated plants	1.92	±0.03	1.68	±0.08	1.92	±0.05	1.85	±0.06
Exp. 2	1999	Monoculture	2.20	±0.11	2.29	±0.08	2.46	±009	2.39	±0.09
		Mixed culture	1.52	±0.13	1.69	±0.10	1.72	±0.13	1.72	±0.13
	2000	Monoculture	1.36	±0.14	1.00	±0.23	0.80	±0.11	1.10	±0.13
		Mixed culture	0.58	±0.18	0.01	±0.17	0.20	±0.23	0.21	±0.12
Exp. 3	2002	Mixed culture	1.91	±0.25	1.91	±0.13	–	–	–	–
	2003	Mixed culture	1.14	±0.22	0.34	±0.18	–	–	–	–
Exp. 4	2005	Monoculture	0.82	±0.07	0.91	±0.11	–	–	–	–
		Mixed culture	1.35	±0.18	0.67	±0.20	–	–	–	–

Data originate from Jungermann (1998), Kozovits et al. (2005a, b), Luedemann et al. (2005, 2009) and Ritter et al. (submitted)

beech, i.e., <0.6, RGR of spruce was larger than that of beech. Conversely, for RGR of beech above 0.9, RGR of spruce was lower than that of beech.

RGR of juvenile beech trees was positively related to the efficiency in aboveground space occupation (see logarithmic fit in Fig. 12.4). Here RGR is understood as a measure of competitive success of a tree and related to its aboveground space occupation, calculated as the ratio of occupied crown space per unit of aboveground biomass (sum of leaf, branch, and stem). The lowest RGR of juvenile beech (<0.5) has been observed in mixture with spruce (open symbols) and corresponds to an occupied crown space of less than 400 $cm^3\ g^{-1}$ biomass. We did not find positive correlations of RGR with aboveground space exploitation, calculated as carbon gain per unit of occupied crown space, or with annual carbon gain in absolute terms. Likewise, neither efficiencies in belowground space occupation and exploitation (i.e., occupied root volume or water uptake per unit of biomass, respectively) nor total resource gain in absolute terms (i.e., annual water uptake) were related to RGR (data not shown). In a similar way in spruce, none of the above-mentioned correlations was significant (data not shown), reflecting the low responsiveness and, hence, low phenotypic plasticity compared to beech.

12.4 Discussion

In this chapter, we aimed at answering two questions. First, whether effects of $+O_3$ and $+CO_2$ and the combination of both ($+O_3/+CO_2$) are modified by the different competitive settings, i.e., growth in isolation, mono- or mixed culture and second, whether the competitive success of a plant is related to an efficient space use, i.e., optimization strategy, or conversely to the maximization of resource gain as such, i.e., maximization strategy.

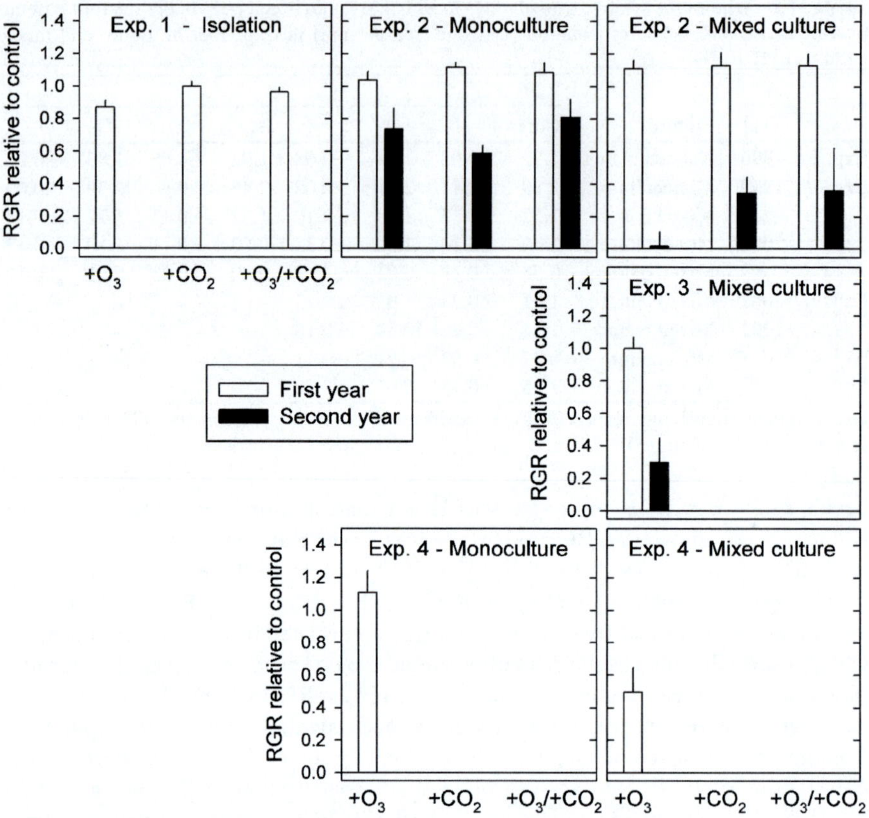

Fig. 12.2 Changes in whole-tree relative growth rate (RGR) of juvenile beech trees under +O_3, +CO_2, and +O_3/+CO_2 relative to the RGR under gaseous control conditions. Data are originating from Experiments 1 (isolated growth in pots: Jungermann (1998)), Experiment 3 (growth in mixed culture: Luedemann et al. (2005, 2009)) and Experiments 2 and 4 (growth in both mono- and mixed cultures: Kozovits et al. (2005a, b) and Ritter et al. (submitted)). *Open* and *closed bars* represent data from the first and second experimental year (see Table 12.1) under the corresponding gaseous treatments

In general, whole-tree RGR of beech was found to be negatively affected by elevated O_3 concentrations, a result consistent with preceding experiments (Langebartels et al. 1997) and recently confirmed for adult beech trees (see Chap. 11, Matyssek et al. 2010a, b). However, the extent of the O_3-related decline in RGR was dependent on the competitive setting and distinctly enhanced in mixture with spruce and at high biomass densities. We therefore conclude that experiments with single potted plants or those grown under low competitive pressure have only limited ecological significance relative to corresponding responses of plant grown in monoculture and, in particular, in mixed culture. This

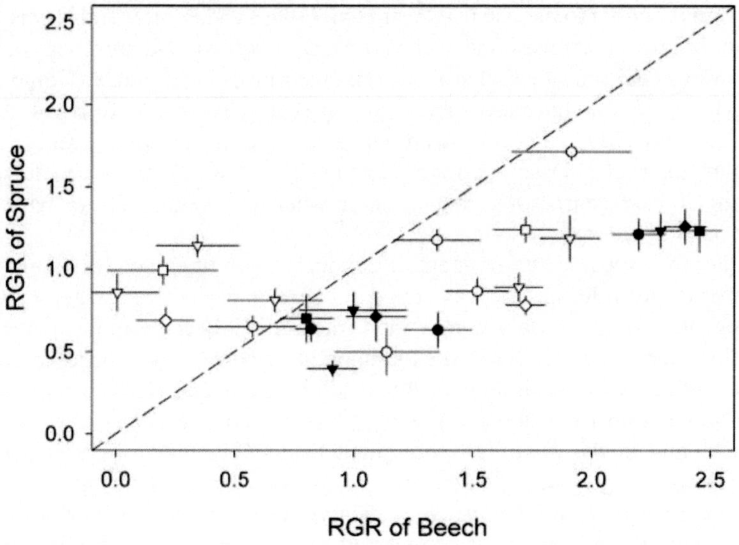

Fig. 12.3 Whole-tree relative growth rate (RGR) of juvenile spruce versus beech grown in mono- or mixed cultures (*closed* and *open symbols*, respectively). *Circles* denote gaseous control, *triangles* +O_3, *squares* +CO_2, and *diamonds* represent the +O_3/+CO_2 treatment. Each symbol gives the RGR of spruce (*y*-axis) versus beech (*x*-axis) under the same treatment and in the same experiment. Data originate from Kozovits et al. (2005a, b), Luedemann et al. (2005, 2009) and Ritter et al. (submitted)

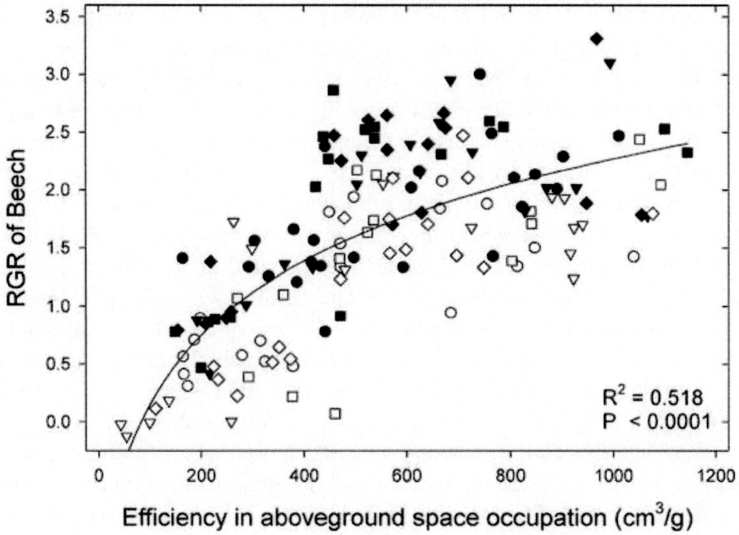

Fig. 12.4 Whole-tree relative growth rate (RGR) of juvenile beech trees correlated with the efficiency in aboveground space occupation, calculated as the occupied crown space per investment of aboveground biomass. *Circles* denote gaseous control, *triangles* +O_3, *squares* +CO_2, and *diamonds* represent the +O_3/+CO_2 treatment. *Closed* and *open symbols* denote growth in mono- and mixed culture, respectively. Data originate from Kozovits et al. (2005a, b)

conclusion is similar to the one by Navas et al. (1999) and Poorter and Navas (2003) for plant biomass responses under elevated CO_2 as growth enhancement in mixed communities could not be scaled from responses of isolated plants (Körner 2006; Millard et al. 2007). Our experiments support such findings on effects of elevated CO_2 as similarly RGR of beech was diminished under intense intra-specific and, in particular, inter-specific competition. Conversely, RGR of spruce benefited from elevated CO_2 concentrations, in particular when grown in mixture with beech (Kozovits et al. 2005a).

Under the specific experimental conditions of the presented phytotron experiments juvenile spruce was found to be a superior competitor compared to beech, in particular when competitive interactions became more intense with increasing plant biomass. Such an experimental outcome may depend on environmental conditions such as light intensity or soil moisture and pH. The dominance of spruce was confirmed when trees were grown in an experiment on acidic soil similar to that in the phytotrons presented here (Körner 2003b; Spinnler et al. 2002). However, growth on calcareous soils favored beech and the competitive advantage of spruce largely vanished. Thus, the better growth performance of spruce compared to beech in the presented study is not the key finding. Instead, focus is on the mechanistic grounds of competitive interactions between the two species. Whole-tree RGR of beech, used as a measure of competitive success, was positively related to the efficiency of aboveground space occupation, i.e., the relation of occupied crown space per unit of biomass investment in leaves, branches, and stem (Fig. 12.4). Only at very high efficiencies of aboveground space occupation, i.e., when the logarithmic fit in Fig. 12.4 starts to saturate, RGR appears to be limited by other factors (e.g., carbon availability). Beech displayed a large range in aboveground space occupation with crown volumes between 50 and 1,100 cm^3 per g of invested aboveground biomass. Apparently, this high phenotypic plasticity enables beech to escape from intense competition with spruce at the expense of less efficient aboveground space occupation. For example in Fig. 12.1d (see arrow), a small crown space is occupied by seven leaves that are supported by a relatively large branch and stem biomass, resulting in a low efficiency of aboveground space occupation. This high phenotypic plasticity of beech and its shift to rather inefficient aboveground space use in mixture was realized as a crucial factor in aboveground competition with spruce in experimental (Grams and Matyssek 2010; Kozovits et al. 2005b) and modeling studies (Gayler et al. 2006; see Chaps. 15 and 17). These findings are challenged by the question of whether optimization or maximization of resource gain is decisive in competitive interactions. In both species, above- or belowground resource gain in absolute terms was not significantly related to whole-tree RGR, indicating minor importance of the resource gain as such. In the case of carbon, this result supports the view that biomass development of trees is not driven by their carbon availability, but that photosynthesis delivers on growth demand—at least as long as surrounding conditions allow for it (Körner 2003a, 2006).

Having identified the efficiency of space occupation as a crucial factor in the competition between juvenile beech and spruce, the question arises of whether space per se is a resource to plants. This has been recently debated by Grams and Lüttge (2011). Their simplest but most straightforward and illustrative example is the space provided for atmospheric bromeliads by a telephone-line wire devoid of any other resources. Hence, they come to the conclusion that indeed sheer space in itself has the function of a resource to plants. Coming back to the competitive interaction between beech and spruce, an example from adult trees may support this conclusion. At the experimental site "Kranzberg Forest" (see Chaps. 11 and 13), studying C balances of branches of adult trees, Reiter et al. (2005) found shaded branches with negative C balances of the foliage to be sustained by the tree for at least 5 years. Having the paradigm of carbon autonomy in mind (Landhausser 2011; Sprugel et al. 1991; Volpe et al. 2008), one might have expected trees to abandon such branches much earlier. This behavior of trees can be interpreted as a "sit-and-wait" strategy (Falster and Westoby 2003; Reiter et al. 2005) as the value of the occupied space may increase with time, e.g., after gap formation by collapsing neighboring trees with resulting increase of irradiance. Such a temporal aspect of space occupation is supported by a belowground example where roots keep occupying soil space of low ecological value (i.e., with low resource availability). If this space is in the vicinity of belowground pathways of rodents (or other animals), the resource availability may improve by more or less frequent urination events (J. F. Cahill, University of Alberta, Canada, personal communication). Thus, competing for and keeping this soil space occupied may pay back over time. In addition to the direct effects of space occupation on the resource budget of a plant, the effects on its competing neighbor should not be overlooked. In particular in the case of unidirectional resources, such as light, that are "pre-emptable" and allow for shading effects (Schwinning and Weiner 1998) successful space occupation may significantly reduce the resource availability to the neighbor. Hence, optimization of above- or belowground space occupation appears to be the mechanistic basis for competitive success—at least in the case of "pre-emptable" resources (Fig. 12.4; Grams and Andersen 2007; Kozovits et al. 2005b). In such a case, the resources gained by a plant may not be raised in absolute terms but be increased relative to a neighbor. Thus, we may conclude that the resources gained relative to its neighbor (i.e., the marginal advantage) is maximized through optimization of space occupation.

12.5 Conclusions

Competitive interactions between plants have the potential to alter abiotic impacts of atmospheric O_3 and CO_2 concentrations. It appears that the more intense the competition is for a limiting resource, the higher the potential becomes to modify the response to other stressors. Hence, responses of plants grown isolated or under low competitive pressure are of only limited relevance for plants grown in mono- or mixed cultures. In particular in situations with intense competitive interactions, the efficient occupation of space represents an effective mechanism to be competitive by increasing the resource accessibility relative to competing neighbors.

Acknowledgements The authors are indebted to Drs. H.-D. Payer and H. K. Seidlitz (Helmholtz Zentrum München—German Research Center for Environmental Health) as well as their teams for excellent and unstinting cooperation and support during the experiments in the phytotron facility. In particular, the Master and PhD-students of this project, i.e., A. Jungermann, Dr. A. R. Kozovits, G. Luedemann and Dr. W. Ritter, is thanked. We gratefully acknowledge the skilful assistance by T. Feuerbach, J. Heckmair, P. Kuba, J. Lebherz, H. Lohner and I. Süß. We thank Drs. C. Andersen (EPA, Corvallis, Oregon, USA), F. Fleischmann (Pathology of Woody Plants, Technische Universität München, Germany) and K.-H. Häberle (Ecophysiology of Plants, Technische Universität München, Germany) for valuable discussions. Dr. U. Lüttge (Institute of Botany, Darmstadt University, Germany) is thanked for valuable comments on an earlier version of this manuscript. The investigation was funded through SFB 607 "Growth and Parasite Defense, Project B5" by the "Deutsche Forschungsgemeinschaft" (DFG).

References

Daigo MJ, Gayler S, Priesack E, Matyssek R, Grams TEE (submitted) The mode of competition for light and water among juvenile beech and spruce under ambient and elevated levels of O_3 and CO_2

Falge E, Graber W, Siegwolf R, Tenhunen JD (1996) A model of the gas exchange response of *Picea abies* to habitat conditions. Trees 10(5):277–287

Falster DS, Westoby M (2003) Plant height and evolutionary games. Trends Ecol Evol 18:337–343

Fowler D, Amann M, Anderson R, Ashmore M, Cox P, Depledge M, Derwent D, Grennfelt P, Hewitt N, Hov O, Jenkin M, Kelly F, Liss P, Pilling M, Pyle J, Slingo J, Stefenson D (2008) Science policy report 15/08: ground-level ozone in the 21st century: future trends, impacts and policy implications. The Royal Society, London, UK, p 132

Garrigues E, Doussan C, Pierret A (2006) Water uptake by plant roots: I – Formation and propagation of a water extraction front in mature root systems as evidenced by 2D light transmission imaging. Plant Soil 283(1–2):83–98. doi:10.1007/s11104-004-7903-0

Gayler S, Grams TEE, Kozovits AR, Winkler JB, Luedemann G, Priesack E (2006) Analysis of competition effects in mono- and mixed cultures of juvenile beech and spruce by means of the plant growth simulation model PLATHO. Plant Biol 8(4):503–514

Grams TEE, Andersen CP (2007) Competition for resources in trees: physiological versus morphological plasticity. In: Esser K, Lüttge U, Beyschlag W, Murata J (eds) Progress in botany, vol 68. Springer, Heidelberg, pp 356–381

Grams TEE, Lüttge U (2011) Space as a resource. In: Esser K, Lüttge U, Beyschlag W, Murata J (eds) Progress in botany, vol 72. Springer, Heidelberg, pp 349–370

Grams TEE, Matyssek R (1999) Elevated CO_2 counteracts the limitation by chronic ozone exposure on photosynthesis in *Fagus sylvatica* L.: comparison between chlorophyll fluorescence and leaf gas exchange. Phyton 39(4):31–39

Grams TEE, Matyssek R (2010) Stable isotope signatures reflect competitiveness between trees under changed CO_2/O_3 regimes. Environ Pollut 158:1036–1042

Grams TEE, Anegg S, Häberle KH, Langebartels C, Matyssek R (1999) Interactions of chronic exposure to elevated CO_2 and O_3 levels in the photosynthetic light and dark reactions of European beech (*Fagus sylvatica*). New Phytol 144(1):95–107

Grams TEE, Kozovits AR, Reiter IM, Winkler JB, Sommerkorn M, Blaschke H, Häberle KH, Matyssek R (2002) Quantifying competitiveness in woody plants. Plant Biol 4(2):153–158

Hunt R, Causton DR, Shipley B, Askew AP (2002) A modern tool for classical plant growth analysis. Ann Bot 90(4):485–488

Jungermann A (1998) Photosynthese, Wasserbedarf und Allokation in Buche (*Fagus sylvatica* L.) unter kombinierten CO_2/O_3-Regimen. Diploma thesis, FH Weihenstephan

Körner C (2003a) Carbon limitation in trees. J Ecol 91(1):4–17

Körner C (2003b) Ecological impacts of atmospheric CO_2 enrichment on terrestrial ecosystems. Philos Transact R Soc Lond A Math Phys Eng Sci 361(1810):2023–2041

Körner C (2006) Plant CO_2 responses: an issue of definition, time and resource supply. New Phytol 172(3):393–411

Kozovits AR, Matyssek R, Blaschke H, Göttlein A, Grams TEE (2005a) Competition increasingly dominates the responsiveness of juvenile beech and spruce to elevated CO_2 and/or O_3 concentrations throughout two subsequent growing seasons. Glob Change Biol 11:1387–1401

Kozovits AR, Matyssek R, Winkler JB, Göttlein A, Blaschke H, Grams TEE (2005b) Above-ground space sequestration determines competitive success in juvenile beech and spruce trees. New Phytol 167(1):181–196

Küppers M (1989) Ecological significance of above-ground architectural patterns in woody plants – a question of cost-benefit relationships. Trends Ecol Evol 4(12):375–379

Landhausser SM (2011) Aspen shoots are carbon autonomous during bud break. Trees 25(3): 531–536

Langebartels C, Ernst D, Heller W, Lütz C, Payer H-D, Sandermann H Jr (1997) Ozone responses of trees: results from controlled chamber exposures at the GSF phytotron. In: Sandermann HJ, Wellburn AR, Heath RL (eds) Ecological studies, vol 127. Springer, Berlin, pp 163–200

Luedemann G, Matyssek R, Fleischmann F, Grams TEE (2005) Acclimation to ozone affects host/pathogen interaction and competitiveness for nitrogen in juvenile *Fagus sylvatica* and *Picea abies* trees infected with *Phytophthora citricola*. Plant Biol 7(6):640–649

Luedemann G, Matyssek R, Winkler JB, Grams TEE (2009) Contrasting ozone × pathogen interaction as mediated through competition between juvenile European beech (*Fagus sylvatica*) and Norway spruce (*Picea abies*). Plant Soil 323:47–60

Matyssek R, Schulze ED (1987) Heterosis in hybrid larch (*Larix decidua* × *leptolepis*) II. Growth characteristics. Trees 1:225–231

Matyssek R, Karnosky DF, Wieser G, Percy K, Oksanen E, Grams TEE, Kubiske M, Hanke D, Pretzsch H (2010a) Advances in understanding ozone impact on forest trees: messages from novel phytotron and free-air fumigation studies. Environ Pollut 158(6):1990–2006

Matyssek R, Wieser G, Ceulemans R, Rennenberg H, Pretzsch H, Haberer K, Löw M, Nunn AJ, Werner H, Wipfler P, Oßwald W, Nikolova P, Hanke DE, Kraigher H, Tausz M, Bahnweg G, Kitao M, Dieler J, Sandermann H, Herbinger K, Grebenc T, Blumenröther M, Deckmyn G, Grams TEE, Heerdt C, Leuchner M, Fabian P, Häberle K-H (2010b) Enhanced ozone strongly reduces carbon sink strength of adult beech (*Fagus sylvatica*) – resume from the free-air fumigation study at Kranzberg Forest. Environ Pollut 158:2527–2532

Millard P, Sommerkorn M, Grelet GA (2007) Environmental change and carbon limitation in trees: a biochemical, ecophysiological and ecosystem appraisal. New Phytol 175(1):11–28

Navas ML, Garnier E, Austin MP, Gifford RM (1999) Effect of competition on the responses of grasses and legumes to elevated atmospheric CO_2 along a nitrogen gradient: differences between isolated plants, monocultures and multi- species mixtures. New Phytol 143 (2):323–331

Payer HD, Blodow P, Köfferlein M, Lippert M, Schmolke W, Seckmeyer G, Seidlitz HK, Strube D, Thiel S (1993) Controlled environment chambers for experimental studies on plant responses to CO_2 and interactions with pollutants. In: Schulze ED, Mooney HA (eds) Ecosystems Research Report No. 6: design and execution of experiments on CO_2 enrichment. Commission European Communities, Brussels, pp 127–145

Poorter H, Navas ML (2003) Plant growth and competition at elevated CO_2: on winners, losers and functional groups. New Phytol 157(2):175–198

Reiter IM, Häberle KH, Nunn AJ, Heerdt C, Reitmayer H, Grote R, Matyssek R (2005) Competitive strategies in adult beech and spruce: space-related foliar carbon investment versus carbon gain. Oecologia 146(3):337–349

Ritter W, Lehmeier CA, Winkler JB, Matyssek R, Grams TEE (submitted) Contrasting responses in carbon allocation of juvenile European beech (*Fagus sylvatica*) and Norway spruce (*Picea abies*) to competition and disturbance by ozone

Rodenkirchen H, Göttlein A, Kozovits AR, Matyssek R, Grams TEE (2009) Nutrient contents and efficiencies of beech and spruce saplings as influenced by competition and O_3/CO_2 regime. Eur J Forest Res 128:117–128

Schwinning S (1996) Decomposition analysis of competitive symmetry and size structure dynamics. Ann Bot 77:47–57

Schwinning S, Weiner J (1998) Mechanisms determining the degree of size asymmetry in competition among plants. Oecologia 113:447–455

Sitch S, Cox PM, Collins WJ, Huntingford C (2007) Indirect radiative forcing of climate change through ozone effects on the land-carbon sink. Nature 448(7155):791–794

Spinnler D, Egli P, Körner C (2002) Four-year growth dynamics of beech-spruce model ecosystems under CO_2 enrichment on two different forest soils. Trees 16(6):423–436

Sprugel DG, Hinckley TM, Schaap W (1991) The theory and practice of branch autonomy. Annu Rev Ecol Syst 22:309–334

Syring KM, Claassen N (1995) Estimation of the influx and the radius of the depletion zone developing around a root during nutrient uptake. Plant Soil 175(1):115–123

Thiel S, Döhring T, Köfferlein M, Kosak A, Martin P, Seidlitz HK (1996) A phytotron for plant stress research: how far can artificial lighting compare to natural sunlight? J Plant Physiol 148:456–463

Volpe G, Bianco LB, Rieger M (2008) Carbon autonomy of peach shoots determined by ^{13}C photoassimilate transport. Tree Physiol 28(12):1805–1812

Chapter 13
Allometry of Tree Crown Structure. Relevance for Space Occupation at the Individual Plant Level and for Self-Thinning at the Stand Level

H. Pretzsch, C. Matthew, and J. Dieler

13.1 Introduction

Allometry in its broader sense is concerned with the size of organisms and its consequences for their shape and functioning. Since the postulation of the allometric equation in the 1930s (Huxley 1932; Teissier 1934), allometry in a narrow sense refers to analysis and modeling of logarithmic transformed bivariate size data by linear regression techniques. Supposing x and y quantify the size of plant organs, the growth x' (dx/dt) and y' (dy/dt) is related to the size x and y as $y'/y = \alpha\, x'/x$. Better known are the integrated ($y = \beta x^{\alpha}$) or logarithmic representations ($\ln y = \ln \beta + \alpha \times \ln x$). Allometry is the relative change of one plant dimension, dy/y (e.g., the relative height growth) in relation to the relative change of a second plant dimension dx/x (e.g., the relative diameter growth). The allometric exponent α can be understood as a distribution coefficient for the growth resources between organs y and x: when x increases by 1%, y increases by α%.

In this chapter the focus is on the allometric parameters β and α, their absolute magnitude, variation and dependency on species and environment as well as on their relevance for tree and stand dynamics. Therefore, we first illustrate the effect of different allometric factors β and exponents α using the relationship between stem diameter, d, and crown cross-sectional area, csa, as a model example (Fig. 13.1a). The more crown size increases with tree diameter, the higher is the resource exploitation of a tree and its repression effect on competitors. Doubling the allometric factor β results in the change from line 1 to line 2; however, an increase of the allometric exponent from 1.33 to 2.0 or 2.67 (lines 2, 3, and 4, respectively) has an exponential

H. Pretzsch (✉) • J. Dieler
Chair of Forest Growth and Yield Science, Technische Universität München, Hans-Carl-von-Carlowitz-Platz 2, 85354 Freising, Germany
e-mail: Hans.Pretzsch@lrz.tum.de

C. Matthew
Institute of Natural Resources, Massey University, Palmerston North, New Zealand

R. Matyssek et al. (eds.), *Growth and Defence in Plants*, Ecological Studies 220,
DOI 10.1007/978-3-642-30645-7_13, © Springer-Verlag Berlin Heidelberg 2012

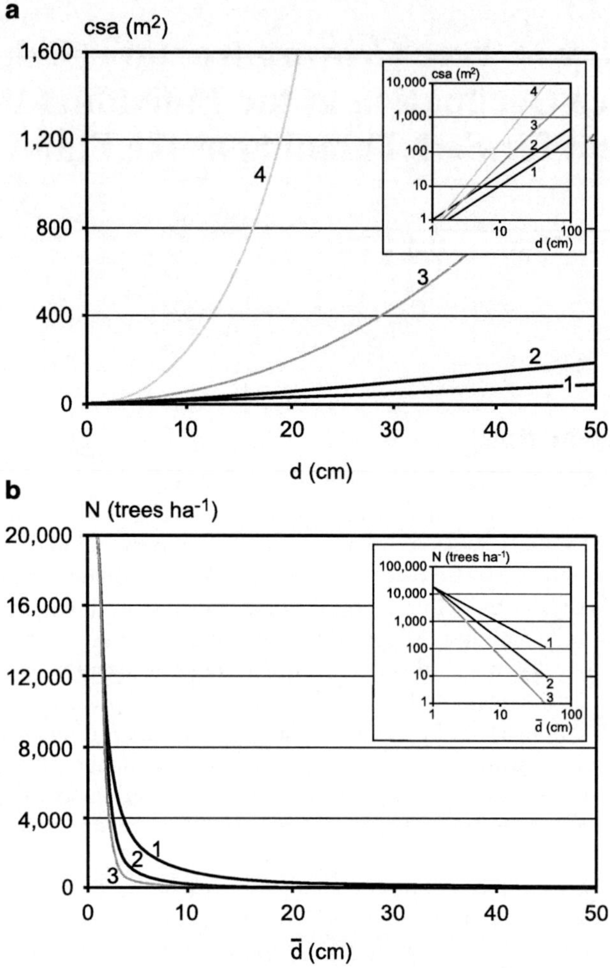

Fig. 13.1 Effect of different allometric factors β and allometric exponents α on (**a**) the relationship between individual tree diameter, *d*, and crown cross-sectional area, csa $= \beta_1 \times d^{\alpha_1}$ (line 1: $\beta_1 = 0.5$, $\alpha_1 = 1.33$; line 2: $\beta_1 = 1.0$, $\alpha_1 = 1.33$; line 3: $\beta_1 = 0.5$, $\alpha_1 = 2.0$; line 4: $\beta_1 = 0.5$, $\alpha_1 = 2.66$) and (**b**) quadratic mean diameter of a stand, $\bar{d}$, and surviving tree number per hectare, $N = \beta_2 \times d^{\alpha_2}$ (line 1: $\beta_2 = 20{,}000$, $\alpha_2 = -1.33$; line 2: $\beta_2 = 20{,}000$, $\alpha_2 = -2.0$; line 3: $\beta_2 = 20{,}000$, $\alpha_2 = -2.66$) in schematic representation. In the *upper right corner* we show the same relationships in the double-logarithmic grid, frequently used for allometric analyses

effect on its space occupation and competitiveness. The same applies for the relationship between mean stem diameter, $\bar{d}$, and tree number per hectare, *N*, which shows a steeper exponential decrease when the exponential exponents are $\alpha = -1.33$, -2.0, and -2.67. Notice how varying degrees of crown expansion cause changes in space occupation (Fig. 13.1a), which result in gradually different decreases in tree number (Fig. 13.1b). We confine discussion here to the elucidation of the parameters' meaning in the allometric equation in principle; when we come back to this figure later on it will become obvious, that the selected exponents $\alpha = -1.33$, -2.0, and -2.67 play a key role in allometric theory. The insertions in the upper right corner of Fig. 13.1a, b show the same relationships in a double-logarithmic grid, a format frequently used for allometric analyses.

Since its beginning, allometric research has been mainly driven by the search for an overarching, universal allometric exponent. Often it was proposed that volume or mass related allometric functions scale with exponents of 1/3 due to the volume dimensionality (von Bertalanffy 1951; Yoda et al. 1963, 1965; Gorham 1979). More recently, West et al. (1997, 2009), Enquist et al. (1998, 2009) and Enquist and Niklas (2001) presented a general theory of allometric scaling with exponents of 1/4 based on fractal networks of transportation systems in organisms.

Mostly, allometric analyses assume geometrical similitude as a starting point and null hypothesis. Geometrical scaling or similitude means that between different linear dimensions $\mathrm{lin}_1, \mathrm{lin}_2 \ldots \mathrm{lin}_n$ (e.g., tree height, h, diameter, d, crown length, cl) applies proportionality: $\mathrm{lin}_1 \propto \mathrm{lin}_2 \ldots \propto \mathrm{lin}_n$. It further assumes that between area-related or quadratic tree attributes (e.g., basal area, ba, leaf area, la, crown cross-sectional area, csa, or growing area, s) and linear dimensions a quadratic relationship applies: quad $\propto \mathrm{lin}^2$, and further between cubic variables (e.g., plant volume, v, crown volume, cv, tree biomass, mt) and linear dimensions, a cubic relationship applies: $\mathrm{cub} \propto \mathrm{lin}^3$. From that it follows that $\mathrm{lin} \propto \mathrm{cub}^{1/3}$ and $\mathrm{quad} \propto \mathrm{cub}^{2/3}$. Since three is in the denominator of the allometric exponent, this corresponds to 1/3 exponent scaling. Application of $\mathrm{quad} \propto \mathrm{cub}^{2/3}$ to the relationship between mean tree volume, $\bar{v}$, ($\equiv$ cubic) and mean growing area, $\bar{s}$, ($\equiv$ quad) yields $\bar{s} \propto \bar{v}^{2/3}$. As $N \propto \bar{s}^{-1}$, we get the $-3/2$ power rule of self-thinning $\bar{v}^{-2/3} \propto N$ or $\bar{v} \propto N^{-3/2}$, which obviously assumes geometrical scaling and isometric form development (Yoda et al. 1963, 1965; Gorham 1979). Note that the $-3/2$ power rule relates to the 1/3 exponent scaling as $\bar{v}$ scales to N with the power of $-2/3$. This relationship is best known in the reverse formulation where N scales to volume with the power of $-3/2$.

Metabolic scaling theory (MST) in contrast predicts that the metabolic rate of an individual plant scales as the 3/4 power of the total body mass, mt, so that la $\propto \mathrm{mt}^{3/4}$ where la denotes leaf area (West et al. 2009). Under this formulation cub $\propto \mathrm{lin}^{\alpha}$ with $\alpha \neq 3$ and a deviation from 1/3 exponent scaling, in other words a fractal scaling. For the relationship between mass, m, and tree stem diameter, d, Enquist et al. (1998) postulate $m \propto d^{8/3}$. With $\mathrm{ba} \propto d^2$, the basal area–mass relationship results in $\mathrm{ba} \propto m^{3/4}$. Obviously, fractal scaling leads to a quarter power (1/4 exponent) scaling rather than 1/3 exponent geometrical scaling. Applied on the mean tree level, Enquist et al. (1998) derive $\bar{m} \propto N^{-4/3}$, which results in a more shallow self-thinning line compared to $\bar{m} \propto N^{-3/2}$ from geometrical scaling (on condition that $m \propto v$). In the following we refer to the MST which assumes common scaling relationships for allometric ideal plants, e.g., $\alpha_{h,d} = 2/3 = 0.67$, $\alpha_{\mathrm{csa},d} = 4/3 = 1.33$ and $\alpha_{v,d} = 8/3 = 2.67$ (West et al. 2009) as well as to geometric scaling which, in contrast, assumes $\alpha_{h,d} = 1$, $\alpha_{\mathrm{csa},d} = 2$ and $\alpha_{v,d} = 3$ (e.g., $\alpha_{h,d}$ means scaling of height, h, over diameter, d). When we compare our results with these common fractal and geometric scaling exponents we do not mean to imply that we are convinced of their generality, but rather to use them as references.

With a few exceptions (e.g., Matthew et al. 1995) most of the works on scaling relationships assume stands with maximum density and self-thinning conditions. In this chapter we analyze a broader range of competitive states and stand densities of individual plants, ranging from solitary growth to maximum density, from poor to rich soil fertility, and from pure to mixed stands. In other words, we analyze tree allometry for a broader range of environmental conditions, so that allometry for self-thinning conditions becomes one special borderline-case in a continuum of growing conditions.

In order to trace allocation pattern from individual to stand level we draw special attention to (1) the species-specific variation of crown scaling exponents as dependent on stand density, intra-, and inter-specific competition and on the relationship between various scaling exponents, (2) the differences in crown scaling between Norway spruce and European beech as well as between the groups of gymnosperm and angiosperm species which each belong to, and (3) the relevance of scaling relationships for space occupation, stand dynamics, especially self-thinning in pure and alien-thinning in mixed-species stands. Finally, we discuss the relevance of the findings for allometric theory, their implications for space occupation at the individual plant level and for self-thinning at the stand level.

13.2 Theory and Methods

13.2.1 Tree and Stand Variables

We base our analysis on two types of tree and stand variables. These are physiological attributes like total tree mass, mt, leaf mass, ml, and leaf area, la, and structure and shape variables like tree diameter, d, tree height, h, crown length, cl, crown cross-sectional area, csa, and crown volume, cv see (Table 13.2). The latter represents the volume under the convex hull when the crown is modeled as a Euclidian geometric body without indentations (following the wrapping approach by Christo und Jeanne-Claude see www.christojeanneclaude.net/wt.shtml). To estimate the total aboveground tree volume, v, we applied species-specific volume functions, which calculate the volume from height and diameter (Pretzsch and Dieler 2012). Specific wood density R (with $R = \mathrm{mt}/v$, for species-specific values see Pretzsch 2009, p. 91) is used to transform volume to mass.

13.2.2 Scaling of Structure

According to Niklas (2004), allometric size-relationships can be quantified by power-law models of the form $y = \beta x^{\alpha}$, where x and y denote the size of an organ or body part of interest, β is the normalization constant and α the allometric scaling exponent. By expressing the power-law function logarithmically as $\ln y = \ln \beta + \alpha \ln x$ it becomes evident that $\alpha = dy/y/dx/x$ is a measure for the relative growth rate of size y as dependent on the relative growth rate of size x.

For all analysis, the allometric exponent α was predicted by ln–ln regression lines. On the individual tree scale, where data from long-term plots are partially autocorrelated due to consecutive measurements (cf. Table 13.1), linear mixed effect models $\ln y_{ik} = \ln \beta + \alpha \ln x_{jk} + a_k \ln x_{jk} + \varepsilon_{jk}$ with individual $= k$ and time $= j$ as a random factor on slope (a_k) were applied (Pinheiro and Bates 2000). The random effect a_k and the error ε_{jk} were assumed to be normally distributed with mean $= 0$ and constant variance. The fixed-effect coefficient α represents the average allometric relationship between x and y and is to be seen as the species-specific mean scaling exponent. This method additionally allows to extract a local subject-specific allometric exponent α^k for each tree k by combining the fixed-effect estimate and the random effect as the intra-specific variation is captured by a_k. The analysis is focusing on this local exponent as it captures the variation in form and development. The mean of α^k equals α and is used for comparisons between species. For fitting the models the function lmer from the R software package "lme4" (Bates and Maechler 2010; R Development Core Team 2009) was chosen.

For allometric analysis of cross-sectional data on stand level, parameters and confidence intervals were estimated by standardized major axis (SMA) regression, also known as reduced major axis (RMA) regression, using the R software package "smatr" (see Sackville Hamilton et al. 1995; Warton et al. 2006; R Development Core Team 2009).

The subscripts of the allometric exponent α indicate which variables are addressed. Note that when reporting $\alpha_{y,x}$, the exponent is calculated on the individual tree scale, whereas $\alpha_{\bar{y},\bar{x}}$ refers to mean tree values. For more details, see Pretzsch and Dieler (2012).

13.2.3 Determination of a Tree's Competitive Status

In order to analyze the effect of competition on the individual tree's allometry the competitive status of each tree was quantified by the tree cover index, tci, and stand density index, sdi, which are explained subsequently. Stand characteristics are analyzed for each individual tree k within a circle of radius r. After having experimented with different search radii, it was assumed that most of a tree's relevant competitors are located within a radius of 2.0 times its mean crown width, which was estimated from tree diameter and height according to Pretzsch et al. (2002). In order to characterize the vertical position, the tree cover index, tci (Fig. 13.2, top), was applied. The first step in deriving tci was to obtain the maximum tree height, h_{max}, within a radius r_k. By setting the height, h_k, of tree k in relation to h_{max}, the tci_k variable can be calculated from $tci_k = 1 - h_k/h_{max}$. Higher tci_k values indicate a suppression of tree k.

Table 13.1 Scaling exponents $\alpha_{h,d}$, $\alpha_{csa,d}$, $\alpha_{v,d}$, and $\alpha_{cv,d}$ observed on the long-term plots of Norway spruce and European beech and correlation between $\alpha^k_{h,d}$ and $\alpha^k_{csa,d}$, $\alpha^k_{h,d}$ and $\alpha^k_{v,d}$, and $\alpha^k_{csa,d}$ and $\alpha^k_{v,d}$ (adapted from Pretzsch and Dieler 2012)

	Norway spruce				European beech			
Characteristics	$\alpha_{h,d}$	$\alpha_{csa,d}$	$\alpha_{v,d}$	$\alpha_{cv,v}$	$\alpha_{h,d}$	$\alpha_{csa,d}$	$\alpha_{v,d}$	$\alpha_{cv,v}$
Expected								
Geometric similitude	1.00	2.00	3.00	1.00	1.00	2.00	3.00	1.00
Allometric ideal plant	0.67	1.33	2.67	0.75	0.67	1.33	2.67	0.75
observed								
n (measurements)	3,668	4,23	3,668	3,528	1,446	1,960	1,446	1,400
n (individuals)	2,566	3,001	2,566	2,425	1,058	1,509	1,058	1,015
mean ± SE	0.63 (±0.017)	1.50 (±0.006)	2.56 (±0.004)	0.80 (±0.006)	0.55 (±0.007)	1.19 (±0.021)	2.54 (±0.006)	0.77 (±0.012)
95 % CI limits	0.62–0.64	1.47–1.54	2.55–2.57	0.79–0.82	0.53–0.56	1.15–1.23	2.53–2.55	0.75–0.80
min α^k–max α^k	0.37–0.78	0.98–2.07	2.41–2.66	0.65–0.91	0.50–0.58	0.87–1.40	2.49–2.59	0.61–0.89
Coeff. var.	0.57	0.74	0.10	0.46	0.51	0.77	0.09	0.58
$r_{Pearson}\alpha^k_{h,d} \times \alpha^k_{csa,d}$	−0.68 ($n = 2{,}425$; $p < 0.001$)				−0.44 ($n = 1{,}015$; $p < 0.001$)			
$r_{Pearson}\alpha^k_{h,d} \times \alpha^k_{v,d}$	+0.95 ($n = 2{,}566$; $p < 0.001$)				+0.99 ($n = 1{,}058$; $p < 0.001$)			
$r_{Pearson}\alpha^k_{csa,d} \times \alpha^k_{v,d}$	−0.62 ($n = 2{,}425$; $p < 0.001$)				−0.43 ($n = 1{,}015$; $p < 0.001$)			

The exponent k refers to the local subject-specific allometric coefficient α^k. Scaling exponents expected under geometric similitude and predicted for the allometric ideal plant serve as reference. $\alpha_{h,d}$, scaling of tree height, h, versus trunk diameter, d; $\alpha_{csa,d}$, crown cross-sectional area, csa, versus trunk diameter, d; $\alpha_{v,d}$, tree volume, v, versus trunk diameter, d; $\alpha_{cv,v}$, crown volume, cv, versus tree volume, v

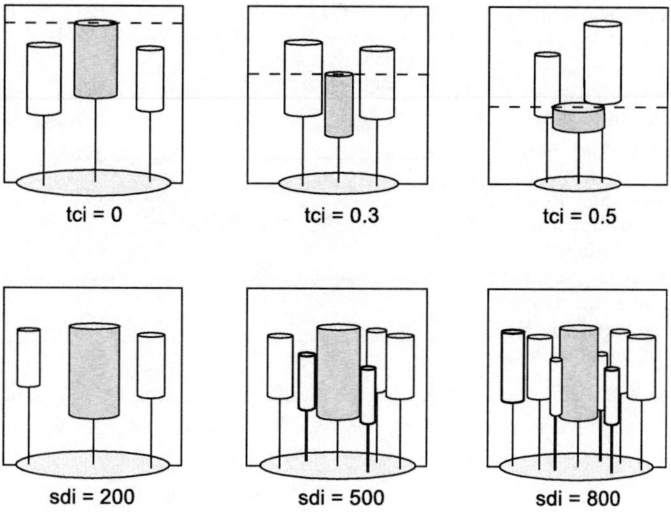

Fig. 13.2 Detection of tree cover index, tci (*top*), and stand density index, sdi (*bottom*), in schematic representation. Tree k is shaded *gray*

For quantifying lateral restriction of tree k we make use of the concept of the stand density index $\mathrm{SDI} = N_{\mathrm{obs}}(25/\bar{d})^{\alpha_{N,\bar{d}}}$ by Reineke (1933), where N_{obs} is the stem number per ha, $\bar{d}$ is the stand's mean stem diameter in centimeters, and $\alpha_{N,\bar{d}}$ is the allometric exponent set to $\alpha_{N,\bar{d}} = -1.605$ (see Reineke 1933). In this common version the SDI is excellent for measuring stand density, by standardizing it and thus making densities of stands in different development phases comparable. For this study we calculate SDI for the sub-collective of trees within radius r and name it sdi_k (Fig. 13.2, bottom). sdi_k is defined as the equivalent trees per hectare at a quadratic mean diameter of 25 cm and is formulated as $\mathrm{sdi}_k = N_k(25/\bar{d}_k)^{\alpha_{N,\bar{d}_k}}$, where N_k is the tree number and $\bar{d}_k$ is the quadratic mean diameter on the respective concentric plot with radius r_k around tree k. For more detailed information on calculation of SDI and sdi_k and their application to measuring density, see Reineke (1933) and Pretzsch and Biber (2010), respectively.

13.3 Evidence

This section starts with scrutiny of 3/4 scaling ($\mathrm{la} \propto \mathrm{mt}^{3/4}$) predicted by MST, then explores the crown structure scaling under various environmental conditions, and reveals the interactions between different crown attributes. We show how species-specific structural allometry and plasticity keeps the tree on the 3/4 power leaf area-plant biomass trajectory. Plasticity of crown structure is high under intra-specific competition but even higher in mixed-species stands. Finally, attention is drawn on the relevance of plastic scaling for space occupation at the individual plant level and for self- and alien-thinning at the stand level.

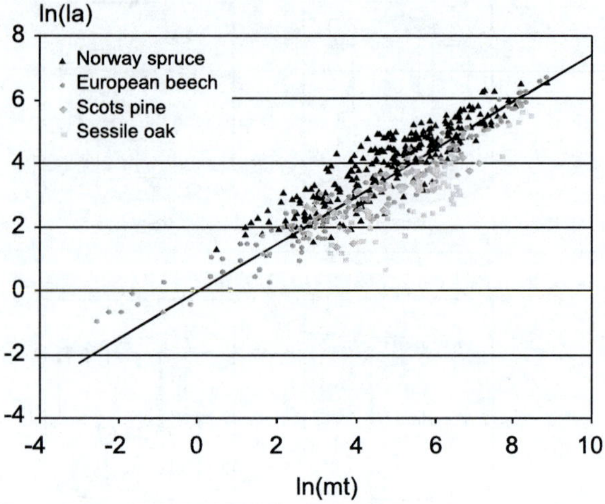

Fig. 13.3 Metabolic scaling of leaf area, la, versus above ground total plant mass, mt, $la \propto mt^{\alpha_{la,mt}}$. Pooled data of Norway spruce ($n = 280$), European beech ($n = 145$), Scots pine ($n = 31$), and Sessile oak ($n = 52$). SMA regression of leaf area, la, versus total aboveground mass, mt, yields $\alpha_{la,mt} = 0.74 \pm 0.016$. The fit of the inserted regression line yielded $\ln(la) = -0.0176 + 0.7418 \times \ln(mt)$

13.3.1 Metabolic 3/4 Scaling of Leaf Biomass versus Total Plant Mass

MST assumes that for the relationship between trunk diameter, d, and plant leaf area, la, $la \propto d^2$, according to the pipe model theory and for the relationship between plant mass, mt, and trunk diameter $mt \propto d^{8/3}$. So, for scaling of leaf area versus plant mass under MST, $la \propto mt^{3/4}$ (Enquist and Niklas 2001; West et al. 1997). This means that when plant mass increases by 1%, leaf area increases by 3/4% rather than the 2/3% expected by geometric scaling. In order to scrutinize this "three-quarter-relationship", which was theoretically derived by West et al. (1997) and assumes fractal geometry for the inner plant structure and functioning, we used 508 records for leaf area and plant mass measurements. Figure 13.3 shows $\alpha_{la,mt} = 0.74 \pm 0.016$ for the relationships between leaf mass and plant mass for the pooled data of Norway spruce, Scots pine, European beech, and Sessile oak. As $\alpha_{la,mt}$ is so close to 0.75, some of the following analyses are founded on the 3/4 scaling, especially the closer scrutiny why the allometry between leaf area and plant mass follows an overarching 3/4 trajectory, but crown structure plant mass allometry can vary in a much wider range.

For analyses and derivations in this study we assume like Enquist and Niklas (2001) and West et al. (1997) that leaf area is proportional to leaf mass ($la \propto ml$). However, thorough analysis of our data (not shown in detail) indicates that scaling of leaf area over plant mass ($\alpha_{la,mt} = 0.74(\pm 0.016)$) may differ from the scaling of leaf mass over plant mass ($\alpha_{ml,mt} = 0.83(\pm 0.020)$), and that scaling of leaf mass over shoot mass lies in between ($\alpha_{ml,ms} = 0.79(\pm 0.017)$). Also works by Niklas et al. (2009) and Price et al. (2010) question the proportionality between leaf area and leaf mass. As the differences seem to be rather small and are not yet sufficiently

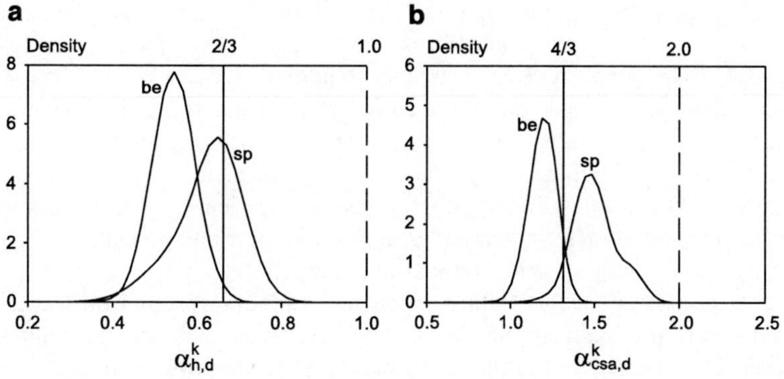

Fig. 13.4 Density distribution of observed scaling exponents **(a)** $\alpha^k_{h,d}$ and **(b)** $\alpha^k_{\text{csa},d}$ for Norway spruce (sp) and European beech (be) based on individual tree measurements on long-term experimental plots in pure stands (adapted from Pretzsch and Dieler 2012). Expected scaling exponents for an allometric ideal plant according to metabolic scaling theory (m.s.) and Euclidian geometric similitude (g.s.) are represented by *solid* respectively *broken bars*. $\alpha^k_{h,d}$, scaling of individual tree height, h, versus individual tree diameter, d; $\alpha^k_{\text{csa},d}$, individual tree crown cross-sectional area, csa, versus individual tree diameter, d

substantiated, we temporarily assume la $\propto$ ml but compile tree data for further scrutiny of differences between $\alpha_{\text{la,mt}}$, $\alpha_{\text{ml,mt}}$, and $\alpha_{\text{ml,ms}}$.

13.3.2 Scaling of Crown Structure in Dependence on Stand Density and Social Status

In order to link the mass-based relationship to plant structure, we consider that leaf area, la, is proportional to crown volume, cv (la $\propto$ cv), according to West et al. (2009). Crown volume, cv, represents the volume under the convex hull when the crown is modeled as an Euclidian geometric body. Assuming constant specific wood density R (with R = mt/v) total mass, mt, is proportional to tree volume, v, (mt $\propto$ v), then la $\propto$ mt$^{3/4}$ can be transformed to cv $\propto v^{\alpha_{\text{cv},v}}$. With further rearrangement, it is possible to separate $\alpha_{\text{cv},v}$ into three components: First, it is considered that scaling between tree height, h, and trunk diameter, d, is represented by the expression $h \propto d^{\alpha_{h,d}}$. According to McMahon and Kronauer (1976), crown length, cl, is expected to be proportional to height yielding the expression $h \propto \text{cl} \propto d^{\alpha_{h,d}}$. Second, it is considered that the scaling relationship between crown cross-sectional area, csa, and trunk diameter is csa $\propto d^{\alpha_{\text{csa},d}}$. As crown volume is the product of crown length and crown cross-sectional area (cv = cl csa), this yields cv $\propto d^{(\alpha_{h,d}+\alpha_{\text{csa},d})}$. Third, it is supposed that $v \propto d^{\alpha_{v,d}}$, so that $d \propto v^{1/\alpha_{v,d}}$. The combination of these three expressions yields cv $\propto v^{(\alpha_{h,d}+\alpha_{\text{csa},d})/\alpha_{v,d}}$ and shows that $\alpha_{\text{cv},v} = (\alpha_{h,d} + \alpha_{\text{csa},d})/\alpha_{v,d}$.

In the following we draw attention on the components of the latter equation as it shows the scaling of the crown structure which provides the holding fixture for the leaf organs. Figure 13.4 shows by way of example the density distribution for the allometric exponents $\alpha^{k}_{h,d}$ and $\alpha^{k}_{\mathrm{csa},d}$ which represent the vertical and lateral crown expansion of Norway spruce and European beech in pure stands (data base see Table 13.1). The graph reveals, that both exponents have a broad intra-specific variation, that the two species differ considerably concerning these two scaling exponents, and that there is no clear correspondence with metabolic scaling (solid vertical lines) or Euclidian geometric scaling (broken vertical lines). In a subsequent section we will see, how a species vertical and horizontal crown expansion in pure stands (indicated by higher allometric exponents in Fig. 13.4) can change under inter-specific competition (Fig. 13.9).

A large proportion of the scaling exponents' variation results from vertical and lateral crown restriction quantified by the tree cover index, tci, and the stand density index, sdi. The effect of tci and sdi on $\alpha^{k}_{h,d}$ and $\alpha^{k}_{\mathrm{csa},d}$ becomes evident by the regression models for Norway spruce and European beech (Fig. 13.5). With increasing competition trees first slightly enhance and then strongly reduce their height growth. With increasing tci, $\alpha^{k}_{\mathrm{csa},d}$ increases as trees develop shade habit. Thus, $\alpha^{k}_{h,d}$ and $\alpha^{k}_{\mathrm{csa},d}$ react contrarily to an increase of competition reflected by the tree cover index, tci. While the former follows a concave (seen from below) trajectory, the latter shows a convex course. The effect of stand density on the two scaling exponents is also inversely related: when stand density increases (sdi $=$ 200 ... 1,000), $\alpha_{h,d}$ increases, but $\alpha_{\mathrm{csa},d}$ declines (Pretzsch and Dieler 2012).

13.3.3 Crown Scaling Patterns of Species and Species Groups

The scaling exponents $\alpha_{h,d}$, $\alpha_{\mathrm{csa},d}$, $\alpha_{v,d}$, and $\alpha_{\mathrm{cv},v}$ for Norway spruce and European beech reveal an intra-specific variation indicated by the range of the 95% CIs, min–max values, and coefficients of variation (Table 13.1). The correlation between the components of $\alpha_{\mathrm{cv},v} = (\alpha_{h,d} + \alpha_{\mathrm{csa},d})/\alpha_{v,d}$ (Table 13.1, bottom) contributes to the stabilization of $\alpha^{k}_{\mathrm{cv},v}$ between 0.79 and 0.82 for Norway spruce and 0.75 and 0.80 for European beech. Analyses of variance (not shown) showed remarkable differences between the two species concerning the scaling exponents $\alpha_{h,d}$, $\alpha_{\mathrm{csa},d}$, $\alpha_{v,d}$, and $\alpha_{\mathrm{cv},v}$ (Pretzsch and Dieler 2012).

In addition to original data from long-term experiments Pretzsch and Dieler (2012) reviewed 126 yield tables and used them for scrutiny of species-specific structural scaling. The frequency distribution of the observed scaling exponents at stand level across 52 species based on yield tables (Fig. 13.6) reveals a high variation of $\alpha_{\overline{\mathrm{csa}},\bar{d}}$ and $\alpha_{\overline{\mathrm{cv}},\bar{v}}$, which encompass lateral crown expansion (Pretzsch and Dieler 2012). In contrast, the variance of $\alpha_{\bar{h},\bar{d}}$ and $\alpha_{\bar{v},\bar{d}}$, which represent scaling of the stem, is more narrow. The means and the 95% CIs (black vertical line with gray peripheral area) amount to $\alpha_{\bar{h},\bar{d}} = 0.830 \pm 0.017$, $\alpha_{\overline{\mathrm{csa}},\bar{d}} = 1.458 \pm 0.030$, $\alpha_{\bar{v},\bar{d}} = 2.820 \pm 0.036$, and $\alpha_{\overline{\mathrm{cv}},\bar{v}} = 0.817 \pm 0.011$. For the group of gymnosperm species the observed scaling exponents, $\alpha_{\bar{h},\bar{d}} = 0.903 \pm 0.021$, $\alpha_{\overline{\mathrm{csa}},\bar{d}} = 1.431 \pm 0.032$, $\alpha_{\bar{v},\bar{d}} = 2.887 \pm 0.047$, and $\alpha_{\overline{\mathrm{cv}},\bar{v}} = 0.837 \pm 0.012$, are 3–23% greater than those for the angiosperm species with $\alpha_{\bar{h},\bar{d}} = 0.733 \pm 0.022$, $\alpha_{\overline{\mathrm{csa}},\bar{d}} = 1.410 \pm 0.055$, $\alpha_{\bar{v},\bar{d}} = 2.732 \pm 0.054$, and $\alpha_{\overline{\mathrm{cv}},\bar{v}} = 0.791 \pm 0.091$. Both

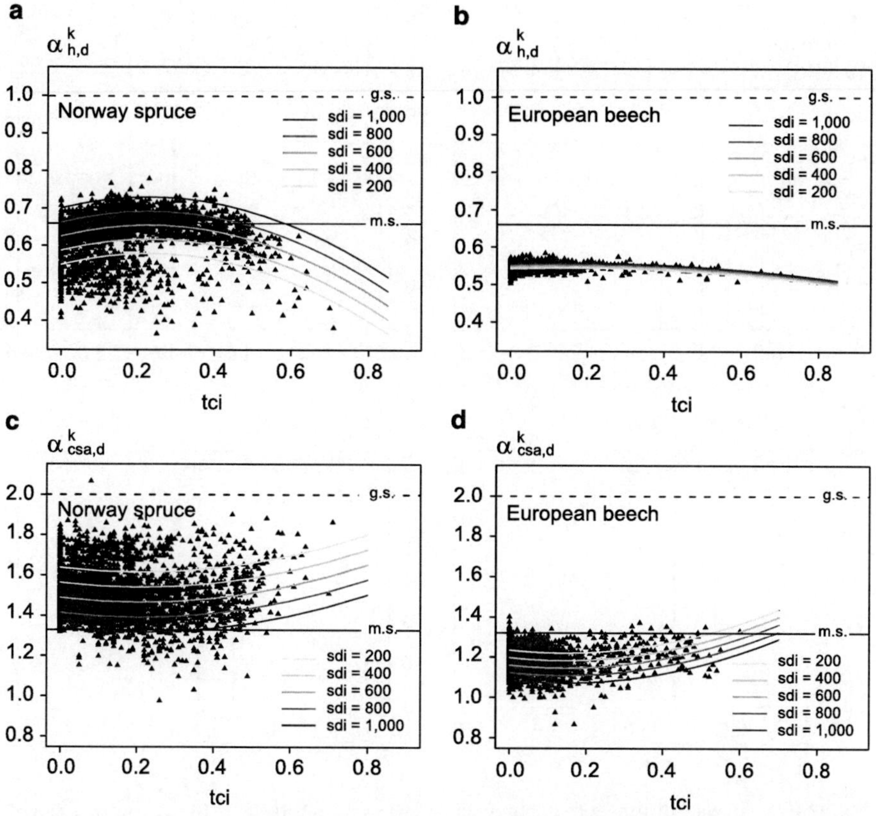

Fig. 13.5 Dependency of allometric scaling exponents on competition displayed for $\alpha^k_{h,d}$ for (**a**) Norway spruce and (**b**) European beech and $\alpha^k_{\text{csa},d}$ for (**c**) Norway spruce and (**d**) European beech (Pretzsch and Dieler 2012). *Horizontal lines* represent scaling expected for geometric similitude, g.s. (*broken line*) and allometric ideal plant, m.s. (*solid line*). *Lines* are ordinary least square regression curves including tci and sdi as covariates. Norway spruce: $\alpha^k_{h,d} = 0.511 + 0.000189$ sdi $+0.258$ tci $- 0.558$ tci^2 ($r^2 = 0.35$, $n = 2{,}566$, $p < 0.001$), $\alpha^k_{\text{csa},d} = 1.713 - 0.000372$ sdi $- 0.227$ tci $+ 0.537$ tci^2 ($r^2 = 0.28$, $n = 3{,}001$, $p < 0.001$). European beech: $\alpha^k_{h,d} = 0.539 + 0.000017$ sdi $+ 0.030$ tci $- 0.096$ tci^2 ($r^2 = 0.16$, $n = 1{,}058$, $p < 0.001$), $\alpha^k_{\text{csa},d} = 1.271 - 0.000186$ sdi -0.201 tci $+ 0.701$ tci^2 ($r^2 = 0.20$, $n = 1{,}509$, $p < 0.001$). In case of the lines in the plots, sdi has been set to values between 200 and 1,000. $\alpha^k_{h,d}$, scaling of height versus trunk diameter; $\alpha^k_{\text{csa},d}$, scaling of cross-sectional area versus trunk diameter; *tci*, tree cover index, *sdi*, stand density index

groups are significantly different in $\alpha_{\overline{h},\overline{d}}$, whereas the confidence intervals of the other allometric exponents overlap. Analysis of the correlation between the components of $\alpha_{\text{cv},v} = (\alpha_{h,d} + \alpha_{\text{csa},d})/\alpha_{v,d}$ reflects the high plasticity by which tree crowns can occupy space (not shown). Pearson's correlation reveals a trade-off between $\alpha_{\overline{h},\overline{d}}$ and $\alpha_{\overline{\text{csa}},\overline{d}}$ indicated by $r = -0.36$***. In contrast, $\alpha_{\overline{h},\overline{d}}$ is positively correlated to $\alpha_{\overline{v},\overline{d}}$ ($r = 0.36$***), and $\alpha_{\overline{\text{csa}},\overline{d}}$ is also positively correlated to $\alpha_{\overline{v},\overline{d}}$ ($r = 0.15$). The trade-off between $\alpha_{\overline{h},\overline{d}}$ versus $\alpha_{\overline{\text{csa}},\overline{d}}$ is mainly responsible for the vertical and

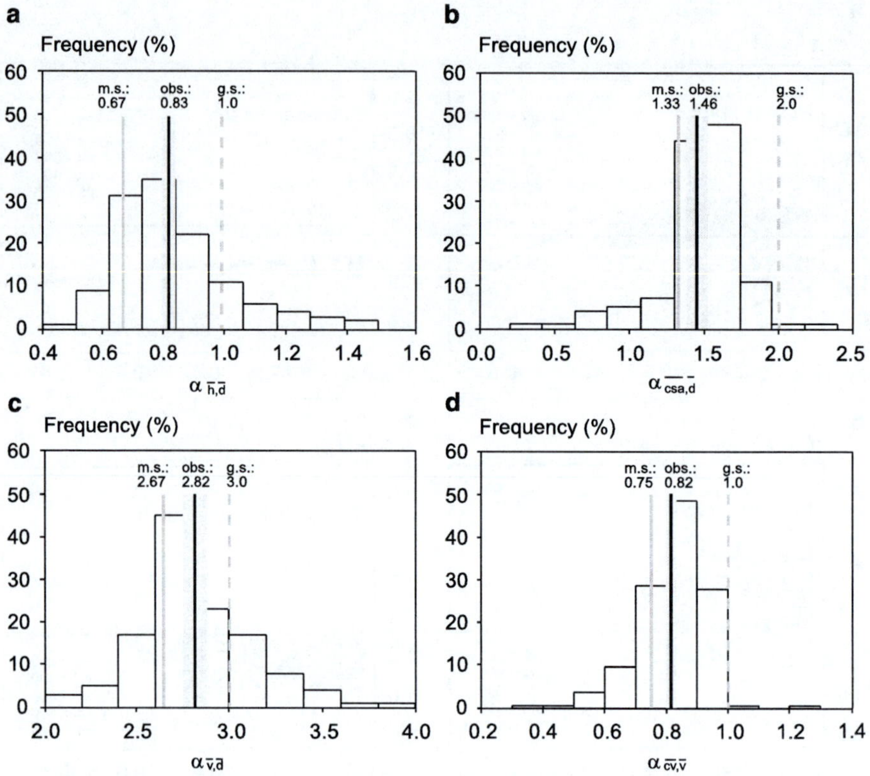

Fig. 13.6 Frequency distributions of observed scaling exponents (**a**) $\alpha_{\bar{h},\bar{d}}$, (**b**) $\alpha_{\overline{csa},\bar{d}}$, (**c**) $\alpha_{\bar{v},\bar{d}}$, and (**d**) $\alpha_{\overline{cv},\bar{v}}$ based on the yield table dataset (Pretzsch and Dieler 2012). Expected scaling exponents for an allometric ideal plant (m.s.) are represented by *solid vertical bars* and for geometric similitude (g.s.) by *broken bars*. The mean observed scaling exponents (obs.) are shown by *black vertical bars*. *Gray bars* around the observed values refer to the confidence limits (95 % CI). $\alpha_{\bar{h},\bar{d}}$, scaling of mean height, $\bar{h}$, versus mean tree diameter, $\bar{d}$; $\alpha_{\overline{csa},\bar{d}}$, mean crown cross-sectional area, $\overline{csa}$, versus mean tree diameter, $\bar{d}$; $\alpha_{\bar{v},\bar{d}}$, mean stem volume, $\bar{v}$, versus mean tree diameter, $\bar{d}$; $\alpha_{\overline{cv},\bar{v}}$, mean crown volume, $\overline{cv}$, versus mean stem volume, $\bar{v}$

lateral crown plasticity. $\alpha_{\overline{cv},\bar{v}}$ shows less variation compared with the variation of its individual components (e.g., $\alpha_{\bar{h},\bar{d}}$, $\alpha_{\overline{csa},\bar{d}}$).

13.3.4 Scaling Under Self-Thinning Conditions in Pure Stands

The above evaluations at the individual plant and mean plant level show how demands on resources and growing space increase when plants grow in size. If resources are no longer sufficient for all individuals, in pure stands self-thinning commences, and the number of plants per unit area, *N*, decreases (Fig. 13.7). Suppose a certain measure of tree size, e.g., the crown cross-sectional area, csa, shows an

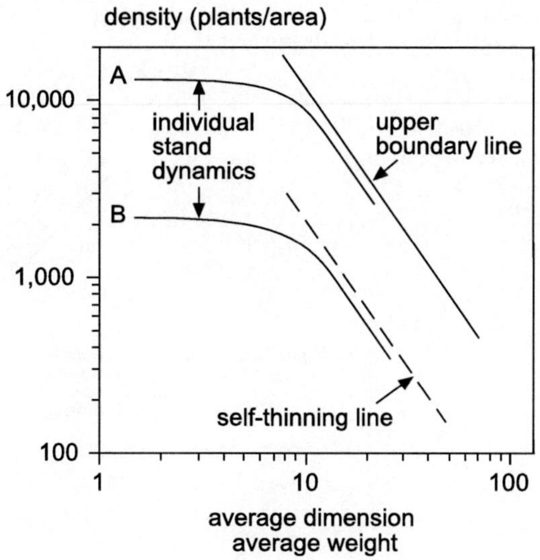

Fig. 13.7 The common principle of Reineke's rule (1933) and the −3/2 power law by Yoda et al. (1963) for even-aged plant populations (Pretzsch and Biber 2005). Both relationships predict that the decrease of plant number (*y*-axis) with growing average plant size or average weight (*x*-axis) follows a *straight line* in a double-logarithmic scale. The *upper solid self-thinning line* represents the "upper boundary line" expected for a species under optimal site conditions, the *lower broken line* represents self-thinning of a stand under resource limitation. Shown are trajectories for a stand with (A) optimal resource supply and (B) poor site conditions. Before crown closure the population density remains rather constant until the limited resources enhance competition and lead to self-thinning

allometric relationship with the size of the total tree, e.g., the total biomass, mt, as $\mathrm{mt} \propto \mathrm{csa}^{\alpha_{\mathrm{mt,csa}}}$. Let us suppose furthermore, that csa is proportional to growing space, s, and resource requirements, r, of a tree ($\mathrm{csa} \propto s \propto r$). Then $\mathrm{mt} \propto \mathrm{csa}^{\alpha_{\mathrm{mt,csa}}}$ equals $m \propto s^{\alpha_{\mathrm{mt},s}}$ or $m \propto r^{\alpha_{\mathrm{mt},r}}$. As an individual's growing space or resource demand is hard to assess, this proposed relationship was mostly analyzed by exploring the relationship of mean tree level $\bar{\mathrm{mt}} \propto \bar{s}^{\alpha_{\bar{\mathrm{mt}},\bar{s}}}$ with mean plant weight, $\bar{\mathrm{mt}}$, and mean plant growing area, $\bar{s}$. The mean growing area, $\bar{s}$ ($\bar{s} = A/N$, with $A =$ unit area, e.g., hectare, $N =$ tree number), is used as surrogate variable for mean lateral crown extension and resource demand of a plant. So, $\bar{\mathrm{mt}} \propto \bar{s}^{\alpha_{\bar{\mathrm{mt}},\bar{s}}}$ represents a linkage between biomass production of the mean tree and the required growing area or resources and thus couples production ecology with population ecology (Zeide 1987). As the average growing area, $\bar{s}$, is the inverse of number of plants, N ($\bar{s} \propto 1/N$), $\bar{\mathrm{mt}} \propto \bar{s}^{\alpha_{\bar{\mathrm{mt}},\bar{s}}}$ can be written as $\bar{\mathrm{mt}} \propto N^{-\alpha_{\bar{\mathrm{mt}},\bar{s}}}$, which is equivalent to $N \propto \bar{\mathrm{mt}}^{-1/\alpha_{\bar{\mathrm{mt}},\bar{s}}}$.

The latter forms the basis of the self-thinning rule which is mostly depicted on a double logarithmic scale (Fig. 13.7). The upper self-thinning, or limiting boundary line (solid line), marks the maximum possible density for a species at a given average plant size, or weight in even-aged pure stands under optimum site conditions. The lower self-thinning line (dashed line) marks the characteristic

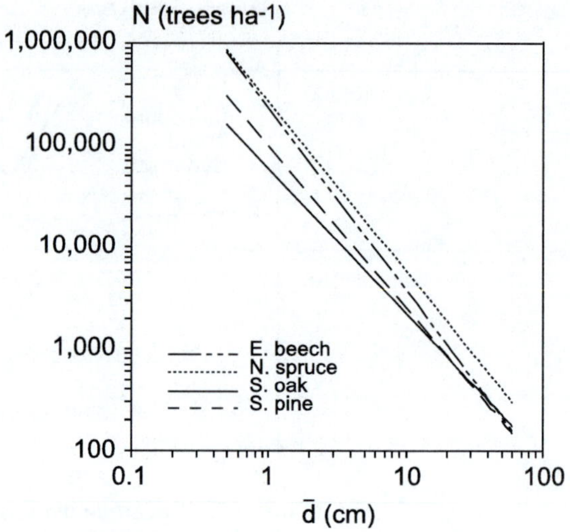

Fig. 13.8 Species-specific ln $(N)-\ln(\overline{d})$-relationships for untreated, fully stocked, pure European beech, Norway spruce, Sessile oak, and Scots pine stands in Bavaria/South Germany under survey since 1870 (Pretzsch and Biber 2005)

boundary relationship for any stand under sub-optimum growing conditions. Given two stands A and B growing under optimum and sub-optimum conditions, respectively, the size–density relationships of each stand initially approximate their stand-specific self-thinning lines, and, subsequently, follow this line. The lines may have different absolute levels, but possess similar gradients (Pretzsch 2009).

Alternative forms of these relationships are seen when self-thinning lines are represented in a Cartesian instead of a double-logarithmic grid (de-logarithmized instead of logarithmized version), when size is represented on the y-axis instead of the x-axis (α instead of $1/\alpha$ exponent), or when plant quadratic mean diameter, $\overline{d}$, is used instead of mean plant mass, $\overline{m}$t, as the size variable (rule of Reineke 1933 instead of Yoda et al. 1963). For the rearrangement and conversion of one to the other representation, see Niklas (1994) or Pretzsch (2010).

In the following we use the self-thinning rule in Reineke's (1933) representation with ln(N) on the y-axis and $\ln(\overline{d})$ on the x-axis as this is based on the variables tree number, N, and quadratic mean diameter, $\overline{d}$, primarily recorded in forest experiments and mostly used. For the relationship between N and $\overline{d}$ in fully stocked, even-aged forest stands Reineke (1933) defined the "stand density rule" $N = b\overline{d}^{-1.605}$. His rule can be represented on the ln–ln scale as a straight line $\ln N = b' - 1.605\ \ln \overline{d}$ with the intercept $b' = \ln b$ and the slope $\alpha = -1.605$. Reineke obtained this scaling rule by plotting $\overline{d}$ and N for untreated forest inventory plots in the USA on a double-logarithmic grid. He found very similar allometric exponents for various tree species, stand structures, and sites, and hence concluded that the rule has a general validity of $\alpha_{N,\overline{d}} \cong -1.605$ for forest stands.

Pretzsch and Biber (2005) have re-evaluated Reineke's rule based on 28 fully stocked pure stands of European beech, Norway spruce, Scots pine, and Sessile oak in Germany, which have been inventoried since 1870. Figure 13.8 shows the

$\ln(N)$– $\ln(\bar{d})$ -relationships for European beech, Norway spruce, Scots pine and Sessile oak with species-specific values of $\alpha_{N,\bar{d}} = -1.789$ for European beech, $\alpha_{N,\bar{d}} = -1.664$ for Norway spruce, $\alpha_{N,\bar{d}} = -1.593$ for Scots pine, and $\alpha_{N,\bar{d}} = -1.424$ for Sessile oak. Physiologically, the species-specific allometric exponent $\alpha_{N,\bar{d}}$ demonstrates how strongly a species enforces self-thinning for a given increase in diameter, or, in the words of Zeide (1985), the species' self-tolerance. According to the results above, European beech exhibits the highest self-thinning, or lowest self-tolerance, and Sessile oak the lowest self-thinning, or highest self-tolerance as defined by Zeide (1985).

Work of Kira et al. (1953) and Yoda et al. (1963) brought the −3/2 power rule of self-thinning to scientific prominence, initiating probably the most extensive discussion that has occurred about a scaling rule. These authors described the relationship between the average shoot mass, $\bar{m}$, and the plant number, N, per unit area in even-aged and fully stocked mono-specific plant populations as $\bar{m} \propto N^{-3/2}$ with the species invariant scaling exponent $\alpha_{\bar{m},N} = -3/2$. Yoda et al. (1963) assumed that plants are simple Euclidian objects, and all plant parts are related to each other isometrically. Effectively, Yoda's allometric coefficient −3/2 is based on the cubic relation between plant diameter, $\bar{d}$, and biomass, $\bar{m}$, $\bar{m} \propto \bar{d}^3$, and the quadratic relation between $\bar{d}$ and occupied growing area, $\bar{s}$, $\bar{s} \propto \bar{d}^2$. As the average growing area, $\bar{s}$, is the inverse of number of plants, N ($\bar{s} = 1/N$), the rearrangement of the above equations yields $\bar{m} \propto (N^{-1/2})^3 \propto N^{-3/2}$.

Harper (1977, p. 183) ascribed the −3/2 power law a validity for annual plants and forests as well. White (1981, p. 479) placed the "empirical generality of the rule ... beyond question" and Long and Smith (1984, p. 195) referred to it as "...a true law instead of the mere rule...". However, from about this time the rule began to be questioned, as indicated in the titles of papers by Hutchings (1983) (Ecology's law in search of a theory) and Lonsdale (1990) (The self-thinning rule: dead or alive?). A quarter of the century after the first euphoria concerning the law Begon et al. (1998, p. 169) pleaded for detection of inter-specific characteristics in allometric scaling studies, indicating a growing dissatisfaction with the law. The difference between the exponents of Reineke and Yoda arises from the different allometry between mt and quadratic mean diameter and mean plant weight. By rearranging Yoda's rule to $N \propto \mathrm{mt}^{1/\alpha_{\bar{\mathrm{mt}},N}}$, and substituting it in Reineke's rule, $N \propto \bar{d}^{\alpha_{N,\bar{d}}}$, one obtains $\bar{\mathrm{mt}}^{1/\alpha_{\bar{\mathrm{mt}},N}} \propto \bar{d}^{\alpha_{N,\bar{d}}} \Leftrightarrow \bar{\mathrm{mt}} \propto \bar{d}^{\alpha_{\bar{\mathrm{mt}},N} \times \alpha_{N,\bar{d}}}$. The original exponents from Yoda and Reineke result in an exponent of $\bar{\mathrm{mt}} \propto \bar{d}^{2.4075}$ (Pretzsch 2009, p. 404) and Reineke's rule becomes just a special case of Yoda's.

13.3.5 Scaling of Crown in Dependence on Intra- and Inter-specific Competition

For European beech in addition to the stand density, measured by sdi, the species composition of a tree's neighborhood significantly influences the scaling of its crown (Dieler and Pretzsch 2012). On inspection of the family of intra-specific

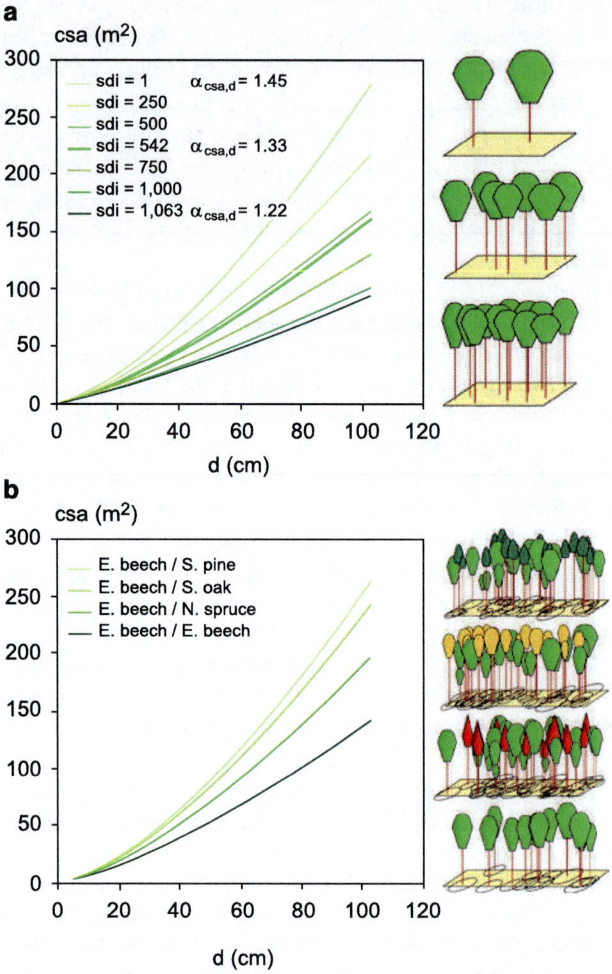

Fig. 13.9 Effect of (**a**) stand density and (**b**) species mixing on the allometric relationship between crown cross-sectional area, csa, and stem diameter, *d*, of European beech (adapted from Dieler and Pretzsch 2012). (**a**) Stand density ranges from SDI = 1 to 1,063 and results for a stem diameter of 50 cm in csa = 98–39 m^2. (**b**) Even under constant stand density of SDI = 463 (mean observed value) in the case of a tree of 50 cm crown cross-sectional area amounts to 55, 72, 86, 91 m^2 (in percent 100, 131, 156, 167) if the neighboring trees are European beech, Norway spruce, Sessile oak, or Scots pine (see Dieler and Pretzsch 2012)

allometric relationships between crown cross-sectional area, csa, and stem diameter, *d*, of European beech (csa $\propto d^{\alpha}$) at varying stand density index, it becomes obvious that Yoda's −3/2 self-thinning is just one special case in the broad range of scaling exponents which vary in dependence on the individual tree's competitive status (Fig. 13.9a). When sdi = 542 the scaling exponent $\alpha_{csa,d}$ amounts to 1.33 predicted by MST. On the included un-thinned plots the maximum sdi is 1,063, which corresponds with $\alpha_{csa,d} = 1.22$.

Even in the case of equal sdi-values, the species composition in the vicinity of a plant has a considerable influence on $\alpha_{csa,d}$. The spread of the crown is most restricted under intra-specific competition, for example, in pure European beech stands (Fig. 13.9b). This restriction diminishes when beech competes with a species like Norway spruce, Sessile oak, or Scots pine, as all transmit a higher proportion of light at a given leaf area and have higher light compensation points. This means that it can be rather beneficial for European beech to forage for light within or besides a Scots pine crown than the other way round. The compensation points increase with ranking European beech<Norway spruce<Sessile oak<Scots pine (Lyr et al. 1967), for example, and have the same ranking as the allometric exponents $\alpha_{csa,d}$.

Such changes of the species-specific scaling exponents from pure to mixed-species stands have a considerable impact on resource supply and capture (Richards et al. 2010), contribute to the understanding of mixing effects, e.g., over- and under-yielding effects (Pretzsch and Schütze 2009), and deliver a quantitative basis for a more realistic modeling of growth, density, and mortality in mixed-species stands (Dieler and Pretzsch 2012).

13.3.6 Scaling Under Alien-Thinning Conditions in Mixed-Species Stands

The slope $\alpha_{N,\bar{d}}$ of Reineke's self-thinning line $N \propto \bar{d}^{\alpha_{N,\bar{d}}}$ reveals the self-tolerance of a tree species growing in pure stands (Zeide 1985). The larger the $\alpha_{N,\bar{d}}$, the lower will be the number of dying trees $\Delta N/N$ for a given diameter increment $\Delta\bar{d}/\bar{d}$ and the greater the self-tolerance of the species in pure stands. The ranking we revealed for the mean species-specific $\alpha_{N,\bar{d}}$-values of European beech (−1.789) < Norway spruce (−1.664) < Scots pine (−1.593) < Sessile oak (−1.424) expresses that in comparison to Norway spruce and European beech, Sessile oak and Scots pine are more tolerant with trees of the same species (Pretzsch and Biber 2005). For instance, in European beech stands, a mean diameter increase of 1% causes a decrease in the number of stems by 1.79%. Given the same diameter increment the decrease in the number of stems is 1.66%, 1.59%, and 1.42% for Norway spruce, Scots pine, and Sessile oak, respectively, or 7%, 11%, and 21% lower than the decrease in stem number for European beech. This underlines the low self-tolerance of European beech and its space consuming investment strategy. The causes for this are its wider and more dynamic lateral crown extension as discussed in the previous section.

For mixed stands on comparable sites mean $\alpha_{N,\bar{d}}$-values determined for European beech, Norway spruce, Scots pine, and Sessile oak were $\alpha_{N,\bar{d}} = -0.40$, −1.02, −1.06, and −2.01, respectively, which indicates a reversal of the ranking in pure stands, e.g., European beech>Norway spruce>Scots pine>Sessile oak (Pretzsch and Biber 2005). Great crown expansion and space occupation abilities under intra-specific conditions evidently confer great assertive power in

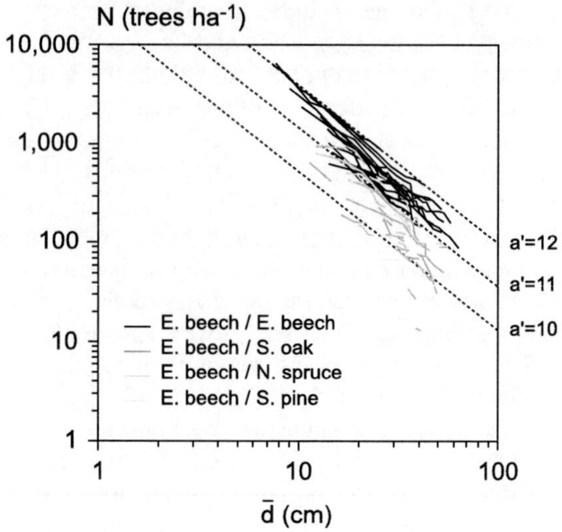

Fig. 13.10 Relationship between quadratic mean tree diameter, $\bar{d}$, and tree number per unit area, N, of European beech in fully stocked pure and mixed stands. The mean slope and standard error of the thinning line $N \propto \bar{d}^{\alpha_{N,\bar{d}}}$ of European beech are $\alpha_{N,\bar{d}} = -1.52$ (± 0.07 SE) in the pure European beech stand and become more shallow in the admixture with Norway spruce (-1.25 ± 0.19 SE), Sessile oak (-1.12 ± 0.14 SE), and Scots pine (-0.51 ± 0.32 SE)

a mixed stand. While intra-specific competition is high and results in steep N–$\bar{d}$-self-thinning slopes in pure beech stands, in mixed-species stands this competitive growth strategy turns against the admixed species and saves beeches from losses. Intra-specific competition turns into inter-specific competition, self- into alien-thinning, and the thinning slope of beech becomes shallower (Fig. 13.10).

13.4 Discussion and Conclusion

13.4.1 Plastic Rather than Fixed Structural Scaling

The tree data analyzed behave according to MST with respect to the relationship between body mass, mt, and leaf area, la, expressed by la $\propto$ mt$^{\alpha_{la,mt}}$ since $\alpha_{la,mt}$ and $\alpha_{ml,mt}$ are both close to 3/4 (Fig. 13.3). However, with respect to the scaling of structure our study provides empirical evidence that plants can deviate from MST as follows:

1. For the allometry between tree height, h, and trunk diameter, d (MST predicts $h \propto d^{2/3}$) observation differs from theory. For Norway spruce analyzed at the individual plant level (Table 13.1) the observed scaling exponent $\alpha_{h,d} < 0.66$ is smaller than predicted but for the group of gymnosperm tree species analyzed at mean tree level $\alpha_{h,d} > 0.66$ (($\alpha_{\bar{h},\bar{d}} = 0.903 \pm 0.021$; mean $\pm$ 95% CI)) exceeds the scaling exponent predicted for the allometric ideal plant by more than 30%. Similar deviations were found for European beech analyzed individually (Table 13.1) and other angiosperm tree species analyzed on the basis of the yield table data ($\alpha_{\bar{h},\bar{d}} = 0.733 \pm 0.022$).

2. In contrast to the predicted scaling of crown cross-sectional area, csa, versus diameter, d (MST predicts csa $\propto d^{4/3}$), higher values were found for Norway spruce and lower values for European beech at the individual level (Table 13.1) and $\alpha_{\overline{csa},\bar{d}} = 1.458 \pm 0.030$ at mean tree level (Fig. 13.6) are observed.
3. Between total volume, v, and tree diameter, d (MST predicts $v \propto d^{8/3}$), $\alpha_{v,d}$ mean values at the individual tree level of Norway spruce and European beech range between $\alpha_{v,d} = 2.54 - 2.56$ (Table 13.1), while $\alpha_{\bar{v},\bar{d}} = 2.820 \pm 0.036$ (Fig. 13.6). In the case of $\alpha_{\bar{v},\bar{d}}$, the real exponent might actually be closer to the allometric ideal plant than the observed values, as yield tables depict stem volume and not total aboveground volume and the brushwood portion is high in the pole stage and becomes smaller in the mature phase (Grundner and Schwappach 1952).
4. The analysis of combined relationships between crown volume, cv, and total tree volume, v (MST predicts cv $\propto v^{3/4}$), yielded $\alpha_{cv,v} = 0.77 - 0.80$(Table 13.1). Additionally, $\alpha_{\overline{cv},\bar{v}}$ extracted from the yield table data with 0.817 ± 0.011 for all species, 0.837 ± 0.012 for the gymnosperms, and 0.791 ± 0.091 for the angiosperms exceeds the value predicted by MST. Frequently discussed error sources, like decrease of crown length, cl, with tree height, h, changes of specific wood density R with size, or increasing portion of physiologically inactive heartwood (Pretzsch 2010) may result in a slight overestimation of the observed scaling exponents, in such a way that a correction (for which appropriate data is lacking) would shift the results somewhat closer to the prediction by MST (Pretzsch 2010).

MST assumes common scaling relationships for the allometric ideal plant, e.g., $\alpha_{h,d} = 2/3, \alpha_{csa,d} = 4/3$, and $\alpha_{v,d} = 8/3$ (West et al. 2009). Insertion of these scaling exponents for an allometric ideal plant in $\alpha_{cv,v} = (\alpha_{h,d} + \alpha_{csa,d})/\alpha_{v,d}$ yields $\alpha_{cv,v} = (2/3 + 4/3)/(8/3) = 3/4$. However, $\alpha_{cv,v} = 3/4$ could also result from diverging components, e.g., $\alpha_{cv,v} = (1/3 + 5/3)/(8/3) = 3/4$. In other words, the revealed deviation in scaling of structure from the allometric ideal plant is not inevitably a contradiction to the core assumption of the 3/4 scaling of MST. Within a broad range, competition can squeeze or stretch the crown and cause the observed broad intra-specific and inter-specific variation in scaling of structure. It is probably this plasticity of the crown, within which branches act as a holding structure for the leaf organs which enables the plant to keep on the 3/4 power leaf area-plant biomass trajectory (Fig. 13.3). Pretzsch et al. (2012) showed a similar plasticity of the root-shoot allometry.

13.4.2 Relevance of Plastic Scaling for Space Occupation and Competition at the Individual Tree Level

The vertical and lateral extension of a tree crown represented by height, h, crown cross-sectional area, csa, and the crown volume, cv, indicate a tree's space occupation and competitive status. We base the following discussion of the dependency of space occupation on crown allometry on the relationship cv $\propto h \times$ csa $\propto v^{\alpha_{h,v}+\alpha_{csa,v}}$, which is equivalent to $v^{(\alpha_{h,d}+\alpha_{csa,d})/\alpha_{v,d}}$ or cv $\propto v^{\alpha_{cv,v}}$. The exponent is assumed to be $\alpha_{cv,v} = 3/4$ for the allometric ideal tree (West et al. 2009) but varies according to empirical observations as reported in Table 13.1 and Fig. 13.6. By setting the occupied space, cv, in relation to the required tree volume, v $(v \propto m)$, we can quantify the efficiency of space occupation $\mathrm{eff_{cv}} = v^{\alpha_{cv,v}}/v = v^{\alpha_{cv,v}-1}$. When we assume according to MST that $\alpha_{cv,v} = 3/4$ we obtain $\mathrm{eff_{cv}} = v^{-1/4}$. In other words, for all species, efficiency of growing space occupation would decrease depending on size to the power of $-1/4$. However, by insertion of the species-specific exponents, reported, e.g., in Table 13.1 ($\alpha_{cv,v} = 0.80$ for Norway spruce and $\alpha_{cv,v}= 0.77$ for European beech), we see that efficiency of space occupation decreases more slowly for spruce $\mathrm{eff_{cv}} = v^{-0.20}$ and more rapidly for beech $\mathrm{eff_{cv}}= v^{-0.23}$. With each percentage increase in size, the efficiency of space occupation decreases by 0.20% and 0.23%, respectively.

For a closer look at the efficiency of vertical and lateral space occupation we can distinguish between the two components of $\mathrm{cv} \propto v^{\alpha_{cv,v}}$, $\alpha_{h,v}$, and $\alpha_{csa,v}$ and derive analogous efficiency ratios for vertical and lateral space occupation which can be formulated as $\mathrm{eff}_h \propto v^{\alpha_{h,v}}/v = v^{\alpha_{h,v}-1}$ and $\mathrm{eff_{csa}} \propto v^{\alpha_{csa,v}}/v = v^{\alpha_{csa,v}-1}$. For an allometric ideal plant it can be shown that $\alpha_{h,v} = 0.25$ and $\alpha_{csa,v} = 0.50$ (West et al. 2009). That means efficiency eff_h decreases with exponent $-0.75(\alpha_{h,v} - 1 = -0.75)$ and $\mathrm{eff_{csa}}$ with $-0.50(\alpha_{csa,v} - 1 = -0.50)$. These differences reflect how species-specific differences in crown allometry determine space sequestration and competition of trees coping with crowding in pure and mixed stands.

13.4.3 Implications for Dynamics at Stand Level

At the individual tree level allometric exponents describe how growth in size is related to the requirement for resources and growing area. At the stand level the exponents describe how many resources or how much growing area an average tree needs to survive and grow in size, both of which are crucial aspects of population dynamics. Several attempts were made to scale from individual tree to average tree allometry in order to link tree physiology with population ecology. Enquist et al. (1998, 2009) transfer the MST, which predicts $\mathrm{la} \propto \mathrm{mt}^{3/4}$ for individual plants in the following way: They suppose a constant resource supply and leaf area on stand level (stand leaf area LA = const.; stand resource supply R = const.; tree number per unit area N), and as $\mathrm{LA} = \mathrm{N} \times \overline{\mathrm{la}}$, $R = N \times \overline{r}$, and $\overline{\mathrm{la}} \propto \overline{r}$, they come to $\overline{\mathrm{la}} \propto \overline{\mathrm{mt}}^{3/4}$, and as $N = \overline{\mathrm{la}}^{-1}$, they derive $\overline{\mathrm{mt}} \propto N^{-4/3}$ equivalent with an overarching self-thinning slope of $\alpha_{\overline{\mathrm{mt}},N} = -4/3$. In other words, they transfer the fractal geometrical scaling based on the tree's inner space filling pipes directly to processes at stand level. This is contradictory to the $-3/2$ power rule of self-thinning developed by Yoda et al. (1963), which results from assumptions on growing area requirements, derived by Euclidean geometry in the following way: Assuming that mean tree growing area and diameter scale as $\overline{\mathrm{csa}} \propto \overline{d}^2$ and mean tree volume and tree diameter as $\overline{v} \propto \overline{d}^3$, then $\overline{\mathrm{csa}} \propto \overline{v}^{2/3}$ and as $\overline{\mathrm{csa}} = N^{-1}$ this yields $N \propto \overline{v}^{-2/3}$, which reveals the self-thinning slope of $-3/2\,(\overline{v} \propto N^{-3/2})$. MST predicts a stronger reduction of tree numbers with size (−0.75% decrease of tree number per 1% of mass growth) compared with the Euclidean geometry approach (−0.67% per 1%).

Table 13.2 Notations and units associated with analyses on individual and mean tree level

Variable	Unit	Definition
$\alpha_{y,x}$		Allometric scaling exponent on population level addressing inter-individual allometry. The subscript expressions show which tree variables are addressed
$\alpha_{y,x}^{k}$		Allometric scaling exponent of tree k related to intra-individual allometric issues. The subscript expressions show which tree variables are addressed
$\alpha_{\bar{y},\bar{x}}$		Allometric scaling exponent on mean tree level. The subscript expressions show which tree variables are addressed
Individual tree level		
d	cm	Trunk diameter measured in 1.3 m stem height
h	m	Tree height
cl	m	Crown length
csa	m^2	Crown cross-sectional area or horizontal crown area
s	m^2	Growing area
cv	m^3	Crown volume
la	m^2	Leaf area
ml	kg	Leaf mass
v	m^3	Total aboveground tree volume
mt	kg	Total plant mass
r		Resource requirement
Mean tree level		
$\bar{d}$	cm	Mean trunk diameter in 1.3 m stem height
$\bar{h}$	m	Mean tree height
$\overline{\text{csa}}$	m^2	Mean crown cross-sectional area
$\bar{s}$	m^2	Mean growing area
$\overline{\text{cv}}$	m^3	Mean crown volume
$\bar{v}$	m^3	Mean trunk volume
N	Trees ha^{-1}	Tree number per hectar
$\bar{r}$		Mean resource requirement per plant

Based on our results both approaches can be reconciled as follows: We start with the overarching MST relationship $\text{la} \propto v^{3/4}$ which predicts that a 1% increase in size is associated with a 3/4% increase in leaf area at the individual tree and mean tree level ($\overline{\text{la}} \propto \bar{v}^{3/4}$). According to the previous section we assume that a given leaf area can require and occupy more or less space or growing area, depending the fractal dimension n of the crown surface area. Suppose, analogous to the previous section that l is the crown extension in one direction, then for the mean crown cross-sectional area csa as well as for the crown surface area, $\text{csa} \propto l^2$ according to Euclidean geometry. Leaf area scales as $\text{la} \propto l^n$, with $n = 2$–3 depending on the pattern of the leaf area arrangement within the crown. Combining the expressions $\text{csa} \propto l^2$ and $\text{la} \propto l^n$ yields the result $\text{csa} \propto \text{la}^{2/n}$ and the combination with $\text{la} \propto v^{3/4}$ yields $\text{csa} \propto v^{3/4 \times 2/n}$. Considering that in a closed stand $\overline{\text{csa}} = N^{-1}$, we come to $v \propto N^{-4/3 \times n/2}$ that is equivalent to a self-thinning slope of $\alpha_{\bar{v},N} = -4/3 \times n/2$. Thus, the slope lies between −4/3 in case of ($n = 2$) and −2 for ($n = 3$).

These theoretically derived ranges of self-thinning slopes correspond well to the empirically observed self-thinning slopes, usually found by Osawa (1995), Pretzsch (2006), Weller (1987), and Zeide (1985) inter alia to fall between −1.3 and −2.0. Reineke's (1933) self-thinning line $N \propto \bar{d}^{\alpha_{N,\bar{d}}}$, originally held to have a widely applicable value $\alpha_{N,\bar{d}} = -1.605$, shows similar species-specific differences (von Gadow 1986; Pretzsch and Biber 2005; Sterba 1981; Zeide 1985).

A consolidated view of all these factors indicates that (in accordance to the MST) all species show the same relative increase in leaf area when growing in size, but, contrary to MST, those species which arrange their leaf area in an umbrella-like shape are more space demanding compared with trees with broom-like crowns. We conclude that observed developments of plant structure and stand self-thinning dynamics seem to result from both a general allometric partitioning, which is inherent in all woody and herbaceous plant species, and a species-specific structural allometry and plasticity which is an adaptation and acclimation to selective pressure.

Acknowledgments We wish to thank the German Science Foundation (Deutsche Forschungsgemeinschaft) for providing the funds for forest growth and yield research as part of the Collaborative Research Centre 607 (Sonderforschungsbereich SFB 607) "Growth and Parasite Defense" and the Bavarian State Ministry for Nutrition, Agriculture and Forestry for permanent support of the project W 07 "Long-term experimental plots for forest growth and yield research". Thanks are also due to Gerhard Schütze for support of the field work and data processing, to Ulrich Kern for the graphical artwork, and to the reviewers and co-editors for their constructive criticism and comments.

References

Bates D, Maechler M (2010) lme4: Linear mixed-effects models using S4 classes. R package version 0.999375-35. http://CRAN.R-project.org/package=lme4

Begon ME, Harper JL, Townsend CR (1998) Ökologie. Spektrum Akademischer Verlag, Heidelberg

von Bertalanffy L (1951) Theoretische Biologie: II. Band, Stoffwechsel, Wachstum, 2nd edn. A Francke AG, Bern

Dieler J, Pretzsch H (2012) Morphological plasticity of European beech (*Fagus sylvatica* L.) in pure and mixed-species stands. Forest Ecol Manage submitted

Enquist BJ, Niklas KJ (2001) Invariant scaling relations across tree-dominated communities. Nature 410:655–660

Enquist BJ, Brown JH, West GB (1998) Allometric scaling of plant energetics and population density. Nature 395:163–165

Enquist BJ, West GB, Brown JH (2009) Extension and evaluations of a general quantitative theory of forest structure and dynamics. Proc Natl Acad Sci USA 106(17):7046–7051

von Gadow K (1986) Observation on self-thinning in pine plantations. S Afr J Sci 82:364–368

Gorham E (1979) Shoot height, weight and standing crop in relation to density of monospecific plant stands. Nature 279:148–150

Grundner F, Schwappach A (1952) Massentafeln zur Bestimmung des Holzgehaltes stehender Waldbäume und Bestände. Verlag Paul Parey, Berlin

Harper JL (1977) Population Biology of Plants. Academic Press, London, New York

Hutchings MJ (1983) Ecology's law in search of a theory. New Sci 98:765–767

Huxley JS (1932) Problems of relative growth. Lincoln Mac Veagh, Dial Press, New York
Kira T, Ogawa H, Sakazaki N (1953) Intraspecific competition among higher plants, I. Competition-yield-density interrelationship in regularly dispersed populations. J Inst Polytech Osaka City Univ Ser D 4:1–16
Long JN, Smith FW (1984) Relation between size and density in developing stands: a description and possible mechanisms. Forest Ecol Manage 7:191–206
Lonsdale WM (1990) The self-thinning rule: dead or alive? Ecology 71:1373–1388
Lyr H, Polster H, Fiedler HJ (1967) Gehölzphysiologie. VEB Gustav Fischer Verlag, Jena
Matthew C, Lemaire G, Sackville Hamilton NR, Hernandez-Garay A (1995) A modified self-thinning equation to describe size/density relationships for defoliated swards. Ann Bot 76:579–587
McMahon TA, Kronauer RE (1976) Tree structures: Deducing the principle of mechanical design. J Theor Biol 59:443–466
Niklas KJ (1994) Plant Allometry. University of Chicago Press, Chicago
Niklas KJ (2004) Plant allometry: is there a grand unifying theory? Biol Rev 79:871–889
Niklas KJ, Cobb ED, Spatz H-CH (2009) Predicting the allometry of leaf surface area and dry mass. Am J Bot 96(2):531–536
Osawa A (1995) Inverse relationship of crown fractal dimension to self-thinning exponent of tree populations: a hypothesis. Can J Forest Res 25:1608–1617
Pinheiro JC, Bates DM (2000) Mixed-effects models in S and S-plus. Springer, New York
Pretzsch H (2006) Species-specific allometric scaling under self-thinning. Evidence from long-term plots in forest stands. Oecologia 146:572–583
Pretzsch H (2009) Forest dynamics, growth and yield. Springer, Heidelberg
Pretzsch H (2010) Re-Evaluation of Allometry: State-of-the-Art and Perspective Regarding Individuals and Stands of Woody Plants. Prog Bot 71:339–369
Pretzsch H, Biber P (2005) A re-evaluation of Reineke's rule and Stand Density Index. Forest Sci 51:304–320
Pretzsch H, Biber P (2010) Size-symmetric versus size-asymmetric competition and growth partitioning among trees in forest stands along an ecological gradient in central Europe. Can J Forest Res 40(2):370–384
Pretzsch H, Dieler J (2012) Evidence of variant intra- and interspecific scaling of tree crown structure and relevance for allometric theory. Oecologia. doi:10.1007/s00442-011-2240-5
Pretzsch H, Schütze G (2009) Transgressive overyielding in mixed compared with pure stands of Norway spruce and European beech in Central Europe: Evidence on stand level and explanation on individual tree level. Eur J Forest Res 128:183–204
Pretzsch H, Biber P, Dursky J (2002) The single tree based stand simulator SILVA. Construction, application and evaluation. Forest Ecol Manage 162:3–21
Price CA, Gilooly JF, Allen AP, Weitz JS, Niklas KJ (2010) The metabolic theory of ecology: prospects and challenges for plant biology. New Phytol 188:696–710
R Development Core Team (2009) R: A language and environment for statistical computing. R Foundation for Statistical Computing, Vienna, ISBN:3-900051-07-0
Reineke LH (1933) Perfecting a stand-density index for even-aged forests. J Agric Res 46:627–638
Richards AE, Forrester DI, Bauhus J, Scherer-Lorenzen M (2010) The influence of mixed tree plantations on the nutrition of individual species: a review. Tree Physiol 30(9):1192–1208
Sackville Hamilton NR, Matthew C, Lemaire G (1995) In defence of the -3/2 boundary rule. Ann Bot 76:569–577
Sterba H (1981) Natürlicher Bestockungsgrad und Reinekes SDI. Centralbl f d ges Forstw 98:101–116
Teissier G (1934) Dysharmonies et discontinuités dans la Croissance. Act Sci et Industr 95 (Exposés de Biometrie, 1). Hermann, Paris
Warton DI, Wright IJ, Falster DS, Westoby M (2006) Bivariate line-fitting methods for allometry. Biol Rev 81:259–291

Weller DE (1987) A Reevaluation Of The -3/2 Power Rule Of Plant Self-Thinning. Ecol Monogr 57:23–43

West GB, Brown JH, Enquist BJ (1997) A general model for the origin of allometric scaling laws in biology. Science 276:122–126

West GB, Enquist BJ, Brown JH (2009) A general quantitative theory of forest structure and dynamics. Proc Natl Acad Sci USA 106(17):7040–7045

White J (1981) The allometric interpretation of the self-thinning rule. J Theor Biol 89:475–500

Yoda KT, Kira T, Ogawa H, Hozumi K (1963) Self-thinning in overcrowded pure stands under cultivated and natural conditions. J Inst Polytech Osaka City Univ D 14:107–129

Yoda KT, Shinozaki K, Ogawa J, Hozumi K, Kira T (1965) Estimation of the total amount of respiration in woody organs of trees and forest communities. J Biol Osaka City Univ 16:15–26

Zeide B (1985) Tolerance and self-tolerance of trees. Forest Ecol Manage 13:149–166

Zeide B (1987) Analysis of the 3/2 power law of self-thinning. Forest Sci 33:517–537

Chapter 14
Principles of Growth Partitioning Between Trees in Forest Stands Under Stress

H. Pretzsch, J. Dieler, and T. Rötzer

14.1 Introduction

While allocation principles within individual plants are thoroughly analyzed (see Landsberg 1986; Mäkelä and Hari 1986; Niklas 1994) and growth dynamics on stand level is well described at the average tree level (see Assmann 1970; Oliver and Larson 1996; Pretzsch 2009), insight into resource and growth distributions between trees of a stand or cohort and their dependency on site fertility is still very limited (Schwinning and Weiner 1998; Weiner et al. 2001). However, the question whether tall, dominant or small, understoried plants suffer more from stress, and whether this size-dependent stress reaction is modified by site fertility, is essential to understand and to manage stands dynamics under steady state conditions and becomes even more relevant under climate change. In the view of growth trends (Spiecker et al. 1996) and stress events (Jentsch et al. 2007; Matyssek and Sandermann 2003; Pretzsch et al. 2010) knowledge of chronic and episodic stress effects on resource and growth partitioning among trees in forest stands is essential for scaling up from individual tree to stand growth, for understanding growth patterns on stand level, and for the prediction of ecosystem dynamics.

Analysis of allocation and tree response on individual plant level might deliver evidence concerning stress response, but conclusions on the relevance of such reactions and the performance on stand level require insight into tree interaction, competition and possible compensation effects between trees under stress. The reaction patterns of stands or cohorts are more than just the sum of individual tree responses and, thus, cannot be derived from trees grown solitarily in greenhouses (Matyssek et al. 2005).

H. Pretzsch (✉) • J. Dieler • T. Rötzer
Chair of Forest Growth and Yield Science, Technische Universität München,
Hans-Carl-von-Carlowitz-Platz 2, 85354 Freising, Germany
e-mail: Hans.Pretzsch@lrz.tum.de

R. Matyssek et al. (eds.), *Growth and Defence in Plants*, Ecological Studies 220,
DOI 10.1007/978-3-642-30645-7_14, © Springer-Verlag Berlin Heidelberg 2012

According to the growth–differentiation balance hypothesis (GDBH) by Herms and Mattson (1992), plant growth and plant defense are subjected to a trade-off in dependence on resource supply. Hence, in high-resource environments carbon is allocated to growth at the expense of defense. Consequently, these plants are predicted to grow more but represent lower levels of defense and secondary metabolites (Stamp 2003). Applied to the trees in a cohort or stand, this would mean that dominant trees with a high resource supply grow considerable more under favorable growing conditions than small trees with a low resource supply. But it also means that limiting site fertility—irrespective of acute or chronic nature—distinctly reduces the superiority of tall trees, and may be less negative or even a benefit for their small neighbors.

This chapter aims at a consistent understanding of the interindividual growth partitioning in forest stands under stress. We scrutinize if the reaction patterns predicted by GDBH at the individual plant level can be found on stand level. We start with the derivation of the growth–size trajectory of open-grown trees as a potential and reference growth and proceed with the revelation of how competition and other stress keep trees below this potential. Subsequently, we show the range of growth distribution patterns among trees within forest stands and analyze how these patterns depend on site-specific stress and limitation. The presented theoretical concept and the evidence from long-term experiments provide new insights by tracing growth distribution patterns from stand to the individual tree level.

14.2 Theoretical Considerations on Interindividual Growth Reaction Patterns

14.2.1 Relation Between Potential Size Growth and Size as Reference

To scrutinize how competition and abiotic stress affect individual tree growth rates in cohorts or stands, we start with the growth–size relationships which can be expected without stress (Fig. 14.1a). In the juvenile phase of trees, anabolism has the upper hand and drives growth exponentially. However, together with size maintenance costs increase and affect culmination of the growth rate which then finally tapers off (Zeide 1993). Therefore, open-grown trees follow an unimodal growth–size trajectory which comprises convex (seen from below) curve sections in the juvenile phase (1–3), concave sections in the middle age (4–6) and again convex sections in the mature phase (7–9) (see Avery and Burkhart 1983, p. 266; Schütz 1989, pp. 4–5). The better the site fertility the higher is the level of the curve, due to the better resource supply per plant for a given size (Pretzsch and Biber 2010). Such potential growth curves can be derived from long-term survey (real time series) or chronosequences (artificial time series) of trees grown under solitary conditions.

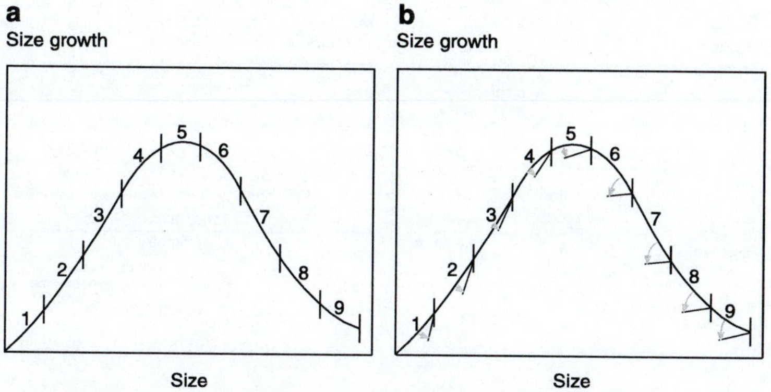

Fig. 14.1 Relationship between size and growth (**a**) without competition effect and (**b**) modification of this relationship by competition in schematic representation (adapted from Pretzsch 2010). (**a**) Potential size–growth trajectory of an open-growing individual tree. Without competition the course of growth in dependence on size consists of convex (segments 1–3 and 7–9) and concave (segments 4–6) parts. (**b**) Unimodal individual growth curve and linear interindividual size–growth relationships. Shown is the relationship between size and growth without competition effect (*thin line*) and growth reduction due to competition (*arrows*). This growth reduction by competition results in linear size–growth relationships with steep positive slopes in early stand phase and increasingly flatter slopes with progressive stand development (segments 1–9)

14.2.2 *Effect of Competition on Growth Partitioning Between Trees*

While growing solitarily mainly size determines the plant's growth–size trajectory. Within a stand especially those trees coping with crowding or stress can fall below this trajectory. For trees within a stand size denotes access to resources, especially to light. The taller a tree, the more privileged is in most cases its access to resources, space occupation and repression effect on neighbors (Biging and Dobbertin 1995; Pretzsch 2009). But size is an ambiguous trait; it can also mean higher susceptibility to windthrow (Peltola 2006; Valinger et al. 1993), drought (Condit et al. 1995; Skov et al. 2004), or bark beetle attacks (Coggins et al. 2010). However, in temperate forests with light as a limiting factor for individual tree growth, taller trees shade and thus reduce growth of their smaller neighbors. In a cohort of even-aged trees growth of smaller trees falls behind the taller ones. Subsequently, it drops more below the potential than that of their fitter neighbors.

This size-dependent growth reduction is behind the phenomenon which forest scientists refer to as diameter increment–diameter line (id–*d* line) and which is often used to describe and model growth–size relationship of even-aged stands in a given developmental phase (e.g., Prodan 1965, pp. 474–476). While Fig. 14.1a shows the potential growth–size trajectory of an individual tree, Fig. 14.1b shows schematically how competition can transform the unimodal curve to linear interindividual growth–size relationships in stands at different ages. Diameter growth

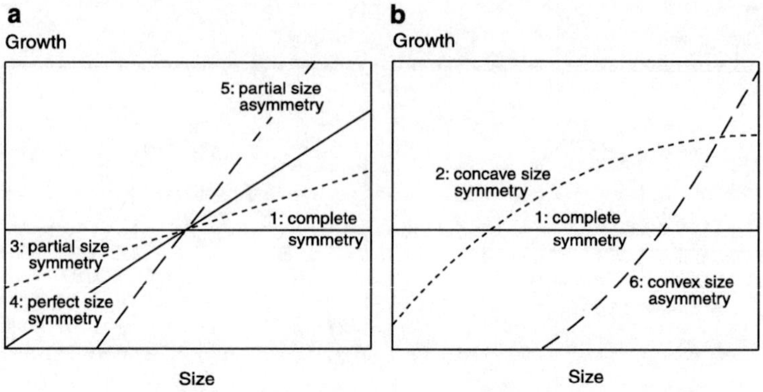

Fig. 14.2 Hypotheses on the relationship between plant size and absolute growth rate. (**a**) Different linear relationships between size and growth and (**b**) nonlinear relationships between plant size and growth (adapted from Pretzsch and Dieler 2011). *Line 1* represents the more theoretical case of complete symmetric size–growth relationship where all plants receive the same budget of growth irrespective of size. *Line 2* displays nonlinear concave size symmetry where growth increases less degressively with size. *Line 3* reflects partial size symmetry where growth increases linearly with size. *Line 4* represents perfect size symmetry and means that growth increases proportionally with size. *Line 5* stands for partial size asymmetry where growth increases linearly with size. *Line 6* represents nonlinear convex size asymmetry as the growth increases progressively with size

plotted above diameter (or volume growth versus volume) yields a straight line with a steep positive slope in early stand phases and an increasingly flatter slope with progressive stand development (Fig. 14.1b, segments 1–9).

Under ceteris paribus conditions, trees on fertile sites can make more use of their privileged position and exert a more negative effect on their neighbors' growth (Wichmann 2001, 2002). This should be reflected by steeper slopes, while nutrient limitation should diminish their superiority and the slope of the growth–size relationship (Pretzsch and Dieler 2011). Supposing that these findings are representative, the growth–size slope might be suitable for indication and further analysis of growth allocation patterns and their dependency on environmental conditions.

14.2.3 Site Fertility as Modifier of Growth–Size Relationship

The absolute growth rate in a defined period, such as 1 year, plotted against plant size at the beginning of the respective period can result in different patterns of growth allocation, representing different modes of competition between trees. Figure 14.2 displays a set of linear (lines 1, 3–5) and nonlinear (lines 2, 6) growth–size relationships (Weiner 1990). A steeper slope gradient indicates a stronger concentration of growth rates and resources on tall trees in the stand. The case of complete size asymmetry, indicated by a line parallel to the *y*-axis (slope = ∞; a sub-cohort of large plants receives all growth), is rare to observe and not integrated in Fig. 14.2. Note that all relationships in Fig. 14.2a are linear. However, only line 4 represents a linear and proportional increase of the absolute growth rate with increasing size, meaning that only in this case is the relative growth rate equal for all individuals.

Complete symmetry (Fig. 14.2a, line 1) would mean that growth and resources, which competitors receive, are independent of their size. Tendency towards complete symmetry (line 1) or partial size symmetry (Fig. 14.2a, line 3; Fig. 14.2b, line 2) is assumed to prevail under limitation by belowground resources (water and mineral nutrients), as they are mobile, diffuse quickly and are difficult to preempt by larger individuals (van Kuijk et al. 2008). Partial or strong size asymmetry (Fig. 14.2a, line 5; Fig. 14.2b, line 6) means that larger individuals obtain a disproportionately higher share of resources and growth. This mode of growth–size relationship can be expected on very fertile sites where light is the limiting factor and, as a vectorial resource, pre-emptible by the larger individuals (Cannell and Grace 1993; Weiner and Thomas 1986).

14.2.4 Tracing Reaction Pattern from Stand to Individual Tree Level

A useful concept for tracing stress-induced growth reductions, above introduced on stand level, to the individual tree level is the potential-modifier model (Mailly et al. 2003; Nagel 1999; Pretzsch 2009, pp. 291–314). This concept links the size-dependent potential growth rate g_{pot} with the expected growth rate, g_{exp}, under a given nonsolitary condition by means of a modifier mod (Fig. 14.3). This modifier is dependent on a competition index (CI) (e.g., Hasenauer et al. 2006; Newnham 1964) which quantifies to what extent the tree is suppressed by neighbors. The expected growth rate can be modeled as $g_{exp} = g_{pot} \times \text{mod}$ with mod $= f(\text{CI})$. The reduction of the potential to the observed growth rate, which yields a rather linear growth–size relationships as shown in Fig. 14.1b, results from the special pattern of individual tree reactions described by the modifier function. Modifier functions used so far are independent from site fertility, i.e., they ignore that a given competition index might reduce growth stronger on fertile site than on poor sites. With other words, a site-dependency of modifier functions on tree level might explain and substantiate different modes of competition sorted out on stand level (Fig. 14.2).

14.3 Empirical Evidence

In the following the focus is on the linkage between potential growth rate of open-grown individuals, growth under competition and growth–size patterns resulting on stand level from different kinds of stress. The empirical considerations base on the

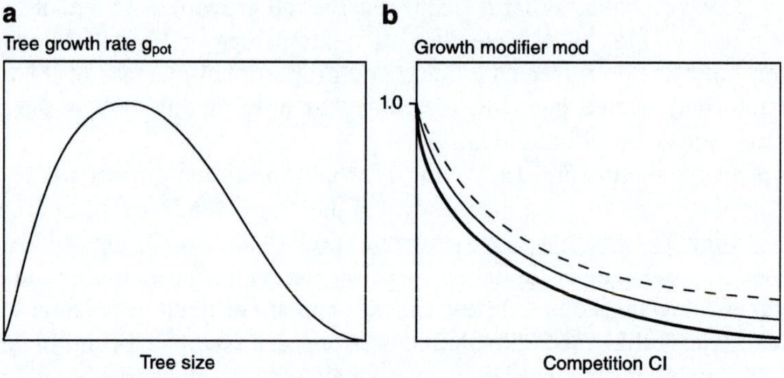

Fig. 14.3 Mainstays for growth estimation by individual tree models are (**a**) potential growth–size relationship and (**b**) relationship between modifier mod and aboveground competitive status of a tree (adapted from Pretzsch and Biber 2010). Potential tree growth rate g_{pot} is estimated in dependence on tree size ($g_{pot} = f(\text{size})$); the modifier is estimated dependent on a competition index (CI) that characterizes the aboveground competitive status of the tree. The expected growth rate g_{exp} results from $g_{exp} = g_{pot} \times \text{mod}$. As a rule models use one fixed relationship mod $= h(\text{CI})$. However, here we assume that site fertility determines the effect of a given CI as follows: the *bold line*, *thin line*, and *broken line*, respectively, indicate rich, mean, and scarce resource supply

following three data sets: (1) individual tree records of size and growth from 120 long-term experimental plots under survey since 1871 in forest stands along an ecological gradient through South Germany (Pretzsch 2010; Pretzsch and Biber 2010), (2) annual measurements of the growth–size relationship, including the extremely dry years of 1976 and 2003, from a mixing experiment of Norway spruce (*Picea abies* (L.) H. KARST.) and European beech (*Fagus sylvatica* L.) (Pretzsch and Dieler 2011), and (3) annual growth rates of spruces and beeches with and without double ambient ozone fumigation between 2000 and 2007 (Wipfler et al. 2009; Matyssek et al. 2010; Pretzsch et al. 2010).

14.3.1 Potential Diameter Increment

For scrutiny of the relationship between tree diameter and potential diameter growth, Pretzsch and Biber (2010) used data from 120 experimental plots in stands of Norway spruce (*Picea abies* (L.) H. KARST.), Scots pine (*Pinus silvestris* L.), European beech (*Fagus sylvatica* L.), and Sessile oak (*Quercus petrea* (MATT.) LIEBL.) and found always unimodal relationships as shown by example in Fig. 14.4. All species' potential growth curves, except European beech, show a distinct dependence on site fertility. Site fertility is indicated by top height, h_o, in meters at age 100. The curves reveal distinctly different species-specific growth rhythms. Scots pine (not shown) culminates at small diameters and drops down quickly after that, while Norway spruce and European beech culminate at slightly larger diameters and do not decline as quickly. Sessile oak (not shown) culminates at even larger diameters but on a lower level of growth rates. Norway spruce

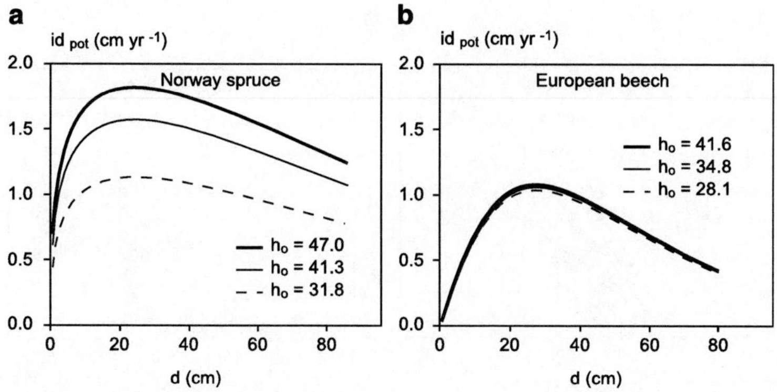

Fig. 14.4 Potential diameter increment model along the diameter for (**a**) Norway spruce and (**b**) European beech (adapted from Pretzsch and Biber 2010). The *bold, thin,* and *broken lines* respectively indicate the highest, medium, and poorest site fertility covered by experimental plots in South Bavaria (Pretzsch and Biber 2010). Site fertility is indicated by the stand top height, h_o, (m) at age 100

(Fig. 14.4a) shows by far the highest potential growth rate and strongest dependency on site fertility; while European beech's trajectories (Fig. 14.4b) lie on a much lower level and depend rather on tree size than on site fertility.

14.3.2 Effect of Stress on the Growth–Size Relationship

In the following, we show how trees fall below the unimodal-shaped growth–size trajectories of open-grown individuals when coping with crowding. For a comprehensive analysis of interindividual growth–size patterns under stress, Pretzsch and Dieler (2011) used long-term plots along an ecological gradient, forest stands under extreme drought stress, and tree cohorts under ozone exposure. The analysis of the growth–size relationship is based on the diameter at breast height, *d*, as the independent size variable (Fig. 14.2, *x*-axis). As the dependent variable (Fig. 14.2, *y*-axis) the mean periodical annual diameter increment, id, was used for the long-term plots, and the current annual increment was used for analyzing the effect of weather conditions and ozone effects. In all cases, the id–*d* relationship is fitted by a straight line $\text{id} = a_0 + a_1 \times d$ and the slope a_1 is applied for quantifying the pattern of interindividual growth partitioning. Steep slopes indicate the superiority of tall trees at the smaller trees' expense. Shallow slopes mean a reduction of the growth rate of tall trees in favor of the small neighbors.

14.3.2.1 Site Fertility as Modifier of the Growth–Size Relationship

Before showing the analysis over all included 64 long-term experimental plots representing an ecological gradient from high to poor sites, we illustrate how slopes

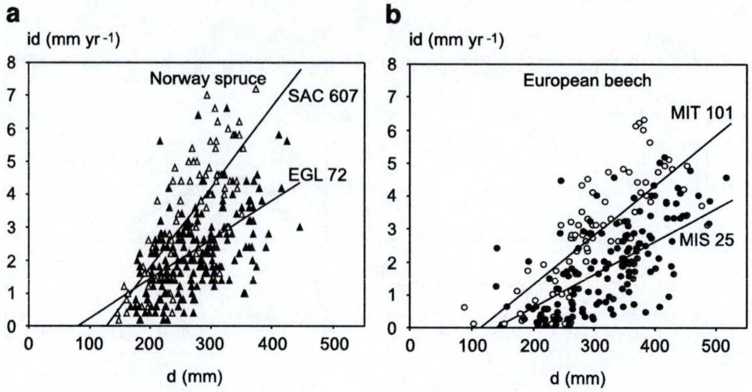

Fig 14.5 Relationship between individual tree diameter, *d*, and diameter growth, id, for (**a**) Norway spruce: Sachsenried 607, plot 7, survey 2001, $\bar{d} = 244$ mm, $h_o = 40.6$ m (*open triangles*; id $= -3.135 + 0.023 \times d, R^2 = 0.593, p < 0.001$) and Eglharting 72, plot 2, survey 1950, $\bar{d} = 252$ mm, $h_o = 31.4$ m (*filled triangles*; id $= -0.941 + 0.012 \times d$, $R^2 = 0.202$, $p < 0.001$) and for (**b**) European beech: Mitterteich 101, plot 2, survey 1960, $\bar{d} = 305$ mm, $h_o = 37.1$ m (*open circles*; id $= -1.487 + 0.014 \times d, R^2 = 0.595, p < 0.001$) and Mittelsinn 25, plot 1, survey 1968, $\bar{d} = 311$ mm, $h_o = 27.5$ m in case of beech (*filled circles*; id $= -1.418 + 0.010 \times d, R^2 = 0.401, p < 0.001$) (adapted from Pretzsch and Dieler 2011)

of growth–size relationships were extracted from the data (Fig. 14.5). The growth–size observations and regression lines are shown for two experimental plots in (a) Norway spruce and (b) European beech with excellent (SAC 607, MIT 101) and low (EGL 72, MIS 25) site fertility but otherwise rather similar stand parameters (Pretzsch and Dieler 2011).

The scrutiny of the relationship between the slope of the growth–size relationship and stand characteristics of all included long-term plots revealed a significant effect of site fertility (represented by top height in meters at age of 100 years, h_o), stand density (represented by the stand density index, SDI, according to Reineke (1933)), and the stand development phase (represented by the quadratic mean diameter, d_q) on the slope a_1 of the id–*d* relationship (Pretzsch 2010).

Figure 14.6 displays the key information of the regression analysis between slope a_1 and site fertility graphically (see Pretzsch and Dieler (2011) for details). The id–*d* relationship is shown to be dependent on h_o, when all other influencing variables are kept constant. The relative growth distribution among trees of different sizes is reflected by the differing slopes. Each of the three straight lines for (a) Norway spruce and (b) European beech represent the growth–size relationship for the highest, medium, and poorest site fertility within the data set. The growth–size relationships (Fig. 14.6) show that size asymmetry increases with increasing site fertility for both species. On poor sites, the relationships indicate a tendency towards a size symmetric growth–size relationship. On medium and fertile sites, perfect size symmetry or even size asymmetry is found. The curve spectrum of both species includes perfect size symmetry (line through the origin with $a_0 = 0$).

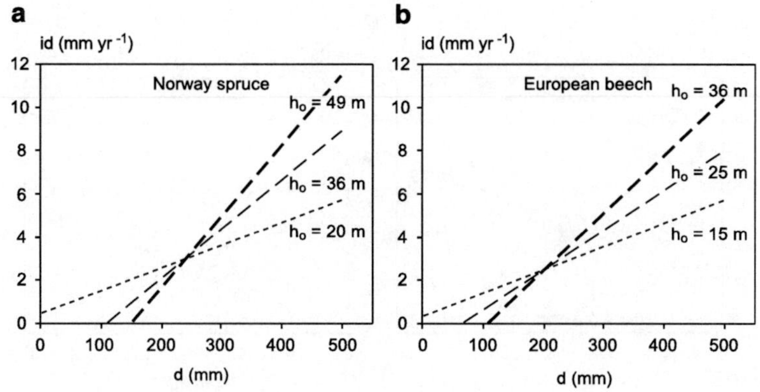

Fig. 14.6 Relationship between diameter growth and stem diameter for the poorest, medium, and highest site fertility for the species (**a**) Norway spruce and (**b**) European beech (adapted from Pretzsch and Dieler 2011). For each species we show the id–*d* relationship for sites with highest fertility (*steepest line*), medium fertility (*medium line*), and lowest site fertility (*shallow line*). Site fertility is indicated by the stand top height, h_o, (m) at age 100. For both species applies, that the steepness of the slope, i.e., the asymmetry of the size–growth relationship increases from poor to fertile sites

14.3.2.2 Drought as Modifier of the Growth–Size Relationship

In order to analyze the effect of annually changing growing conditions and drought on the growth–size relationships, we used data of the long-term experimental plot FRE 813/1 (Pretzsch et al. 1998) which is not included in the analysis in the last section. Basis data are the annual sequence of the growth–size relationships between 1972 and 2007, including the extremely dry years 1976 and 2003. By example, Fig. 14.7 displays the reaction patterns for the extremely dry and warm year 1976 in comparison to the rather moist and temperate year 1978. Both species reveal a change in their mode of growth–size relationship; after a rather shallow slope in the dry/warm year 1976, which indicated size symmetry, the allocation pattern changes within 2 years towards a far more size asymmetric relationship in 1978. In the case of Norway spruce, the slopes a_1 in 1976 and 1978 are significantly different at the level $p < 0.05$, whereas European beech shows no significant change.

14.3.2.3 Effect of Ozone Stress

For a further assessment of how growth–size relationships in stands are affected by stress, data of the Kranzberg ozone fumigation experiment (Matyssek et al. 2010; Werner and Fabian 2002) between 2000 and 2007 were compared (Pretzsch and Dieler 2011). The comparison between mean periodic diameter increment of trees with and without ozone fumigation yields neither a significantly different intercept nor slope for Norway spruce. In the case of European beech, ozone fumigation

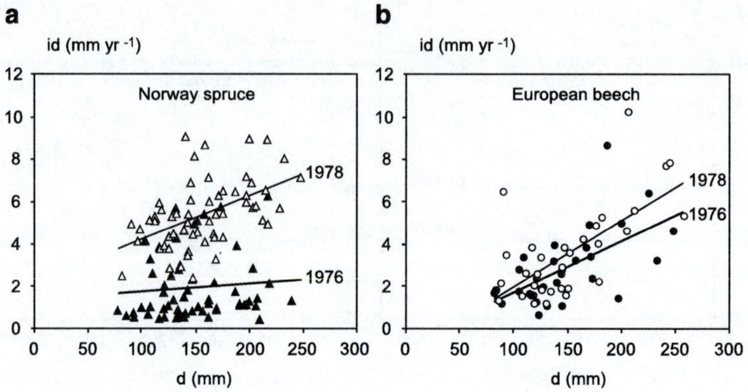

Fig. 14.7 Annual size–growth relationship in years with contrasting climate condition for (**a**) Norway spruce and (**b**) European beech (adapted from Pretzsch and Dieler 2011). Year 1976 (OLS fit: spruce 1.368 + 0.004 × *d*, $p > 0.05$; beech −0.683 + 0.024 × *d*, $p < 0.001$; *filled symbols*; *bold lines*) represents an extremely dry year whereas 1978 (OLS fit: spruce 2.139 + 0.021 × *d*, $p < 0.001$; beech −1.156 + 0.031 × *d*, $p < 0.001$; *open symbols*; *thin lines*) is characterized by excellent growing conditions

results in a significantly shallower slope ($a_1^{1\times O_3} = 0.033$; $a_1^{2\times O_3} = 0.011$) of the id–*d* relationship. The difference is significant on the level $p < 0.05$. This indicates that ozone stress reduces the growth of dominant European beeches more than their smaller neighbors, which are the relative winners of the ozone stress. The id–*d* line representing the growth–size relationship changes from size asymmetry under ambient ozone exposition (1 × O_3) to size symmetry under ozone stress (2 × O_3).

In addition, the combined effect of summer drought and ozone on the id–*d* line was analyzed for the extreme dry year 2003 (Pretzsch and Dieler 2011). Figure 14.8 shows the annual diameter increment over the initial diameter of trees exposed to 1 × O_3 and 2 × O_3 conditions in 2003. Again, it was found that ozone fumigation reduces the slope considerably, which is equivalent to a shift from size asymmetric to a more size symmetric growth–size relationship under combined drought and ozone stress in 2003 (see Fig. 14.2 for basic growth–size distribution patterns and terminology). Norway spruce shows a stronger flattening of the slope under ozone fumigation than European beech. For spruce, the flattening of the slope is significant (Fig. 14.8a) whereas beech shows no significant change (Fig. 14.8b). This suggests that spruce reacts more sensitively to 2 × O_3 in years with poor water supply than beech.

14.3.3 Tracing Site-Dependent Stress Effects from Stand to Individual Level

So far, the hypothesis that the mode of competition and biomass partitioning among plants is determined by the site fertility—above all the dominating limiting factor—stems mainly from analyses of herbaceous plants on stand level (e.g., Müller et al. 2000; Weiner et al. 2001). However, long-term experiments in forest stands with

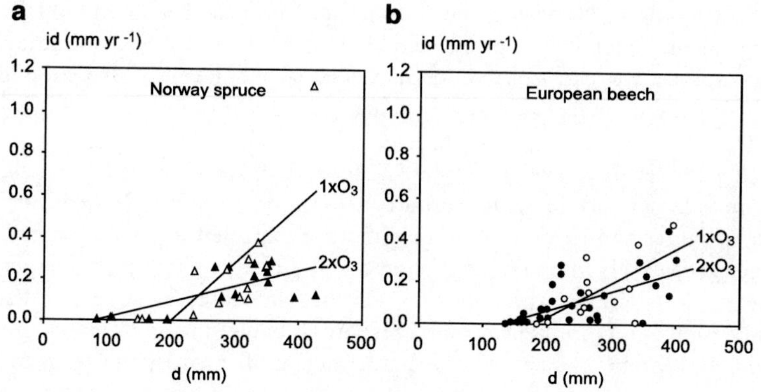

Fig. 14.8 Annual diameter increment in the extremely dry year 2003 over initial diameter of (**a**) Norway spruce and (**b**) European beech trees without ozone fumigation (1 × O_3: *open symbols*) and with double ambient ozone fumigation (2 × O_3: *filled symbols*) (adapted from Pretzsch and Dieler 2011). The OLS fit of the size-increment relationship is represented by *thin lines* in case of 1 × O_3, and by *bold lines* in case of 2×O_3. In case of spruce the slope for 2 × O_3 is significantly flatter than for 1 × O_3

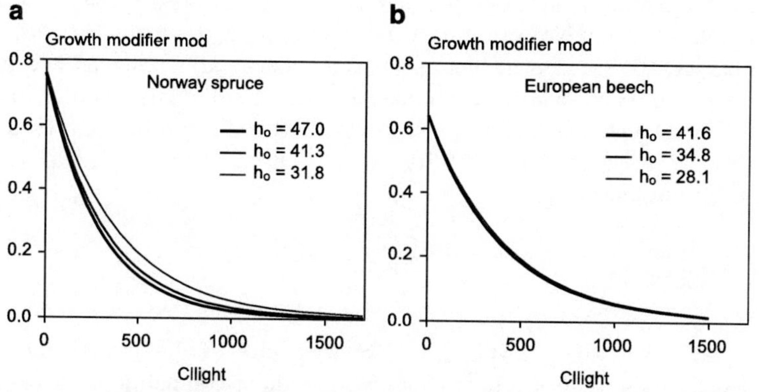

Fig. 14.9 Site-dependent modifier functions exp($f(x)$) for (**a**) Norway spruce and (**b**) European beech (adapted from Pretzsch and Biber 2010). The modifier function $f(x)$ is formulated in dependence on site fertility indicated by stand top height, h_o, at age 100. The *bold*, *thin*, and *broken lines*, respectively, indicate how site fertility determines the effect of aboveground competition on tree growth (see Fig. 14.3)

spatially explicit records of growing space, competition, size and growth enables tracing the growth–size relationship from the stand to the individual plant level. A suitable approach is to analyze whether the relationship between competition index (CI) and potential modifier (Fig. 14.3b) changes with site fertility.

The study by Pretzsch and Biber (2010) along an ecological gradient reveals that for most of the analyzed species, a given competition index reduces diameter growth more on high-quality sites than on poor sites. Figure 14.9 shows that this applies distinctly for Norway spruce but not for European beech. Site fertility was indicated again by top height of the stand in meters at age 100 years. In the case of Norway spruce we can observe both a rise in the level of potential growth (Fig. 14.4a) and an increase in size asymmetric competition (Fig. 14.9a), which is equivalent to a resource allocation in favor of the tall members of a population with increasing site fertility. The poorer the site fertility is, the lower is the potential and the more size symmetric is the resource allocation. A given competition index caused by larger and shade casting neighbors is expected to have a greater effect (stronger growth reduction on neighbors) on high-quality sites and a smaller effect (weaker growth reduction) on poor sites. For the latter case, the larger neighbors are certainly competitors for belowground resources, but on the high-quality sites they additionally become competitors for light, causing disproportional growth reduction. Hence, on poor sites the modifier–CI relationship is expected to be shallower (broken upper line in Fig. 14.3b), while on good sites it is expected to be steeper (bold lower line in Fig. 14.3b).

European beech seems to be less dependent on biotic stress; neither the potential growth (Fig. 14.4b) nor the relationship between competition index and potential modifier (Fig. 14.9b) shows a distinct dependency on site fertility.

For pine and oak, Pretzsch and Biber (2010) found a maximum difference between competition-induced growth reduction on the poorest and highest sites of approximately 0.1. For Norway spruce, a difference is also evident, but with a value of only about 0.05 (see Fig. 14.9a, growth modifier mod). European beech is the only exception, where additional information about site quality hardly improves diameter increment estimation. This corresponds with our findings of a lower sensibility of beech to site fertility (Fig. 14.6b), drought and ozone stress (Fig. 14.7b) on stand level.

14.4 Discussion

Starting point was the growth curve-like trajectory of the growth rate, id, in dependence on the size of a tree, *d*, growing without competition. Subsequently, we revealed how this unimodal curve is transformed by competition to the well-known linear interindividual id–*d* relationships and how this transformation depends on site fertility and stress. We showed how stress, no matter whether induced by competition, drought, or ozone, distinctly reduces the superior growth rate of tall trees in relation to smaller neighbors. A lower degree of stress increases steepness and size asymmetry of the growth–size relationship, i.e., it favors the superiority of tall trees at the smaller trees' expense.

Insight into these patterns is of particular interest because tree growth and stand dynamics are increasingly affected by trends of changing growing conditions (Spiecker et al. 1996) and stress events (Jentsch et al. 2007; Matyssek and Sandermann 2003; Matyssek et al. 2010). The effects of site fertility, disturbances, and stress events on tree as well as stand growth have been the subject of many empirical analysis (Röhle 1987; Schweingruber et al. 1986), as well as being summarized (Pretzsch 1999) and predicted by models (Bugmann et al. 1997;

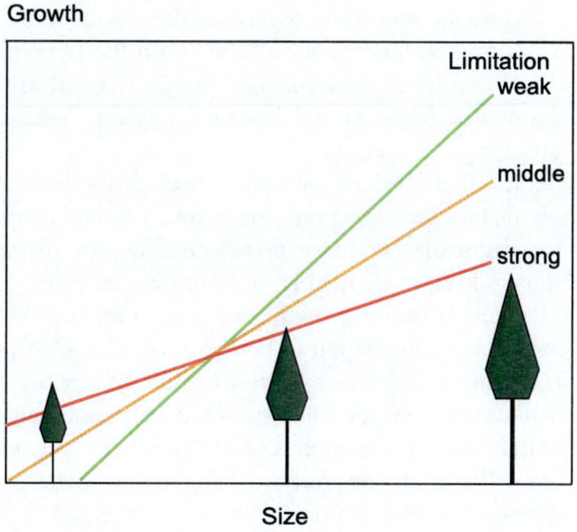

Fig. 14.10 Working hypothesis on the effect of stress, e.g., caused by limitation of belowground resources or ozone exposition, on the relationship between size and growth in schematic representation

Pretzsch et al. 2008; Rötzer et al. 2009). The reaction on individual tree level reaches from growth acceleration (Pretzsch 1999) to a gradual decrease in vitality caused by long-term deposition (Elling 1993), abrupt growth losses (Utschig 1989), and even dieback (Röhle 1987). Numerous investigations have contributed to a differentiated understanding of the effect of stress on tree or stand growth, but how stress changes the resource and growth distributions between trees of a cohort or stand was still open for debate.

14.4.1 Implications for the Growth–Differentiation Balance Theory

The relatively stronger growth reduction of tall, dominant trees in relation to small, overtopped trees in a stand in declining growing conditions appears as an overarching reaction pattern (Fig. 14.10). This pattern which appears along an ecological gradient from fertile to poor sites, in moist versus dry years, and ambient to double-ambient ozone concentrations can be interpreted as an internal trade-off in tree metabolism between growth and defensive mechanisms. It equips the dominant trees with superior ability to grow but with a low defensive capacity, whereas the opposite applies for small trees. Furthermore, a cause of this stress-dependent growth distribution pattern might be that the growth reduction of tall trees lowers their ability to pre-empt light and compete with their smaller neighbors, thus allowing their smaller counterparts to profit from the redistribution of resources and biomass growth. Finally, in the case of drought and ozone stress, the reaction pattern might simply be the result of a stronger impact of stress (radiation, wind, ozone concentration) on the more exposed dominant trees compared with the more sheltered and overtopped small trees.

While the reasoning regarding the causes remains speculative, the pattern itself is informative and in accordance with the hypothesis of reaction patterns at individual tree level. According to the growth–differentiation balance theory (GDBH, Herms and Mattson 1992), which applies to the individual plant level, there is a physiological trade-off between growth and secondary metabolism. The theory predicts that nutrient (or water) availability will have a parabolic effect on secondary metabolite concentrations, resulting in a unimodal optimum curve with a maximum of secondary metabolites at low to medium nutrient availability (see Matyssek et al. 2012, Fig. 1.1). In source-limited plants, a positive correlation is predicted between growth and secondary metabolism. In sink-limited plants the correlation is predicted to be negative: Carbohydrates are invested more and more into growth and less into constitutive secondary metabolism with rising nutrient availability. The revealed growth distribution pattern—which favors the tall plants relatively to the small ones under low stress and abundant resource supply, or rather favors the small ones relatively to tall ones under high stress and scarce resources—corresponds with the prediction of the GDBH.

Our finding that growth decreases from tall to small trees in a stand is not a new concept, but the finding that this gradient is more pronounced on rich sites and less pronounced on poor sites, in dry years or under stress appears as a general reaction pattern.

14.4.2 Ecological Implications

Similar analyses are partially available for herbaceous stands (Cannell and Grace 1993; Müller et al. 2000; Weiner and Thomas 1986, 1992), more seldom for juvenile woody stands (Hara 1993; van Kuijk et al. 2008; Thomas and Weiner 1989; Weiner and Thomas 1986), and even rarer for mature forest stands (Wichmann 2001, 2002). Only forest stands under survey for decades to centuries were used in this study, where the growth of individuals has been measured repeatedly without artifacts caused by interferences in the stand structure. Furthermore, the stands represent a broad range of ages, stand phases and site fertilities and are positioned along an ecological gradient, and thus reflect time series with differing growth and stress conditions.

In all three reported evaluations, the slope of the relationship between growth and size shows a distinct dependency on site fertility. The general reaction pattern, which can be assumed as a working hypothesis, is shown schematically in Fig. 14.10. On poor sites, under drought, or even under ozone stress the slope is shallow, meaning that the difference between the growth of small and large trees is smaller, as occurs on high-quality sites or in the absence of stress. In stands with unfavorable growing conditions, growth increases less than proportionally with plant size. As growth is less concentrated on the large and dominant trees, even their small neighbors grow well and survive. Under medium site fertility, growth increases proportionally to tree size resulting in size symmetric competition. With increasing resource supply and total stand growth, large trees will tend to succeed over their smaller neighbors. Optimal resource supply results in a disproportional increase of growth with tree size, i.e., size asymmetric competition (Fig. 14.10).

This general reaction pattern is in agreement with, among others, the empirical findings of Wichmann (2001, 2002) in adult forest stands, of van Kuijk et al. (2008) for tree dominated communities in the succession phase, of Thomas and Weiner (1989) respectively Weiner and Thomas (1986) for herbaceous stands, of Pretzsch (2010) respectively Pretzsch and Biber (2010) for forest stands, and the model scenarios from Hara (1993) or Rötzer et al. (2012) for forest stands. In searching for a causal explanation, it is speculated that under favorable site fertility and without additional stress, dominant trees profit disproportionally more as they make use of their privileged access to light. By pre-empting light they raise their own growth rate and reduce that of their smaller neighbors. In contrast, under unfavorable conditions, where growth is restricted by limitation of water or mineral nutrient supply, or under drought or ozone stress, dominant trees cannot make use of their superior position, causing the relationship between growth and size to become shallower. In the sense of Darwinian fitness, the *relative* advantage compared with the competitors rather than the *absolute* growth rate is decisive for success under selective pressure and fitness. The relationship between growth and size is a particularly informative indicator, as it distinctly reflects this relative difference between the growth of small and large members of a community.

14.5 Conclusion

14.5.1 Understanding and Modeling

Growth models for temperate forests consider primarily competition for light and include mainly just larger trees into their algorithm for estimation of competition and individual tree growth (Pretzsch et al. 2008). They apply a given competition index in the same way for different site fertilities. In view of our results, generalization of a function for the growth reduction in individual-tree models is not appropriate. On poor sites and under the same competition indices and sizes, trees still grow better than on rich sites.

When models are developed and applied to a narrow range of site fertilities and for short-term simulations, the practical relevance of the presented findings is somewhat limited. In such cases, the resource limitation and mode of competition varies only within a small range, and the prediction of resource and biomass distribution among trees of a stand is most likely reasonable. But if such models are used for spatial or temporal extrapolation beyond their range of parameterization, the following severely implausible results may occur: The conventional model underestimates the overall level of growth, in particular the growth of small understory trees, especially on poor sites. For the latter case and in unfavorable years, competition becomes more size symmetric. A more size symmetric distribution of resources and growth means that understory trees grow more than predicted, that their probability to die is lower than predicted, and that the stand structure remains more vertically structured than predicted. The compound interest effect of such an underrating may become considerable in the course of a long-term model run, owing to the feedback between structure and growth. An underestimation of the understory tree's growth results in an overestimation of the dominant tree's share of resources, and in rather mono-layered stand structure as small trees slow down their growth rate or even die due to the overestimated asymmetric competition. If a model should be applicable to a broader range of site fertilities, such as for the prediction of growth reactions under climate change (shift from light limitation to belowground resource limitation), then the hitherto site-independent reduction function in individual-tree models (e. g. Biging and Dobbertin 1995) requires site-dependency. Pretzsch and Biber (2010) proposed an improvement of existing statistical growth models and a more flexible algorithm for the reduction of the potential growth by considering different trajectories of modifier–CI relationships depending on the shape of the growth–size relationship.

14.5.2 Silvicultural Treatment

The id–*d* slope reflects the resource supply and affects the structural diversity on stand level. Shallow slopes occur on poor sites and indicate adequate growth conditions for small members of the stand, while steep slopes indicate self-thinning of the understory trees as their dominant neighbors take the larger share. On high-quality sites, those with sufficient resource supply with the exception of light, growth can be more easily concentrated on a limited number of selected trees by silvicultural thinning operations, while understory trees are difficult to keep. On poor sites, by contrast, understory trees still grow proportionally to their size, and vertical structure is easier to maintain. This is because water and nutrients are limiting factors and restrict the competition and superiority of dominant trees.

References

Assmann E (1970) The principles of forest yield study. Pergamon, Oxford, 506 p

Avery TE, Burkhart HE (1983) Forest measurements, 3rd edn. McGraw-Hill, New York, 331 p

Biging GS, Dobbertin M (1995) Evaluation of competition indices in individual tree growth models. Forest Sci 41:360–377

Bugmann H, Grote R, Lasch P, Lindner M, Suckow F (1997) A new forest gap model to study the effects of environmental change on forest structure and functioning. In: Mohren GMJ, Kramer K, Sabaté S (eds) Impacts of global change on tree physiology and forest ecosystems, Forestry sciences. Kluwer, Wageningen, pp 255–261

Cannell MGR, Grace J (1993) Competition for light: detection, measurement, and quantification. Can J Forest Res 23:1969–1979

Coggins SB, Coops NC, Wulder MA (2010) Estimates of bark beetle infestation expansion factors with adaptive cluster sampling. Int J Pest Manage 57(1):11–21

Condit R, Hubbell St P, Foster RB (1995) Mortality rates of 205 neotropical tree and shrub species and the impact of severe drought. Ecol Monogr 65:419–439
Elling W (1993) Immissionen im Ursachenkomplex von Tannenschädigung und Tannensterben. AFJZ 48(2):87–95
Hara T (1993) Mode of competition and size-structure dynamics in plant communities. Plant Species Biol 8:75–84
Hasenauer H, Kindermann G, Steinmetz P (2006) The tree growth model MOSES 3.0. In: Hasenauer H (ed) Sustainable forest management. Growth models for Europe. Springer, Heidelberg, pp 64–70
Herms DA, Mattson WJ (1992) The dilemma of plants: to grow or to defend. Q Rev Biol 67:283–335
Jentsch A, Kreyling J, Beierkuhnlein C (2007) A new generation of climate change experiments: events, not trends. Front Ecol Environ 5(7):365–374
Landsberg JJ (1986) Physiological ecology of forest production. Academic, London, 198 p
Mailly D, Turbis S, Pothier D (2003) Predicting basal area increment in a spatially explicit, individual tree model: a test of competition measures with black spruce. Can J Forest Res 33:435–443
Mäkelä A, Hari P (1986) Stand growth model based on carbon uptake and allocation in individual trees. Ecol Model 33:205–229
Matyssek R, Sandermann H (2003) Impact of ozone on trees: an ecophysiological perspective. Prog Bot 64:349–404
Matyssek R, Agerer R, Ernst D, Munch JC, Oßwald W, Pretzsch H, Priesack E, Schnyder H, Treutter D (2005) The plant's capacity in regulating resource demand. Plant Biol 7:560–580
Matyssek R, Wieser G, Ceulemans R, Rennenberg H, Pretzsch H, Haberer K, Löw M, Nunn AJ, Werner H, Wipfler P, Oßwald W, Nikolova P, Hanke DE, Kraigher H, Tausz M, Bahnweg G, Kitao M, Dieler J, Sandermann H, Herbinger K, Grebenc T, Blumenröther M, Deckmyn G, Grams TEE, Heerdt C, Leuchner M, Fabian P, Häberle KH (2010) Enhanced ozone strongly reduces carbon sink strength of adult beech (*Fagus sylvatica*) – resume from the free-air fumigation study at Kranzberg Forest. Environ Pollut 158(8):2527–2532
Matyssek R et al (2012) The balance between resource sequestration and retention: a challenge in plant science. In: Matyssek R et al (eds) Growth and defence in plants. Sprimger, Heidelberg
Müller I, Schmid B, Weiner J (2000) The effect of nutrient availability on biomass allocation patterns in 27 species of herbaceous plants. Perspect Plant Ecol Evol Syst 3(2):115–127
Nagel J (1999) Konzeptionelle Überlegungen zum schrittweisen Aufbau eines waldwachstumskundlichen Simulationssystems für Nordwestdeutschland. Schr Forstl Fak Univ Göttingen u Niedersächs Forstl Versuchsanst, vol 128. JD Sauerländer's Verlag, Frankfurt am Main, 122 p
Newnham R (1964) The development of a stand model for Douglas-fir. PhD thesis, University of British Columbia, Vancouver, 201 p
Niklas KJ (1994) Plant allometry. University of Chicago Press, Chicago, IL
Oliver CD, Larson B (1996) Forest stand dynamics. Wiley, New York, 520 p
Peltola HK (2006) Mechanical stability of trees under static loads. Am J Bot 93:1501–1511
Pretzsch H (1999) Waldwachstum im Wandel, Konsequenzen für Forstwissenschaft und Forstwirtschaft. Forstw Cbl 118:228–250
Pretzsch H (2009) Forest dynamics, growth and yield. From measurement to model. Springer, Heidelberg, 664 p
Pretzsch H (2010) Zur Verteilung des Zuwachses zwischen den Bäumen eines Bestandes und Abhängigkeit des Verteilungsschlüssels von den Standortbedingungen. AFJZ 181(1/2):4–13
Pretzsch H, Biber P (2010) Size-symmetric versus size-asymmetric competition and growth partitioning among trees in forest stands along an ecological gradient in central Europe. Can J Forest Res 40:370–384
Pretzsch H, Dieler J (2011) The dependency of the size-growth relationship of Norway spruce (*Picea abies* [L.] Karst.) and European beech (*Fagus sylvatica* [L.]) in forest stands on long-term site conditions, drought events, and ozone stress. Trees 25:355–369

Pretzsch H, Kahn M, Grote R (1998) Die Fichten-Buchen-Mischbestände des Sonderforschungsbereiches "Wachstum oder Parasitenabwehr?" im Kranzberger Forst. Forstw Cbl 117:241–257

Pretzsch H, Grote R, Reineking B, Rötzer T, Seifert S (2008) Models for forest ecosystem management: a European perspective. Ann Bot 101:1065–1087

Pretzsch H, Dieler J, Matyssek R, Wipfler P (2010) Tree and stand growth of mature Norway spruce and European beech under long-term ozone fumigation. Environ Pollut 158:1061–1070

Prodan M (1965) Holzmeßlehre. JD Sauerländer's Verlag, Frankfurt am Main, 644 p

Reineke LH (1933) Perfecting a stand-density index for even-aged forests. J Agric Res 46:627–638

Röhle H (1987) Entwicklung von Vitalität, Zuwachs und Biomassenstruktur der Fichte in verschiedenen bayerischen Untersuchungsgebieten unter dem Einfluß der neuartigen Walderkrankungen, vol 83, Forstl Forschungsber München. Frank, München, 122 p

Rötzer T, Seifert T, Pretzsch H (2009) Modeling above and below ground carbon dynamics in a mixed beech and spruce stand influenced by climate. Eur J Forest Res 128:171–182

Rötzer T et al (2012) Effects of stress and defence allocation on tree growth: simulation results at the individual and stand level. In: Matyssek R et al (eds) Growth and defence in plants. Sprimger, Heidelberg

Schütz JPh (1989) Der Plenterbetrieb. Fachbereich Waldbau, Zürich, 54 p

Schweingruber FH, Albrecht H, Beck M, Hessel J, Joos K, Keller D, Kontic R, Lange K, Niederer M, Nippel C, Spang S, Spinnler A, Steiner B, Winkler-Seifert A (1986) Abrupte Zuwachsschwankungen in Jahrringabfolgen als ökologische Indikatoren. Ber Eidg Anst forstl Versuchswesen, Dendrochronologia 4:125–182

Schwinning S, Weiner J (1998) Mechanisms determining the degree of size asymmetry in competition among plants. Oecologia 113:447–455

Skov KR, Kolb TE, Wallin KF (2004) Tree size and drought affect ponderosa pine physiological response to thinning and burning treatments. Forest Sci 50(1):81–91

Spiecker H, Mielikäinen K, Köhl M, Skovsgaard JP (eds) (1996) Growth trends in European forests. European Forest Institute Research Report No 5, Springer, Heidelberg, 372 p

Stamp N (2003) Out of the quagmire of plant defense hypotheses. Q Rev Biol 78:23–55

Thomas SC, Weiner J (1989) Growth, death and size distribution change in an Impatiens Pallida population. J Ecol 77:524–536

Utschig H (1989) Waldwachstumskundliche Untersuchungen im Zusammenhang mit Waldschäden. Auswertung der Zuwachstrendanalyseflächen des Lehrstuhles für Waldwachstumskunde für die Fichte (Picea abies (L.) Karst.) in Bayern, vol 97, Forstl Forschungsber München. Frank, München, 198 p

Valinger E, Lundquist L, Bondesson L (1993) Assessing the risk of snow and wind damage from tree physical characteristics. Forestry 66(3):249–260

van Kuijk M, Anten NPR, Oomen RJ, van Bentum DW, Werger MJA (2008) The limited importance of size-asymmetric light competition and growth of pioneer species in early secondary forest succession in Vietnam. Oecologia 157:1–12

Weiner J (1990) Asymmetric competition in plant populations. Trends Ecol Evol 5:360–364

Weiner J, Thomas SC (1986) Size variability and competition in plant monocultures. Oikos 47:211–222

Weiner J, Thomas SC (1992) Competition and allometry in three species of annual plants. Ecology 73(2):648–656

Weiner J, Stoll P, Muller-Landau H, Jasentuliyana A (2001) The effects of density, spatial pattern, and competitive symmetry on size variation in simulated plant populations. Am Nat 158 (4):438–450

Werner H, Fabian P (2002) Free-air fumigation of mature trees – a novel system for controlled ozone enrichment in grown-up beech and spruce canopies. Environ Sci Pollut Res 9:12–117

Wichmann L (2001) Annual variations in competition symmetry in even-aged sitka spruce. Ann Bot 88:145–151

Wichmann L (2002) Competition symmetry. In: Modelling the effects of competition between individual trees in forest stands. PhD thesis, University of Forestry, Copenhagen, pp 67–77
Wipfler P, Seifert Th, Biber P, Pretzsch H (2009) Intra-annual growth response of adult Norway spruce (*Picea abies* [L.] KARST.) and European beech (*Fagus sylvatica* L.) to an experimentally enhanced, free-air ozone regime. Eur J Forest Res 128:135–144
Zeide B (1993) Analyses of growth equations. Forest Sci 39(3):594–616

Part III
Unifying Theory in Resource Allocation?

Introduction

In continuation of Part II, which comprehended available knowledge on the topic of this book as derived from empirical evidence, Part III considers theoretical grounds. In preparing for elucidating the predictability and its enhancement of plant resource allocation, and whether elaborated knowledge allows for posing a unifying theory on the subject beyond the current status (Chap. 19), Part III introduces into theoretical approaches conducive to knowledge consolidation. Such approaches are borne by modelling. Chapter 15 explores capacities of numeric modelling for mathematically describing and synopsizing cause–effect-based knowledge, and of serving as integrated hypothesis constructs amenable to validation in comparison with empirical evidence. In particular, model falsification augments knowledge. To this end, Chaps. 17 and 18 demonstrate novel models, PLATHO and BALANCE, respectively, which mirror the transition of individual plant resource allocation and its regulation between growth and defence towards a stand-level assessment of resource turnover under the perspective of long-term development. Explored is the potential of mathematical simulation in view of multi-factorial complexity inaccessible to experimentation. Chapter 16 introduces into novel approaches of statistical modelling, fathoming capacities for assessing universality of findings, in particular, for unifying evidence from diverse environmental scenarios. Means of guiding empirical research are elucidated while highlighting constraints on empirical and theoretical approaches. Part III sets the stage for the conclusions and perspectives presented in Chap. 20 (Part IV) on the book subject.

R. Matyssek et al. (eds.), *Growth and Defence in Plants*, Ecological Studies 220,
DOI 10.1007/978-3-642-30645-7, © Springer-Verlag Berlin Heidelberg 2012

Chapter 15
Mechanistic Modelling of Soil–Plant–Atmosphere Systems

E. Priesack, S. Gayler, T. Rötzer, T. Seifert, and H. Pretzsch

15.1 Introduction

Mathematical models use mathematical methods to describe, to investigate and to explain real-world phenomena. These models are applied to test conjectures or hypotheses on the phenomena they represent and in particular provide the scientific means to make predictions about the real world. Since the reality or the real system of interest is in the majority of cases far too complex to be entirely modelled, models have to be designed for a specific aim and consequently consider only the most important aspects or parts of the real system in order to deal with the designated model objectives. For example, plant growth models may compile reliable knowledge about single aspects of plant or stand growth and integrate the selected parts to form an abstracted picture of the whole system. The systems

E. Priesack (✉)
Institute of Soil Ecology, Helmholtz Zentrum München, Ingolstädter Landstr. 1, 85764 Neuherberg, Germany
e-mail: priesack@helmholtz-muenchen.de

S. Gayler
Institute of Soil Ecology, Helmholtz Zentrum München, Ingolstädter Landstr. 1, 85764 Neuherberg, Germany

Center for Applied Geoscience, Water & Earth System Science Competence Cluster, c/o University of Tübingen, 72074 Tübingen, Germany

T. Rötzer • H. Pretzsch
Chair of Forest Growth and Yield Science, Technische Universität München, Hans-Carl-von-Carlowitz-Platz 2, 85354 Freising, Germany

T. Seifert
Chair of Forest Growth and Yield Science, Technische Universität München, Hans-Carl-von-Carlowitz-Platz 2, 85354 Freising, Germany

Forest and Wood Science, Stellenbosch University, Private Bag X1, 7602 Matieland, South Africa

R. Matyssek et al. (eds.), *Growth and Defence in Plants*, Ecological Studies 220,
DOI 10.1007/978-3-642-30645-7_15, © Springer-Verlag Berlin Heidelberg 2012

model, with its underlying system elements and chains of cause and effect, may then be regarded as a hypothesis about the structure and behaviour of the entire system.

Model evaluations are nowadays usually achieved by numerical simulations using a computer system, i.e. by numerical calculations to evaluate the model structure or the mathematical equations that form the model. By these results of quantitative simulation the scientific knowledge which is integrated in the model can be compared to experimental findings. Therefore, the model can be seen as a quantitative expression of the scientific hypotheses we have about the real-world phenomena the model represents. In the case of plant growth models hypotheses can be tested at study sites for which the initial conditions, environmental factors, resource availability, etc. as well as the stand dynamics are known. If the simulation results are plausible or agree with the reality, it enhances the reliability of the underlying chain of hypotheses in the model. In this case, the model is the hypothesis. For falsification trials, the observations and model outcomes are compared and checks for model plausibility and consistency as well as validation and verification of software codes should be performed. Besides the general aims to investigate research questions and to predict system dynamics which are addressed by *research models*, another prevalent general model purpose is the aid in decision making for which *management models* are developed.

Models can be further distinguished by their outcome: *Deterministic models* deliver always the same outcome for a given initial condition or starting point, whereas *stochastic models* consider random variations and rather predict distributions of possible outcomes (Fig. 15.1). Models may also be classified by the considered hierarchy of organisation levels of the modelled system. A typical scheme of such a hierarchy in plant biology ranges from the level $i + 1$ describing the canopy/stand: over the level i for the individual plant, to the level $i - 1$ of plant organs, and further down to the levels of tissues, cells, organelles, and finally to the level of molecules and atoms (level $i - n$). A more complex model based on a considerable number of theoretical assumptions generally attempts to describe the behaviour at the level i by considering mechanisms, understanding and explanations at the lower levels. These models are called *mechanistic models*, since they explicitly take account of the inner structures and mechanisms which cause changes in the modelled system. In contrast, *empirical models* are essentially descriptions of observational data and do not account for the inner structure and underlying mechanisms by which changes occur. These models attempt to describe system behaviour at an organisation level as expressed by observational data in terms of attributes of this level alone. Therefore, empirical models often are statistical representations of the observational data based on equations or functional relationships determined by regression analysis.

To support a better understanding and an extended view of plant strategies of resource allocation as outlined in the "growth–differentiation balance theory"

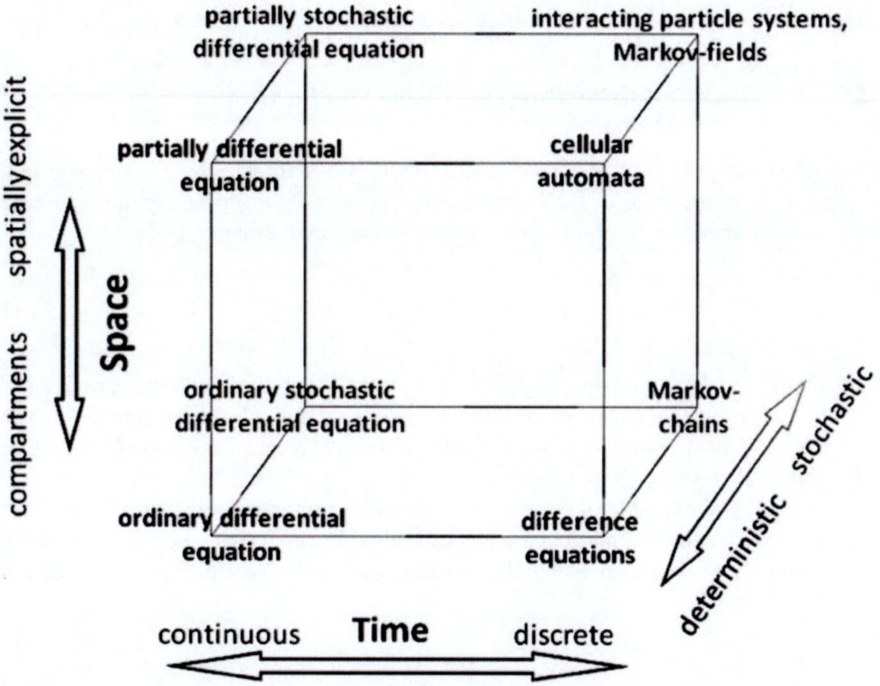

Fig. 15.1 Classification of mathematical models to represent dynamical systems

(Herms and Mattson 1992), mathematical modelling was applied not only to statistically analyse measured data sets, but also to integrate spatial–temporal processes across hierarchical dimensions in structure and time from the cell level to the levels of organs, plants and stands (c.f. Preface and Chap. 1). For this purpose besides new and flexible statistical approaches (c.f. Chap. 16) two mechanistic models have been developed to analyse resource allocation in plants. One model focuses on the description of processes from the organ level to the level of the whole plant (c.f. Chap. 17), whereas the other model takes the perspective of the whole stand to simulate tree–tree interaction and resource competition based on individual tree growth (c.f. Chap. 18). In this chapter we first point out the principles of mechanistic mathematical models of growth processes and elucidate their capability to analyse cause–effect interactions in complex systems. We then go into more detail focusing on mechanistic plant growth models based on eco-physiological processes. Finally, we refer to Chaps. 17 and 18 exemplifying the potential of mechanistic modelling of soil–plant–atmosphere systems to get insights into the complex and dynamic behaviour of resource allocation in plants which would not be attained by experiments alone.

15.2 Dynamic Deterministic Mathematical Models

15.2.1 Description of Processes and Mechanisms

A single process of a system may be described by a state variable x representing a property or an attribute of the system, which is described further by the way how this state changes with time t. This may be written mathematically by

$$\frac{\mathrm{d}x}{\mathrm{d}t} = f(x,p) \tag{15.1}$$

the use of a functional relationship $f(x, p)$ between the variable x and the parameter p. Given the values of $x(t_0)$ at the starting point t_0 and of the parameter p, the development of the state variable x is determined and may be calculated for a finite time interval from times t_0 to t_1.

The classical example of such a process description is the model for the motion of a particle or centre of mass of a material body that moves with a constant velocity v covering the distance s between the starting position $s(t_0)$ at time t_0 and its position at the later time t:

$$\frac{\mathrm{d}s}{\mathrm{d}t} = v. \tag{15.2}$$

The movement is conceived to be sustained on a straight line and only an extraneous cause can lead to a change of the motion. By this concept of inertial motion, which was first formulated in this generality by the French philosopher René Descartes, motion (at a certain constant velocity v) or rest (motion with velocity $v = 0$) is seen as a primitive state of the material body whose cause needs no further explanation (Hund 1996).

Another example is given by the description of the process of biomass growth. Since growth of a living system is based on the division and replication of cells which constitute the biomass of the system, the newly formed biomass of the next generation of cells is proportional to the biomass of the parent cells. Therefore, the change in biomass B of the living system may be described as directly proportional to the original biomass B

$$\frac{\mathrm{d}B}{\mathrm{d}t} = \mu B \tag{15.3}$$

by use of a constant factor of proportionality μ which defines the specific growth rate. The development of the growing biomass is then determined by giving a value for the initial biomass $B(t_0)$ at the starting time t_0. As for the state of constant velocity of a particle in the first example, the state of growth of the living system and its cause is not further explained. It is only assumed that the newly built biomass is proportional to the growing biomass B and that the growth continues with the same constant specific growth rate μ. This leads to the description of unlimited or optimal growth, which most often can be sustained only for a rather small time interval. To obtain a more realistic description of the usually limited growth process, additional properties of the growing system and its interaction with or limitation by its environment need to be considered, which can cause a change of the specific growth rate. The impact of environmental conditions on the growth process thus can be regarded in analogy to the consideration of the action of forces on a moving particle that can change its constant velocity and cause an acceleration of the particle.

A mechanism of a system is usually modelled by considering the coupling or interaction of different processes of the system. If the state of the system at the time t is defined by the n independent state variables $x_1,\ldots,x_n$ that represent properties or attributes of the system (e.g. biomass dry matter content, leaf area, root nitrogen content), then the model of the mechanism may be defined by n first-order ordinary differential equations that describe the change of the n state variables with time t (Thornley and Johnson 1990). These can be written formally as

$$\begin{aligned}\frac{dx_1}{dt} &= f_1(x_1, x_2, \ldots, x_n; p; q)\\ \frac{dx_2}{dt} &= f_2(x_1, x_2, \ldots, x_n; p; q)\\ &\vdots\\ \frac{dx_n}{dt} &= f_n(x_1, x_2, \ldots, x_n; p; q).\end{aligned} \tag{15.4}$$

The $f_1,\ldots,f_n$ denote functions of the state variables, of the parameters indicated by p and of the environmental quantities indicated by q. The equations define how the rates of change of the system state variables explicitly depend on the current values of the state variables (Thornley and Johnson 1990).

An example for such a model which describes a mechanism expressing the cause–effect interaction of two different state variables is the more realistic description of biomass growth. For that purpose we model the relation between substrate use and biomass growth. If B denotes the weight of the growing biomass and S the amount of available substrate needed for growth (e.g. carbon or nitrogen), the following two ordinary differential equations provide a rather simple example:

$$\frac{dB}{dt} = \mu\frac{S}{S+K_S}B - \sigma B \quad \text{and} \quad \frac{dS}{dt} = -\frac{1}{Y}\left(\mu\frac{S}{S+K_S}B\right) + \phi(S,q) \tag{15.5}$$

with parameters μ denoting the specific growth rate, σ the specific decay rate of the biomass, K_S the half saturation constant of substrate use, Y the yield coefficient and ϕ the source term depending on the substrate S and the environmental quantity q. The half saturation relation or Michaelis–Menten relation $\frac{S}{S+K_S}$ can be seen as an expression for substrate utilisation kinetics based on a reaction which is driven by an enzyme substrate complex (Bailey and Ollis 1986). The first of the two differential equations is also known as the bacterial growth model of Monod (Monod 1949), in combination with the second equation this growth model is usually used to simulate microbial growth in a chemostat (Bailey and Ollis 1986).

15.2.2 Mechanistic Models

The last example which describes the mechanism of the effect–cause relationship between growth and substrate utilisation can therefore be regarded as simple mechanistic model of microbial growth in a chemostat by defining the upper level $i + 1$ to be the chemostat level, where, if we use the source-sink term $\phi(S,q) = \sigma(S_0 - S)$, the input of dissolved substrate takes place at an inflow rate of σS_0 with a constant input concentration of substrate $q = S_0$ and the dilution rate σ. The outflow is given by the rates σS and σB for the current concentrations of substrate S and microbial biomass B in the chemostat solution. The lower level i may then be represented by the two components, substrate and microbial biomass, and their interaction, which defines the level i mechanism.

Although the mechanistic models that integrate processes and mechanisms at different organisation levels often are deterministic models and their outcome is determined by the prescribed initial conditions, the model integration can give rise to new knowledge. This new knowledge may be gained by new and eventually unexpected system understanding, because model behaviour based on a certain hierarchy of interacting mechanisms can show emergent properties, which are more than the mere sum of the underlying process components and yet can be explained in terms of the components and their interactions (Thornley and Johnson 1990). Often feedback mechanisms between processes at the same or different hierarchy levels determine the dynamical system behaviour which cannot be derived from the isolated process components alone (Pretzsch 2009). Also interaction between many particles can result in an emergent phenomenon such as superconductivity investigated in condensed matter physics (Laughlin 2006).

15.2.3 Complex Systems and Modular Modelling

The series of equations and algorithms defining the mechanistic model (i.e. as represented by Eq. 15.4) can be built up in a way that single components as given by certain sub-series of equations describe a single natural process, e.g. transport of a chemical or growth of a plant organ. This component-wise composition of the model system based on models describing single processes defines the modularity of the model, since the single process models can be considered as elementary modules from which sub-models, e.g. the water flow model or the crop growth model, and, finally the total model can be constructed.

This modularity of the model allows a thorough model system analysis starting with the validation of the single process model, often by using laboratory experiments, and ending by the comparison of simulation results with experimental data from field experiments to test the mutual couplings of processes and related feed back loops that determine the total model. Furthermore, model modularity facilitates model extension, since the model can be easily expanded by adding further components. To strengthen this extensibility, the model can be designed as an open model, which allows the software programming user to insert his self-defined and self-programmed sub-models. Such an open and modular model concept was realised by the development of the model system EXPERT-N (Engel and Priesack 1993; Priesack 2006), resulting in one of the first soil–plant system models with an open and modular model architecture (Abrahamsen and Hansen 2000). Because of consequently implementing several different sub-models that describe the same single process given by different soil–plant system models such as e.g. CERES (Jones and Kiniry 1986; Ritchie et al. 1987) or LEACHM (Hutson and Wagenet 1992), from the beginning of the EXPERT-N model development, attention had to be paid to the exchangeability of single process models. Basis of the model development was a documentation and review of known agricultural models and modelling approaches (Engel et al. 1993). This overview led to an improved model structure with the partitioning into modular model groups for water flow, heat transfer, solute transport, plant growth and land use management and further model group divisions into single process components. On the one side the open modular model concept of EXPERT-N facilitates the integration and testing of new modelling approaches, on the other side the model already includes for each of the model components several different exchangeable sub-models, which are well documented in the literature. The modular structure thus supports component-wise model comparison or model system analysis and strengthens the model adaptability to different application objectives. Therefore the EXPERT-N model was especially suited for the development of the individual-based plant growth model PLATHO (Gayler and Priesack 2007) by providing flexible sub-models to represent the atmosphere and soil compartments needed to simulate growth of woody or herbaceous plants in phytotrons, in lysimeters and at different field sites.

15.3 Mechanistic Plant Growth Models Based on Eco-Physiological Processes

15.3.1 Growth, Decay and Matter Cycling

Plants grow at the interface between the soil and the atmosphere and hence their growth processes are influenced by below- as well as above-ground environmental conditions and gradients (e.g. temperatures, water contents). Vice versa, plant growth processes impact soil and atmospheric conditions and shape their nearest environment to gain favourable growth conditions. Therefore, mechanistic plant growth models have not only to consider the direct environmental effects on plant growth, but also the change of the environmental conditions produced by growing plants including the feedback of this change on growth processes. In particular, mechanistic plant growth models describing the growth of plant biomass based on carbon (sometimes also on nitrogen and water) mass balances have to consider the change of matter contents in the different parts of the plant and to balance the incoming and outgoing matter fluxes (e.g. induced by carbon gain through photosynthesis and by carbon losses through respiration and senescence). In this case, to include environmental feedback mechanisms, the matter balances of the neighbouring environmental compartments have to be assessed which regard the fluxes at the plant–atmosphere and plant–soil interfaces (Fig. 15.2).

In most of these process and mass-balance based models the atmospheric conditions are represented by state variables related to available observational data on weather conditions such as global solar radiation, air temperature, precipitation rate, air humidity and wind speed at hourly or daily resolution. These state variables are used to simulate the plant–atmosphere interactions based on the spatial distribution of the leaf area either at the individual plant or the canopy level by applying sub-models of

- Absorption and extinction of light
- Gas exchange of CO_2, H_2O and possibly O_3
- Interception and transpiration

The plant–soil interactions are considered by sub-models which describe

- Plant litter fall including reproductive organs
- Root release of carbon by root exudation, root respiration and root death and
- Root uptake of water and nutrients (including mineral nitrogen) based on the spatial distribution of the root area

For the mass balances in the soil compartment the corresponding soil processes are modelled. They are needed to estimate the water and nutrient availability for the plants and comprise

- Soil water flow including evaporation and drainage
- Heat transfer and solute transport, the latter closely linked to
- Soil organic matter turnover, which depends on soil moisture and soil temperature and often involves the turnover of soil microbial biomass and possibly of mulch layers at the soil surface

In combination with the sub-models for the interactions with atmosphere and soil compartments plant growth models are typically built up by sub-models for

- Phenological development
- Photosynthesis at the leaf, individual plant or possibly canopy level
- Allocation of assimilates and nutrients to plant organs usually represented by roots, stem, leaves and fruits
- Biomass growth of the plant organs, respectively, of the whole plant or canopy

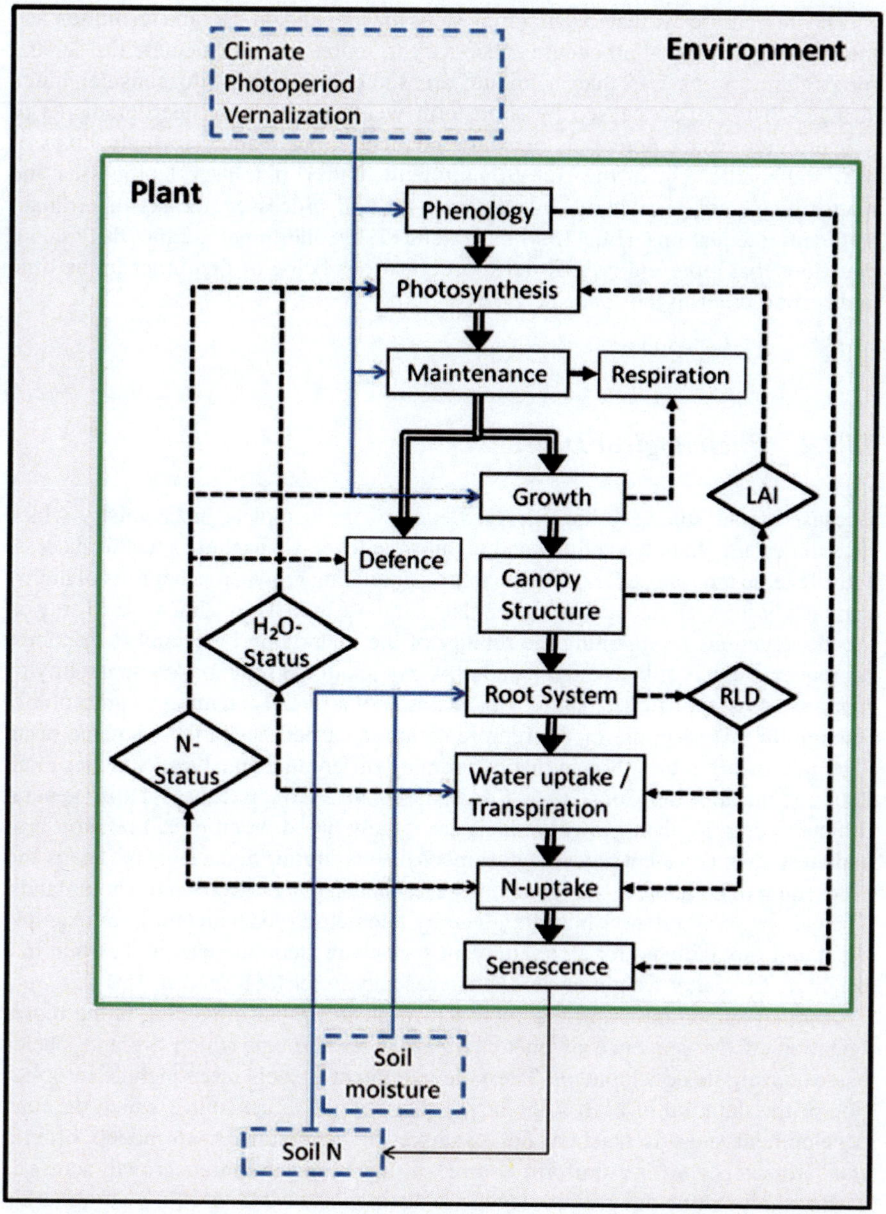

Fig. 15.2 Flow chart of a typical plant growth model with its major feedback loops (*dashed arrows*) and environmental impacts (*blue arrows*)

- Formation of leaf and root area at the different above-ground height or below-ground depth levels either for individual plants or for the whole stand
- Senescence and mortality

The models are evaluated using time steps that depend on the time resolution and the numerical solution procedures necessary to represent and calculate the driving forces. Time steps less than a minute are sometimes needed to consider quick changes that may be caused by short-term weather events or steep gradients of temperatures, water potentials or solute concentrations leading to high fluxes of heat and matter e.g. in the soil compartment. Often, if transport processes and matter fluxes are considered, in particular for soil processes, the set of ordinary differential equations (Eq. 15.4) is extended by additional partial differential equations that represent typical transport equations being of first order in the time and second order in the space derivative.

15.3.2 Phenological Development

Because of our still very limited knowledge of the complete biochemistry which determines growth at the cellular and organismic level, we cannot expect to successfully develop mechanistic models capable of simulating growth mechanisms of highly complex living systems such as higher plants, if these models are almost based only on kinetic equations representing the totality of the underlying biochemical reactions. Moreover, there is still an ongoing controversy about the principal computability of living systems, even of very simple organisms or of artificial and minimal units of life. Though, in this case, the model representation is carried out in the language of an algebraic theory rather than in that of ordinary differential equations (Letelier et al. 2006; Munteanu and Solé 2006; Cárdenas et al. 2010). Because of this general difficulty, current plant growth models are largely based on empirical relationships and most often represent physiological mechanisms starting at the level of organs and proceeding to higher levels as given by single plants, plant cohorts and whole stands. Typical empirical relationships are given by allometric models relating for example shoot and root biomass for a certain plant species, or stem diameter at a certain tree height to dry matter weight of stem, branches and leaves (see also Eq. 15.11).

Another important empirical approach in plant growth modelling is the representation of the sequence of phenological stages through which a plant species passes during its development. These development models often include information of the duration of each stage or provide critical values above which the next development stage is reached. For example, in agricultural crop models often a standardised scale for a uniform coding of phenological similar growth stages is used and nine principal growth stages are distinguished (Meier 1997): germination or bud development, leaf development, formation of side shoots, stem elongation or rosette growth, development of harvestable vegetative plant parts or vegetatively propagated organs, inflorescence emergence, flowering, development of fruit, ripening or maturity of fruit and seed, senescence or beginning of dormancy. In case of modelling the wheat phasic development Ritchie (1991) defines for each developmental stage a thermal time-duration which identifies the thermal time t_d the plant needs to pass the specified development stage. The thermal time has the unit C°d, degree-days and is defined by

$$t_d = \sum_{i=1}^{n} (\bar{T}_a - T_b), \tag{15.6}$$

where $\bar{T}_a$ is the daily mean air temperature, T_b is the base temperature at which development stops and n is the number of days of temperature observations used in the summation.

In general, by designating criteria when to switch from one stage to the next stage during plant growth we are able to model the development of the growing plant. Thus by determining the thermal time intervals of the phenological stages, we know when to switch between the different systems of ordinary differential equations, if each system describes plant growth for a different phenological stage. For example, if plant growth is modelled by differential equations representing the growth of different plant organs in a winter wheat growth model, the equation for ear growth may be added to the initial equation system, when the stage of booting ends and the stage of inflorescence emergence begins (Schröder and Richter 1993).

15.3.3 Photosynthesis and Allocation

For carbon-based mechanistic plant growth models the starting point to calculate dry matter production is the determination of the photosynthesis at the leaf level. The applied leaf photosynthesis models range from simple approaches such as light response curves to the explicit consideration of biochemical processes (differently for C_3 and C_4 plants) to assess the interaction of absorbed radiation, atmospheric and intercellular CO_2 concentration, leaf temperature and leaf nitrogen content (Farquhar et al. 1980; Yin and van Laar 2005; Yin and Struik 2009). However, most often leaf photosynthesis is modelled by applying a light response curve and considering the impacts of environmental conditions such as atmospheric CO_2 concentration, temperature, stomatal conductivity and nutrient supply by a related modification of the maximal photosynthesis rate (Pretzsch 2009).

To up-scale leaf photosynthesis to the stand level different concepts of integration are pursued (Priesack and Gayler 2009). The simplest approach, the so-called big leaf model, treats the canopy surface as one leaf and calculates the leaf energy balance and the leaf photosynthesis for the conditions prevailing at the top of the canopy. The more realistic but also far more complex multi-layer concept was introduced by Goudriaan and van Laar (1978, 1994) by considering direct and diffusive radiation components, shaded and sunlit leaves and also different spatial layers within the canopy (Grote and Pretzsch 2002). Moreover, unlike most models, which compute gross assimilation at a daily time step, the diurnal variation of incident radiation is explicitly considered. A simplification of the multi-layer approach avoiding the over-simplification of the big-leaf concept is the two-leaf model (De Pury and Farquhar 1997; Wang and Leuning 1998) in which the canopy leaves are split into sunlit and shaded fractions and each leaf fraction is treated by a one-layered leaf model. The two-leaf model proved to be computationally about ten times more efficient than the multi-layer model (Leuning et al. 1995), but resulted at nearly the same prediction of canopy photosynthesis (De Pury and Farquhar 1997).

By substracting the carbon consumption for crop maintenance respiration from the up-scaled gross crop photosynthesis, the pool of assimilated carbohydrates which is available for the growth of the different plant organs can be quantified and partitioned to the plant organs. To model the partitioning of assimilates four main modelling concepts have been indentified (Thornley and Johnson 1990; Lacointe 2000; Le Roux et al. 2001; Génard et al. 2008) reaching from empirical partitioning coefficients, teleonomic approaches including functional balance and optimisation principles and models that explicitly describe within-plant transport processes and source-sink dynamics.

A typical example of an empirical approach is the crop and variety specific assimilate partitioning scheme applied in the crop model SUCROS (Goudriaan and van Laar 1994; Priesack and Gayler 2009). Following Penning de Vries et al. (1989) the daily dry matter biomass growth rates of the different plant organs $\mu_{B,\mathrm{plo}}^{\mathrm{day}}$ (kg ha^{-1} day^{-1}), i.e. of leaves (plo = lvs), stems (plo = stm), storage organs (plo = sto) and roots (plo = rts) depend directly on the daily pool of carbohydrates $C_{\mathrm{asm}}^{\mathrm{day}}$ [kg (CH_2O) ha^{-1} day^{-1}] that is available for growth. This pool will be partitioned to the plant organs according to their growth rates:

$$\mu_{B,\mathrm{plo}}^{\mathrm{day}} = C_{\mathrm{asm}}^{\mathrm{day}} f_{\mathrm{ap,plo}} \varsigma_{B,\mathrm{crp}}^{-1}, \tag{15.7}$$

where the efficiency coefficient of glucose to dry matter transformation $\varsigma_{B,\mathrm{crp}}$ (–) for the entire canopy is given by

$$\varsigma_{B,\mathrm{crp}} = f_{ap,\mathrm{lvs}} \varsigma_{B,\mathrm{lvs}} + f_{\mathrm{ap,stm}} \varsigma_{B,\mathrm{stm}} + f_{\mathrm{ap,sto}} \varsigma_{B,\mathrm{sto}} + f_{\mathrm{ap,rts}} \varsigma_{B,\mathrm{rts}} \tag{15.8}$$

resulting from the efficiency coefficients $\varsigma_{B,\mathrm{plo}}$ (–) and the partitioning coefficients $f_{\mathrm{ap,plo}}$ (–) of the single plant organs. Whereas the efficiency coefficients are given by constant empirical values, the partitioning coefficients can depend on the actual development stage s_{dev} via fixed tabular functions $f_{\mathrm{plo}}(s_{\mathrm{dev}})$ for the shoot (plo = sht), the leaves (plo = lvs) and the stems (plo = stm):

$$f_{\mathrm{ap,sht}} = \varsigma_\theta f_{\mathrm{sht}}(s_{\mathrm{dev}}), \quad f_{\mathrm{ap,rts}} = 1 - f_{\mathrm{ap,sht}}, \tag{15.9}$$

$$\begin{aligned} f_{\mathrm{ap,lvs}} &= f_{\mathrm{ap,sht}} f_{\mathrm{lvs}}(s_{\mathrm{dev}}), \quad f_{\mathrm{ap,stm}} = f_{\mathrm{ap,sht}} f_{\mathrm{stm}}(s_{\mathrm{dev}}), \\ f_{\mathrm{ap,sto}} &= f_{\mathrm{ap,sht}} - f_{\mathrm{ap,lvs}} - f_{\mathrm{ap,stm}}. \end{aligned} \tag{15.10}$$

The additional factor ς_θ (–) is a water stress factor, which represents the ratio of actual to potential transpiration and increases the proportion of assimilates assigned to the root to promote root growth under conditions of water shortage. Although the partitioning coefficients can vary during the season and can be modulated by external conditions such as water shortage, they can be applied only in the range of conditions for which they have been measured. However, if these conditions are actually met, the empirical allocation scheme can be very efficient and usually leads to reasonable predictions (Le Roux et al. 2001).

This approach to model carbon allocation was extended in the individual-based plant growth model PLATHO (Gayler and Priesack 2007) by including a pool of defence-related carbon-based secondary compounds (Gayler et al. 2008, see also Chap. 17). The partitioning of assimilates to defence-related compounds is simulated depending on crop stage and environmental factors such as nitrogen supply and water supply. This approach is based on the consideration that the partitioning of the mostly limited assimilates in plants is determined by the balance between the different, sometimes conflicting demands for maintenance and growth of organs on the one side and for defence against biotic and abiotic stresses on the other side (Matyssek et al. 2005; Priesack and Gayler 2009).

Another empirical approach is to model assimilate partitioning by the so-called allometric formulas assuming that during growth certain allometric proportions are maintained between the organs of an individual plant: If x and y quantify the size (e.g. of volume or surface area) of plant organs or of a total plant, then the allometric formula may be given by

$$y = ax^{\alpha} \tag{15.11}$$

or in differential form by

$$\frac{1}{y}\frac{\mathrm{d}y}{\mathrm{d}t}\left(\frac{1}{x}\frac{\mathrm{d}x}{\mathrm{d}t}\right)^{-1} = \frac{\mathrm{d}y}{y}\left(\frac{\mathrm{d}x}{x}\right)^{-1} = \frac{\mathrm{d}y}{\mathrm{d}x}\frac{x}{y} = \alpha \tag{15.12}$$

for a constant allometric exponent α (Pretzsch 2009). From the last equations α can be interpreted as denoting the ratio between the relative growth rates of x and y. Therefore, according to the allometric formula a plant allocates its resources to the organs x and y in such a proportion that α remains constant (Pretzsch 2009). For example in the physiological growth model BALANCE this approach is applied to model carbon allocation in trees (Rötzer et al. 2009; see also Chap. 18).

15.3.4 Transport Processes

During the last decades two concepts, the cohesion–tension theory (Tyree and Zimmermann 2002) and, based on it, the electrical circuit analogy have been applied to model water transport in plants using resistances and capacitances, respectively, functional relationships to characterise plant hydraulic properties (Cruizat et al. 2002). By applying these concepts the description of plant hydraulic architecture has made a strong improvement towards a more realistic and comprehensive vision of plant water relationships (Cruizat et al. 2002). Only recently, this approach was improved by substituting the electrical circuit analogy by a porous media description for sap flow based on Darcy's law, which includes a water capacity term to better account for the dynamic behaviour of the hydraulic storage in plants such that mass conservation is directly calculated (Arbogast et al. 1993; Früh and Kurth 1999; Kumagai 2001; Aumann and Ford 2002; Bohrer et al. 2005; Chuang et al. 2006). This approach avoids simulations of negative water contents and unlimited water withdrawal from the plant which can occur in applications of the electrical circuit model (Chuang et al. 2006). By using a three-dimensional representation of the plant architecture and the physically based representation of tree hydrodynamics, questions can now be addressed of how the variability in canopy structure (i.e. inter-specific, age or ecosystem dependent structure) would lead to different transpiration responses (Bohrer et al. 2005). In particular, by considering the stomatal response to leaf xylem water potentials, the stomatal closure, the corresponding change in leaf water conductance and the resulting actual transpiration rate can be simulated. Moreover, the consideration of conductivity loss with decreasing leaf xylem water potentials allows the representation of the vulnerability to cavitation at critical, low potentials (Hölttä et al. 2005).

Since water transport from the soil through the plant into the atmosphere takes place in a soil–plant–air continuum that is interconnected by a continuous film of water, modelling of plant water flow has to consider both the water exchange at the leaf–air interface and the water flux at the soil–root interface. Therefore, also root water uptake including root hydraulic architecture has to be described to integrate the plant hydro-dynamic model into ecosystem models (Doussan et al. 2003). This is a step still to be taken to replace current models of plant water dynamics, e.g. big leaf or resistor–capacitor approaches (Bohrer et al. 2005) that are often coupled to one-dimensional effective root water uptake models (Cowan 1965; Gardner 1960; Nimah and Hanks 1973; Feddes et al. 1978; Campbell 1985). New root water uptake models are available that describe root architecture and related soil–plant processes in three dimensions by explicitly considering the three-dimensional distribution of the uptake (Clausnitzer and Hopmans 1994; Somma et al. 1998; Vrugt et al. 2001) and also consider water flow in the root system (Doussan et al. 2006; Javaux et al. 2008). They can be based on root growth models that allow the integration of a great diversity of environmental conditions and their impact on root system development (Dunbabin et al. 2002). To simulate water flow in the total soil–plant system Janott et al. (2011) extended the above-ground xylem water flow model of Bohrer et al. (2005) by a corresponding root water flow model also based on the porous media equation. The new model was then coupled to a second porous media equation describing the soil water flow. In this way the observed similarities between water flow in the soil and in the xylem are exploited and a unified description of hydrodynamics in the soil–plant continuum is obtained. Moreover, the water storage capacity of the stem can now be directly considered by modelling its function in the regulation of diurnal transpiration and its potential to provide supplementary water to the leaves, if excessive water deficits occur.

Based on Münch theory which is now agreed to explain the mass flow in the phloem, non-steady-state transport models can be developed to study the long distance phloem transport (Hölttä et al. 2006; Thompson and Holbrook 2003; Daudet et al. 2002). If we assume the pressure-flow model, the active loading of solutes at the mature leaves lowers the water potential of the sieve elements, with their semi-permeable outer membranes. The arising gradient of water potentials then draws in water from the surrounding tissue and causes the hydrostatic water pressure, i.e., turgor pressure, to rise. Reversely, unloading of solutes at the sink areas raises the water potential of the sieve elements and water is pushed out. "This pressure gradient drives the bulk water flow in the phloem from the solute sources to sinks, and solutes move by convection along with the water stream" (Hölttä et al. 2006). Up to now the two conduit pathways, the xylem and phloem, have mainly been studied as separate systems, only recently Hölttä et al. (2006) developed a mathematically simple conceptual model to simultaneously treat flow and transport in both systems including the exchange of water and solutes between xylem and phloem. In this way the model represents a first approach to simulate the full "Münch circulation". However, the model is still very simple, as it represents plant architecture only by an above-ground cylinder and does not describe roots. A slightly more complex but still very simple model using a network of 18 cylinders to represent a branched architecture is applied by Lacointe and Minchin (2008) to simulate xylem and phloem flows and their interaction at the presence of one carbon source in the leaves and of two carbon sinks in two different roots of the considered plant system.

15.4 The Models PLATHO and BALANCE

Two representatives of the more complex mechanistic models are given by the models PLATHO (Gayler and Priesack 2007) and BALANCE (Grote and Pretzsch 2002; Rötzer et al. 2009), which are applied to yield the simulations presented in Chaps. 17 and 18. The model PLATHO focuses on the modelling of growth and allocation processes in individual plants by explicitly considering different defence mechanisms at the organ level at an hourly time resolution. Mechanisms of competition between neighbouring plants are only described in a simple and compact way, for example without differentiating impacts of the three-dimensional crown architecture on the light capture of the individual tree. In contrast the model BALANCE simulates the interaction of the light regime with the three-dimensional stand and crown architecture to describe tree growth and stand structure development on a monthly and yearly basis (Rötzer et al. 2009), but considers the distribution of net assimilates and nitrogen gains in a more simple way. In this way the two models enable us to analyse experiments at different organisational levels, that of the individual plant and that of the whole stand. In this way we can approach our central question about the balancing of plants between growth and defence by theoretical considerations from both levels.

15.5 Conclusions

Mechanistic models are tools to represent and evaluate scientific hypotheses about cause and effect relationships by providing the possibility for a quantitative comparison of results obtained by derivations from theoretical considerations and by experimental findings. Through integration of empirical relationships and explicit descriptions of well-understood processes also more complex systems including highly nonlinear interactions and feedbacks can be studied. These methods made plant growth models available that simulate such complex systems as agricultural and silvicultural systems and are successfully applied for management purposes in plant production systems. The given examples of different models to describe carbon allocation during plant growth range from representations of empirical relationships to the explicit consideration of transport processes within plants and show the recent effort to gain more mechanistic understanding of the source and sink relationships between different plant organs and their dependence on plant architecture and environmental conditions. In particular, the two new mechanistic models PLATHO and BALANCE successfully helped to improve our understanding of resource allocation both at the level of the whole plant and at the level of the whole stand.

References

Abrahamsen P, Hansen S (2000) Daisy: an open soil-crop-atmosphere system model. Environ Model Softw 15:313–330

Arbogast T, Obeyesekere M, Wheeler MF (1993) Numerical methods for the simulation of flow in root-soil systems. SIAM J Numer Anal 30:1677–1702

Aumann CA, Ford DE (2002) Modeling tree water flow as an unsaturated flow through a porous medium. J Theor Biol 219:415–429

Bailey JE, Ollis DF (1986) Biochemical engineering fundamentals. McGraw-Hill, New York

Bohrer G, Mourad H, Laursen TA, Drewry D, Avissar R, Poggi D, Oren R, Katul GG (2005) Finite element tree crown hydrodynamics model (FETCH) using porous media flow within branching elements: a new representation of tree hydrodynamics. Water Resour Res 41:W11404. doi:10.1029/2005WR004181

Campbell GS (1985) Soil physics with BASIC, transport models for soil-plant systems, vol 14, Developments in soil science. Elsevier, New York

Cárdenas ML, Letelier J-C, Gutierrez C, Cornish-Bowden A, Soto-Andrade J (2010) Closure to efficient causation, computability and artificial life. J Theor Biol 263:79–92

Chuang YL, Oren R, Bertozzi AL, Phillips N, Katul GG (2006) The porous media model for the hydraulic system of a conifer tree: linking sap flux data to transpiration rate. Ecol Model 191:447–468

Clausnitzer V, Hopmans JW (1994) Simultaneous modeling of transient three-dimensional root growth and soil water flow. Plant Soil 164:299–314

Cowan IR (1965) Transport of water in the soil-plant-atmosphere system. J Appl Ecol 2:221–239

Cruizat P, Cochard H, Améglio T (2002) Hydraulic architecture of trees: main concepts and results. Ann Forest Sci 59:723–752

Daudet FA, Lacointe A, Gaudillère JP, Cruiziat P (2002) Generalized Münch coupling between sugar and water fluxes for modelling carbon allocation as affected by water status. Journal of Theoretical Biology 214:481–498
De Pury DGG, Farquhar GD (1997) Simple scaling of photosynthesis from leaves to canopies without the errors of big-leaf models. Plant Cell Environ 20:537–557
Doussan C, Pagès L, Pierret A (2003) Soil exploration and resource acquisition by plant roots: an architectural and modelling point of view. Agronomie 23:419–431
Doussan C, Pierret A, Garrigues E, Pages L (2006) Water uptake by plant roots: II. Modelling of water transfer in the soil root-system with explicit account of flow within the root system – comparison with experiments. Plant Soil 283:99–117
Dunbabin VM, Diggle AJ, Rengel Z (2002) Simulation of field data by a basic three-dimensional model of interactive root growth. Plant Soil 239:39–54
Engel T, Priesack E (1993) Expert-N, a building block system of nitrogen models as a resource for advice, research, water management and policy. In: Eijsackers HJP, Hamers T (eds) Integrated soil and sediment research: a basis for proper protection. Kluwer, Dordrecht, pp 503–507
Engel T, Klöcking B, Priesack E, Schaaf T (1993) Simulationsmodelle zur Stickstoffdynamik. Verlag Eugen Ulmer, Stuttgart
Farquhar GD, Caemmerer S, Berry JA (1980) A biochemical model of photosynthetic CO2 assimilation in leaves of C3 species. Planta 149:78–90
Feddes RA, Kowalik PJ, Zaradny H (1978) Simulation of field water use and crop yield, Simulation monographs. Pudoc, Wageningen
Früh T, Kurth W (1999) The hydraulic system of trees: theoretical framework and numerical simulation. J Theor Biol 201:251–270
Gardner WR (1960) Dynamic aspects of water availability to plants. Soil Sci 89:63–73
Gayler S, Priesack E (2007) PLATHO – A dynamic plant growth model considering competition between individuals and allocation to carbon-based secondary compounds. In: Fourcaud T, Zhang XP (eds) PMA06 – Plant growth modeling and applications. IEEE Computer Society, Los Alamitos, CA, pp 85–92
Gayler S, Grams TEE, Heller W, Treutter D, Priesack E (2008) A dynamical model of environmental effects on allocation to carbon-based secondary compounds in juvenile trees. Ann Bot 101:1089–1098
Génard M, Dauzat J, Franck N, Lescourret F, Moitrier N, Vaast P, Vercambre G (2008) Carbon allocation in fruit trees: from theory to modelling. Trees 22:269–282
Goudriaan J, van Laar HH (1978) Calculation of daily totals of the gross CO_2 assimilation of leaf canopies. Neth J Agric Sci 26:373–382
Goudriaan J, van Laar HH (1994) Modelling potential crop growth processes. Textbook with exercises. Kluwer, Dordrecht
Grote R, Pretzsch H (2002) A model for individual tree development based on physiological processes. Plant Biol 4:167–180
Herms DA, Mattson WJ (1992) The dilemma of plants: to grow or defend. Q Rev Biol 67:283–335
Hölttä T, Vesala T, Nikinmaa E, Perämäki M, Siivola E, Mencuccini M (2005) Field measurements of ultrasonic acoustic emissions and stem diameter variations. New insight into the relationship between xylem tensions and embolism. Tree Phys 25:237–243
Hölttä T, Vesala T, Sevanto S, Perämäki M, Nikinmaa E (2006) Modeling xylem and phloem water flows in trees according to cohesion theory and Münch hypothesis. Trees 20:67–78
Hund F (1996) Geschichte der physikalischen Begriffe. Spektrum, Heidelberg
Hutson JL, Wagenet RJ (1992) LEACHM: Leaching estimation and chemistry model: a process-based model of water and solute movement, transformations, plant uptake and chemical reactions in the unsaturated zone. Version 3.0. Cornell University, Ithaca, NY
Janott M, Gayler S, Gessler A, Javaux M, Klier C, Priesack E (2011) A one-dimensional model of water flow in soil-plant systems based on plant architecture. Plant and Soil 341:233–256
Javaux M, Schröder T, Vanderborght J, Vereecken H (2008) Use of a three-dimensional detailed modeling approach for predicting root water uptake. Vadose Zone J 7:1079–1088

Jones CA, Kiniry JR (1986) CERES-Maize: a simulation model of maize growth and development. Texas A&M University Press, Temple, TX

Kumagai T (2001) Modeling water transportation and storage in sapwood: model development and validation. Agric Forest Meteorol 109:105–115

Lacointe A (2000) Carbon allocation among tree organs: a review of basic processes and representation in functional-structural tree models. Ann Forest Sci 57:521–533

Lacointe A, Minchin PEH (2008) Modelling phloem and xylem transport within a complex architecture. Funct Plant Biol 35:772–780

Laughlin RB (2006) A different universe: reinventing physics from the bottom down. Basic, New York

Le Roux X, Lacointe A, Escobar-Gutiérrez A, Le Dizèsa S (2001) Carbon-based models of individual tree growth: a critical appraisal. Ann Forest Sci 58:469–506

Letelier J-C, Soto-Andrade J, Guíñez Abarzúa F, Cornish-Bowden A, Cárdenas ML (2006) Organizational invariance and metabolic closure: analysis in terms of (M, R) systems. J Theor Biol 238:949–961

Leuning R, Kelliher FM, De Pury DGG, Schultze E-D (1995) Leaf nitrogen, photosynthesis, conductance and transpiration: scaling from leaves to canopies. Plant Cell Environ 18:1183–1200

Matyssek R, Agerer R, Ernst D, Munch JC, Oßwald W, Pretzsch H, Priesack E, Schnyder H, Treutter D (2005) The plant's capacity in regulating resource demand. Plant Biol 7:560–580

Meier U (1997) Growth stages of mono- and dicotyledonous plants, BBCH-monograph. Blackwell, Berlin

Monod J (1949) The growth of bacterial cultures. Annu Rev Microbiol 3:371–394

Munteanu A, Solé RV (2006) Phenotypic diversity and chaos in a minimal cell model. J Theor Biol 240(3):434–442

Nimah MN, Hanks RJ (1973) Model for estimation of soil water, plant, and atmospheric interrelations: I. Description and sensitivity. Soil Sci Soc Am Proc 37:522–527

Penning de Vries FWT, Jansen DM, ten Berge HFM, Bakema A (1989) Simulation of ecophysiological processes of growth in several annual crops. Pudoc, Wageningen

Pretzsch H (2009) Forest dynamics, growth and yield. Springer, Berlin

Priesack E (2006) Expert-N Dokumentation der Modell-Bibliothek. Hieronymus, München

Priesack E, Gayler S (2009) Agricultural crop models: Concepts of resource acquisition and assimilate partitioning. In: Lüttge UE, Beyschlag W, Murata J (eds) Progress in botany, vol 70. Springer, Heidelberg, pp 195–222

Ritchie JT (1991) Wheat phasic development. In: Hanks J, Ritchie JT (eds) Modeling plant and soil systems. ASA, CSSA, SSSA, Madison, WI, pp 31–54

Ritchie JT, Godwin DC, Otter-Nacke S (1987) CERES-Wheat - A simulation model of wheat growth and development Texas A&M University Press, College Station, TX, pp. 185

Rötzer T, Seifert T, Pretzsch H (2009) Modelling above and below ground carbon dynamics in a mixed beech and spruce stand influenced by climate. Eur J Forest Res 128:171–182

Schröder U, Richter O (1993) Parameter estimation in plant growth models at different levels of aggregation. Model Geobiosph Process 2:211–226

Somma F, Hopmans JW, Clausnitzer V (1998) Transient three-dimensional modeling of soil water and solute transport with simultaneous root growth, root water and nutrient uptake. Plant Soil 202:281–293

Thornley JHM, Johnson IR (1990) Plant and crop modelling. A mathematical approach to plant and crop physiology. Clarendon, Oxford

Thompson MV, Holbrook NM (2003) Application of a single-solute non-steady-state phloem model to the study of long-distance assimilate transport. Journal of Theoretical Biology 220:419–455

Tyree MT, Zimmermann MH (2002) Xylem structure and the ascent of sap, 2nd edn, Springer series in wood sciences. Springer, Berlin

Vrugt JA, van Wijk MT, Hopmans JW, Simunek J (2001) One-, two-, and three-dimensional root water uptake functions for transient modeling. Water Resour Res 37:2457–2470

Wang Y-P, Leuning R (1998) A two-leaf model for canopy conductance, photosynthesis and partitioning of available energy I: model description and comparison with a multi-layered model. Agric Forest Meteorol 91(1–2):89–111

Yin X, Struik PC (2009) C3 and C4 photosynthesis models: an overview from the perspective of crop modelling. NJAS Wageningen J Life Sci 57:27–38

Yin X, van Laar HH (2005) Crop systems dynamics. Wageningen Academic, Wageningen

Chapter 16
Learning from Various Plants and Scenarios: Statistical Modeling

W. zu Castell, R. Matyssek, A. Göttlein, F. Fleischmann, and A. Staninska

16.1 Complex Phenomena in Nature

In a book on his view on science and its perspectives for the future, the physicist and Nobel laureate Robert B. Laughlin comments on the claim that doing fact-based science, theories are neither necessary nor helpful (Laughlin 2005). Within the debate between theoretical versus experimental approaches such a claim is occasionally expressed. Laughlin counters that theories are inevitable to design proper experiments (Chap. 13 in Laughlin 2005). Theories clearly pass through different stages of maturity (see Chap. 1). But even being mature and widely accepted in a scientific community, theories have to be constantly questioned and probed towards their explanatory quality.

There is an underlying question, guiding the following exposition: Are there laws in biology? Interestingly, this question has recently gained much interest in philosophy and the theory of sciences. It is deeply rooted in the overwhelming success of ancient Greek reductionism which has guided our way of thinking about what science is over centuries and led us to explain and control the world around us.

W. zu Castell (✉) • A. Staninska
Scientific Computing Research Unit, German Research Center for Environmental Health, Helmholtz Center Munich, Ingolstädter Landstraße 1, 85764 Neuherberg, Germany
e-mail: castell@helmholtz-muenchen.de

R. Matyssek
Chair of Ecophysiology of Plants, Technische Universtität München, Hans-Carl-von-Carlowitz-Platz 2, 85354 Freising, Germany

A. Göttlein
Forest Nutrition and Water Resources, Technische Universtität München, Hans-Carl-von-Carlowitz-Platz 2, 85354 Freising, Germany

F. Fleischmann
Pathology of Woody Plants, Technische Universtität München, Hans-Carl-von-Carlowitz-Platz 2, 85354 Freising, Germany

R. Matyssek et al. (eds.), *Growth and Defence in Plants*, Ecological Studies 220,
DOI 10.1007/978-3-642-30645-7_16,

But life at its core has yet defied all our ambitious intentions, leaving us in a world of speculation.

Some biologists (e.g., Kauffman 1995; Solé and Goodwin 1962) argue that the reason for persisting uncertainty lies in complex phenomena, showing emergent properties, i.e., properties of systems which cannot be explained from the components of the system. Beyond doubt, biological systems are complex. Highly nonlinear behavior is coupled with the overall presence of feedback loops. Furthermore, the aim to capture just a few key agents soon leads to high-dimensional data. Highly variable performance is characteristic to biological systems. But behind the obvious difficulties of capturing complex phenomena, the question about the existence of biological laws remains and answering it has immediate consequences for the theories we develop and the design of our experiments.

We will argue in this exposition that the complexity of biological systems, in particular of plant systems, forces us to think beyond classical reductionism in its pure form. Rigorous theoretical concepts are available, which lead us to allow for more contingency in our statements. But this insight has consequences for the methodological toolkit which we apply to confirm that the experimental data derived are consistent with the hypotheses underlying the experimental design in an appropriate way. We will show how this can actually be achieved using examples from the work being addressed in this book. This way, statistics does not only tell us something about our data, at least, as far as the biological core of information is concerned. Conversely, the experimental work and its appraisal also teach us about how to refine and re-design our statistical approaches, when preparing for the next step in empirical assessment.

16.2 Contingency of Scientific Laws

Beatty established what he calls the *Evolutionary Contingency Thesis*:

> All generalizations about the living world:
> a) are just mathematical, physical, or chemical generalizations (or deductive consequences of mathematical, physical, or chemical generalizations plus initial conditions),
> b) are distinctively biological, in which case they describe contingent outcomes of evolution.
> (Beatty 1995, p.46f)

While the first case just acknowledges the fact that biology is contingent on physics and chemistry, leaving open the question whether there are indeed proper laws in those disciplines, the second case implies that there are no laws in biology. Thinking about the appropriateness of such statements, one soon is lead with the challenge to properly define what is meant by the word “law.” We will cite a definition below. Before doing so, we first want to consider some aspects helping to better understand the context of the above cited statements.

Beatty picks up on Gould's (1989) mind game of thinking about what the world would look like if the tape could be rewound and evolution would be started again. Newton's law of gravity would still hold, i.e., evolution will not be able to result in a world where masses will not be attracted with a force proportional to the product of the two masses and the reciprocal of the squared distance of the masses. Conversely, it seems to be much easier to "*imagine circumstances in which unequal segregation of alleles among gametes would prevail than to imagine circumstances which would favor evolution by natural selection of 50:50 segregation ratios à la Mendel*" (Beatty 1995, p. 54).

Beatty's arguments have been source of inspiration for many other ideas since then, with contributions supporting and questioning his line of thought (Sober 1997; Mitchell 2000; Woodward 2001; Mitchell 2009). Some of the suggestions to overcome apparently contradictive conclusions concerning the existence of laws in biology aim at clarifying the meaning of the word "law." Following Mitchell, "*there is general agreement that laws allow us to explain, predict, and successfully intervene in the world. The features which are supposed to allow them to accomplish these functions are:*

1. *logical contingency (have empirical content),*
2. *universality (cover all space and time),*
3. *truth (exceptionless), and*
4. *natural necessity (not accidental)*" (Mitchell 2000, p. 246).

From the attempt to classify biological generalizations according to these criteria, the conclusion that there are no laws in biology seems obvious. It is in this sense that Beatty concludes saying that "*biological generalizations are evolutionary contingent is to say that they are not laws of nature—they do not express any natural necessity*" (Beatty 1995, p. 52). But, as Mitchell observed, this fact does not distinguish biology from other sciences (Mitchell 2000). Note that whatever is left as not evolutionary contingent will then be considered as existing "by chance," leading to an indirect definition of chance. We might be tempted to conclude that the debate on the existence of laws in biology is a philosophical one and as such of little relevance for biological experimentalists.

On the other hand, if the debate were intrinsically biological, would this imply that there are no other laws in biology than those directly inherited from chemistry and physics?

Putting arguments together implies that emergent phenomena in biology either need to be defeated due to philosophical reasons—as sketched above—or might never be extracted, since, per definition, they do not reside in the "theory of the parts," i.e., chemistry or physics.

Therefore, we prefer to take up a pragmatic position, as laid out by Mitchell and others (see Skyrms 1980; Waters 1998; Woodward 2001; Mitchell 2002). Rather than postulating a dichotomy of law versus chance, Mitchell suggests to span a "*continuum of contingency*" (Mitchell 2000). What is indeed needed is a framework which allows classifying generalizations obtained from experiments according to their explanatory value. One of Mitchell's "dimensions" of scientific laws is stability, which "*is roughly the extent to which a generalization is contingent on conditions that are stable across space and time*" (Woodward 2001, p. 2).

Woodward prefers invariance over stability: "*a generalization is invariant if and only if it would continue to hold under some range of physical changes involving interventions*" (Woodward 2001, p. 4). Resiliency would be another one (Skyrms 1980). "*Resiliency of a proposition has to do with the extent to which its subjective probability remains stable or unchanged (or changes only by some small amount) as one conditionalizes on other truth functional propositions in some family, all of which are consistent with the original proposition and its denial*" (Woodward 2001, p. 17), i.e., sensitivity with respect to limited alterations of the assumptions.

Whatever property one prefers, from a pragmatic point of view they hint into the same direction: they force us to re-think our strategy to derive generalizations with explanatory value. Either way, reductionist or non-reductionist, setting up a hypothesis and carefully thinking over an experimental design which allows tackling the precise core of the hypothesis and at the same time blending out as much of the blurring side effects as possible, remains to be at the core of scientific progress. Additional thought has to be spent considering the proper scale (e.g., spatial, temporal, organizational) of the phenomenon to be probed, the stability and therefore contingency of the experiment, and its inherent limitations. Even more, proper validation through repetition and confirmation through independent observation are inevitable.

Laughlin and Pines extend the claims also to physics. Although it is not yet commonly appreciated, emergence is present in physics, too. "*The emergent physical phenomena regulated by higher organizing principles have a property, namely their insensitivity to microscopics, that is directly relevant to the bread question of what is knowable in the deepest sense of the term*" (Laughlin and Pines 2000, p. 29). They conclude: "*the central task of theoretical physics on our time is no longer to write down the ultimate equations but rather to catalogue and understand emergent behavior in its many guises, including potentially life itself. . . . For better or worse we are now witnessing a transition from the sciences of the past, so ultimately linked to reductionism, to the study of complex adaptive matter, firmly based in experiment, with its hope of providing a jumping-off point for new discoveries, new concepts, and new wisdom*" (Laughlin and Pines 2000, p. 30).

16.3 Adequate Mathematical Modeling: Classical Statistics

So far we have seen arguments towards a pluralist strategy in the theory of science. One key concept is the notion of contingency, allowing us to formulate statements of varying level of lawfulness. Despite unavoidable insufficiencies in experimental setup, the possible existence of emergent phenomena forces us to broaden our approach. We do not want to dwell deeper on the question of emergent properties in biology, here, but rather turn our attention to the question, what an appropriate mathematical answer would be. Why this is necessary?

Consider an example presented in Chap. 3. Analyzing the response of European beech (*Fagus sylvatica* L.) to an infection with the root pathogen *Phytophthora plurivora* at the transcript and protein level led to the conclusion that the pathogen is indeed able to suppress defense reactions of the plant (also see Valcu et al. 2009; Schlink 2010). However, the answer at the transcript or protein level is neither specific, i.e., similar to wounding, nor very strong. It is therefore reasonable to expect that such an unspecific, weak answer will be reflected in a likewise unspecific effect at the organ level. But, as Oßwald et al. (Chap. 3) observed, "*there was no indication of a trade-off between growth and defense in surviving plants. However, these plants were characterized by higher net assimilation rates per unit leaf area, by a smaller root system, but unchanged total biomass, and showed an increased number of root tips per root biomass compared to healthy control trees*" (Sect. 3.4.4; see also Fleischmann et al. 2010). First discrepancies between different levels of observation were already found between transcripts and proteins. Although the general plant response was similar at both levels, only two direct overlaps of distinctly regulated transcripts/proteins were found (Sect. 2.3).

With a classical statistical approach, one would have chosen a stochastic model, describing the expected relation between root growth and/or root biomass and carbon/nitrogen levels in leaves. Adding further assumptions, which are commonly needed for the mathematical treatment of the model, but can rarely be truly justified in reality, we would have designed a test and, in this case, found that the relation is weak. But what does such a result tell us about the plant's defense system? Even more, would our a priori expectation of the sequential signaling cascade from gene expression over the protein to the metabolic level have made us design a statistical test which includes the possibility of effects by treatment on the higher scale without a proper effect on the "supporting" lower scale? Assume for the moment being, a clear effect can be identified on the higher organ level, resulting from multiple, small effects on the signaling level, each on its own too weak to be clearly identified. Searching for the significant effect, would we ever have considered these many small causes? This option should clearly always be considered in any form of scientific argumentation, nevertheless, our classical thinking in reductionism constantly forces us to simplify and reduce, blurring our senses for emergent phenomena.

Breiman (2001) describes the two cultures of statistical modeling. Classical statistics in the sense of Fisher (1956) starts with an appropriate probability model for the data. The data then are assumed to be independently drawn samples from this model. Then parameters of the model are estimated and the model is used for inference. The problem is that "*standard probability model assumptions in statistical theory are strong, unverifiable idealizations*" (Beran 2008, p. 218). Nevertheless, almost all work in statistics as a scientific discipline is of this type.

The second culture also considers data to be independently drawn, but from an unknown probability measure on the space of inputs and outputs. The goal then is to find a function, to be precise an algorithm, mimicking the behavior of the unknown distribution. The problem now is to consistently argue how the way the algorithm behaves indeed sheds light on the underlying biological black box system. In this respect, the classical approach is more appealing to theoreticians because it provides a sound framework for the statements to be deduced. Critics argue that in turn statistics based on unverifiable idealizations led to irrelevant theory and questionable scientific conclusions and, what might even weigh worse, prevent statisticians from working on relevant problems (Breiman 2001). Hereby, relevant is referring to problems resulting from experimental work rather than mathematical modeling. The second approach conversely needs high computing power and new ways of dealing with data and asking questions.

Bruggeman et al. (2002) call the second approach the *anti-reductionist strategy* in contrast to the classical *reductionist strategy*. They argue that none of them is suited to satisfy all needs we have in order to analyze, understand, and even control complex biological systems. It is a coexistence of both of them, using the strengths of either one and supporting the other on its weaknesses which carries the highest potential for future progress in finding biological explanations.

What is important for us here is the question of what the results tell us about the underlying biological phenomenon the data were taken from? The classical modeling approach seems indeed better suited for deducing properties of the underlying stochastic model, whereas the data-driven approach gives us this awkward feeling of never being able to know, whether what we see is indeed biologically relevant, or just an artifact produced by the way we look at the data.

We do not want to side with either position here. In any case, we will not be able to guarantee universal validity using mathematics/statistics. What we might instead do is to add to the explanatory power extracting the logical core out of the biological interpretation. Doing so we frame our current hypothesis about the biological system to be observed into a language which makes it accessible for mathematical computation. Following this line, a p-value on its own has no higher value of natural necessity than a set of clusters found by a hierarchical clustering algorithm. Even more, we would normally neither claim natural necessity, nor universality. It is the way we use the modeling within our line of reasoning which makes the rule deduced appear less accidental.

If at the end of the day, we cannot decide whether the stochastic model or the data-driven approach is more feasible, what is the difference? It appears to be a matter of taste to accept the paradigm that observations from experiments with complex systems in nature can be described with an a priori fixed parametric model. But there is the trap of relying on a classical goodness-of-fit criterion, or any other type of statistical error control without ever questioning the model chosen a priori. Keep in mind that a model means reduction and thereby comes along with blindness for explanations outside of the model's scope.

Let us give an example supporting the model-driven approach in statistics. The free-air ozone fumigation facility in the Kranzberger Forst (Werner and Fabian 2002; Matyssek et al. 2007, 2010) was based on the assumption that ozone being released through small openings in fumigation tubes may stay locally defined within a discrete canopy volume, from where diffusion quickly occurs into the surrounding space above and below canopy. This physical intuition is important for the design of the fumigation facility, since the number of fumigation tubes per unit canopy volume and the distance to the control zone are crucial parameters, as are the side conditions, given by the presence of trees and scaffolding.

Assuming second-order stationarity, the O_3 data can be modeled as samples drawn according to the distribution of a spatial process. As a consequence of the assumption, the covariance structure of the spatial process, i.e., its variogram, can be modeled using a parametric model. Computing the experimental variogram then allows gaining confidence in the usefulness of this assumption. Doing so, we can see that data points indeed show a behavior which one would expect for a variogram. A suitable variogram model can then be chosen, capturing the essential properties of the experimental variogram. Parameters can be fit using maximum likelihood or any other algorithm of choice, and the resulting variogram be used in a kriging process to estimate the spatial distribution. Recall that kriging constructs the best linear unbiased predictor by minimizing the variance of the expected error. Figure 16.1 shows the kriging estimate together with a plot of the kriging variance for a particular day. And indeed, this type of analysis confirms our physical intuition and raises confidence that the experimental design be appropriate to study biological phenomena in plants exposed to different ozone regimes. Note that the graph shows the kriging estimate rather than values derived from conditional simulation. If a concise prediction of ozone level at a particular day and location was needed, conditional simulation might be more appropriate.

The assumption needed to use the described statistical approach is stationarity. Note that there are different notions of stationarity, leading to different technical adoptions, but this question will not be addressed here. Roughly speaking, stationarity says that the correlation between the random variables at two distinct points x_1 and x_2 in space is a function of the vector $x_1 - x_2$ only. The assumption thus guarantees that the stochastic phenomenon is in some sense independent of the choice of the coordinate system. Again, whether this mathematical property is true in nature, remains—at the end of the day—a matter of belief. The fact that there also exists a deterministic approach to kriging-type predictors supports the claim that mathematical assumptions are needed to support the mathematical procedure, rather than describe properties of natural phenomena. To give an example, such methods have successfully been used to estimate volumes of tree trunks (see Fig. 16.2) based on computer tomography imaging (zu Castell et al. 2005; Beatson and zu Castell 2011).

To stress the important point again, stationarity is an assumption which is necessary for methodological reasons. There is not much in geostatistics beyond this assumption. Therefore, this assumption is often needed for practical reasons and as often criticized due to the difficulty to provide a proper justification.

Let us now consider a less supportive example. Working with scaffolding in a mature forest for about one decade, as it was the case at Kranzberger Forst, may pose specific problems. Soil compaction can become relevant in the vicinity of scaffolding, bearing the risk of affecting belowground parameters. Aboveground, scaffolding may influence rainfall and radiation patterns. Besides the risk of mechanical abrasion in the canopy introduced by scaffolding, the latter can act like a prop for trees, restricting the aboveground movability, which has been shown,

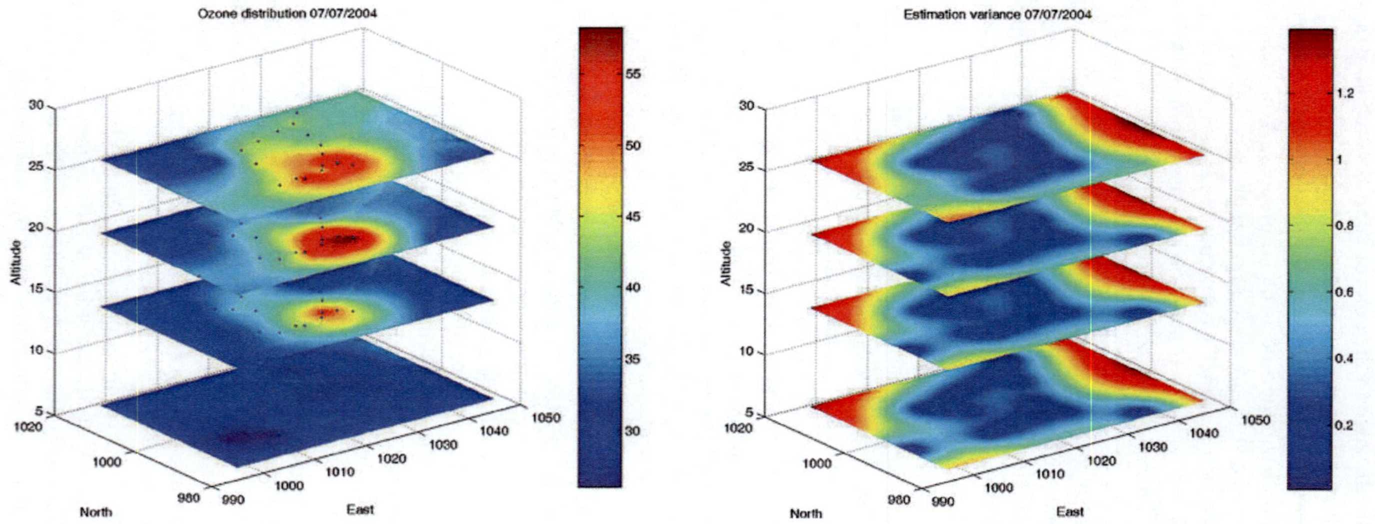

Fig. 16.1 Kriging estimation (*left*) and kriging variance (*right*) of the spatial ozone distribution in Kranzberger Forst using a spherical variogram model (*source*: SFB 607 database)

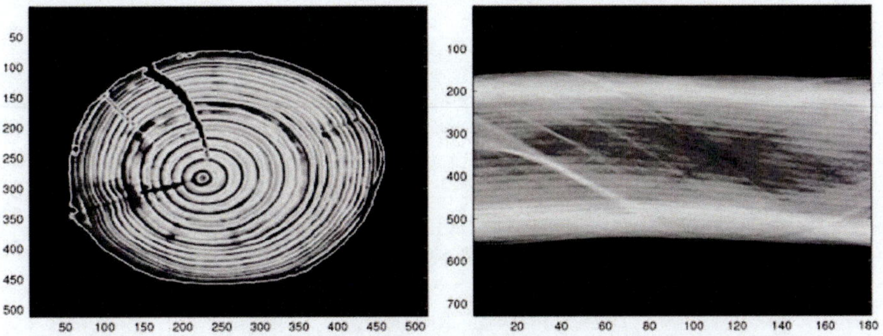

Fig. 16.2 Computer tomography image of a tree trunk (*left*) and corresponding sinogram (*right*). The sinogram shows the values of integrals along *lines* through the original image in a coordinate system determined by a rotation angle of the line and its distance to the origin

together with tangency effects, to influence tree growth (De Boeck et al. 2008). Apart from minor impact of abrasion, perceivable or measurable indications of scaffolding did not emerge at Kranzberger Forst to interfere with the O_3 responsiveness of the trees. Nevertheless, some potential lower bound for the size of physiological treatment effects may not be ruled out. But it does not make sense trying to search for effects in the data, whatever source there might be, which are of similar order or even smaller. Thus, there will always be a lower bound on the size of an effect which can statistically be identified from the data. Such a bound will be determined by the noise level present in the data and the sample size.

Returning to statistical assumptions, let us consider another aspect. Due to the scaffolding being zinc (Zn) plated, deposition to the soil occurs in the area of this installation through weathering. The question is, whether the Zn concentration in the soil after several years of the experiment has an impact on the plants growing there. Zn as a constituent of vital enzymes is an essential nutrient. However, because Zn is a micronutrient, elevated concentrations may soon reach toxic ranges. Although there are databases available on adverse effects of elevated heavy metals in soil (for a compilation see Riek and Wolff 2005) and trees (van den Burg 1985, 1990), measured Zn concentrations in combination with the respective threshold values have to be interpreted with caution. Especially in soil the respective threshold levels considered in practice have rather been set in view of high safety margins for precautionary policy reasons. To evaluate the influence of the scaffolding and other installations on soil chemical properties, soil samples were taken in grid-like patterns of different grid sizes underneath and in the vicinity of the installations (Fig. 16.3). Locally elevated Zn levels below the scaffolding turned out to exceed suggested critical concentrations for humus layers, which are set to 500–600 μg/g with respect to soil biological parameters (Tyler 1992).

From the chemical point of view, there is no reason to believe that Zn is subject to major horizontal drift in the soil. Thus, the phenomenon apparently is local, although some spatial effect may occur due to spills by rainwater around the scaffolding.

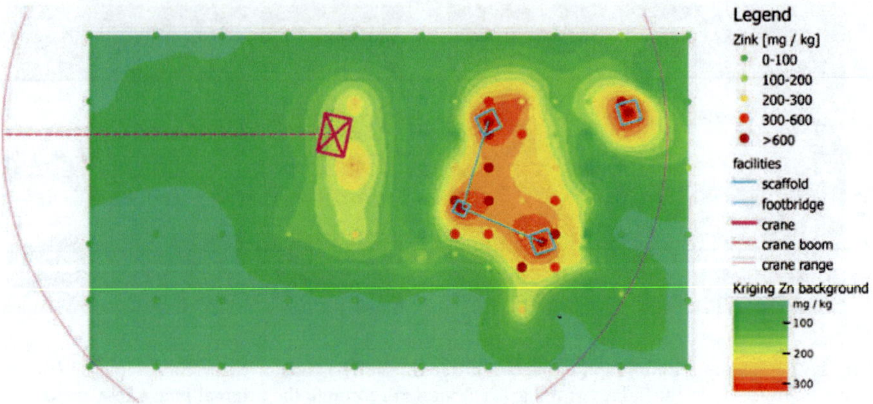

Fig. 16.3 Background Zn distribution in the humus layer of Kranzberger Forst as calculated from all values ≤300 mg/kg using a kriging estimate based on a spherical model. Grid points of soil sampling are indicated, the ones exceeding 300 or 600 mg/kg being highlighted by *larger circles* (*source*: based on unpublished data from SFB 607)

Addressing the question from the geostatistical point of view, the attempt to analyze the experimental variogram fails, because stationarity is not a proper assumption in this context. Independently of the chosen bin size, the calculated point cloud does not show any behavior that would allow for presuming stationarity. Measurement shows high Zn levels close to the scaffolding, whereas they are inconspicuous outside the critical boundary. In order to allow for an estimation of the spatial pattern, a geostatistical analysis can be performed based on all measured Zn data below a certain level (in our case the threshold was set to 300 mg/kg). Upon doing so, the experimental variogram becomes acceptable. Figure 16.3 was computed based on a Whittle–Matern model, fit to the experimental variogram. Nevertheless, for the kriging estimate all data points were used. It has recently been shown that such a procedure can be theoretically justified (Scheuerer 2009). The resulting map shows a realistic distribution of Zn concentrations in the humus layer of the Kranzberger Forst, but does not deliver any explanatory value to the biological aspect of Zn accumulation at the site. In summary, we obtained a visualization technique, rather than statistical support to analyze hypothesized biotic effects.

Nevertheless, biology can provide its own answers in this case. As exemplified for poplar (Todeschini et al. 2011), Zn from the soil eventually arrives at the leaves, i.e., the endpoint of the transpiration stream. There it is deposited mainly in the cell walls of the vascular tissue and bundle sheet cells, and becomes more enriched in the foliage than in the other tree organs. Morphological and ultrastructural changes may occur, having potential significance also in mitigating photosynthesis (Hermle et al. 2007). In sun leaves of the beech trees at Kranzberger Forst, Zn levels did not provide evidence for such an observation (Fig. 16.4). The trees without ozone treatment, which are in tendency located closer to the scaffolding (see Fig. 9.1) had higher Zn

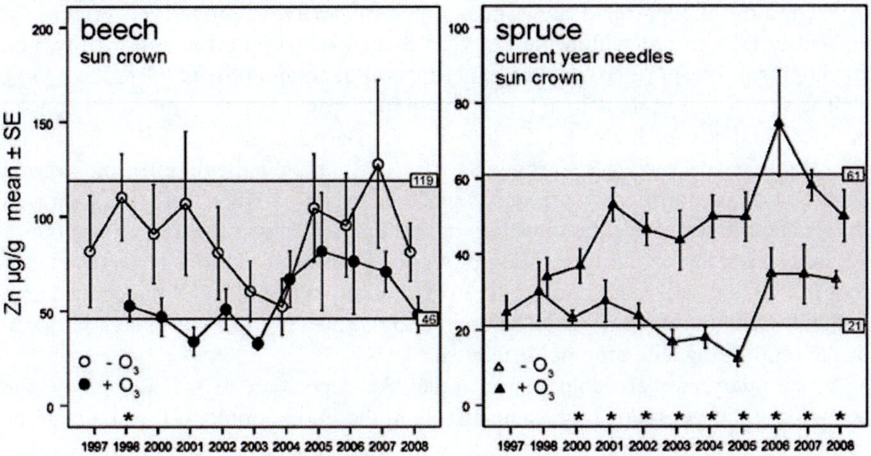

Fig. 16.4 Time course of Zn concentrations in beech leaves and spruce needles of the intensively measured trees of the ozone experiment at the Kranzberger Forst; the *gray shaded* region indicates the concentration range of normal nutrition according to statistical evaluation of threshold values compiled by van den Burg (1985, 1990) (*source*: based on unpublished data from SFB 607)

values, although the difference was significant only for the year 1998. Zn levels were mostly confined between two thresholds as derived from a statistical evaluation of the threshold values given by van den Burg (1985, 1990), with the lower one claiming the transition to latent Zn deficiency, and the upper one indicating the beginning of luxury Zn supply. Thus, the region between the two lines can be attributed as the normal range for Zn nutrition of beech and spruce. No indication of bias by Zn emerged in the beech leaves of the intensively measured trees, neither regarding morphology nor photosynthesis (Kitao et al. 2009), in particular, in view of interference with the O_3 impact. As elevated Zn tends to distinctly lower photosynthesis and water-use efficiency (WUE) while enhancing the CO_2 concentration in the intercellular leaf space (c_i, Hermle et al. 2007), opposite effects were found for WUE and c_i each under the enhanced O_3 regime in Kranzberger Forst, with the photosynthesis showing minor limitation only (Kitao et al. 2009). Similar to beech, the Zn levels of spruce were predominantly located between the threshold values for normal nutrition (see Fig. 16.4), although a general trend of increasing Zn concentrations was reflected under the $+O_3$ regimes, which were operated close to the scaffolding (see Fig. 9.1). Spruce sun needles did not display O_3 effects on the gas exchange (Matyssek et al. 2010). Neither was annual stem growth limited by the enhanced O_3 regime, as opposed to findings, however, in beech (Pretzsch et al. 2010). One can therefore conclude that if Zn effects existed, they were not resolvable and sufficiently low in biological terms to stay irrelevant at Kranzberger Forst in relation to the trees' responses to O_3 impact. This conclusion is consistent with a re-evaluation of the data in the context of this chapter.

Regarding aboveground assessments, careful analyses proved Zn effects—if present at all—to fall within the range of noise caused by other site factors, i.e., staying insignificant relative to the experimental treatment effects related to ozone.

The situation is more demanding belowground, regarding the immediate contact of roots and mycorrhizae to Zn in the soil. Because of potentially inhibiting effects of Zn on enzymes (Lingua et al. 2008), biochemical tests for enzyme activities of mycorrhizae were not considered when sampled soil cores showed Zn concentrations above the threshold of 600 μg/g. This was the case in one out of 12 cores from the ozone fumigation plot only (Pritsch 2011). For analyses of fine root growth and soil respiration (Nikolova et al. 2010), samples of the control area within the central plot have been compared to samples from a control area far away of the scaffolding, showing no differences.

In summary, we are able to show that Zn deposition to the soil below the scaffolding did not incite assessable effects in the O_3 experiment. But it should be pointed out here that using statistics alone we would not have been able to provide a suitable pathway for explanation. There is a balance to be kept between trusting the ability to quantify uncertainty in the data using statistics on the one side and treating experimental data as outcome of a particular incidence on the other side. The final conclusion will always be derived from a synopsis of both points of view.

The situation is not always such an ambiguous one. In many cases, classical statistical methods with a sound theoretical foundation can be generalized and adapted to situations, where the original assumptions might not be fully satisfied. Within the context of the experiments described in this book, periodogram techniques, i.e., techniques for time series analogously to spatial processes, have been adapted to non-equally spaced time series data. In another case, simultaneous confidence bands were used to incorporate neighborhood information in sequential data. Thus, there is a large zoo of statistical methods which cover a range of applications.

Not all of them are based on the same theoretical foundation. Brillinger and Tukey conclude: "*Here there are at least three successively larger regions, namely: An inner core of proven quality (usually quite unrealistically narrow) ...; a middle-sized region of understanding, where we have a reasonable grasp of our technique's performance ...; a third region, often much larger than the other two, in which the techniques will be used*" (cited in Beran 2008).

The worst thing one can do is to fix the statistical toolkit a priori and not allow for other methods. Unfortunately, this seems often to be the case, due to convenience and familiarization—or mere inexperience—within scientific communities. Instead we should consider using various statistical approaches, taking advantage of their differing power to deal with complex phenomena (cf. Fig. 16.5). The prize to pay is to deal with varying amounts of contingency, but it is not clear whether this is a restriction at all.

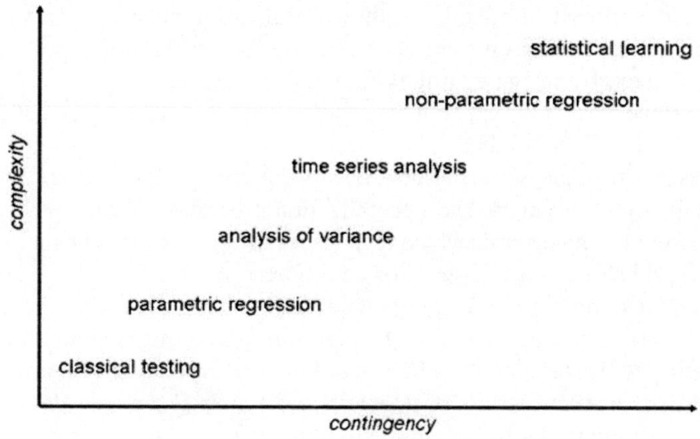

Fig. 16.5 Statistical methods can be arranged according to their contingency and the level of complexity they can address

16.4 Adequate Mathematical Modeling: Data-Driven Methods

Let us now turn to data-driven methods for statistical analysis. As mentioned above, the stochastic assumptions for those methods are relatively weak. We just assume the data to be independently drawn samples according to an underlying, but unknown probability distribution. Typical examples of such methods are neural networks or tree-based clustering. The basic idea is to minimize a given risk functional based on a sample drawn according to the unknown probability distribution (*empirical risk minimization*).

Clearly, it does not make much sense to ask for classical tests to be derived for those methods. Therefore, the quality criteria had to be thought over again. Goodness-of-fit tests have been replaced by prediction accuracy. The underlying paradigm of these approaches is the agreement that data have been produced by a black box, providing outputs $y_1,\ldots,y_n$ for certain inputs $x_1,\ldots,x_n$. Again, the goal is to compute a function, f say, capturing the relation between input and output. The function f is then called the *hypothesis*. The new quality criterion now aims at $f(x)$ providing a good prediction for a yet unforeseen input x. The idea is to *learn* the relation from a set of given labeled data, e.g., function evaluations (*supervised learning problem*). Alternatively, there is a quality functional measuring the quality of the current hypothesis f and punishing or rewarding guesses made using the hypothesis (*unsupervised learning problem*). Note that the learning approach is much closer to the way we start to explore our environment as children, than a modeling approach based on a theoretically founded stochastic model.

For historical reasons, much of the theory for such approaches has not been developed within statistics. Observe that this historical development is a consequence of the critique addressed above that statistical theory has been kept from addressing relevant problems.

Vapnik (1995) derived bounds for the generalization error of binary classification algorithms depending on a capacity measure for the class of hypotheses H, the function f is taken from. Let us briefly sketch the basic idea.

Clearly, if the class H is very rich, i.e., contains many different types of functions, every possible finite set of labeled data can be reproduced exactly, using an element of the class H. In other words, H contains an interpolating function. But such an element will commonly have bad generalization properties. This means, our algorithm will provide a hypothesis which reproduces the given training data exactly, but with a high probability for a large error on unforeseen data.

To give an example, think of polynomial regression. Assume we have ten data values with corresponding inputs on the real line, which might be obtained from a linear relation with some noise. There will be an interpolating polynomial, if the class H consists of polynomials of degree at most nine. For higher degree polynomial classes, an interpolating solution will still exist, but no longer be unique.

Now add one additional unforeseen data point and ask for prediction. An interpolating polynomial will commonly do a very bad job in giving the correct prediction.

Clearly, one would normally not suggest interpolation given noisy data, but this is not the point here. What is important is the *principle of parsimony* meaning the postulate to tend towards the simplest explanation possible in favor of complicated alternatives being difficult to be justified. Following the principle we shall go for the simplest model, explaining the data. Thus, we need to reduce the complexity to allow for a proper justification of our approach. Note that there are several other reasons why the principle of parsimony is adequate. We do not dwell on this topic here, but just mention that there are many empirical arguments in favor of it. Choosing the appropriate scale of complexity is the crucial task in this context. The capacity term in Vapnik's error bounds quantifies to what extent the class H contains complex elements.

The other criterion to take care of is accuracy in the sense of the ability to reproduce training data. In the supervised problem we can easily quantify accuracy. But will the quantity obtained tell us something about the accuracy of the derived relation? In neural networks, there is the well-known problem of over-fitting the training data. Roughly speaking, in favor of higher accuracy on the training data, we choose a network which shows bad generalization performance, because it tends to stay "to close to the data." Accuracy and simplicity therefore have to be balanced out.

One of the most successful algorithms in machine learning are *support vector machines*. This algorithm has been developed by Vapnik and coworkers having the theoretical bounds in mind and aiming at an optimal algorithm obeying the lessons they learned from statistical learning theory.

Consider a binary classification problem in the two-dimensional space, i.e., red and blue points in a plane. Assume the data are linearly separable, i.e., there is a line such that all blue points lie on one side of the line and all red points on the other. The support vector machine algorithm computes a separating line which is furthest

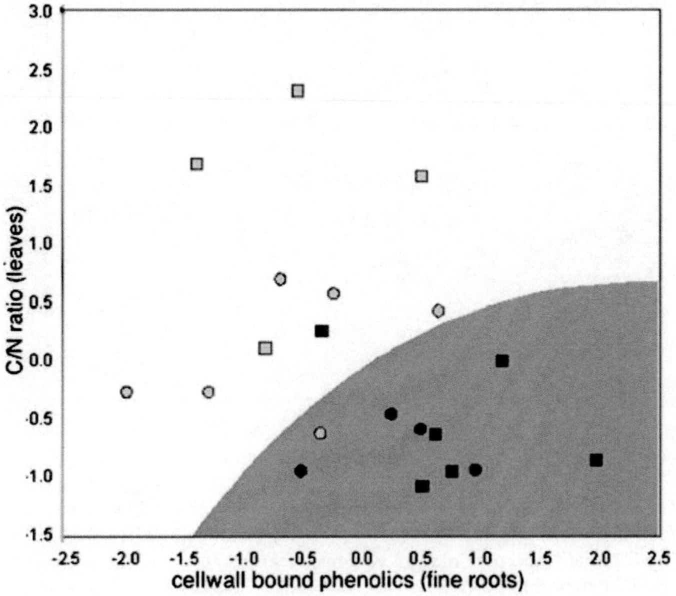

Fig. 16.6 Two-dimensional projection of a nonlinear separating hyperplane onto the plane spanned by the variables cell wall bound phenolics and *C/N* ratio, derived from a support vector machine algorithm (*source*: SFB 607 database)

away from each set of points. This is the so-called *maximal margin separating hyperplane*.

Separating the points, the hyperplane does the job; we can use it for further classification. If a point coming in lies on the side with the blue points, we predict a blue label, otherwise the point is labeled red. Being at maximal distance to each group of points, we guarantee that even with some perturbation of the points, the hyperplane will still be the same. This is what we want for good generalization performance.

The drawback of the described algorithm is the assumption of linear separability. This is clearly a very restricting property. But the algorithm itself is geometrical; it is computing a separating hyperplane. We can compute the same object in another space, where we can embed the data in. If this embedding is nonlinear, we can likewise solve nonlinear classification problems (see Fig. 16.6). This is what the so-called *kernel-trick* is doing. Kernels are nothing, but suitable embeddings.

Let us consider an example. Given juvenile beech trees in a phytotron experiment and in a similar experiment in the greenhouse, we can ask for changes in allocation behavior of the plants between growth-related and defense-related metabolism as suggested by the GDB under elevated and ambient CO_2, respectively. Doing such an experiment, various biomass variables had been measured, as well as phenolic substances in the leaves and the fine roots.

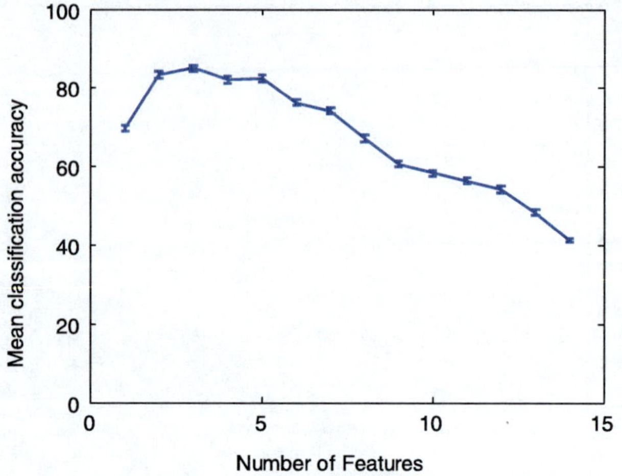

Fig. 16.7 Greedy variable selection for the discrimination of ambient CO_2 versus elevated CO_2 in phytotron experiments. Maximal classification accuracy (estimated) 85.13 % $\pm$ 0.75 % based on training set with 19 data sets containing 15 variables (features) from two independent experiments (*source*: SFB 607 database)

The experiment itself directly poses a binary classification problem: which of the measured variables show differences best between the two CO_2 scenarios? Splitting the data set into a randomly chosen training set and a remaining test set, we can train a support vector machine on the training set and use it to predict labels for the vectors of variables in the test set. Comparing the predicted labels with the actual ones, we can estimate the misclassification error. Repeating this cross-validation type procedure over several, randomly chosen subsets of data improves the accuracy of the estimated error.

In order to determine the importance of the variables for the classification task, we perform, in a first step, the described procedure with each variable alone. Taking the variable with the best prediction accuracy, we can then try every combination of the chosen variable with any other one from the remaining set of variables. We then proceed by adding one variable at a time to the set of so far chosen ones, each time voting for the combination with highest prediction accuracy. Solving an optimization problem by optimizing locally within each step is called a *greedy approach*. Upon doing so, we end up with a list of variables and a corresponding list of achieved accuracies, obtained with the corresponding selection of variables (Fig. 16.7). Commonly, prediction accuracy first rises during the process and then decreases again, if too many variables have been chosen. This has to do with additional noise being introduced and depending on variables, which do not carry additional information.

The rough sketch of a procedure using support vector machine classification in a greedy way to determine an ordered list of variables for a given classification task combines several aspects which have been mentioned in the first part of this chapter.

First of all, we do not start from an a priori fixed hypothesis/model. Thus, we might learn something from the data. But the use of whatever we learned lies outside of the statistical approach. Second, we aim at a stable relation, which is what the maximal margin hyperplane tries to compute. We do not want a classification rule which completely changes with perturbations of our data. Robustness is also a reason for applying cross-validation techniques. Averaging out results from various iterations using randomly chosen subsets commonly stabilizes the generalization rule we are aiming at.

The algorithm further focuses on the scale which seems to be best for the given classification task. Further variables on other scales might be present, but their influence will be ruled out. The maximal margin separating hyperplane is determined by the vectors in each group of data being closest to the plane. Others do not play a role in the mathematical optimization problem behind. In this sense, support vector machines implement some kind of sparsity principle. This is a principle often observed in data. It is based on the observation that the features we try to extract from the data are often encoded in just a few of the data points given. If we knew in advance which ones to consider, we would have much more efficient algorithms. Using regularization terms in functionals for empirical risk minimization sparsity can be enforced algorithmically, as can other structural information given a priori (see Slawski et al. 2010).

16.5 Lessons Learned

Dealing with complex biological systems, we are facing natural limits of our ability to capture the system. Since there is no one-size-fits-all answer, statistical approaches have to provide enough flexibility to adjust to the questions being asked.

From the methodological side, we should never close our toolbox of algorithms. Every problem posed by experimentalists carries its own challenges to be addressed. Being open for new questions raised in the empirical sciences, we will shift our focus from the discussion of the optimality of a method towards exploring chances and limits of its use in practical applications. We need to be aware of the fact that there will always be a need for proper balancing among several aims.

Concerning the insight we can provide, it indeed seems to be appropriate to share Mitchell's pluralistic approach. We can trade higher contingency for being able to capture higher complexity. Choosing the optimal balance is as much a consequence of the question being addressed as is the choice of the optimal resolution. Being aware of the existence of emergent properties of complex systems, we should use multiple approaches. While a data-driven, nontargeted search through the experimental outcome may lead to hypotheses otherwise not being considered, a modeling approach provides a rigorous formulation of the state of our knowledge which can be explored in its implications through mathematical simulation. Robustness tests and sensitivity analysis hereby govern our sense of quality, since appropriateness might always be subject to discussion. Imitating Karl Popper, a model is a valid hypothesis as long as our experiments have not yet falsified it. But it is the latter which provides progress in science.

References

Beatson RK, zu Castell W (2011) Scattered data interpolation of Radon data. Calcolo 48:5–19

Beatty J (1995) The evolutionary contingency thesis. In: Wolters G, Lennox JG (eds) Concepts, theories, and rationality in the biological sciences. University of Pittsburgh Press, Pittsburgh, pp 45–81

Beran R (2008) Discussion: approximating data. J Korean Stat Soc 37(3):217–219

Breiman L (2001) Statistical modeling: the two cultures. Stat Sci 16:199–231

Bruggeman FJ, Westerhoff HV, Boogerd FC (2002) BioComplexity: a pluralist research strategy is necessary for a mechanistic explanation of the ‘live’ state. Philos Psychol 15:411–440

De Boeck HJ, Liberloo M, Gielen B, Nijs I, Ceulemans R (2008) The observer effect in plant science. New Phytol 177:579–583

Fisher RA (1956) Statistical methods and scientific inference. Hafner, New York

Fleischmann F, Raidl S, Oßwald WF (2010) Changes in susceptibility of beech (*Fagus sylvatica*) seedlings towards *Phytophthora citricola* under the influence of elevated atmospheric CO_2 and nitrogen fertilization. Environ Pollut 158:1051–1060

Gould SJ (1989) Wonderful life. Norton, New York

Hermle S, Vollenweider P, Günthardt-Goerg MS, McQuattie CJ, Matyssek R (2007) Leaf responsiveness of *Populus tremula* and *Salix viminalis* to soil contamination by heavy metals and rainwater acidity. Tree Physiol 27:1517–1531

Kauffman S (1995) At home in the universe – the search for the laws of self-organization and complexity. Oxford University Press, New York

Kitao M, Löw M, Heerdt C, Grams TEE, Häberle KH, Matyssek R (2009) Effects of chronic elevated ozone exposure on gas exchange responses of adult beech trees (*Fagus sylvatica*) as related to the within-canopy light gradient. Environ Pollut 157:537–544

Laughlin RB (2005) A different universe – reinventing physics from the bottom down. Basic Books, New York

Laughlin RB, Pines D (2000) The theory of everything. Proc Natl Acad Sci USA 97:28–31

Lingua G, Franchin C, Todeschini V, Castiglione S, Biondi S, Burlando B, Parravicini V, Torrigiani P, Berta G (2008) Arbuscular mycorrhizal fungi differentially affect the response to high zinc concentrations of two registered poplar clones. Environ Pollut 153:137–147

Matyssek R, Bahnweg G, Ceulemans R, Fabian P, Grill D, Hanke DE, Kraigher H, Oßwald W, Rennenberg H, Sandermann H, Tausz M, Wieser G (2007) Synopsis of the CASIROZ case study: carbon sink strength of *Fagus sylvatica* L. in a changing environment – experimental risk assessment of mitigation by chronic ozone impact. Plant Biol 9:163–180

Matyssek R, Wieser G, Ceulemans R, Rennenberg H, Pretzsch H, Haberer K, Löw M, Nunn JJ, Werner H, Wipfler P, Oßwald W, Nikolova P, Hanke D, Kraigher H, Tausz M, Bahnweg G, Kitao M, Dieler J, Sandermann H, Herbinger K, Grebenc T, Blumenröther M, Deckmyn G, Grams TEE, Heerdt C, Leuchner M, Fabian P, Häberle KH (2010) Enhanced ozone strongly reduces carbon sink strength of adult beech (*Fagus sylvatica*) – resume from the free-air fumigation study at Kranzberg Forest. Environ Pollut 158:2527–2532

Mitchell S (2000) Dimensions of scientific law. Philos Sci 76:242–265

Mitchell SD (2002) Ceteris paribus – an inadequate representation for biological contingency. Erkenntnis 57:329–350

Mitchell SD (2009) Unsimple truths – science, complexity, and policy. University of Chicago Press, Chicago

Nikolova PS, Andersen CP, Blaschke H, Matyssek R, Häberle K-H (2010) Belowground effects of enhanced tropospheric ozone and drought in a beech/spruce forest (Fagus sylvatica L./Picea abies [L.] Karst.). Environ Pollut 158:1071–1078

Pretzsch H, Dieler J, Matyssek R, Wipfler P (2010) Tree and stand growth of mature Norway spruce and European beech under long-term ozone fumigation. Environ Pollut 158:1061–1070

Pritsch K (2011) Private communication

Riek W, Wolff B (2005) Bodenkundliche Indikatoren für die Auswertung der Bodenzustanderhebung im Wald (BZE II). Bundesministerium f. Verbraucherschutz, Ernährung u. Landwirtschaft, Bonn

Scheuerer M (2009) A comparison of models and methods for spatial interpolation in statistics and numerical analysis. PhD thesis, University of Göttingen

Schlink K (2010) Down-regulation of defense genes and resource allocation into infected roots as factors for compatibility between Fagus sylvatica and Phytophthora citricola. Funct Integr Genomics 10:253–264

Skyrms B (1980) Causal necessity. Yale University Press, New Haven

Slawski M, zu Castell W, Tutz G (2010) Feature selection guided by structural information. Ann Appl Stat 4:1056–1080

Sober E (1997) Two outbreaks of lawlessness in recent philosophy of biology. Philos Sci 64:458–467

Solé R, Goodwin B (1962) Signs of life: how complexity pervades biology. Basic Books, New York

Todeschini V, Lingua G, D'Agostino G, Carniato F, Roccotiello E, Berta G (2011) Effects of high zinc concentration on poplar leaves: a morphological and biochemical study. Environ Exp Bot 71:50–56

Tyler G (1992) Critical concentrations of heavy metals in the mor horizon of Swedish forests. Swedish Environmental Protection Agency, Report 4078, pp 1–38

Valcu C-M, Junqueira M, Shevchenko A, Schlink K (2009) Comparative proteomic analysis of responses to pathogen infection and wounding in *Fagus sylvatica*. J Proteome Res 8:4077–4091

van den Burg J (1985) Foliar analysis for determination of tree nutrient status – a compilation of literature data. Rijksinstituut voor onderzoek in de bos- en landschapsbouw "de Dorschkamp", Wageningen

van den Burg J (1990) Foliar analysis for determination of tree nutrient status – a compilation of literature data; 2. Literature 1985–1989. "de Dorschkamp" Institute for Forestry and Urban Ecology, Wageningen

Vapnik VN (1995) Statistical learning theory. Wiley, New York

Waters CK (1998) Causal regularities in the biological world of contingent distributions. Biol Philos 13:5–36

Werner H, Fabian P (2002) Free-air fumigation of mature trees. Environ Sci Pollut Res 9(2):117–121

Woodward J (2001) Law and explanation in biology: invariance is the kind of stability that matters. Philos Sci 68:1–20

zu Castell W, Schrödl S, Seifert T (2005) Volume interpolation of CT images from tree trunks. Plant Biol 7:737–744

Chapter 17
Modelling the Defensive Potential of Plants

S. Gayler, E. Priesack, F. Fleischmann, W. Heller, T. Rötzer, T. Seifert, and R. Matyssek

S. Gayler (✉)
Institute of Soil Ecology, Helmholtz Zentrum München, Ingolstädter Landstr. 1, 85764 Neuherberg, Germany

Water & Earth System Science (WESS) Competence Cluster, c/o University of Tübingen, Hölderlinstr. 12, 72074 Tübingen, Germany
e-mail: sebastian.gayler@uni-tuebingen.de

E. Priesack
Institute of Soil Ecology, Helmholtz Zentrum München, Ingolstädter Landstr. 1, 85764 Neuherberg, Germany

F. Fleischmann
Pathology of Woody Plants, Technische Universität München, Hans-Carl-von-Carlowitz-Platz 2, 85354 Freising, Germany

W. Heller
Institute of Biochemical Plant Pathology, Helmholtz Zentrum München, Ingolstädter Landstr. 1, 85764 Neuherberg, Germany

T. Rötzer
Chair for Forest Growth and Yield Science, Technische Universität München, Hans-Carl-von-Carlowitz-Platz 2, 85354 Freising, Germany

T. Seifert
Department of Forest and Wood Science, Stellenbosch University, Bosman Street, 7599 Stellenbosch, South Africa

R. Matyssek
Chair of Ecophysiology of Plants, Technische Universität München, Hans-Carl-von-Carlowitz-Platz 2, 85354 Freising, Germany

R. Matyssek et al. (eds.), *Growth and Defence in Plants*, Ecological Studies 220,
DOI 10.1007/978-3-642-30645-7_17,

17.1 Introduction

Plants distribute their limited resources between different, sometimes conflicting demands such as organ growth and different biochemical functions. Carbohydrates, nutrients and energy are needed for feeding growth-related metabolism as well as defence-related metabolism, and for maintenance-related processes. The latter comprise re-synthesis of enzyme proteins, ribonucleic acids, and membrane lipids that undergo turnover in the normal process of metabolism, maintaining ion gradients across cellular membranes, and energy support for metabolic activities. As the total demand of all processes cannot be met simultaneously in most cases, resource allocation on the whole plant level is determined by competition for the plant internal resources and energy. After fulfilling the demand for process which are indispensable to life, plants have to trade-off the demand for growth to be competitive against neighbours and the demand for defence against pathogens (Matyssek et al. 2005). Many carbon-based "secondary" compounds (CBSCs), in particular products of the phenylpropanoid metabolism, such as soluble phenolic compounds, flavanoids, and condensed tannins, are known to be involved in different defence responses of plants (Dixon 2001). As environmental factors such as nutrient supply, temperature, light conditions, or atmospheric carbon dioxide concentrations can influence the level of these compounds in plant tissues, partitioning of carbohydrates between growth-related metabolism and defence-related metabolism has been the subject of ecological research since decades. Patterns and variations in concentration of defence-related metabolites in plant tissues have been explained by means of different complementary and, in some cases, contradictory plant-defence hypotheses, such as the carbon–nutrient-balance hypothesis (Bryant et al. 1983), the growth–differentiation-balance hypothesis (Loomis 1932; Herms and Mattson 1992) and the protein-competition model (Jones and Hartley 1999). These hypotheses, which are discussed in detail in Chap. 1, are conceptual approaches to predict the impact of environmental resource availability on the balance of whole-plant resource allocation between the more growth-related metabolism and the more defence-related metabolism. They provide a theoretical framework from which different working hypotheses can be derived considering impacts of experimental treatments on the availability of carbohydrates and nitrogen for competing demands within plants and in the end on the concentration of defence-related metabolites in plant tissues. Under the different approaches, the growth–differentiation balance theory (in the following abbreviated as "GDB"), in particular in its expanded form (Herms and Mattson 1992; Matyssek et al. 2002, 2005) was identified as the most theoretically mature of the plant defence hypotheses (Stamp 2003a; Matyssek et al. 2005). The GDB predicts for resources, whose increasing shortage reduces actual growth rate to a greater degree than photosynthesis, a parabolic relation between resource availability and allocation to defence-related compounds in plant tissues (see Fig. 1.1). This prediction is founded by the consideration that the utilisation of carbohydrates for growth processes (cell division and enlargement) has priority over the utilisation for differentiation processes, i.e. chemical and morphological changes leading to qualitative differences between cells. Consequently, allocation of carbohydrates to defence-related metabolites like the group of CBSCs takes place predominantly in situations where their availability exceeds the demand for growth processes. Considering different levels of resource availability, therefore the highest fraction of carbohydrates should be allocated to CBSCs in plants growing in an environment with a moderate shortage of resources, if growth is already limited but photosynthesis is still less affected resulting in an accumulation of photosynthates. Typical resources of this type are water and nitrogen. An exception is the resource of light, which influences photosynthesis more than growth. In this case, with increasing light, CBSCs will increase proportionally with growth (Herms and Mattson 1992).

Many studies were conducted to investigate the effect of different levels of resource availability on growth and allocation to defence-related metabolism. In many cases, the results agree with the predictions of GDB. Herms (2002) showed that the effect of fertilisation on insect resistance of woody ornamental plants is consistent with GDB. The growth of apple trees was favoured by N-fertilisation but flavonoid biosynthesis was inhibited resulting in a decreased pathogen resistance (Rühmann et al. 2002; Leser and Treutter 2005). Plant growth was negatively correlated with foliar phenolic concentrations in a study investigating effects of nutrient availability on constitutive herbivore resistance of poplar (Glynn et al. 2003). Glynn et al. (2007) also observed in a study across five fertility levels that effects of nutrient availability on total phenylpropanoid levels in two willow species largely conformed to predictions of GDB. A study carried out by Blodgett et al. (2005) showed that fertilisation decreases resistance of red pine. Stefanelli et al. (2010) looked within a review at the effect of nitrogen and water supply on quality of a wide range of fruits and legumes. They showed that the optimal nitrogen application to maximise carbon-based secondary metabolite concentrations corresponds to or is near by the minimum of the range of applied N in the examined studies. For water, in the same review, it was shown that deficit irrigation can increase the level of CBSCs in different plant species. Further examples which substantiate the predictions of GDB are given in Mattson et al. (2005) for birch, in Mittelstraß et al. (2006) for potato, and in Le Bot et al. (2009) for tomato. But there are also many studies which yielded results which do not agree with predictions of GDB (Herms and Mattson 1992; Lindroth et al. 1993; Koricheva et al. 1998; Koricheva 2002). As a consequence of these contradictory results considerable dissatisfaction with the importance and usefulness of GDB was raised, which is discussed in Stamp (2003a, b, 2004) (cf. Chap. 1). In particular the question was discussed if it is resource availability at all which determines shifts in the balance between growth-related and defence-related metabolism (Hamilton et al. 2001).

In this chapter, we discuss constraints in the evaluation of GDB by experimental approaches and point out the potential of the newly developed plant growth model PLATHO to assist experimentalists in evaluating biological hypotheses. The transfer of new empirical evidence as described in Chaps. 3 and 12 into a complex simulation tool is demonstrated. In particular, it will be shown how the GDB hypothesis can be embedded into a dynamic plant growth simulation model, where it can be complemented by the protein-competition model (Jones and Hartley 1999), which also describes environmental impacts on phenolic allocation in plants in a conceptual way. The latter plant-defence hypothesis complements GDB by the assumption that for synthesis of CBSCs, which contain no nitrogen, nevertheless a minimum amount of free nitrogen is needed due to the nitrogen demand of precursory compounds (cf. Chap. 1). When included into a dynamic plant growth model, simulation allows quantifying the effect of fluctuating environmental conditions on the variability of source and sinking strengths of the plant internal resources within different phenological growth stages and thus enables the study of resource allocation in the plant with a high time resolution. In this sense PLATHO can be seen as a dynamic extension of GDB which enables the evaluation of this theory in the dynamic context of plant growth.

17.2 Constraints in the Evaluation of the Growth–Differentiation Balance Theory by Experimental Approaches

There are a number of reasons which make it difficult to test the explanatory power of GDB directly by empirical experiments. Six most important constraints of hypothesis evaluation can be identified.

1. It is hardly possible to quantify whether the plant-internal resource availability in a special experiment is high, intermediate or low in the sense of GDB. Consequently, we do not know whether we are below, at, or above the optimum for resource allocation to defence-related metabolism. Since allocation to the different biochemical demands of a plant is a matter of plant-internal competition for common resources, carbohydrate pools and nutrient levels in the plant should be measured rather than plant external availabilities of light, water and nutrients. Due to competition effects between individuals in a stand or due to constrained functionality of the resource capturing organs, plant-internal and plant-external resource availability can differ from each other considerably.
2. In general, it is hardly possible in experiments to change the availability of a single resource without changing the relative availability of the other resources (Stamp 2004). But complex interactions of different factors hamper the precise analysis of the effect of a single resource on allocation to defence-related metabolism. The clearing of interactions between two or more factors is difficult, as for different treatment factors different non-linear effects are predicted by GDB.
3. Fluctuating dynamics of carbon partitioning between growth-related and defence-related metabolism takes place at different temporal scales. Under the influence of circadian rhythms fluctuating transcript abundance of genes encoding enzymes in the phenylpropanoid pathway was observed in *Arabidopsis*

(Harmer et al. 2000). Diurnal variations of secondary metabolism were also shown at the metabolite level in leaves of wild-type tobacco plants under nitrogen-deficient conditions (Fritz et al. 2006). In leaves of beech trees, a continuous decrease of concentrations of phenolic compounds was observed in each season over a time span of three growing seasons for adult trees (Bahnweg et al. 2005) and over a time span of four growing seasons for juvenile trees in a lysimeter study (see Sect. 17.5.2). This indicates variable patterns of resource partitioning between growth-related and defence-related metabolism during different phenological growth stages at the organ level. Moreover, some of the secondary metabolites are continually catabolised and re-synthesised, and thus measurements of metabolite concentrations are just snapshots of the products of defence-related metabolism which do not reflect the real costs of allocation to defence-related metabolism, except the turnover rates of all single metabolites are known. Consequently, measuring directly total allocation to defence-related metabolism over a time period of weeks or month seems impossible, because measurements of metabolite concentrations are restricted to net values of resource allocation which neglect metabolite turnover.

4. Since a parabolic dependency of allocation on defence from resource availability is expected, a minimum of five levels of resource availability spread out along a gradient is advisable (Stamp 2004). However, in most experiments only two or three levels of resource supply are realised. An exception is the study of Pizarro and Bisigato (2010), in which the GDB was tested on the response of allocation of photo-assimilates to five water supply regimes. In general, it is difficult precisely to address the range of the resource gradient at which an experiment takes place. Statements about the validity of the GDB are therefore problematic, because different plant behaviour is predicted by this hypothesis for severe resource shortage compared to moderate resource shortage. Glynn et al. (2007) therefore point out that for testing GDB a priori knowledge of the physiological status of the plant and soil nutrient availability is required.
5. If results from different experiments are compared to evaluate the coherence of GDB over a wide range of experimental conditions, the problem arises to assess the resource availability levels for the specific experimental situations of the individual studies. Due to the non-linearity of the expected relation between resource availability and allocation rates to defence-related metabolism, we otherwise could not account for the magnitude of the differences in the applied resource supply in the single studies and would not be able to assess the observed treatment effects. This precept amounts to determine for each experimental treatment, which is considered in such a study, the position of both the control plants and the experimental plants along the resource availability gradient (the x-axis in the conceptual model of Herms and Mattson (1992); see Fig. 1.1 in this book). Moreover, this has to be done separately for nutrients, water and light, if it cannot be excluded that the availability of one of these factors is constant over all experimental situations. Consequently, a meta-analysis, which compares treatment effects between different experiments carried out under different environmental conditions, different intensities of experimental treatments, over

different time periods, and maybe with different plant species, should be done only if it is possible to determine the degree of limitation of each resource for the single experiments under consideration.
6. Plant-defence hypotheses are mainly related to constitutive defence at the whole plant level and neglect to a large extent induced defence mechanisms. Whole leaf level concentrations of CBSCs are critical for the resistance of the plant only for specific host/pathogen systems. This was shown by Ros et al. (2004) for different cultivars of potato in combination with the pathogen *Phytopthora infestans*, and Leser and Treutter (2005) for two apple tree cultivars which were infested with *Venturia inaequalis*. However, in many cases it is not the concentration of constitutive defensive compounds at the whole plant level, which is decisive for susceptibility or resistance of a plant against a specific parasite or pathogen. The induction of synthesising defensive compounds may be restricted to the specific site of the pathogen attack and is therefore hardly measurable at the whole plant level and even at the level of a single organ (cf. Chap. 3).

Some of these limitations in evaluating GDB could be accomplished with a considerable effort by more elaborate designs of the experiments, as pointed out by Stamp (2004). However, barely any empirical study was performed which fulfils these required standards. Therefore Stamp (2003a) and Matyssek et al. (2005) conclude that there are often difficulties in experimental designs or failures to identify and test assumptions in experiments which lead to contrasting and sometimes premature conclusions about the validity of GDB and other plant-defence hypotheses. Consequently, approaches are needed which provide quantitative estimations of the resource availability levels within the plants over the time periods of the experiments.

17.3 Concepts for Modelling Allocation to Defence-Related Metabolism

Only few attempts have been undertaken to develop mathematical models which describe the growth-defence trade-off in plants. Coley et al. (1985) as well as Rötzer et al. (see Chap. 18) use an abstract percentage term of defence investment to analyse opportunity cost and benefits of resource usage for defence instead of growth. In contrast, Gayler et al. (2004, 2008) and Le Bot et al. (2009) present models which simulate explicitly allocation of carbohydrates to a pool of carbon-based defence-related compounds depending on environmental factors.

The one equation model (Eq. 17.1) present by Coley et al. (1985) was developed to explain the relation between growth rate, allocation to defence and biomass loss due to herbivory. This model, named the Growth Rate Hypothesis, is a modification of the basic exponential growth equation:

$$\frac{dW}{dt} = r_{\max} \cdot W \cdot (1 - k \cdot D^{\alpha}) - (H - m \cdot D^{\beta}), \tag{17.1}$$

where W (g) is the plant biomass, r_{max} (day^{-1}) is the maximum inherent growth rate of a species, and D [g (defence) · g^{-1} (biomass)] is the defence investment, which can vary due to genotypic variations between individuals. k [g (biomass) · g^{-1} (defence)] and α (–) are constants that relate an investment in defence to a reduction of growth. H [g (biomass) · day^{-1}] is the potential loss caused by herbivores in the absence of defence, m [g (biomass) · g^{-1} (defence) · day^{-1}] represents the effectiveness of defence and β (–) determines the shape of the defence effectiveness curve. With some restrictions in relations to the parameters of the model to avoid defence investment exceeding inherent growth rate or negative herbivory, $k{\cdot}D^{\alpha} < 1$ represents the percentage term of reduction in realised growth due to investment into defensive compounds and the term $(H - m{\cdot}D^{\beta}) > 0$ represents the reduction in the realised growth due to herbivory. In this model it is expressed that plants, which allocate more of their available resources towards the defence-related metabolism, would reduce their inherent growth rate to a larger degree than plants with a lower investment for defence. However, a higher investment into defensive compounds reduces the potential damage by herbivores and consequently plants could benefit from this investment in habitats with a high herbivore pressure. The predictive power of the model depends on the extent to which the assumption holds that herbivores consume a fixed proportion of plant tissue and not of plant productivity (Coley et al. 1985).

Although the model has further constraints on its parameters ($\alpha > \beta$) (Basey and Jenkins 1993) and the assumption that all resources which are not allocated to defence are available for growth is a strong simplification, it provides a very clear look at the relationship between costs and benefits of allocation to defence at the whole plant level. For example, the model predicts the existence of an optimum value of the investment rate into defence, D_{opt}, which maximises the realised growth rate dW/dt under a given stress scenario H. Further on, the model predicts a decreasing value of D_{opt} with an increasing maximum inherent growth rate r_{max} of a plant. Consequently, the constitutive defensive potential of fast growing plants species with a high turnover of their organs should be lower compared to that of slow growing species with a low turnover of their organs. A comparison of saplings of 41 tree species, which were grown in light gaps of a rain forest, confirmed these predictions of the model. A negative correlation between growth rate and tannins was observed as well as a positive correlation between growth rate and the rate of herbivore damage to leaves (Coley 1988). Other examples cited in Stamp (2003a) do not agree with the Growth Rate Hypothesis. Cates (1996) for instance observed in fast-growing early successional pine species higher constitutive levels of defence compared to slower-growing late successional fir species. However, the potential for induced defence reactions after wounding was markedly higher in fir compared to pine.

Le Bot et al. (2009) present a simulation model for tomato plants, in which a pool of carbon-based defence-related compounds is explicitly considered. This model is designed to analyse concentrations of carbon-based defence-related compounds in leaves in response to nitrogen nutrition levels. The allocation to defence-related metabolism is linked to the plant internal availability of carbohydrates, which on its turn depends on the demands for maintenance, growth and defence. At increasing nitrogen availability, more carbohydrates can flow into growth and maintenance, due to the re-synthesis of degraded organic nitrogen. However, due to the simplifications made in this model, fluctuating environmental conditions and the dynamic of plant growth are not considered in this approach.

17.4 Modelling Carbon Allocation to Defence-Related Metabolism Regarding "Source–Sink" Relations

An approach to simulate dynamically environmental impacts on resource allocation to carbon-based defensive metabolites and the implication of their function in defence against parasites is realised in PLAnts as Tree and Herb Objects (PLATHO), a comprehensive plant growth simulation model (Gayler et al. 2004, 2008). The idea behind this model is to estimate the volatile dynamics of source and sink strengths of carbon and nitrogen within plants during different phenological growth stages and to simulate resource allocation between biochemical pools at the whole plant level as functions of these variables. Four main pools are considered: *assimilates* (temporarily existing products of photosynthesis and reserve remobilisation, handled as glucose), *reserves* (which can be mobilised if required), *carbon-based secondary compounds* (CBSCs) and *structural biomass*. The simulation of resource partitioning between theses pools is described in detail in Gayler et al. (2008). Equation 17.2 shows the function, which describes allocation of carbohydrates to the pool CBSCs, G_S (g g^{-1}), depending on two key variables, the plant internal carbon availability factor φ_C (–) and the plant internal nitrogen availability factor φ_N (–):

$$G_S(\varphi_C, \varphi_N) = \begin{cases} \sigma \cdot \varphi_N^\delta & \text{if} \quad \varphi_C \geqslant \frac{\sigma}{\left[1-(1-\sigma)\cdot\varphi_N^\beta\right]} \\ \left[1-(1-\sigma)\cdot\varphi_N^\beta\right] \cdot \varphi_N^\delta \cdot \varphi_C & \text{else} \end{cases}, \quad (17.2)$$

where σ is a species-specific parameter describing the maximum allocation rate of carbohydrates from the dispensable pool to CBSCs and β, δ are parameters determining the shape of this function (Fig. 17.1). For β, a value of 0.5 is assumed to reflect the fact that the growth rate is increased in a non-linear way by additional nitrogen supply. Metabolism of carbon-based secondary compounds is less affected by nitrogen deficiency than growth processes are, since phenylalanine is continuously regenerated from a limited nitrogen pool during phenolic biosynthesis (Mattson et al. 2005). The exponent δ must therefore be lower than β and is set to 0.33 in the model. Thus the function G_S reflects a parabolic behaviour of allocation to CBSCs with respect to the availability of the resource "nitrogen", but a linear

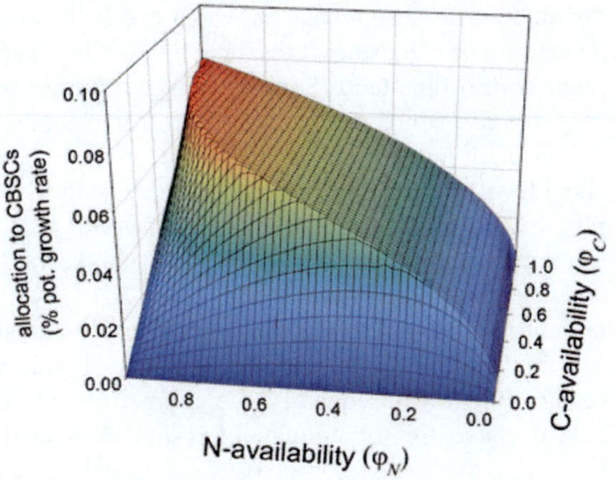

Fig. 17.1 Shape of the function $G_S(\varphi_C, \varphi_N)$ for simulating allocation to the pool of carbon-based secondary compounds. A *blue colour* of the surface indicates low allocation rates; an *orange colour* indicates high allocation rates to this pool

behaviour with respect to the availability of the resource "carbon" up to a certain level where a saturation level is reached. As plant internal availability of carbon is closely related to light supply as long as plant is operating below light saturation of photosynthesis, both is in accordance with the predictions of GDB. Consequently, Eq. 17.2 is a two-dimensional extension of the conceptual model illustrated by the line for defence-related metabolism in the figure of Herms and Mattson (1992); see Fig. 1.1. Moreover, and this is the main progress compared to the conceptual model of Herms and Mattson, the numerical simulation model allows to quantify the exact position along the resource gradients, given the boundary conditions of an experiment are known.

Resource availability factors $\varphi_C(t)$ and $\varphi_N(t)$ are calculated as the ratios between source strength and sink strength of carbon and nitrogen in the plant (Eq. 17.3).

$$\varphi_C(t) = \min\left\{1; \frac{\text{"C - source"}}{\text{"C - sink"}}\right\}; \quad \varphi_N(t) = \min\left\{1; \frac{\text{"N - source"}}{\text{"N - sink"}}\right\}, \qquad (17.3)$$

where source and sink strengths of plant internal resources result from resource uptake rates and potential growth rates, which are on their turn determined by the availabilities of the resources in the environment, by climatic factors and by the spatial distribution of the resource capturing organs of the plant individuals. The sink strengths of both resources are defined as the total amounts of C and N respectively, which are needed for supporting maintenance, for fulfilling the potential growth rates of plant organs and for satisfying the demand for the synthesis of defensive compounds. During a given time step, they are calculated from actual biomass of plant organs and its biochemical composition, from the phenological stage of the plant, from abiotic factors like temperature and, if present, from additional demands for defence reactions caused by pathogen diseases or for ozone detoxification (Gayler et al. 2004, 2009). The source strengths are defined as the amounts of carbon and nitrogen available within the plant during the considered time step for supporting all sinks. Their values are calculated as the amount of these resources which can be captured by the plant within this time depending on the external resource availability and the competitive situation of the plant within the stand. Source strengths further include the amount of C and N respectively, which can be mobilised from mobile pools in the plant.

For plant individuals growing in stands, the external resource availability at the stand level does not necessarily coincide with the resource uptake rate. For example, spatial distribution of resources within a stand of individuals of different sizes can cause asymmetries between uptake rates of single plants (Schwinning and Weiner 1998; Pretzsch and Biber 2010), see also Chaps. 12 and 14. Plants are able to respond to resources not homogeneously distributed within a stand by locating their resource capturing organs to sites with high resource supply to a certain extent. Consequently, the competitive situation of a plant within a stand is also decisive for the amount of resources available for metabolic processes. In PLATHO, growth-related processes are largely based on well-tested modules from other plant growth models (Jones and Kiniry 1986; Bossel 1996; van Ittersum et al. 2003), but are further extended by some newly developed process descriptions taking into consideration inter- and intra-specific plant–plant interaction between competing neighbours (Gayler et al. 2006). Assimilation rates as well as uptake rates of water and nitrogen of individual plants are calculated depending on size and position of a plant within the stand. Competition effects between neighbouring plants are estimated from the overlap of "zones of influence" assigned to the individual plants, where occupied crown and soil volumes of each individual are defined as a cylinder. The applicability of the approaches to simulated growth and competition between individuals was tested under different experimental conditions for juvenile beech (*Fagus sylvatica* [L.]) and spruce (*Picea abies* [L.] Karst.) in Gayler et al. (2006).

The key assumptions underlying the module for simulating carbohydrate allocation to the pool of CBSCs are adopted from the GDB and from the protein-competition model. The demand for maintenance therefore takes priority over all other processes in the model, growth takes priority over defence and photosynthesis is less affected by moderate nitrogen deficiency than growth. This is in accordance with the GDB, which predict that additional assimilates may be converted to CBSCs if carbohydrates accumulate in excess over growth demands or if availability of nitrogen is lower than the nitrogen demand required for growth processes. Further on it is assumed that the formation of carbon-based defensive compounds, even though they contain no nitrogen, requires sufficient levels of available nitrogen in the plant. This is due to the requirements for biosynthesis of precursory compounds (Jones and Hartley 1999). Based on a study of Häberle et al. (2009), it is assumed that some CBSCs are always needed for plant tissue and only a restricted amount of assimilates is relocatable between growth-related and defence-related metabolism (cf. Chap. 11). Consequently, a constant minimal baseline of carbohydrates is always allocated to the pool of CBSCs in the model whereas allocation of additional carbohydrates to CBSCs depends on plant internal carbon and nitrogen availability. Finally, recent photosynthetic assimilates are at first directed to energy-consuming processes before reserves are remobilised (e.g. Lötscher and Gayler 2005).

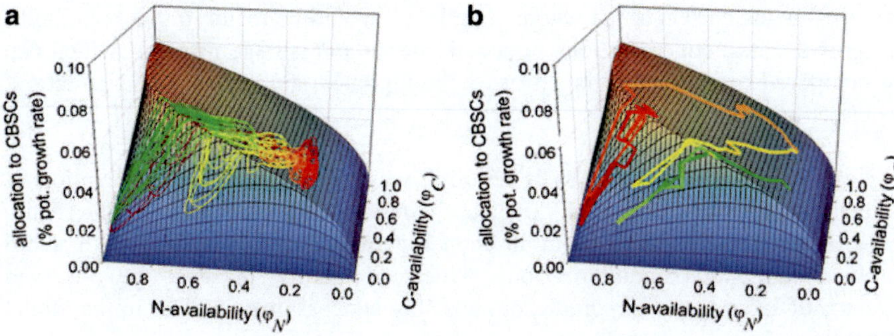

Fig. 17.2 Simulated dynamic of plant internal C- and N-availability factors and allocation to the pool of carbon-based secondary compounds. (**a**) Simulation of an experimental scenario with apple trees, high temporal resolution; (**b**) simulation of an experimental scenario with juvenile beech trees, where simulated values are aggregated to daily mean values. Each scenario is starting in May and ending in September. The time course of the simulations is represented in by the colour of the *line* from *green* (May) to *red* (September). A *blue* background colour indicates low allocation rates to the pool of carbon-based secondary compounds; an *orange colour* indicates high allocation rates to this pool

Simulations show that $\varphi_C(t)$ and $\varphi_N(t)$ can strongly fluctuate at different temporal and spatial scales. During the course of 1 day, the variability of sources and sinks of carbohydrates is mainly determined by the light/darkness cycle and differences in temperature. At the scale from several months to one or a few growing seasons, processes like phenological development, competition between individuals of different sizes (e.g. shading) or fluctuating supply of water and nutrients (e.g. precipitation, fertilisation) strongly influence source and sink strengths. This is illustrated in Fig. 17.2, where two examples of simulated $\varphi_C(t)$ and $\varphi_N(t)$ and predicted allocation to the pool of CBSCs are given at different temporal resolution. In Fig. 17.2a, the time course of allocation to the pool of carbon-based secondary compounds is simulated for an experimental scenario with apple trees for one growing season at high temporal resolution. Daily loops of the availability factors are visible, which are the result of the daily light/darkness cycles. Figure 17.2b shows an example with juvenile beech where simulated values are aggregated to daily mean values, resulting in a smoother curve.

17.5 Evaluation of the Modelling Approach Developed in PLATHO

How it is pointed out in Sect. 17.2, it is rather intricate to get all the data in one experiment needed for a direct and thorough evaluation of a modelling approach like the one which is used in the model PLATHO. High-resolution concentrations time series of metabolites and of other variables, which point to source and sink strengths of nitrogen and carbon at the whole plant level, would be required. Such variables are growth rates, concentrations of sugars, starch and storage proteins, which can be compared to simulated pool sizes of assimilates and reserves and structural biomass, but also uptake rates of nitrogen, water and carbon dioxide. Further on detailed information about forcing functions (climate, light supply, soil properties and experimental management) is required to get an accurate representation of boundary conditions and initial values in the model. This contributes to the reduction of the degrees of freedom in model parameterisation and thus to avoid ambiguous model parameterisations, which admittedly all meet the observed patterns of secondary compounds, but possibly conceal inadequacies in the model structure.

Unsurprisingly, such complete data sets were not available during the development of PLATHO. However, more and more new data sets became available, which complemented one another and provided the information needed for model calibration and later on for model testing. Consequently, the single processes of the model were evaluated step by step. The model was run for many different experimental scenarios, in each case considering the specific situations of the experiments in which the data were acquired. So the model was repeatedly confronted with new situations and new experimental results. The adequacy of the model was so tested by the ability of the model to reproduce treatment effects on growth on defensive compounds in the different specific situations. If required, model corrections or re-parameterisations were done, followed by new tests of the modified model. Consequently, the final structure and parameterisation of the model is the result of an iterative process of model testing and model refinement.

17.5.1 Varying Experimental Scenarios and Treatments

The first evaluation of the modelling approach for simulating resource allocation to defence-related metabolism was done with a data set originating from an experiment with apple trees of the cultivar Golden Delicious (Gayler et al. 2004). Based on five measurements per treatment during the first 4 weeks of leaf development, it was shown that the model is able to simulate the time course of phenylpropanoid concentrations in the leaves of this cultivar and to reproduce the effect of nitrogen supply on allocation rates to defence-related metabolism in this experiment. The successful transfer of the modelling approach to another apple cultivar (cv. Rewena) and even to other plant species (beech and spruce) was shown in a subsequent study (Gayler et al. 2008). The capability of the model to match treatment effects and the impact of varying experimental boundary conditions on total allocation to defence-related metabolism is shown in Fig. 17.3, in which simulated and measured concentrations of carbon-based secondary compounds are compared over different experiments with beech and spruce. Experimental factors, which were changed between and within the different experiments, are the level of nitrogen supply, the levels of atmospheric concentrations of carbon

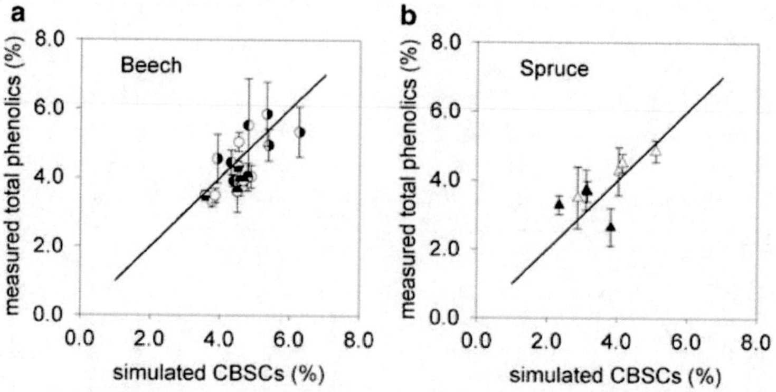

Fig. 17.3 Measured total phenolic vs. simulated carbon-based secondary compound concentrations (% of dry matter) in leaves and fine roots of juvenile (**a**) beech and (**b**) spruce under different experimental scenarios and treatments. Measured values are mean values of $n = 6$–12 individuals. *Error bars* represent standard deviations of measurements. In all cases, the time point of sampling was at the end of the vegetation period, but before onset of leaf senescence. Thus, measured values represent rather net allocation to defence-related metabolism over the lifespan of the respective organ than a snapshot of actual allocation rate to these compounds at the time of measurement. Data originate from SFB 607 (http://www.sfb607.de), personal communication

dioxide and ozone, plant age (2–7 years), and stand density (4–100 trees m^{-2}, depending on age). Furthermore, the trees were grown either in mono-specific or in mixed cultures of both species. The different experimental scenarios had a strong impact on the level of observed carbon-based secondary compounds, which vary between 3 and 6 % of leaf dry weight in case of beech and between 2 and 5 % in case of spruce respectively. This can be largely reproduced by the model. The simulations were carried out without changing the species-specific parameters of the model, which describe the ecophysiological and genetic properties of beech and spruce respectively. Only the input data describing the climatic conditions, planting density, fertilisation and irrigation management or soil properties were adapted to the specific experimental condition. Thus simulation results show that varying concentrations of CBSCs in plant tissues can be explained to a large extent by the simulated plant internal availabilities of carbon and nitrogen.

17.5.2 Simulation of Plant Internal Carbon Pools in a Lysimeter Study with Juvenile Beech Trees

The most comprehensive evaluation of the model PLATHO was achieved using a data set originating from a 4-year outdoor lysimeter study with juvenile beech trees (Winkler et al. 2009). In this experiment, half of the plants were fumigated during four vegetation periods at double compared to the ambient ozone concentration.

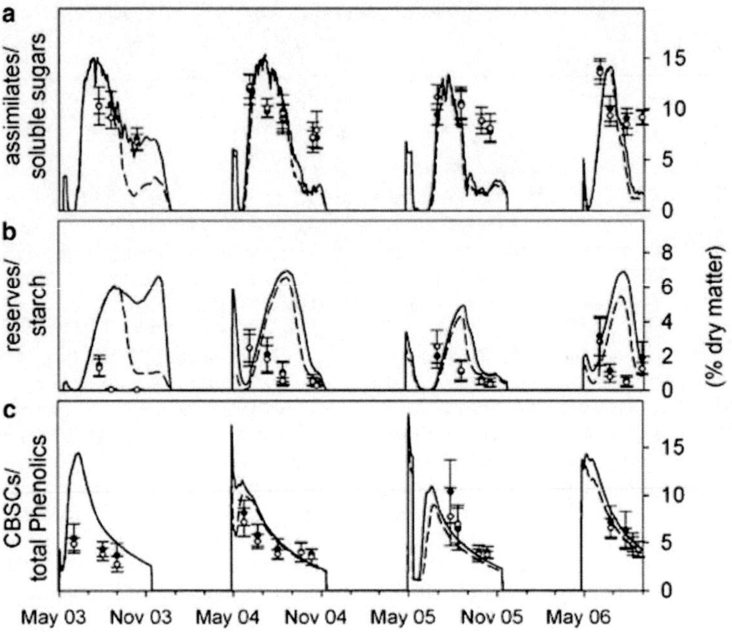

Fig. 17.4 Simulation of carbon pools in leaves of beech saplings during a 4-year lysimeter experiment in comparison with measured concentrations of corresponding compounds: (**a**) simulated "assimilates" and measured soluble sugars, (**b**) simulated "reserves" and measured concentrations of starch, and (**c**) simulated carbon-based secondary compounds and measured total phenolics. *Lines* are simulations and represent mean values over all trees in a treatment; the *symbols* illustrate measurements, which also represent mean values over one treatment. *Error bars* represent standard deviations of measurements. *Solid lines* and *closed symbols* represent trees grown under ambient atmospheric concentrations of ozone; *dotted lines* and *open symbols* represent trees which were fumigated with a doubled concentration of ozone. Data originate from SFB 607 (http://www.sfb607.de), personal communication

The ability of the model PLATHO to reproduce the effect of fumigation as well as the role of differences in soil properties and in initial biomass of trees on plant growth rates were analysed in detail in Gayler et al. (2009). Here we present the dynamic of the simulated carbon pools of the model, and compare these data with measured time series of concentrations of metabolic products, which are representative for the pools of the model. So the pool of assimilates is compared to measured concentrations of soluble sugars (sucrose, glucose and fructose). The simulated pool of reserves is compared to measured concentrations of starch, and the model pool of carbon-based secondary compounds is compared to measured phenolic compounds in leaves (see Fig. 17.4). For all three pools, in each season the highest concentrations were observed at the beginning of the vegetation period followed by a continuous decrease during the course of the season. This could be reproduced by the model for the pool of assimilates and for the pool of carbon-based secondary compounds (Fig. 17.4a, c). In case of simulated reserves and measured starch concentrations, a discrepancy between model and measurements is obvious (Fig. 17.4b). This can be explained by the fact that carbon reserves comprise other sources than starch. Thus amounts of starch cannot be regarded as representative for this pool in the model. However, a strong decrease of reserves due to translocation to stem and roots and deposition at the end of each vegetation period is visible as well in measurements as in the simulations. Nearly no effect of doubled atmospheric ozone concentrations on the level of soluble sugars and starch was measured during the first three vegetation periods of the experiment. However, in the last vegetation period, in 2006, the fumigation with ozone resulted in a significant decrease in concentration of sucrose as well as of starch (Fleischmann et al. 2009). This indicates probably a long-term effect of an increased consumption of carbohydrates and energy under elevated ozone for the detoxification of ozone entering leaves, where Rubisco damaged by ozone has to be repaired or replaced. The reduced availability of carbohydrates results also in slightly reduced allocation rates to carbon-based secondary compounds in ozone-fumigated trees during the first month of any single vegetation period. The model behaves in a similar way and simulates the effect of high ozone concentrations by decreased carbon pool sizes in the plant. In particular, however with except for the first vegetation period, the dynamic of and the treatment effect on the pool of carbon-based secondary compounds is well represented by the model.

Fumigation with doubled atmospheric ozone concentrations had no significant effect on concentration of nitrogen in plant tissues, which were measured once per growing season (simulations and data not shown). Mean values of nitrogen concentrations in leaves of control trees varied between 1.74 and 2.37 % of dry weight, in leaves of fumigated trees between 1.72 and 2.32 %. Simulations are within the range of the measurements and show also no effect of elevated ozone on nitrogen uptake. Simulated nitrogen concentrations at the level of the single plant organs, measured in detail at the end of the experiment, were also in the range of the measurements.

17.6 Model Applications

In this section, we focus on applying the model to support the interpretation of experimental results, on interrelating results from different experiments, and on analysing the complex interaction of experimental treatments and constraints in a scenario simulation going beyond the specific situation of a real experiment.

17.6.1 Model Simulations Support the Interpretation of Experimental Results

PLATHO simulations were carried out to support the interpretation of experimental results in a corresponding experiment. In this experiment, 3 years old beech

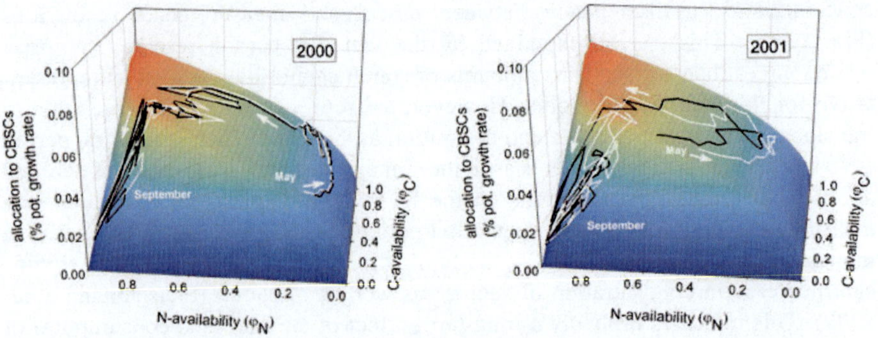

Fig. 17.5 Simulated time course of the plant-internal availabilities of carbon (φ_C) and nitrogen (φ_N) and corresponding allocation to the pool of carbon-based secondary compounds (CBSCs) in a phytotron experiment with juvenile beech trees during the vegetation periods 2000 and 2001. The *black lines* represent trees grown under ambient concentrations of atmospheric carbon dioxide (350 μl l^{-1}), *white lines* represent trees, which were exposed to an elevated concentration of atmospheric carbon dioxide (700 μl l^{-1}). A *blue* background colour indicates low allocation rates to the pool of carbon-based secondary compounds; an *orange colour* indicates high allocation rates to this pool

saplings grown for two vegetation periods either under ambient or under elevated concentrations of atmospheric carbon dioxide, concentrations of phenolic compounds in fine roots were measured repeatedly. Different effects of the treatment were observed at the end of each vegetation period. After the first vegetation period, i.e. in 2000, a slight, but not statistically significant, increase of concentrations of phenolic compounds was observed in the trees' fine roots under elevated carbon dioxide. 2.6 % of dry weight was measured under elevated CO_2 concentrations compared to 2.5 % under ambient concentrations. However, after the second vegetation period (2001), CBSCs concentrations in fine roots were significantly decreased under elevated CO_2 (2.5 % compared to 3.2 % under ambient concentrations). This was a surprising finding, because it was expected that increased CO_2 should increase photosynthesis and would lead to a higher amount of assimilates available for allocation to defence-related metabolism. Figure 17.5 shows the result of the PLATHO simulation of carbon and nitrogen availability in trees throughout two vegetation periods at both treatments and the resulting allocation rates to the pool of carbon-based secondary compounds. The time course of the system behaviour is indicated by arrows and start and end points of the simulations. In the simulations, during the first vegetation period elevated CO_2 had nearly no effect on plant internal availabilities of carbon and nitrogen. This could be a hint that that there was possibly not enough free nitrogen available within the plants to convert the additional external CO_2 to further carbohydrates. However, there is a different situation for the second vegetation period. After the dormant period, the simulated relative nitrogen availability within the plants is higher, due to depletion of the pool of assimilates in maintenance respiration. Now the additional external CO_2 supply would be used to enhance growth rates.

This would shift the availability curve to a lower relative N-availability, mainly during the first weeks of vegetative growth. This might be interpreted as a dilution effect resulting in decreased nitrogen concentrations, as it was observed in other experiments with elevated CO_2 (e.g. Bloom et al. 2010). Overall, the simulated curves are shifted towards decreased allocation to carbon-based secondary compounds, corresponding to the decreased concentrations of phenolic compounds measured in this experiment.

17.6.2 *Cross-sectional Analysis over Different Experimental Scenarios*

According to the GDB, increased carbon availability at constant nutrient supply should increase the plant internal pool of carbohydrates followed by increased allocation rates to carbon-based defensive compounds (see Sect. 17.1). However, in some cases experimental results do not fit on this expected behaviour, as shown in the previous section. In order to answer the question if there are patterns in the reaction of the plants related to their specific growing situations, we used the model PLATHO to conduct a cross-sectional study of a data set which covers different experimental scenarios and treatments affecting the carbon economy of 3–7 years old beech trees (Kozovits et al. 2005a; Luedemann et al. 2005; Winkler et al. 2009). The aim of the study was to analyse under which circumstances plants did enhance allocation to defence-related metabolism as predicted by GDB and to identify factors causing a deviating reaction of plants.

Only treatments were examined, where an increase in the carbon pool of trees compared to the pool of free nitrogen in the plants could be expected. An increase of carbohydrates was expected if plants were exposed to elevated atmospheric carbon dioxide concentrations and under increased light supply. After exposure to elevated concentrations of atmospheric ozone ("+O_3"), a decrease of the pool of available carbohydrates is to be expected due to enhanced demand for energy and assimilates for detoxification processes and repair of damaged Rubisco. Consequently, compared to a "+O_3" treatment, an analogous treatment with ambient ozone concentrations corresponds to an increase of carbohydrate availability in the plant. An increase in light supply of beech trees was caused in these experiments indirectly by the planting scheme: Beech trees grown in mixed culture were shaded by dominating spruce trees, whereas higher irradiance was measured for beech trees grown in mono-culture (Kozovits et al. 2005a).

The first step of the analysis was to quantify the plant internal availabilities of carbon and nitrogen under the different experimental conditions and for the respective treatments. For this purpose simulation runs with PLATHO were done for all experimental scenarios using the parameterisation of the model gained in previous studies. In each simulation scenario, the experimental growth conditions such as climate, stand density, fertilisation and soil properties were considered as inputs

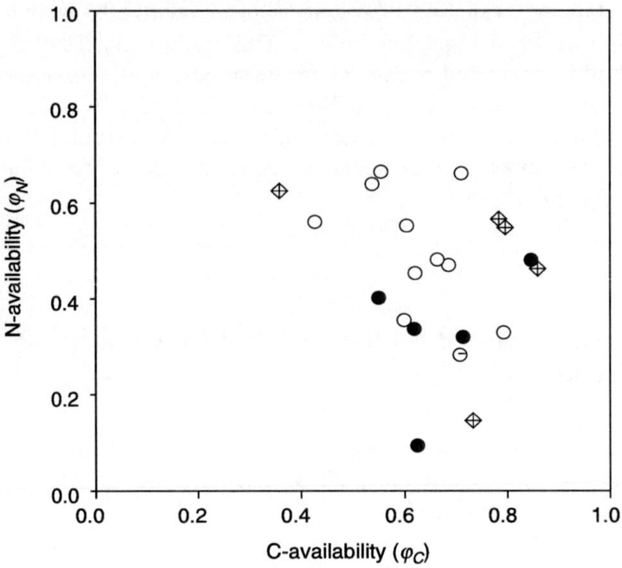

Fig. 17.6 Simulated position of different experimental scenarios with juvenile beech in the C- and N-availability space. The reaction of plants on treatments which potentially enhance the availability of carbohydrates is indicated by different symbols. *Open circles*: defence and growth increased; *diamonds*: defence increased, growth not increased; *half-filled circle*: defence not increased, growth increased; *closed circles*: neither defence nor growth is increased

and constraints. Mean values $\bar{\varphi}_C$ and $\bar{\varphi}_N$ (Eq. 17.4) were calculated from simulated carbon- and nitrogen-availability factors $\varphi_C(t)$ and $\varphi_N(t)$, where t_b [days] and t_e [days] denote the time points of beginning and end of the period of averaging, respectively:

$$\bar{\varphi}_C = \frac{1}{(t_e - t_b)} \int_{t_b}^{t_e} \varphi_C(t) \mathrm{d}t; \quad \bar{\varphi}_N = \frac{1}{(t_e - t_b)} \int_{t_b}^{t_e} \varphi_N(t)\ \mathrm{d}t. \tag{17.4}$$

Simulation results for plants, which were treated in a way that availability of carbohydrates was expected to be increased, were compared to the respective "controls", that means to the simulated plant behaviour in the corresponding scenario without this treatment. In the second step of the analysis, the simulated results of resource availability in the "control" and in the "treatment" of each experimental scenario were aggregated again to mean values, $\tilde{\phi}_C = (\bar{\varphi}_{C,\mathrm{control}} + \bar{\varphi}_{C,\mathrm{treatment}})/2$ and $\tilde{\phi}_N = (\bar{\varphi}_{N,\mathrm{control}} + \bar{\varphi}_{N,\mathrm{treatment}})/2$. This allowed getting the coordinates of a point in a two-dimensional space of resource availability. All points, each representing a realised experimental scenario, were plotted in a diagram which represents this space (see Fig. 17.6). In the third step of the analysis, measured effects of the respective treatments on allocation to growth-related metabolism and to defence-related metabolism were then assessed from plant growth rates and concentrations of phenolic compounds in plant tissues (soluble and cell wall bound). Measurements were done in all cases at the end of the vegetation periods before senescence of the leaves. The observed effects of the enhanced availability of carbohydrates on both growth- and defence-related metabolism were divided into the two categories: "increased" and "not increased". Four different responses of the plants could be expected: (1) increase of growth and defence, (2) growth increased, defence not increased, (3) increase of defence but no increase of growth, (4) neither growth nor defence are increased. Finally, the outcome of step 3 of the analysis is indicated by different symbols in Fig. 17.6.

The analysis was done repeatedly using varying simulation time spans in step 1 ranging from a few weeks up to the whole vegetation period. From these variations, it was observed that it is the plant internal resource availability during the early phase of the vegetation period from leaf unfolding till the end of leaf development that is decisive for the observed plant response. This seems reasonable because most of carbon-based secondary compounds are probably synthesised in this early phase and underlie a turnover, decay or irreversible deposition processes in the following period. The decrease of concentrations of phenolic compounds which follows from this assumption is for example documented in Bahnweg et al. (2005) for adult trees. The same effect was also observed in the lysimeter study mentioned above. The concentrations of phenolic compounds measured in plant tissues at the end of the vegetation period should therefore be seen as integrative values which reflect the history of the plant during the whole vegetation period.

Figure 17.6 shows the result of the analysis based on simulated $\tilde{\varphi}_C$ and $\tilde{\varphi}_N$ for the early vegetation period until completion of leaf development. Tracing in this figure a fictive path from the upper left to the lower right corner, this represents a gradient from growing situations, where resource allocation in the plants is determined rather by carbon limitation than by nitrogen limitation, towards growing situations, where nitrogen limitation dominates carbon limitation. In the analysed scenarios, an enhancement of defence-related metabolism could be induced by experimental treatments predominantly in situations where allocation is dominated rather by low carbon availability than by nitrogen shortage. In situations with low relative nitrogen availability compared to the carbon availability, plants are in most cases not able to increase defence-related metabolism after experimental treatments which are expected to increase carbohydrate availability. Concurrently, varying reactions to these treatments were observed with respect to growth. Considering Fig. 17.6 together with Fig. 17.1, the results of the analysis may now be interpreted in the following way: increasing availability of carbohydrates forces allocation to CBSCs, but only up to a distinct saturation level. Beyond this level, additional carbohydrates cannot be used for further CBSC formation. The produced surplus of carbohydrates remains in the pool of assimilates or will be used for reserve formation. The effect of saturation is reached earlier in situations with low nitrogen availability in the plant, because in this case the level of the amino acid phenylalanine, a precursor compound needed for both, biosynthesis of growth-related as well as defence-related metabolites, is also very low and consequently additional carbohydrates can be used neither for growth processes nor for allocation to secondary metabolism (closed symbols). However, if nitrogen availability is high enough, both growth- and defence-related metabolism can be met (open symbols). Nevertheless, in some cases only allocation to CBSCs is increased by enhanced carbohydrate supply but not growth. Probably this reflects situations in which growth is depressed by other reasons than nitrogen limitation.

In summary, the model-based analysis shows that the different responses of the trees can be explained to a large extent by the experimental conditions which lead to varying relative availabilities of carbon and nitrogen in the plants. Whether an experimental treatment, which enhances carbohydrate availability within the plant, induces increased concentrations of CBSCs in pant tissues or not, depends to a large extent on the question whether growth conditions during the decisive phase of leaf development were rather dominated by carbon-limitation or by nitrogen limitation.

17.6.3 Scenario Simulation: Impact of Stand Density and Nitrogen Fertilisation on Allocation to Defence-Related Metabolism in a Mixed Canopy of Juvenile Beech and Spruce

The complex interaction of different experimental factors affecting partitioning of resource allocation to growth- or defence-related metabolism is shown in Fig. 17.7. Starting from two experiments with mixed stands of juvenile beech and spruce trees (Kozovits et al. 2005a; Luedemann et al. 2005), scenario simulations were carried out, which go beyond the situation of the actual experiments. In the empirical experiments, trees were grown within a phytotron for two vegetation periods. The planting density was comparable in both experiments (about 100 trees m^{-2}, 50 of each species), but the age of the trees differed between the experiments. After two vegetation periods, spruce outcompeted beech. This was more pronounced in the older stands. This means that plant–plant interaction was more prominent due to the higher space occupation of trees. Previous studies showed that the model allows to describe the competitive behaviour of the two species under the specific constraints of the actual experiments. Also the direction of treatment effects on growth and allocation to defence-related metabolism in both species was simulated adequately by the model (Gayler et al. 2006, 2008). Here, scenario simulations are presented to investigate the interaction of nitrogen supply and stand density in the specific situation of such experiments. In the simulations presented in the following, stand density was varied within a range of 10–200 trees m^{-2}, and nitrogen-fertilisation rate was varied simultaneously from no fertilisation to 28 g(N) m^{-2}, twice the amount of the fertilisation rate applied in the real experiment. All other factors and experimental constraints were in accordance with the conditions of the actual experiment.

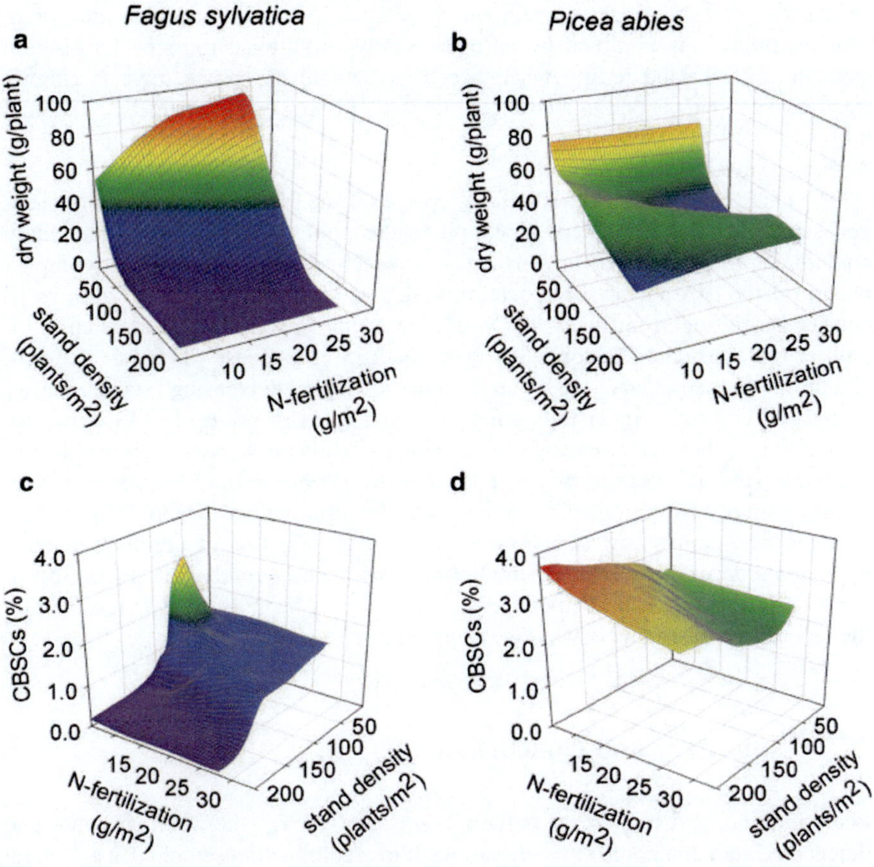

Fig. 17.7 Simulated impact of stand density and N-fertilisation on growth (**a**, **b**) and concentrations of carbon-based secondary compounds (**c**, **d**) in leaves of beech (**a**, **c**) and spruce (**b**, **d**) in a mixed canopy. *Dark blue colours* of the surfaces indicate low values of biomass and carbon-based secondary compounds, respectively; *orange colours* indicate high values of these variables

The model predicts an increasing competitive advantage for spruce either if nitrogen fertilisation is decreased or if stand density is increased (Fig. 17.7a, b). If trees grow under high fertilisation rates or with low stand density, beech is predicted to out-compete the spruce trees. This can be explained by the lower nitrogen demand of spruce compared to beech and by the time span until canopy closure (Kozovits et al. 2005b; Luedemann et al. 2005). At lower stand densities, under the specific conditions applied in the experiments, beech would have an advantage over spruce, because beech plants can grow for a longer time without competing for resources with neighbouring individuals. When the canopy closes and competition for light increases, beech is in a dominant position in relation to spruce. Conversely, at high plant densities, competition for nitrogen is more intense. Also competition for light starts earlier and spruce is not overgrown by beech as long as the canopy is not closed. This relationship is modified by nitrogen supply. Additional nitrogen fertilisation shifts the critical stand density, where beech can outcompete spruce, towards a higher stand density. This is due to a stronger increase of beech growth rates by additional nitrogen fertilisation than that of spruce, which leads to a dominant position of the former species at canopy closure.

The changing competitive situation superposes the effects of shading and nitrogen supply on the simulated allocation to the pool of carbon-based secondary compounds (Fig. 17.7c, d). In particular for beech trees, this results in a complex pattern of the dependency of allocation to defence-related metabolism from stand density and nitrogen fertilisation. Depending on the specific situation, changes of both of the considered factors can either result in an increase or in a decrease of carbon-based secondary compounds. For example decreasing stand density (corresponding to a higher light supply) results in an increase of CBSCs in case of beech only, but in decreasing CBSCs concentrations in leaves of spruce. Moreover, the parabolic dependency of allocation to defence-related metabolism from nitrogen supply is only visible for beech in case of a medium stand density. This example now shows how strong the acceptance or rejection of the plant defence hypotheses depends on experimental conditions. Consequently, the possibility of contrasting responses in falsification tests of a theory such as GDB is resolved by the model in a mechanistic way (cf. Chap. 19).

17.7 Summary and Conclusions

Plant internal partitioning of carbohydrates and energy between growth- and defence-related metabolism is the subject of ecological research since several decades. Different hypotheses about environmental impacts on constitutive defence in plants were developed during this period to explain patterns and variations in the concentration of carbon-based secondary compounds in plant tissues. Under these conceptual frameworks, the growth–differentiation balance theory (GDB) was identified as that one which is most mature from aspects of science theory (cf. Chap. 1). However, testing the explanatory capacity of this theory by experimental approaches is difficult, and results seem to be contradictory in some cases.

The dynamic plant growth model PLATHO supports the analysis of experimental results, assumptions and predictions of the GDB. This model offers the possibility to estimate the plant-internal relative availabilities of carbohydrates and nitrogen across different experimental scenarios and treatments. It further allows capturing dynamic variations of resource supply and resource demand, which can result from management, competition with neighbours and changing different growth stages during an experiment. Thus the model allows for the first time an evaluation of consequences of the conceptual framework of the GDB under fluctuating experimental constraints and to consider the impact of different experimental factors simultaneously. Consequently, PLATHO can be considered as a dynamic extension of the GDB.

Different tests have shown that PLATHO can reproduce the impact of experimental constraints and treatments on allocation to growth- and defence-related metabolism in plants. Based on this, applications of the model are presented in Sect. 17.6 which demonstrate the potential of numerical simulations in supporting empirical approaches. By combining measured findings with simulated availability factors for carbon and nitrogen, deeper insights into the complex process of resource allocation are possible, as it would be possible by experiments alone (cf. Chap. 19). In particular, a cross-sectional analysis like that one presented in Sect. 17.6.2 for juvenile beech goes beyond the possibility of previous experimental tests of the GDB. In this example across different experimental scenarios and treatments, it was shown that the observed pattern of the impact of experimental treatments on CBSCs concentrations could be explained to a large extent by the relation of relocatable carbohydrates and nitrogen within plants during the period of leaf development. This would not be visible without combining empirical data analysis with a dynamic simulation model. From this we conclude that also in other comparable studies the impact of resource availability on the defensive potential of plants is possibly more pronounced than it could be shown by the experimental approaches alone, even though resource availability is not the only factor which is responsible for variations in allocation rates to defence-related metabolism. A dynamic simulation model like PLATHO can help to detect hidden relationships and underlines the importance of resource availability concepts as formulated in the GDB.

Acknowledgements We are grateful to the Deutsche Forschungsgemeinschaft which funded this study within the frame of Sonderforschungsbereich 607 Growth and Parasite Defence – Competition for Resources in Economic Plants from Forestry and Agronomy. We further gratefully thank Thorsten Grams for providing growth parameters of beech and spruce trees from different experimental studies and Axel Göttlein for providing data of nutrient concentrations in organs of juvenile beech trees.

References

Bahnweg G, Heller W, Stich S, Knappe C, Betz G, Heerdt C, Kehr RD, Ernst D, Langebartels C, Nunn AJ, Rothenburger J, Schubert R, Wallis P, Müller-Starck G, Werner H, Matyssek R, Jr HS (2005) Beech leaf colonization by the endophyte *Apiognomonia errabunda* dramatically depends on light exposure and climatic conditions. Plant Biol 7:659–669

Basey JM, Jenkins SH (1993) Production of chemical defenses in relation to plant growth rate. Oikos 68:323–328

Blodgett JT, Herms DA, Bonello P (2005) Effects of fertilization on red pine defense chemistry and resistance to *Sphaeropsis sapinea*. Forest Ecol Manage 208(1–3):373–382

Bloom AJ, Burger M, Asensio JSR, Cousins AB (2010) Carbon dioxide enrichment inhibits nitrate assimilation in wheat and arabidopsis. Science 328:899–903

Bossel H (1996) TREEDYN3 forest simulation model. Ecol Model 90:187–227

Bryant JP, Chapin FS, Klein DR (1983) Carbon/nutrient balance of boreal plants in relation to vertebrate herbivory. Oikos 40:357–368

Cates RG (1996) The role of mixtures and variation in the production of terpenoids in conifer-insect-pathogen interactions. In: Romeo JT, Saunders JA, Barbosa P (eds) Recent advances in phytochemistry. Plenum, New York, pp 179–216

Coley PD (1988) Effects of plant growth rate and leaf lifetime on the amount and type of anti-herbivore defense. Oecologia 74:531–536

Coley PD, Bryant JP, Chapin FS (1985) Resource availability and plant antiherbivore defense. Science 230:895–899

Dixon RA (2001) Natural products and plant disease resistance. Nature 411:843–847

Fleischmann F, Winkler JB, Oßwald W (2009) Effects of ozone and *Phytophthora citricola* on non-structural carbohydrates of European beech (*Fagus sylvatica*) saplings. Plant Soil 323:75–84

Fritz C, Palacios-Rojas N, Feil R, Stitt M (2006) Regulation of secondary metabolism by the carbon-nitrogen status in tobacco: nitrate inhibits large sectors of phenylpropanoid metabolism. Plant J 46:533–548

Gayler S, Leser C, Priesack E, Treutter D (2004) Modelling the effect of environmental factors on the "trade-off" between growth and defensive compounds in young apple trees. Trees 18:363–371

Gayler S, Grams TEE, Kozovits A, Luedemann G, Winkler B, Priesack E (2006) Analysis of competition effects in mono- and mixed cultures of juvenile beech and spruce by means of the plant growth simulation model PLATHO. Plant Biol 8(4):503–514

Gayler S, Grams TEE, Heller W, Treutter D, Priesack E (2008) A dynamic model of environmental effects on allocation to carbon-based secondary compounds in juvenile trees. Ann Bot 101(8):1089–1098

Gayler S, Klier C, Mueller CW, Weis W, Winkler JB, Priesack E (2009) Analysing the role of soil properties, initial biomass and ozone on observed plant growth variability in a lysimeter study. Plant Soil 323:125–141

Glynn C, Herms DA, Egawa M, Hansen R, Mattson WJ (2003) Effects of nutrient availability on biomass allocation as well as constitutive and rapid induced herbivore resistance in poplar. Oikos 101(2):385–397

Glynn C, Herms DA, Orians CM, Hansen RC, Larsson S (2007) Testing the growth–differentiation balance hypothesis: dynamic responses of willows to nutrient availability. New Phytol 176(3):623–634

Häberle K-H, Nunn AJ, Reiter IM, Werner H, Heller W, Bahnweg G, Gayler S, Lütz C, Matyssek R (2009) Variation of defence-related metabolites in the foliage of adult beech and spruce – a conceptual approach to approximating trade-off carbon. Eur J Forest Res 128:99–108

Hamilton JG, Zangerl AR, DeLucia EH, Berenbaum MR (2001) The carbon-nutrient balance hypothesis: its raise and fall. Ecol Lett 4:86–95

Harmer SL, Hogenesch JB, Straume M, Chang H-S, Han B, Zhu T, Wang X, Kreps JA, Kay SA (2000) Orchestrated transcription of key pathways in Arabidopsis by the circadian clock. Sci Cult 290:2110–2113

Herms DA (2002) Effects of fertilization on insect resistance of woody ornamental plants: reassessing an entrenched paradigm. Environ Entomol 31(6):923–933

Herms DA, Mattson WJ (1992) The dilemma of plants: to grow or defend. Q R Biol 67(3):283–335

Jones CG, Hartley SE (1999) A protein competition model of phenolic allocation. Oikos 86:27–44

Jones CA, Kiniry JR (eds) (1986) CERES-Maize. A simulation model of maize growth and development. Texas A&M University Press, College Station, TX

Koricheva J (2002) Meta-analysis of sources of variation in fitness costs of plant antiherbivore defenses. Ecology 83(1):176–190

Koricheva J, Larsson S, Haukioja E, Keinänen M (1998) Regulation of woody plant secondary metabolism by resource availability: hypothesis testing by means of meta-analysis. Oikos 83:212–226

Kozovits AR, Matyssek R, Blaschke H, Göttlein A, Grams TEE (2005a) Competition increasingly dominates the responsiveness of juvenile beech and spruce to elevated CO_2 and O_3 levels throughout two subsequent growing seasons. Glob Change Biol 11:1387–1401

Kozovits AR, Matyssek R, Winkler JB, Göttlein A, Blaschke H, Grams TEE (2005b) Above-ground space sequestration determines competitive success in juvenile beech and spruce trees. New Phytol 167:181–196

Le Bot J, Benard C, Robin C, Bourgaud F, Adamowicz S (2009) The 'trade-off' between synthesis of primary and secondary compounds in young tomato leaves is altered by nitrate nutrition: experimental evidence and model consistency. J Exp Bot 60(15):4301–4314. doi:10.1093/jxb/erp271

Leser C, Treutter D (2005) Effects of nitrogen supply on growth, contents of phenolic compounds and pathogen (scab) resistance of apple trees. Physiol Plant 123:49–56

Lindroth RL, Kinney KK, Platz CL (1993) Responses of deciduous trees to elevated atmospheric CO_2: productivity, phytochemistry, and insect performance. Ecology 74(3):763–777

Loomis WE (1932) Growth differentiation balance vs. carbohydrate-nitrogen ratio. Proc Am Soc Hortic Sci 29:240–245

Lötscher M, Gayler S (2005) Contribution of current photosynthates to root respiration of non-nodulated *Medicago sativa*: effects of light and nitrogen supply. Plant Biol 7(6):601–610

Luedemann G, Matyssek R, Fleischmann F, Grams TEE (2005) Acclimation to ozone affects, host/pathogen interaction, and competitiveness for nitrogen in juvenile Fagus sylvatica and Picea abies trees infested with *Phytophthora citricola*. Plant Biol 7:640–649

Mattson WJ, Julkunen-Tiitto R, Herms DA (2005) CO_2 enrichment and carbon partitioning to phenolics: do plant responses accord better with the protein competition or the growth-differentiation balance model? Oikos 111:337–347

Matyssek R, Schnyder H, Elstner E-F, Munch J-C, Pretzsch H, Sandermann H (2002) Growth and parasite defence in plants: the balance between resource sequestration and retention. Plant Biol 4(2):133–136

Matyssek R, Agerer R, Ernst D, Munch JC, Osswald W, Pretzsch H, Priesack E, Schnyder H, Treutter D (2005) The plant's capacity in regulating resource demand. Plant Biol 7:560–580

Mittelstraß K, Treutter D, Pleßl M, Heller W, Elstner EF, Heiser I (2006) Modification of primary and secondary metabolism of potato plants by nitrogen application differentially affects resistance to *Phytophthora infestans* and *Alternaria solani*. Plant Biol 8(5):653–661

Pizarro LC, Bisigato AJ (2010) Allocation of biomass and photoassimilates in juvenile plants of six Patagonian species in response to five water supply regimes. Ann Bot 106:297–307

Pretzsch H, Biber P (2010) Size-symmetric versus size-asymmetric competition and growth partitioning among trees in forest stands along an ecological gradient in central Europe. Can J Forest Res 40:370–385

Ros B, Thümmler F, Wenzel G (2004) Analysis of differentially expressed genes in a susceptible and moderately resistant potato cultivar upon *Phytophthora infestans* infection. Mol Plant Pathol 5:191–201

Rühmann S, Leser C, Bannert M, Treutter D (2002) Relationship between growth, secondary metabolism, and resistance of apple. Plant Biol 4(2):137–143

Schwinning S, Weiner J (1998) Mechanisms determining the degree of size asymmetry in competition among plants. Oecologia 113:447–455

Stamp N (2003a) Out of the quagmire of plant defense-hypotheses. Q R Biol 78:23–55

Stamp N (2003b) Theory of plant defensive level: example of process and pitfalls in development of ecological theory. Oikos 102(3):672–678

Stamp N (2004) Can the growth-differentiation hypothesis be tested rigorously. Oikos 107:439–448

Stefanelli D, Goodwin I, Jones R (2010) Minimal nitrogen and water use in horticulture: effects on quality and content of selected nutrients. Food Res Int. doi:10.1016/j.foodres.2010.04.022

van Ittersum MK, Leffelaar PA, van Keulen H, Kropff MJ, Bastiaans L, Goudriaan J (2003) On approaches and applications of the Wageningen crop models. Eur J Agron 18:201–234

Winkler JB, Lang H, Graf W, Reth S, Munch JC (2009) Experimental setup of field lysimeters for studying effects of elevated ozone and below-ground pathogen infection on a plant-soil-system of juvenile beech (*Fagus sylvatica L.*). Plant Soil 323:7–19

Chapter 18
Effects of Stress and Defence Allocation on Tree Growth: Simulation Results at the Individual and Stand Level

T. Rötzer, T. Seifert, S. Gayler, E. Priesack, and H. Pretzsch

18.1 Introduction

The long life span of trees implies that they are more or less frequently confronted with different biotic and abiotic stress situations during their lives. However, biotic stress such as attacks by herbivores or pathogens and abiotic stress such as frost or drought could strongly vary in frequency, intensity, duration, time of occurrence as well as in the involved tissues. This urged trees to develop flexible defence mechanisms during their evolution ensuring a high probability of survival to regenerate successfully. Based on an analysis of existing literature on plant response to herbivory, McNaughton (1983) concludes that "...the yield of the tissue affected and other tissues is not affected in proportion to the amount of tissues damaged by the herbivore", referring also to Lee and Bazzaz (1980) and Neilsen (1981). McNaughton presents a set of alternative patterns about the effect

T. Rötzer (✉) • H. Pretzsch
Chair for Forest Growth and Yield Science, Technische Universität München, Hans-Carl-von-Carlowitz-Platz 2, 85354 Freising, Germany
e-mail: thomas.roetzer@lrz.tum.de

T. Seifert
Department of Forest and Wood Science, Stellenbosch University, Private Bag X1, 7602 Matieland, South Africa

S. Gayler
Water & Earth System Science Competence Cluster, University of Tübingen, Keplerstr. 17, 72074 Tübingen, Germany

E. Priesack
Institute of Soil Ecology, Helmholtz Zentrum München, Ingolstädter Landstr. 1, 85764 Neuherberg, Germany

R. Matyssek et al. (eds.), *Growth and Defence in Plants*, Ecological Studies 220,
DOI 10.1007/978-3-642-30645-7_18, © Springer-Verlag Berlin Heidelberg 2012

of herbivory; at low herbivory levels growth reactions of individual plants range from immediate decline to a certain compensation. At moderate herbivory even a positive effect and overcompensation can be the result. These described patterns are in line with the phases of stress response defined by Larcher (2003).

Under the conditions of limited resources, a tree has the "choice" either to invest in growth to stay competitive against its neighbours, or to adapt to biotic and abiotic stress to keep the gained resources (see Chap. 1). Koricheva (2002) classified the costs involved in the context of herbivory in three major groups, which are also applicable to other pathogens and stress situations: allocation costs, ecological costs and opportunity costs. Allocation costs define the internal trade-offs between growth, defence and reproduction within the individual plant (Chew and Rodman 1979; Rhoades 1979; Bazzaz et al. 1987; Simms 1992; Seifert and Müller-Starck 2009), while ecological costs are external trade-offs caused by defence investments in other than defence or growth-related properties, such as increased susceptibility against other pathogens or abiotic stress. Other ecological costs are negative effects of the induced defence on pollinators, predators and parasitoids (Simms 1992; Rausher 1996). Opportunity costs are costs incurred through an investment in defence instead of investing in growth. In this sense these are costs of resources to defend against herbivores and pathogens, based on what could have been earned if the resources had been used for growth. Based on published results of Coley et al. (1985), Gulmon and Mooney (1986), and Koricheva (2002) and first simulation studies of Seifert (2007) with an empirical growth model, it could be assumed that opportunity costs have a substantial influence on the consequences of an allocation strategy of a tree and may be even more important than the actual diversion of resources. Opportunity costs involve similarly tree and stand level, and imply a decline of competitiveness of the subject tree compared to neighbouring trees. This effect is caused by the fact that the subject tree cannot use the invested products of the assimilation for growth since they were allocated to defence. This can lead to a feedback cycle that consistently increases the loss of competitiveness over time (Fig. 18.1).

It is obvious that an adapted defence strategy is only one determinant of the allocation strategy that ensures a tree's long-term survival. Growth and space occupation strongly determine the access to above- and below-ground resources, and thus define the absolute amount of resources that can be acquired (e.g. Grams et al 2002; Gayler et al. 2006; Rötzer et al. 2009). The consequence is that social position and tree size in relation to the competing trees are highly correlated with tree growth (e.g. Seifert 2007; Seifert and Müller-Starck 2009). As a result a tree has to adjust its allocation in a way that it allows sufficient defence without losing competitiveness regarding space occupation and resource acquisition. "Allocation" can be defined as biomass partitioning, in the sense of translation of resources to structural components of a plant (e.g. Niklas and Enquist 2002). Defence-related allocations are only partly structural; they involve chemical compounds in a non-structural sense.

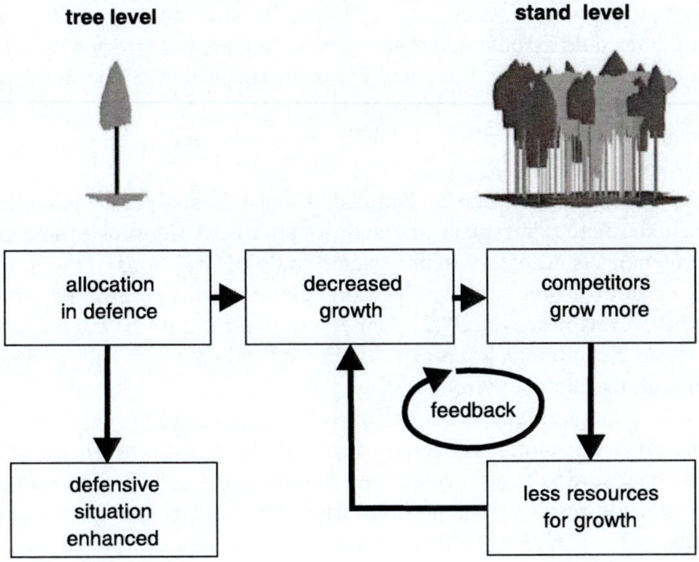

Fig. 18.1 Concept of opportunity costs of a tree as a consequence of defence allocation

During the past years allocation patterns of trees have gained increasing scientific attention. Particularly the trade-off questions of growth and defence, subject to the resource availability of the tree, have been analysed (e.g. Häberle et al. 2009; Chap. 1). Several competing hypotheses have been presented and were highly discussed (Hamilton et al. 2001; Stamp 2003). Most of the empirical and theoretical work is focussed on the allocation pattern within the individual plant. With regards to forest trees, it seems especially rewarding to extend the focus in order to have a closer look at the effects of stress-induced allocation patterns in a competitive environment. This opens up a way to evaluate the long-term success of certain allocation strategies. This approach includes the examination of the effects at the tree and at the stand level alike, to take into consideration that forests are highly complex spatially and temporally determined systems.

There is strong evidence for a feedback of the resource availability on the allocation pattern in defence- or growth-related metabolism (Coley et al. 1985; Herms and Mattson 1992; see also Chap. 17). This interdependence of defence- and growth-related allocation can lead to trade-off situations where the tree grows less in favour of strengthening its defence. The correlation between environmental stimuli and internal allocation processes are still not understood well enough.

Further lack of knowledge becomes evident if the focus of interest is moved from the individual tree to the stand level. Due to the intense interaction of the individuals, the stand cannot be explained from reaction of solitary trees. For example if a certain proportion of trees in a stand has decreased growth in consequence of herbivore attack, or an allocation shift from growth to defence metabolism, one could assume that other trees, which are not affected, will take the opportunity and try to occupy the space. These compensation effects are dependent on the spatial pattern of trees and are especially regarded to be attenuated in mixed species stands, which contribute to their resilience and economic feasibility (e.g. Knoke and Seifert 2008).

From the above stated it can be concluded that stressor–plant interactions are inevitably linked with plant–plant interaction patterns. A rigorous examination of those dependencies is necessary to understand trade-off reactions of the individual. As a result of an extensive meta-analysis on costs of plants confronted with stress as, e.g. herbivores, Koricheva (2002) pointed out that most of the compiled studies were aimed at the internal trade-offs of the individual plant and excluded the interaction with the biotic environment.

To overcome the problems described above an alternative way was followed to evaluate the effects of biotic and abiotic stress on the growth, defence and allocation patterns by resorting to a growth model-based approach. Provided the tree–tree interaction and the resource competition are reproduced plausibly in the model, scenario analysis can give most desired stress scenarios for various types of tree stands with defined spatial stand structures.

In the following a simulation study is presented to analyse the effects of biotic and abiotic stress on growth, allocation and defence at tree and stand level. The scenario simulations were based on the eco-physiological individual-tree growth model BALANCE and include three scenarios with pure and mixed stands. The objectives were to investigate

1. The magnitude of opportunity costs
2. The effects of different intensities and frequencies of defoliation stress and
3. The effects of drought stress.

18.2 The Eco-Physiological Growth Model BALANCE

Growth models–particularly eco-physiological growth models–have been proven suitable tools for analysing the effects of environmental changes on the growth of entire forest stands, individual trees or even specific tree compartments such as roots or stems, as they integrate a wide range of system knowledge (Pretzsch et al. 2008; Fontes et al. 2010). They simulate forest growth on the basis of generally accepted eco-physiological principles (see Chap. 15). Well-known physiological models were developed by Running and Coughlan (1988), Kellomäki et al. (1993), Mohren and van de Veen (1995), Landsberg and Waring (1997), Bossel (1996), and Bartelink (2000).

Apart from the evaluation and understanding of a tree's internal resource allocation and related "decisions" on regulation, these kinds of models can also be used as versatile tools for revealing the effects of defence allocation on the long-term growth and competitive success of trees.

Dedicated spatial and eco-physiologically based tree-growth models, such as the newly developed model BALANCE (Grote and Pretzsch 2002; Rötzer et al. 2005, 2010), allow for scenarios confronting different forest stands with biotic and abiotic stress. The competitive capacity of a tree can be quantified, i.e. the amount of resources which has to be allocated to defence instead of to growth. Eco-physiologically based models give answers about the consequences and implications of these trade-offs. BALANCE is a functional–structural model, which accounts for the influence of competition, stand structure, species mixture, and management impacts on individual tree growth. Tree development is calculated depending on the explicitly spatial environmental conditions of each tree. In turn, each tree is influencing its microenvironment in terms of, e.g. shading or water consumption.

The three-dimensional development of the individual trees in a forest stand is estimated based on the annual increase of biomass. The simulation of the carbon, water and nutrient balances of the individual trees form the core processes of the model. Each tree of a forest stand is divided into crown and root layers, which are in turn subdivided into eight crown- and eight root sectors. For each layer or each sector respectively, the micro-climatic conditions and the water balance are computed daily. Assimilation, respiration, nutrient uptake, growth, senescence and allocation are calculated in 10-day periods from the aggregated weather and water balance variables. This way, the approach takes into account the physiological responses to weather conditions, CO_2 concentration, water and nitrogen availability, as well as biotic and abiotic stress. The seasonal development of foliage is explicitly modelled to include light availability and radiation absorption changes depending on leaf biomass (see also Chap. 8). The beginning of bud burst is predicted based on temperature sums (Rötzer et al. 2004), while foliage senescence is estimated depending on the respiration sum (Rötzer et al. 2010). The individual tree approach of BALANCE facilitates the simulation of pure and mixed stands of various age structures and patterns of species mixtures.

For the initialisation of BALANCE, tree position, stem diameter, crown dimensions and stem height for every tree are required as well as a basic description of the soil (field capacity, wilting point, nutrient status, rooting depth). Daily meteorological values of temperature, radiation, wind speed, humidity and precipitation are used as driving forces of the simulations. Additionally, CO_2 concentration of the air and N-deposition data can be considered. Output is obtained from daily up to annual values in a spatial resolution of individual tree compartments and can be aggregated to tree and stand values. Apart from growth data such as diameter, height or carbon content, data describing stand micro-climate and water balance can also be obtained.

Growth model evaluation was an integral part of the development of BALANCE. The basic processes, i.e. micro-climate within the stand, water balance, photosynthesis and phenology, have been evaluated for the site "Kranzberger Forst" (Rötzer et al. 2010). Further on, numerous validation studies for different sites in Central Europe, for different tree species and different climate regimes have proven that

BALANCE is suitable for growth simulations and analyses after adapting the model to the site (Grote 2002; Grote and Pretzsch 2002; Rötzer et al. 2004, 2005, 2010).

18.3 Scenario Definition

To elucidate the reactions and feedback reactions of growth, defence and competition different scenarios simulating biotic stress are presented: Defined investment proportions for the defence against a pathogen, specific loss of biomass to herbivores, and abiotic drought stress. Both biotic and abiotic stresses were varied with respect to intensity, duration, frequency and time of occurrence. Within the simulations not only individual trees, but also entire stands, different stand structures, i.e. species mixtures and their spatial formations were considered. Each scenario required defining an initial stand with a certain structure. These stands were then confronted with defined biotic and abiotic stress scenarios over the simulation period of 14 years.

18.3.1 Initial Stand Structure

Virtual tree stands (30 m $\times$ 30 m) of different mixing ratios and spatial structures of adult Norway spruce (*Picea abies* (L.) Karst.) and European beech (*Fagus sylvatica* L.) were created based on the structure generator "STRUGEN" (Pretzsch 1997). The mixing structure varied to represent mixture in groups and clusters of beech and a random individual tree mixture (Fig. 18.2).

18.3.2 General Site Conditions

The site conditions used for the scenario simulations were based on data of the SFB 607 site "Kranzberger Forst", located in Southern Germany (48.420°N, 11.662°E, elevation 490 m a.s.l.). For the simulation period of 1994–2007 the mean annual air temperature was 8.1°C, the mean annual radiation 1,096 J/cm and the mean annual precipitation sum 842 mm. Thus, temperature as well as precipitation were at the upper limit of the long-term regional averages (BayFORKLIM 1996), which ranged between 7 and 8°C (period 1951–1980) resp. between 750 and 850 mm (period 1961–1990). Radiation, on the other hand, was somewhat higher than the long-term mean with 1,015 J/cm for the period 1976–1989 (BayFORKLIM 1996). The growing season extended from mid April through the end of October. The soil of the study site—a Luvisol—is described in more detail in Chap. 9. For the calculation of the water balance, we distinguish four soil layers with a maximum rooting depth of 1.2 m.

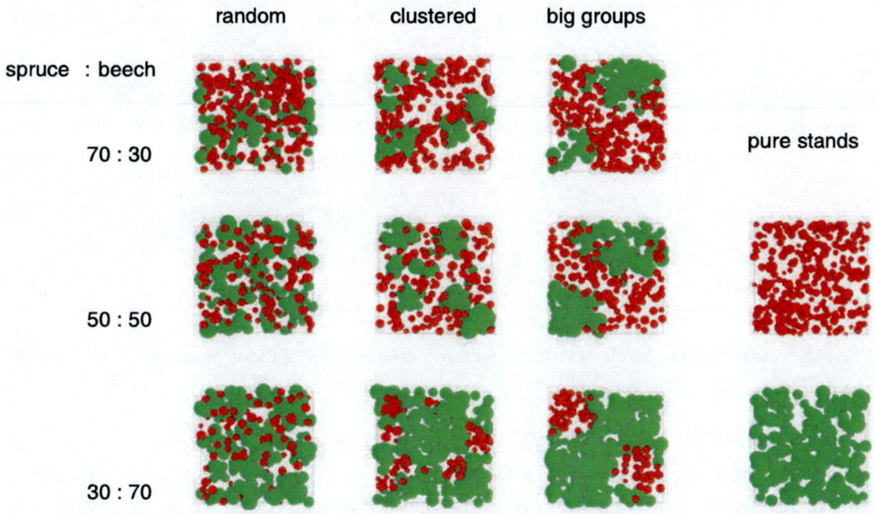

Fig. 18.2 Initial stands for the scenario simulation. The stands differ in the proportion of spruce and beech in the crown projection and the degree of clustering of the tree species (*green circles* represent beech, *red circles* spruce trees)

18.3.3 Stress Scenarios

Regardless whether biotic or abiotic, stressors can occur and lead to disturbances in the growth of individual trees and forest stands. Growth rates may change competition of trees or allocation patterns to a great extent according to the stress intensity, the duration of the stress (short, long or even continuous) the stress frequency (received as a single or repeated pulse), and the phase of stress occurrence (early or late within the growing period) (Fig. 18.3).

For the simulation studies three stress scenarios were defined that varied according to the intensity, duration and frequency of the stress (Table 18.1):

Scenario I (opportunity costs) was set up to investigate opportunity costs of beech in a mixed stand. A constant defence investment rate was deducted from the available net resources for all beech trees while spruce trees were not subjected to defence investments.

In *Scenario II (consequences of defoliation)*, different rates of defoliation were simulated for a pure beech stand. Defoliation was simulated by a proportional removal of leaf biomass equally over all crown parts. A partial replacement of leaves by proleptic sprouting as found in beech was not simulated. Four levels of defoliation intensity were compared with the control. These were combined with three different frequencies of the years of leaf loss in a 10-year period. In the last 5 years no defoliations were simulated.

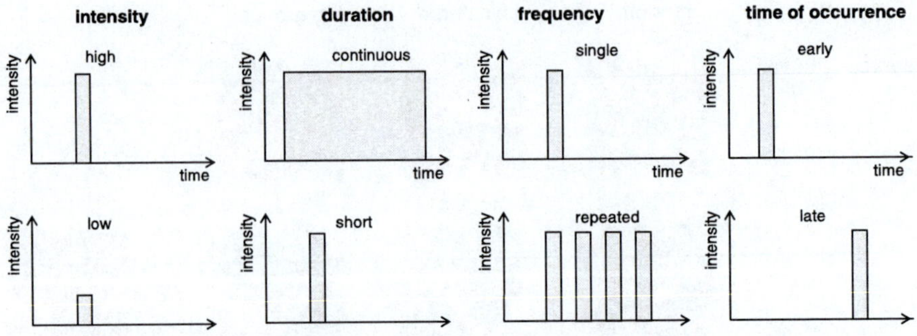

Fig. 18.3 Contrasting levels of stress patterns (intensity, duration, frequency and time of occurrence)

In *Scenario III (effects of drought stress)* years with different intensities of drought periods as well as years with optimal growth conditions were chosen to analyse the effects of water deficit on the growth of individual trees and entire pure and mixed stands. The influence of drought stress on allocation patterns was studied.

18.4 Analysis Methods

The response variables to characterise stress effects were production rates of leaf biomass, fine root biomass, free C-pool (reserve), total structural biomass (above- and below-ground), tree diameter and net primary production. Additionally, allometric relationships were calculated to look for changes in the partitioning. Pretzsch (2009) proposed a set of different ways for empirical diagnosis of growth disturbances of forest trees. Two main methods were applied to analyse the simulation results: a pairwise comparison of affected and unaffected trees and a comparison of the average response with variation on the plot with a control plot.

The first method is the method of choice if the initial stand is homogenous. Compared to empirical studies, the advantage of simulation studies is that the determination of equal pairs of trees is simple since initially identical trees can be simulated *ceteris paribus* with different treatments (Seifert 2007; Pretzsch 2009).

In the case that stand structure is used as an independent variable, the initial stand structure is different. That favours the second method with the average and variance of trees affected and unaffected by stress of a plot to be contrasted with the values of the control plot.

In most cases relative changes of the stress scenarios to the control were chosen to analyse the patterns because the main focus of the simulation study was to reveal the reaction rather than providing absolute magnitudes. The main response variables were the allocation patterns of the trees, and tree growth, both analysed on the individual tree and stand level.

Table 18.1 Stress scenarios that vary in intensity, frequency, duration and time of occurrence of the stress

Scenario	Initial stands	Intensity	Frequency		Duration	Time of occurrence (month)	Number of simulation runs
Scenario I							
"Opportunity costs"	Pure beech	0, 10, 20, 30, 40, 50 % for defence investment	Single		Continuous	–	6
	All (only beech was stressed)	0, 10, 20, 30, 40, 50 % for defence investment	Single		Continuous	–	60
Scenario II							
"Consequences of defoliation at tree level"	Pure beech	% Defoliation	Repetition	Year			
		0	0×	–	–	–	1
		25, 50, 75, 99	1×	1	Pulse	5	4
		25, 50, 75, 99	3×	1, 6, 11	Pulse	5	4
		25, 50, 75, 99	4×	1, 4, 7, 10	Pulse	5	4
Scenario III							
"Drought stress and stand structure"	All	No drought and selected "drought" years	–	–	–	–	11

18.5 Effects of Stress on Growth and Allocation Pattern

18.5.1 Scenario I: Biotic Stress on Stand Level

In the first scenario, a response to a species-specific pathogen was simulated for a mixed stand. The beech trees in the stand were subjected to a constant defence investment while the competing spruce trees were not. For each beech tree the available free carbon pool was reduced by 10–50% to simulate different defence rates. The objectives of the scenario simulation were the quantification of opportunity costs and the investigation of the structural stand variables that influence the opportunity costs. Additionally, a closer look was taken at adaptive changes of structural allometric relationships as a consequence of defence allocations.

In Scenario I, it was assumed that a tree incurred opportunity costs if the predicted growth losses were higher than indicated by the specified defence investment. For example in case of incurred opportunity costs a tree with a defence investment of 30% of the available net resources should lose more than 30% of its growth, compared to the control tree, which was not subjected to that investment. Over time compound interest effects have to be considered.

18.5.1.1 Quantification of Opportunity Costs

Economic theory can be employed to analyse investment patterns in plants (Bloom et al. 1985). An adequate response variable to assess possible opportunity costs is the total biomass of the tree and its changes because it is unaffected by shifts in the partitioning between the tree compartments. A cross-sectional analysis of all Scenario I simulations revealed a significant drop in biomass production after 14 years of simulation (Fig. 18.4). The results indicate an effect beyond the relative reduction of net resources. At 10% defence rate beech trees showed a decrease in total biomass of 16.1% on average compared to the control. At 50% defence rate, 67.4% of the biomass was lost. These results strongly suggest the presence of compound interest effects caused by losses of competitiveness in space occupation and resource acquisition of trees investing in defence instead of growth, which leads to opportunity costs.

The spruce trees profit from the decreased growth of the beech trees, but from a beech defence investment rate of 30% upwards, spruce trees were not able to compensate for the decreased growth of beech trees any more. However, we have to take into account the "compound interest effect". A quantification of the annual costs instead of the accumulated costs over the 14 years is therefore necessary to investigate the presence of opportunity costs. In the simulated early growth phase trees grow with nearly constant increment rates. Thus in approximation the well-known concept of the compound interest rate can be adapted from economics (Eq. 18.1)

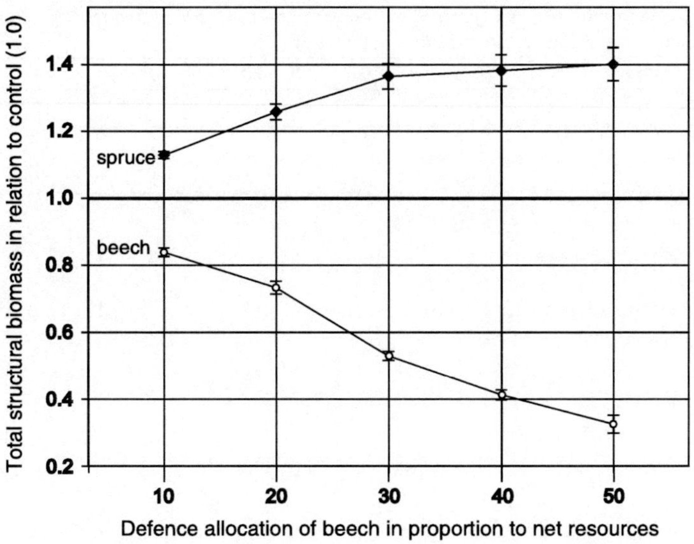

Fig. 18.4 Predicted deviation of total structural biomass of beech and spruce in a simulated mixed stand in relation to the control (1.0 line): Beech trees are subjected to different defence investment proportions (*lower line* with *circles*). They are compared to the reaction of the non-investing spruce trees in the same stand (*upper line* with *diamonds*)

$$i = (\mathrm{FV}/\mathrm{PV})^{1/n} - 1, \tag{18.1}$$

where i is the compound interest rate, PV is the present value, FV the future value, and n the time in years.

Translated to biological terms PV is the tree biomass at the beginning and FV the biomass at the end of the simulation period which will be denoted by x and y resp. further on. The compound interest rate i represents the annual biomass increment. In our case two effects influence biomass increment: the given defence investment d, and c, the true opportunity costs that are incurred by the loss in competition. It is important to note that the opportunity costs c are incurred additionally to the defence investment d. In order to obtain c Eq. 18.1 has to be extended (Eq. 18.2)

$$y_\mathrm{d} = x \cdot (1 + i \cdot (1 - d) \cdot (1 - c))^n, \tag{18.2}$$

where y_d is the final biomass with defence investments considered according to the simulation; x is the initial biomass, i can be obtained from the control simulation without defence investment (Eq. 18.3) and d is the given defence investment percentage divided by 100; n represents the simulation period (14 years)

$$y_\mathrm{c} = x \cdot (1 + i)^n, \tag{18.3}$$

where y_c is the final biomass of the control simulation, without defence investment.

Table 18.2 Opportunity costs expressed as competition-based growth reductions c in relation to direct defence investments d (given in the first column)

Defence investment	%	10	20	30	40	50
Opportunity costs	Mean	1.40	1.35	1.87	2.63	3.08
95% Confidence interval	Lower	1.25	1.23	1.79	2.39	2.98
	Upper	1.57	1.51	1.97	3.03	3.18

By obtaining i from Eq. 18.3 and inserting it in Eq. 18.2, Eq. 18.2 can be solved for c (Eq. 18.4)

$$c = 1 - \frac{\sqrt[n]{\frac{y_d}{x}} - 1}{i \cdot 1 - c}. \tag{18.4}$$

The results are presented in Table 18.2, where the opportunity costs (competition effects c) are calculated in proportion to the defence investments d.

The scenario results suggest that the relative importance of opportunity costs compared to allocation costs increase in a nonlinear fashion with increasing defence rates, ranging from 1.5 at 10% defence investment to more than 3.0 at 50% defence investment. This means that opportunity costs caused by competition effects were always higher than the direct defence investments.

18.5.1.2 Influence of Stand Structure on the Opportunity Costs

A particularly relevant question in mixed stands is how stand structure and interspecific competition influence the opportunity costs, because this determines the long-term competitiveness of a tree species. It also gives an indication for practical forestry on how stand composition impacts growth, in situations where one species is stressed in a mixed stand. An analysis of the simulation results of Scenario I was carried out based on multiple linear regression. The objective was to determine the influence of the beech proportion and the spatial distribution of beech (mixing pattern) within the stand on biomass growth. As a response variable, the biomass at the end of the simulation period in relation to the control was selected (Eq. 18.5)

$$\text{Biomass}_{\text{proportion}} = a + b \cdot \text{defence}_{\text{proportion}} + c \cdot \text{beech}_{\text{proportion}} + d \cdot \text{mixing pattern}, \tag{18.5}$$

where a, b, c and d are fitted regression parameters.

Again beech was the tree species subjected to different degrees of defence allocation.

The results in Table 18.3 indicate for beech that the deviation of biomass from the control is significantly influenced by the proportion of beech in the stand.

Table 18.3 Multiple linear regression analysis of simulation results in Scenario I: Influence of proportionally invested resources in defence (Defence proportion), beech proportion in the stand and spatial species distribution (Cluster) on the total biomass in relation to the control

		Unstandardised coefficients		Standardised coefficients		
Species	Model parameter	*B*	Std. error	Beta	*t*	Sig.
Spruce	*a* (Constant)	1.276	0.037		34.787	0.000
	b Defence proportion	0.007	0.001	0.280	11.623	0.000
	c Beech proportion	0.000	0.001	0.017	0.716	0.474
	d Cluster	−0.099	0.010	−0.236	−9.786	0.000
Beech	*a* (Constant)	0.826	0.019		42.427	0.000
	b Defence proportion	−0.014	0.000	−0.733	−46.953	0.000
	c Beech proportion	0.002	0.000	0.213	10.916	0.000
	d Cluster	0.008	0.004	0.037	1.891	0.059

The spatial distribution of the mixture (random, cluster, group) is only significant at the 0.1 level. An increased beech proportion leads to less severe growth reductions, because the interactions with spruce trees were decreased. For spruce, which was not subjected to defence investments, the beech proportion in the mixture was non-significant while the distribution of species had a highly significant influence. Spruce trees seem to profit from a random individual tree mixture much more and are able to gain competitiveness over the beech in our scenario. The bigger the beech groups, the less attenuated the advantage of spruce.

18.5.1.3 Changed Allometric Relations as a Result of Defence and Different Stand Structures

It is important to consider the cause of the changes at the individual tree level. Changes in structural partitioning become obvious if the allometric relationships of the trees are changed. Based on the results of Scenario I, allometric functions were calculated that relate the diameter at breast height (dbh) to the total tree biomass. Based on a double logarithmic function: $\ln(\text{biomass}) = a + b \cdot \ln(\text{dbh})$ the influence of the beech proportion and the way the trees are mixed spatially (mixing pattern) on the allometric slope parameter b were assessed for each of the stands in Scenario I. The database for the analysis was again the situation at the end of the 14 years simulation period.

The regression fits of the simulation results with a second-degree polynomial suggest that beech trees subjected to defence allocation react to increasing inter-specific competition with a changed allometry. The slope parameter showed a correlation to the beech proportion (Fig. 18.5, left). In the stand with the 70:30 spruce/beech proportion, the allometric relation had the lowest slope parameters indicating the lowest biomass per dbh. The biomass per stem diameter in beech is reduced with increasing amount of spruce in the stand. Interestingly, the allometric

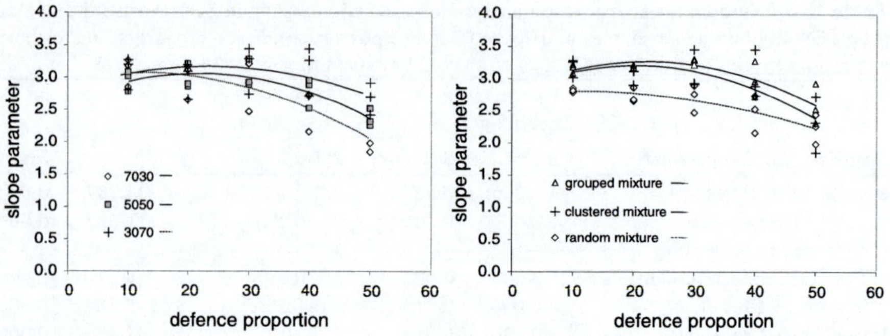

Fig. 18.5 Influence of species' mixing ratios (*left*) and spatial mixing structure (*right*) on the allometric relationship of beech between dbh and total biomass (the first two digits in the species code indicate the proportion of spruce, the second two the proportion of beech; e.g. 7030 codes for 70 % spruce and 30 % beech)

slope values peek according to the proportion of net resources that are allocated to defence. At 70:30 spruce/beech proportion the peek is around 15% defence proportion, 50:50 proportion led to a peak at around 20% and with 30:70 spruce/beech proportion in the stand the peek was around 25%. The influence of the interspecific mixture of the two species is similar (Fig. 18.5, right). The more intensively the individual trees of the two species intermingle spatially, the more the allometry is changed and the biomass per dbh of beech is reduced. The statements made on the nonlinearity of the correlation and the maximum of the slope parameter at the examination of the beech proportion are equally valid for the clustering structure as for the independent variable.

18.5.1.4 Scenario I: Discussion

Our results suggest that opportunity costs in their compound interest effect are a relevant factor for trees, confirming results obtained by Koricheva (2002) and Seifert (2007). The observed pattern suggests that competition-related opportunity costs have a bigger effect on annual biomass increment than the defence investments themselves. Moreover, opportunity costs increased over-proportionally with increasing defence investments. In the mixed scenario decreased increments of the defence investing beech trees were compensated for by the non-investing spruce trees, if defence investments of beech did not exceed 20–30%. The proportion and type of mixture of affected trees in a stand influenced growth losses and compensation, and were reflected in changed allometric relationships between stem diameter at breast height and biomass of the affected species. The allometric changes indicate an adaptation of the allometric pattern of beech according to defence investment and the amount of beech in the stand with an interaction between both variables. Spruce trees were strong competitors for resources because they were not subjected to a defence allocation in this scenario. With an increasing proportion of spruce trees in the stand, the maximum biomass per diameter of beech trees is

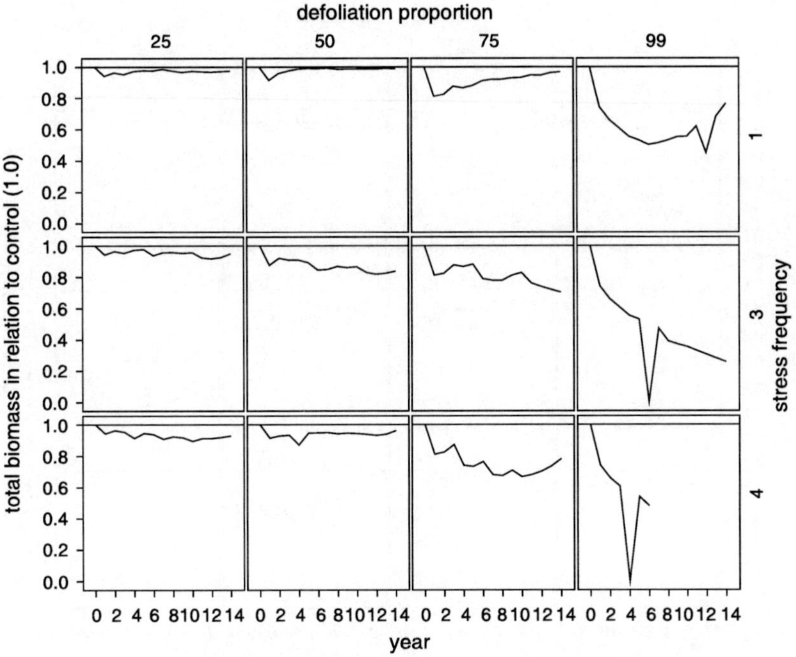

Fig. 18.6 Effects of proportion of defoliated leaves and the frequency of the defoliation on the total biomass in relation to the control

reached at increasingly smaller defence rates. This reaction pattern shown in the simulations is supported by the fact that intensive intermingling of the species resulted in decreased biomass per dbh of the beech trees.

18.5.2 Scenario II: Biotic Stress on Tree Level

18.5.2.1 Influence of Defoliation Patterns

The second scenario aims at investigating the influence of different defoliation patterns on tree growth and partitioning at the individual tree level. Using the total biomass in relation to the control as a response variable, the simulation results show that low intensities of defoliation like 25 and 50% at low frequencies (only once in the simulation period) led to a recovery but no overcompensation of biomass (Fig. 18.6). Also three consecutive defoliations with 25% leaf loss were followed by a near complete recovery. Higher defoliation rates, especially in combination with higher frequencies, resulted in a decline in total biomass compared to the control which only recovered gradually after defoliation was stopped, or otherwise lead to a permanent decline (Fig. 18.6.).

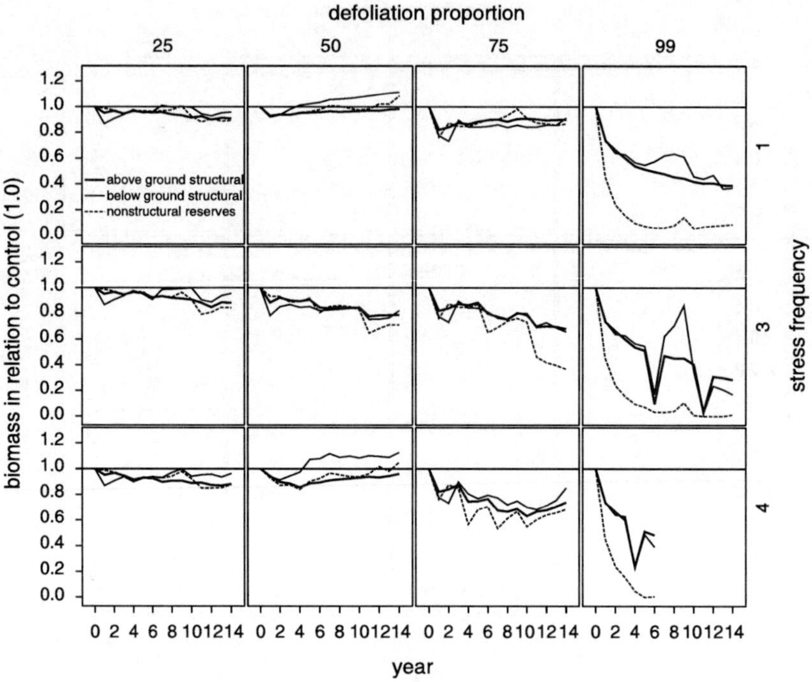

Fig. 18.7 Effects of proportion of defoliated leaves and the frequency of the defoliation on the above-ground biomass (*bold line*), below-ground biomass (*thin line*) and reserves (*dotted line*) in relation to the control

A closer look at the pattern of above-ground, below-ground, and reserve allocation reveals that the simulation results propose a more complex reaction (Fig. 18.7). While defoliation at any intensity and frequency was detrimental to the predicted relative above-ground biomass, below-ground biomass and free C-pool showed a distinctive overcompensation (values greater 1.0) at certain combinations of defoliation intensities and frequencies, as a result of internal interactions of the model.

18.5.2.2 Scenario II: Discussion

The results obtained in Scenario II would favour McNaughton's second hypothesis (McNaughton 1983) that suggests compensation to a certain extent, but not overcompensation at low levels of herbivory. Decreases in diameter and volume increments were often described as proportional to the needle losses in conifers (Craighead 1940; Bruce 1956; Chalupa 1965) while other authors negate proportionality in a strict sense (Fournier et al. 2010). Brubaker (1978) stated that no significant decrease in ring width occurred if more than 50% of needles were remaining, which can be seen as an argument for a nonlinear relationship, where small needle losses do not lead to proportional diameter growth losses and with increasing defoliation the growth losses increase progressively. A similar reaction pattern was apparent in our simulation results for beech. Our findings confirm the empirical results of Krause and Raffa (1996), who compared the growth and recovery rates of 10-year-old deciduous and evergreen conifers after defoliation. They found a quick recovery for deciduous *Larix decidua* in contrast to evergreen *Pinus resinosa* trees. Pines with 66% defoliation rates never recovered with regards to above-ground biomass. They attributed the differences partially to the plastic architectural response of *Larix* but also to the differences of nutrient and carbon distribution in evergreen and deciduous conifers. Beech may be more comparable with *Larix* as it is also a deciduous tree. There is ample evidence from defoliation studies that leaf and needle losses are nonlinearly related to diameter growth and above-ground biomass (Kulman 1971).

The overcompensation of below-ground biomass, predicted by BALANCE, requires some interpretation. BALANCE does not allow for efficiency changes in the chemical pathways of photosynthesis itself, but is sensitive to changed leaf areas, modified root–shoot ratios and higher light and water availability. However, the observed effects suggest a complex interaction on the eco-physiological levels covered by the model, which includes extrinsic and also intrinsic mechanisms to a certain extent as discussed by McNaughton (1983).

18.5.3 Scenario III: Abiotic Stress (Drought)

In the Scenario III simulations, the influence of abiotic stress on growth and competition of individual trees and entire forest stands was analysed. Abiotic stress that changes tree and stand growth includes, for example, storms, drought or heat. In this study, we concentrate on the influence of drought. The analysis provides answers to the questions: (1) What is the influence of drought stress on biomass growth? (2) Does drought stress change the allocation patterns of trees (free C-pool, above and below-ground biomass)? and (3) Do stand structure and interspecific competition influence the growth of individual trees resp. entire stands in drought periods?

18.5.3.1 Quantification of Drought Stress

To assess drought quantitatively and qualitatively for the single years of the period from 1999 to 2007, several drought indicators were analysed. As a first indicator a meteorologically based index, i.e. the number of days with precipitation sums lower than 0.1 mm within the summer months June to August (n_{NIE0}), was used. In Fig. 18.8a, the annual deviations of n_{NIE0} from the mean of the years 1999–2007 are given. The year with the highest positive deviation, i.e. the year with the most days without precipitation, was 2003. Other dry years were 2002 and 2006, while

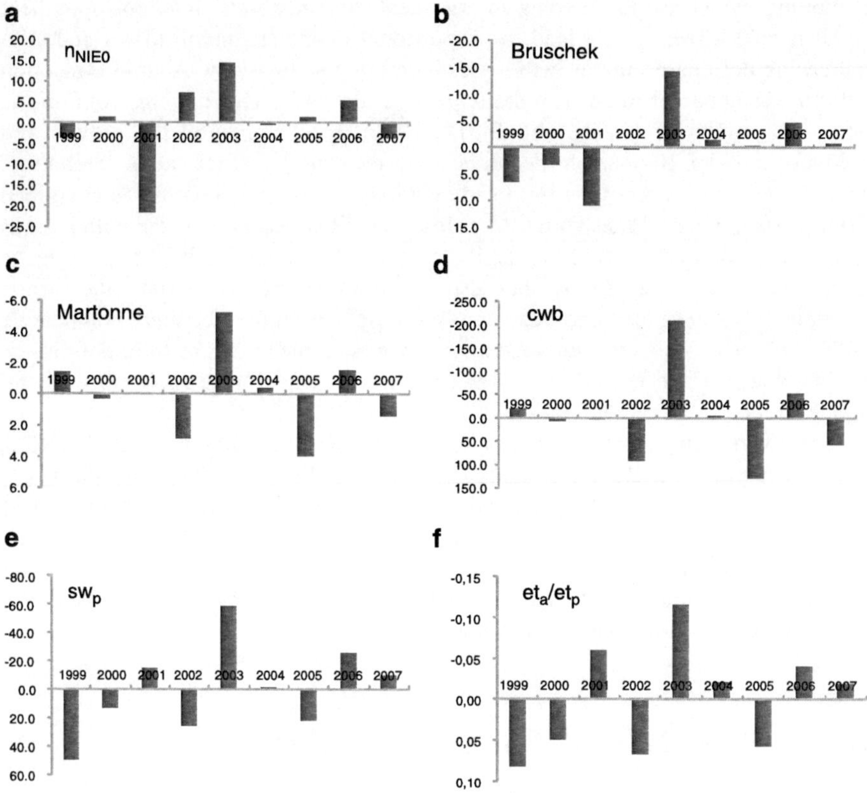

Fig. 18.8 Anomalies of different drought indices (**a** = n_{NIE0}, **b** = Bruschek, **c** = Martonne, **d** = cwb, **e** = sw_p, **f** = et_a/et_p) for the years 1999–2007 from the average of 1999–2007 for the site Kranzberger Forst

2001 was the year with the lowest number of n_{NIE0}. If, however, other climate parameters that influence the water balance of a site are additionally taken into account, the ranking of years with drought is changing. The drought indices of Bruschek or De Martonne consider temperature as well as precipitation. While the Bruschek index is based on the annual precipitation sum and the number of summer days, i.e. the number of days with temperatures above 25°C (Bruscheck 1994), the De Martonne index is calculated from the precipitation sum and the temperature of the summer months June, July and August (De Martonne 1926). Both indicators show the highest drought stress in 2003 and a lighter drought stress in 2006, followed by the years 2004 (Bruschek) and 1999 (De Martonne) (Fig. 18.8b, c).

A more sophisticated index is the climatological water balance cwb, calculated from the difference between precipitation sum and potential evapotranspiration. Based on the values for the summer months, the order of the drought stress intensity is again 2003 followed by 2006 and 1999 (Fig. 18.8d).

By using the plant available soil water sw_p as an indicator for drought, site conditions other than the meteorological conditions are also considered (Fig. 18.8e). The annual anomalies of the summer month June to August against the average were highest for the year 2003 and 2006. If the ratio of actual to potential evapotranspiration et_a/et_p is used 2003 indicated the strongest drought (Fig. 18.8f). Hereby the water balance parameters $sw_{p,}$, et_a and et_p were simulated with BALANCE for the beech stand using the above-mentioned site and weather conditions. In contrast to the above drought indices, positive anomalies based on sw_p and the et_a/et_p ratio were also found for the years 2001 and 2007. This is because the favourable growth conditions in these 2 years increased growth and in consequence also actual transpiration, which lowered et_a/et_p ratios and sw_p values.

Based on these results the year 2001 was selected as a year with sufficient water availability, the year 2006 as a year with a moderate drought and the year 2003 with a severe drought in order to analyse the influence of drought on tree growth and competition.

18.5.3.2 Influence on Individual Tree and Stand Level

To compare total biomass development of single years, simulated total biomasses of 10-day periods were related to the initial biomass of each year. Figure 18.9 presents the results for the pure beech stand.

Starting in May biomass development of the year 2001 increased stronger than the mean values of the period 1999–2007. Already in the spring months growth rates were smaller compared to the average for the moderate dry year 2006. In the severe dry year 2003 beech tree growth in spring was somewhat higher than on average. However, in the mid of June growth rates dropped severely. At the end of the dry years 2006 and 2003, there was no detectable increase in the stand biomass compared to the initial biomass.

Compared to the average biomass increment (1999–2007) the biomass increment of the beech stand was increased by 1.8% in the year 2001 while it was clearly lower in the drought years with 4.5% (2006) respectively 4.9% (2003). Spruce, on the other hand, was more affected by the year 2003 with a decrease of 7.1% compared to the mean growth rate. In 2006, biomass growth was lowered by 1.5%. In 2001, a year with good growing conditions, the growth rates of spruce were 4.1% higher compared to the average.

An essential factor of the production of forest stands is the condition of each tree, i.e. its individual light regime, nutrient status and water supply. Figure 18.10 illustrates that at the individual tree level the influence of droughts on the growth rates of trees varied strongly. In dependence on the tree size, denoted as dbh, net primary productivity (npp) increased clearly with tree size.

As can be seen from Fig. 18.10, particularly in the year 2001 with good growth conditions individual trees developed huge differences in npp. Small-sized trees

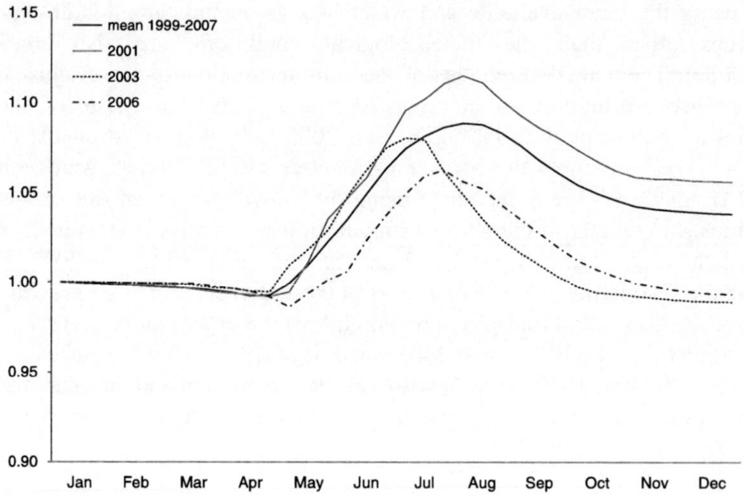

Fig. 18.9 Relative total biomass development of a pure beech stand at the site Kranzberger Forst for single years and on average

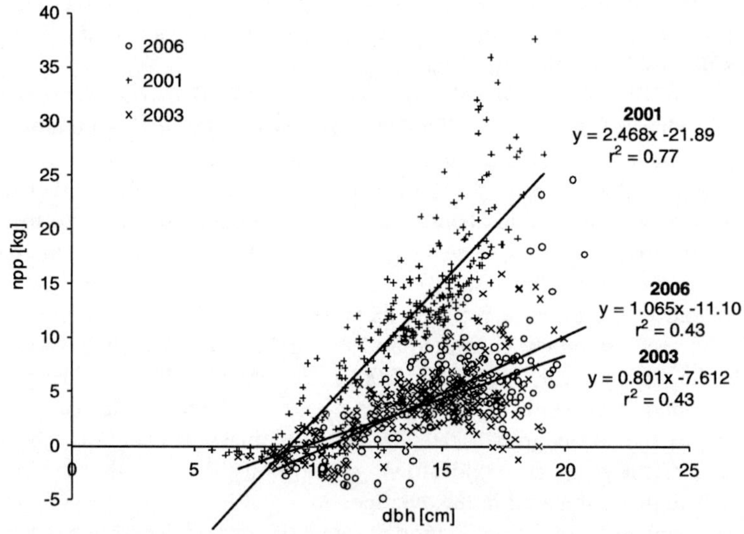

Fig. 18.10 Net primary productivity of individual beech trees at the site Kranzberger Forst for the years 2001, 2003 and 2006 in dependence on their sizes (dbh); *lines* denote the linear regressions of the 3 years

had npp values smaller than 5 kg, big-sized trees, on the other hand, had rates above 15 kg. In years with drought, the slope of the regression line clearly dropped. While small trees still grew with the same increment rates as calculated for 2001, big trees reduced their growth rates severely (Fig. 18.10).

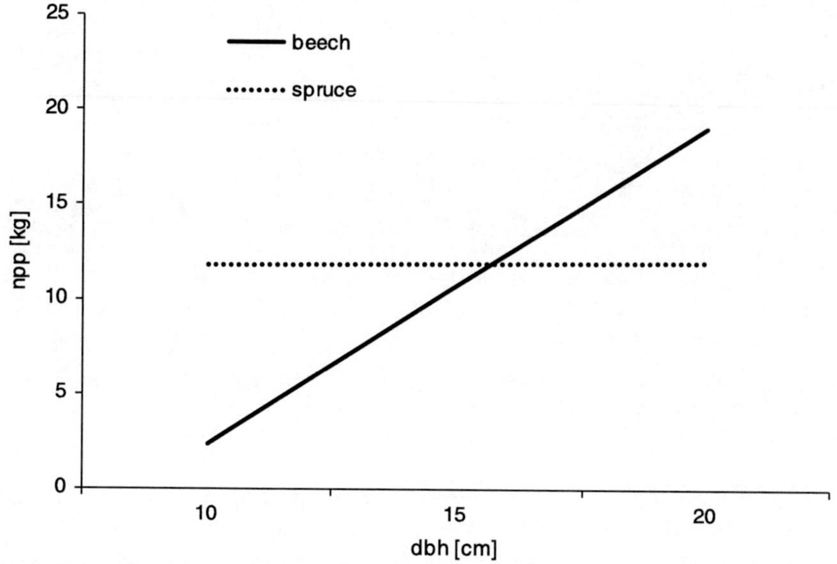

Fig. 18.11 Regression line of the npp differences between the year with favourable growing conditions (2001) and the drought year 2003 for beech and spruce trees at the site Kranzberger Forst depending on tree size (dbh)

To elucidate the loss of npp based on tree size in a drought year compared to a year with favourable growing conditions, the differences of the linear regression lines of the years 2001 and 2003 were calculated (Fig. 18.11). While the spruce trees of the Kranzberger Forst lost approximately 12 kg per tree in all size classes, small-sized beech trees of about 10 cm in dbh decreased npp in 2003 by 2.5 kg per tree compared to 2001. The npp of big beech trees with a dbh of 20 cm, on the other hand, decreased by 19 kg per tree.

18.5.3.3 Influence on Individual Tree and Stand Level: Discussion

Higher growth rates of spruce compared to beech in years with suitable growing conditions are well known (e.g. Pretzsch 2009). This is reflected in our simulation results since the growth of spruce is reduced more than the growth of beech in intensive drought years. These findings confirm the results reported by Beierkuhnlein and Foken (2008). Additionally, Pichler and Oberhuber (2007) found for the year 2003 a 30% reduction of the radial tree growth for spruce compared to the reference period (1998–2002). The annual energy flux measurements of Grünwald and Bernhofer (2007) for the spruce forest Tharandt show a distinctive lower net ecosystem productivity of 395 g C/m in 2003 compared to 2001 with 559 g C/m. Based on the long-term mean for the years

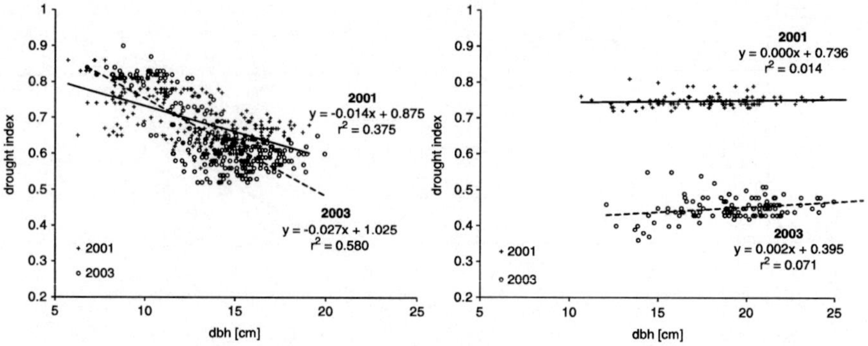

Fig. 18.12 Drought index et_a/et_p of beech (*left*) and spruce (*right*) at the site Kranzberger Forst for 2001 and 2003; *lines* denote the linear regressions for the single years

1996–2005 this is a 30% reduction in 2003, but no reduction in 2001. An increase of the above-ground biomass with tree size is also reported by Ogaya and Penuelas (2007) for a holm oak forest using initial above-ground biomass as a representative for tree size.

The growth patterns for different sized trees under stress and no stress conditions can also be seen in the studies shown in Chap. 14, where size–growth relationships in forest stands are analysed, amongst others under drought events. Pichler and Oberhuber (2007) found at a dry mesic site for dominant spruces the strongest growth reductions in the year 2003 compared to the 1998–2002 period. The results of Ogaya and Penuelas (2007) were similar to our results: The above-ground biomass increment was clearly lower under drought conditions averaged over all species; the slope of the regression line for the above-ground biomass increment based on tree size was higher for non-drought conditions compared to the drought conditions averaged over the species. The response of the single tree species, however, was quite different (high for *Arbutus unedo* and *Quercus ilex*, low for *Phillyrea latifolia*).

The simulation results for npp based on tree size in drought years (Fig. 18.11) were in accordance with findings of Peterken and Mountford (1996) that 30-year-old beeches (small trees) were largely unaffected by droughts, while the growth of 75- to 105-year-old beeches (big trees) declined or ceased. Further on, Dohrenbusch et al. (2002) reported that spruces of the dominant social classes reacted more on low water supply than suppressed trees.

A reason for the different growth patterns of beech and spruce depending on tree size is related to their water supply. The et_a/et_p ratio can serve as an indicator for the water supply of a tree (Fig. 18.12). For beech trees a clear decrease of the et_a/et_p ratio with increasing tree size (dbh) for the year 2003 and 2001 is evident, which means that big trees have higher water deficits.

Spruce trees, however, showed no correlation of the drought index to tree size. For the year 2001 the ratio oscillated around a value of 0.75, which denotes good water supply for all size classes. In 2003 the et_a/et_p ratios ranged between 0.36 and 0.44, which indicates severe drought stress for all size classes. For beech trees such low ratios were only found for big trees in the year 2003. The 2003 drought stress for all tree sizes of spruce caused low npp values in 2003, while in 2001 the high et_a/et_p ratios for all tree sizes (indicating no water stress) caused equally high npp values. This implies that drought stress changed growth rates of the individual trees, of the entire stands, but also of the tree species to a different extent. Longer stress periods can therefore change stand structure and stand growth (e.g. Peterken and Mountford 1996; Dohrenbusch et al. 2002; Ogaya and Penuelas 2007). The assumption of Dohrenbusch et al. (2002) that a worsening of the environmental conditions, for example drought, mainly affects the dominant trees of a stand compared to the dominated was confirmed for beech, but not for spruce. It seems that the 2003 drought intensity was so strong that all tree sizes of the spruces were similarly influenced.

18.5.3.4 Effects on Allocation Patterns

Up to now, the influence of abiotic stress is shown on individual tree and on stand level. But there is also evidence (e.g. Polomski and Kuhn 1998; Frank 2007; Pichler and Oberhuber 2007; Rötzer et al. 2009) that stress such as droughts change the allocation pattern of plants. In two examples, BALANCE simulations demonstrate allocation changes of a beech stand.

Figure 18.13 presents the relative annual development of the free C-pool averaged over the years 1999–2007 as well as for the single years 2001, 2003 and 2006. In 2001, a year with good growing conditions, from mid May onwards the free C-pool increased more than the long-term mean. In the moderately dry year 2006 with the beginning of May, the free C-pool increased slower than on average.

In the dry year 2003, the development of the free C-pool nearly matched the average values in the first months. Since the beginning of July, however, the free C-pool had clearly dropped. At the end of the year 2003, the C-pool was 19% lower than at the end of the year 2001.

Fine root biomass was also clearly influenced by drought stress. Figure 18.14 depicts the development of the fine root growth again for the long-term mean and the three chosen years. Higher growth rates for the fine roots than on average can be found in all years. In the course and at the end of the year, the highest amount of fine root biomass was available in the severe dry year of 2003. However, there was only a small difference in the fine root growth rate at the end of the year 2003 compared to 2001, a year with favourable growing conditions.

If the fine root to leaf biomass ratio of the beech trees of the single years were combined with a drought stress indicator like the et_a/et_p ratio, a close relationship ($r^2 = 0.59$) is obvious (Fig. 18.15). Increasing drought stress, i.e. decreasing et_a/et_p values, stimulated fine root to leaf biomass ratio positively.

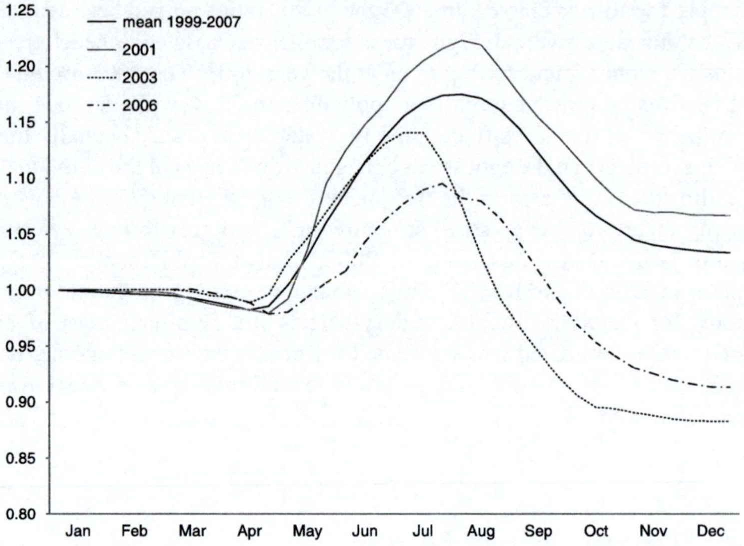

Fig. 18.13 Relative course of the free C-pool of beech at the site Kranzberger Forst

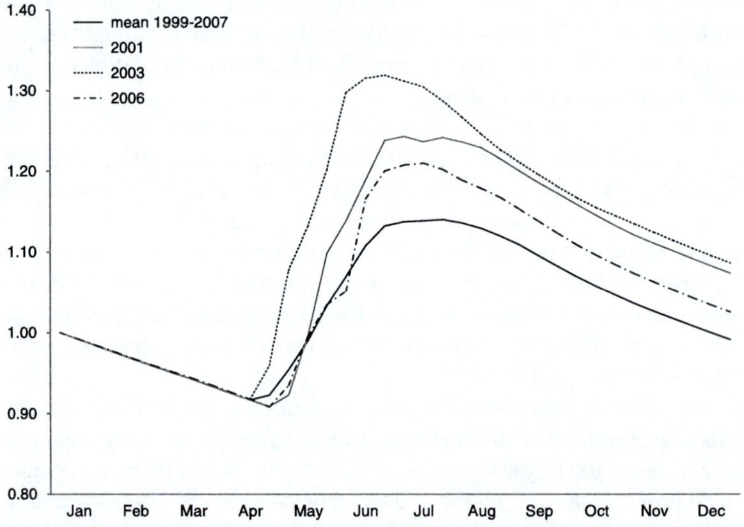

Fig. 18.14 Relative course of the fine root growth of beech at the site Kranzberger Forst

18.5.3.5 Effects on Allocation Patterns: Discussion

Nikolova et al. (2009) found for the mixed beech/spruce stand in the Kranzberger Forst similar levels of fine root production and fine root recovery for beech in the year 2003 compared to 2002. For spruce, on the other hand, these values were

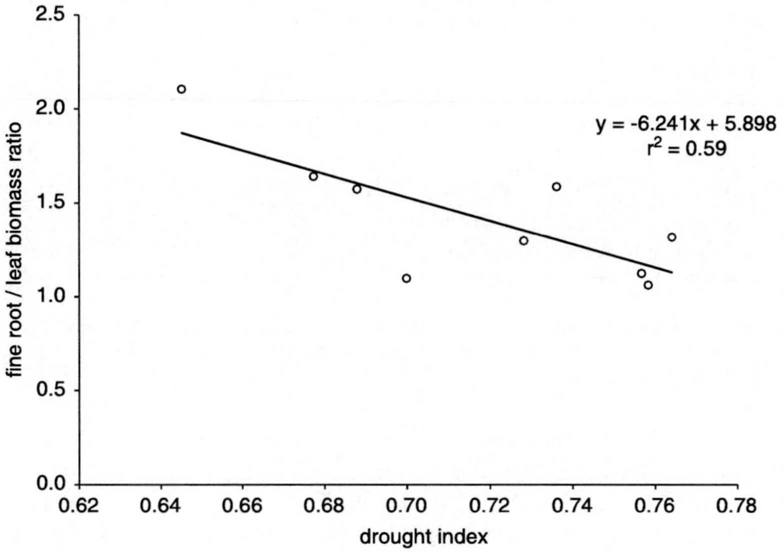

Fig. 18.15 Simulated fine root/foliage ratio and drought index et_a/et_p of the vegetation period for beech

significantly lower in 2003. Regarding that both years 2001 and 2002 show quite low drought indices (Fig. 18.8b–f), the results for beech of Nikolova et al. (2009) are consistent with the simulation results. Konopka et al. (2007) found that fine root biomass of Japanese cedar decreased by drought treatment. They refer to Eissenstat (1997), who stated that "plant root responses to drought are diverse, ranging from shedding of roots to the stimulation of root production". Konopka et al. (2007) further argued that these strategies are modified by the intensity and duration of a drought. This is in accordance with our results (Fig. 18.14), and with the conclusion of Mainiero and Kazda (2006) who found no evidence that fine root formation of beech counter-balances short-term soil water shortages.

Leuschner et al. (2001) stated that fine root growth is more vulnerable to soil water shortage than leaf expansion and photosynthesis, causing (inter-annual) shifts in the carbon allocation patterns of trees. This suggestion is confirmed by the simulation results as shown in the Figs. 18.13 and 18.14 for the free C-pool and the fine root biomass. Despite the fact that fine root biomass greatly varies among tree species, forest type and climate (e.g. Vogt et al. 1996; Noguchi et al. 2007), there are also significant year to year changes. Our results (Figs. 18.9, 18.13 and 18.14) support these findings.

Increased root growth as observed in the simulations in response to drought (Fig. 18.15) was reported by several authors (e.g., Cienciala et al. 1994; Leuschner et al. 2001; Frank 2007). This is in conformity with the plant allocation theory. This theory states that above- and below-ground biomass is allocated in order to

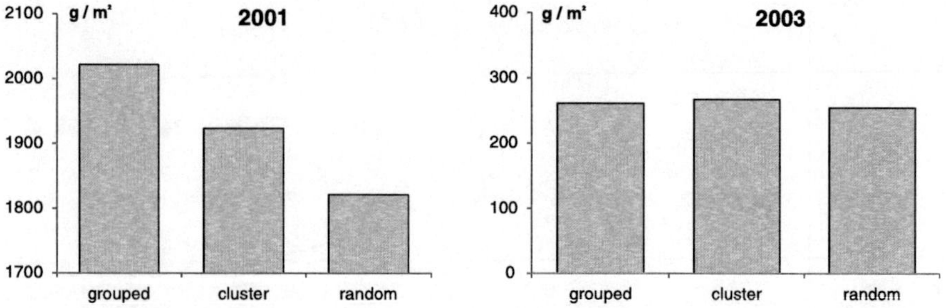

Fig. 18.16 Net primary production of the entire stand for different mixing structures (average of mixing ratios) in the year 2001 with good growing conditions (*left*) and in the dry year 2003 (*right*) at the site Kranzberger Forst

minimise resource limitation resp. to maximise resource capture (e.g. Chapin 1980). Vivin and Guehl (1997), however, report that the root to shoot biomass ratio of *Quercus robur* seedlings decreased by drought, which contradicts the theory that under limiting water supply increased root growth would be advantageous for the acquisition of water (Tyree and Alexander 1993).

In dry periods more carbon was invested in root growth of beeches compared to leaf growth (Fig. 18.15) indicating that more fine root biomass is formed in the trees' search for resources (water). These changes in the carbon allocation from above- to below-ground guarantee the water supply during drought (Cermak et al. 1993; Leuschner et al. 2001). Noguchi et al. (2007) found that the ratio of fine root biomass to total biomass of different tree species clearly increased in drought periods pointing out a biomass allocation to the fine roots.

18.5.3.6 Influence of Stand Structure

Other factors that influence growth of forest stands in drought periods are stand structure and interspecific competition. Based on three different mixing ratios (spruce to beech 30:70, 50:50 and 30:70) and three different mixing structures (random, clustered, grouped), the growth of mixed stands was simulated. Figure 18.16 shows the net primary production of the mixed stands for the three mixing proportions averaged over the mixing ratios for the year 2001 and 2003.

Again, distinctively smaller net primary production values in the drought year 2003 compared to the 2001 values were evident. In 2003 only about 14% of the npp of 2001 was gained by the different forest stands. When the effects of the different mixing structures on npp were compared, decreasing values were obvious in 2001, starting with a grouped mixing structure over a clustered structure to randomly distributed beeches and spruces. The low npp of the randomly structured stand was due to the low npp of spruce that cannot be compensated by the higher npp values of beech in the randomly structured stand. In 2003, all three mixing structures produced nearly the same (250–270 g/m). Hereby, the spruces had the highest

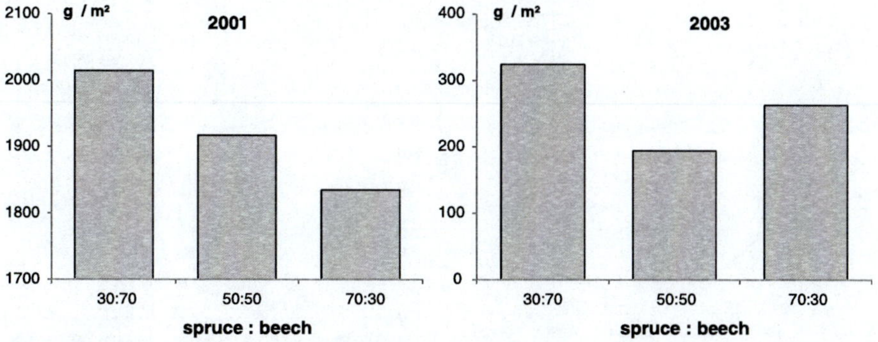

Fig. 18.17 Net primary production of the entire stand for different mixing ratios (average of mixing structures) in the year 2001 with good growing conditions (*left*) and in the dry year 2003 (*right*) at the site Kranzberger Forst

npp reductions (in relation to the 2001 values), particularly the grouped mixing structure. Because there was also a clear drop in the npp of the beech trees in all mixing structures, npp of the year 2003 fell to a low level showing no differences between the mixing structures.

If the mixing ratios were analysed (Fig. 18.17), in the year 2001 with good growing conditions the stand with a spruce beech ratio of 30:70 had the highest npp, followed by the 50:50 ratio and the 70:30 ratio. The differences between the mixing ratios, however, were small, less than 9% based on the maximum. In the dry year 2003 the lowest npp was found for the 50:50 mixture, which was 40% less compared to the 30:70 mixture and 26% less compared to the 70:30 mixture. Generally, npp rates were clearly lower in 2003 than in 2001.

Based on npp and on the et_a for the different mixed forest stands the water use efficiencies (wue), i.e. the dry matter production per kg of water used for evapotranspiration, were calculated for the years 2001 and 2003 (Fig. 18.18).

The lower the proportion of beeches in the stand the lower is the wue. For the different mixing structures, the efficiencies of the year 2001 ranged from 6.1 g/kg H_2O for grouped mixtures to 4.3 g/kg H_2O for randomly distributed trees. The sharp decline in wue from pure beech to pure spruce stands for the year 2001 could not be found for the dry year 2003. Although wue of pure beech was relatively high with 2.5 g/kg H_2O, the efficiencies of all mixing ratios and mixing structures and of the pure spruce stand varied between 0.2 and 1.0 g/kg H_2O.

18.5.3.7 Influence of Stand Structure: Discussion

Stand structure can influence the growth of mixed beech–spruce stands to a great extent (see Chap. 13). Productivity and efficiency of a stand was changed by the mixing structure (random, cluster, group) and mixing proportion, as well as by drought stress. The simulations revealed comparatively small efficiency differences between the mixing ratios respectively the mixing structures in years with drought

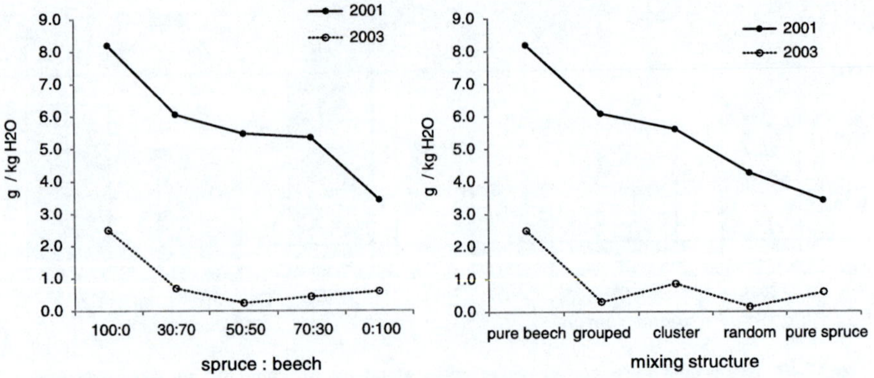

Fig. 18.18 Water use efficiency of mixed beech–spruce stands for different mixing ratios (*left*, averaged over all mixing structures) and different mixing structures (*right*, averaged over all mixing ratios) for the years 2001 and 2003 at the site Kranzberger Forst

stress but distinct differences in years with good water supply. This result was consistent for pure stands as well as for stands with different mixing structures of beech and spruce. It can be deduced that the wue of a tree species is reduced under drought stress. The mixing structure of a forest stand altered the water use efficiency. The efficiency differences are more pronounced under favourable growing conditions than under drought influence. The findings of Guehl et al. (1995), Barr et al. (2007) and Menon et al. (2007) underline this assumption of lower water use efficiencies under stress.

18.6 Conclusions

The scenario simulations with the eco-physiologically based, individual tree growth model BALANCE proved feasible for analysing the influence of environmental changes on growth, productivity and efficiency of forest stands. The simulation results were plausible and in accordance with empirical studies. From the simulations, the following conclusions can be drawn:

Defoliation of beech led to a decline of total biomass growth as expected. The intensity and the frequency of the defoliation stress determined whether trees recovered fully, partially to a lower biomass level, or died.

Even without changing physiological parameters like photosynthetic efficiency explicitly in BALANCE, intrinsic changes of allocation patterns led to an overcompensation of the growth of certain tree compartments under low defoliation rates. The recovery time between the defoliation events was more important than the number of defoliations.

Opportunity costs were incurred by beech trees in mixed stands that invested in defence of a species-specific pathogen, which did not attack their interspecific competitors. Competition effects between investing and non-investing trees caused losses that ranged from 1.5 to 3.1 of the direct defence investments. The competing non-affected spruce trees were able to increase their total biomass at the cost of the beech trees, to 140% when compared to the control, but were only able to compensate for the losses of beech if the defence investments of beech did not exceed 20–30%. This showed the limits of the buffering capacity in mixed stands.

Additionally to the defence rate of beech, stand structure influenced the growth of beech and spruce significantly. The simulation results revealed a differential pattern considering the two species. Spruce significantly profited from an individual tree mixture but could not compensate for the growth of beech in clustered and grouped situations. In contrast, beech reacted more to the absolute proportion of spruce in the stand than to the distribution of the spruce trees.

The simulation results suggest that the allometry between dbh and above-ground biomass was changed nonlinearly. The factors influencing allometry were the proportion of net resources invested in defence, the proportion of a tree species at the total crown area of a stand, and the degree of clustering of the tree species. The findings suggest a higher biomass to dbh ratio for beech at higher proportions of beech in the stand. The more beech was spatially intermingled with spruce the lower the biomass to dbh ratio of beech trees.

Compared to dominated, small trees, productivity of dominant big trees was influenced more by decreasing resource availability (e.g. water). Under severe drought stress as for example for the year 2003, however, for spruce a nearly constant loss of productivity for all size classes was obvious.

Drought stress clearly changed the allocation patterns of trees. Total biomass growth was decreased particularly for spruce under intensified drought. While in years with favourable growing conditions, the free C-pool of the trees increased, more severe drought caused increasing losses of the free C-pool. Fine root biomass of beech, on the other hand, increased with increasing drought. This is in conformity with the plant allocation theory, that above and below-ground allocation is adjusted to minimise resource constraints resp. maximise resource acquisition.

Stand structure characterised by the kind of mixing and/or tree species composition was shown to change productivity and efficiency of a forest stand.

Overall, climate sensitive, physiologically based growth models like BALANCE are useful tools for analysing influences of environmental changes on productivity and efficiency. They can help to map out adaption strategies to avoid negative consequences of environmental changes. Furthermore, such models can provide interesting hypotheses for subsequent empirical or experimental testing and can thus support theory building.

References

Barr AG, Black TA, Hogg EM, Griffis TJ, Morgenstern K, Kljun N, Theede A, Nesic Z (2007) Climatic controls on the carbon and water balances of a boreal aspen forest, 1994-2003. Glob Change Biol 13:561–576

Bartelink HH (2000) A growth model for mixed forest stands. Forest Ecol Manage 134:29–43

BayFORKLIM (1996) Klimaatlas Bayern. Fachbuchhandlung Kanzler, München

Bazzaz FA, Chiariello NR, Coley PD, Pitelka LF (1987) Allocating resources to reproduction and defense. Bioscience 37:58–67

Beierkuhnlein C, Foken T (2008) Klimawandel in Bayern: Auswirkungen und Anpassungsmöglichkeiten. Bayreuther Forum Ökologie, vol 113, University Bayreuth, 501 p

Bloom AJ, Chapin FS III, Mooney HA (1985) Resource limitation in plants – an economic analogy. Ann Rev Ecol Syst 16:363–392

Bossel H (1996) TREEDYN3 forest simulation model. Ecol Model 90:187–227

Brubaker LB (1978) Effects of defoliation by Douglas-fir tussock moth on ring sequences of Douglas-fir and Grand fir. Tree Ring Bull 38:49–60

Bruce D (1956) Effect of defoliation on growth of longleaf pine seedlings. Forest Sci 2:31–35

Bruscheck GJ (1994) Waldgebiete und Waldbrandgeschehen in Brandenburg im Trockensommer 1992. PIK Rep 2:245–264

Cermak J, Matyssek R, Kucera J (1993) Rapid response of large, drought-stressed beech trees to irrigation. Tree Physiol 12:281–290

Chalupa V (1965) Influence of the reduction of leaves on the beginning and course of radial growth. Commun Inst Forest Czech 4:61–73, Forest Abstr 27 no. 4580

Chapin FS III (1980) The mineral nutrition of wild plants. Annu Rev Ecol Syst 11:233–260

Chew FS, Rodman JE (1979) Plant resources for chemical defense. In: Rosenthal GA, Janzen DH (eds) Herbivores: their interaction with secondary plant metabolites. Academic, New York, pp 271–307

Cienciala E, Lindorth A, Cermák J, Hällgren JE, Kucera J (1994) The effects of water availability on transpiration, water potential and growth of Picea abies during a growing season. J Hydrol 155:57–71

Coley PD, Bryant JP, Chapin FS III (1985) Resource availability and plant herbivore defense. Science 230:895–899

Craighead FC (1940) Some effects of artificial defoliation on pine and larch. J Forest 38:885–888

De Martonne E (1926) Une nouvelle fonction climatologique: L'indice d'aridité. La Meteorologie 2:449–458

Dohrenbusch A, Jaehne S, Bredemeier M, Lamersdorf N (2002) Growth and fructification of a Norway spruce (Picea abies L. Karst) forest ecosystem under changed nutrient and water input. Ann Forest Sci 59:359–368

Eissenstat DM (1997) Trade-offs in root form and functioning. In: Jackson LE (ed) Ecology in agriculture. Academic, San Diego, CA, pp 173–179

Fontes L, Bontemps JD, Bugmann H, van Oijen M, Gracia C, Kramer K, Lindner M, Rötzer T, Skovsgaard JP (2010) Models for supporting forest management in a changing environment. Forest Syst 19:8–29

Fournier C, Bauce E, Dupont A (2010) Wood losses and economical threshold of Btk aerial spray operation against spruce budworm. Pest Manag Sci 66:319–324

Frank DA (2007) Drought effects on above- and belowground production of a grazed temperate grassland ecosystem. Oecologia 152:131–139

Gayler S, Grams TEE, Kosivits AR, Winkler JB, Luedemann G, Priesack E (2006) Analysis of competition effects in mono- and mixed cultures of juvenile beech and spruce by means of the plant growth simulation model PLATHO. Plant Biol 8:503–514

Grams TEE, Kozovits AR, Reiter IM, Winkler JB, Sommerkorn M, Blaschke H, Häberle K-H, Matyssek R (2002) Quantifying competitiveness in woody plants. Plant Biol 4:153–158

Grote R (2002) Estimation of crown radii and crown projection area from stem size and tree position. Ann Forest Sci 60:393–402

Grote R, Pretzsch H (2002) A model for individual tree development based on physiological processes. Plant Biol 4(2):167–180

Grünwald T, Bernhofer C (2007) A decade of carbon, water and energy flux measurements of an old spruce forest at the Anchor Station Tharandt. Tellus 59B(3):387–396

Guehl JM, Fort C, Ferhi A (1995) Differential response of leaf conductance, carbon isotope discrimination and water use efficiency to nitrogen deficiency in maritime pine and pedunculate oak plants. New Phytol 131:149–157

Gulmon SL, Mooney HA (1986) Costs of defense and their effects on plant productivity. In: Givnish TJ (ed) On the economy of plant form and function. Cambridge University Press, Cambridge, pp 681–698

Häberle K-H, Nunn AJ, Reiter IM, Werner H, Heller W, Bahnweg G, Gayler S, Lütz C, Matyssek R (2009) Variation of defence-related metabolites in the foliage of adult beech and spruce: a conceptual approach to approximating trade-off carbon. Eur J Forest Res 128:99–108

Hamilton JG, Zangerl AR, DeLucia EH, Berenbaum MR (2001) The carbon-nutrient balance hypothesis: its rise and fall. Ecol Lett 4:86–95

Herms DA, Mattson WJ (1992) The dilemma of plants: to grow or to defend. Q Rev Biol 67:283–335

Kellomäki S, Väisinen H, Strandman H (1993) FINNFOR: a model for calculating the response of boreal forest ecosystems to climate change. Research notes, no. 6. University of Joensuu, Joensuu, Finnland, 120 p

Knoke T, Seifert T (2008) Integrating selected ecological effects of mixed European beech – Norway spruce stands in bioeconomic modelling. Ecol Model 210:487–498

Konopka B, Noguchi K, Sakata T, Takahashi M, Konopkova Z (2007) Effects of simulated drought stress on the fine roots of Japanese cedar (Cryptomeria japonica) in a plantation forest on the Kanto Plain, eastern Japan. J Forest Res 12:143–151

Koricheva J (2002) Meta-analysis of sources of variation in fitness costs of plant antiherbivore defenses. Ecology 83(1):176–190

Krause SC, Raffa KF (1996) Differential growth and recovery rates following defoliation in related deciduous and evergreen trees. Trees 10:308–316

Kulman HM (1971) Effects of insect defoliation on growth and mortality of trees. Annu Rev Entomol 16:289–324

Landsberg JJ, Waring RH (1997) A generalised model of forest productivity using simplified concepts of radiation-use efficiency, carbon balance and partitioning. Forest Ecol Manage 95:209–228

Larcher W (2003) Physiological plant ecology. Springer, Berlin, 513 p

Lee TD, Bazzaz FA (1980) Effects of defoliation and competition on growth and reproduction in the annual plant Abutilon theophrasti. J Ecol 68:813–821

Leuschner Ch, Backes K, Hertel D, Schipka F, Schmitt U, Terborg O, Runge M (2001) Drought responses at leaf, stem and fine root levels of competitive *Fagus sylvatica L.* and *Quercus petraea (Matt.) Liebl.* trees in dry and wet years. Forest Ecol Manage 149:33–46

Mainiero R, Kazda M (2006) Depth-related fine root dynamics of Fagus sylvatica during exceptional drought. Forest Ecol Manage 237:135–142

McNaughton SJ (1983) Compensatory plant growth as a response to herbivory. Oikos 40:329–336

Menon M, Hermle S, Günthardt-Goerg MS, Schulin R (2007) Effects of heavy metal soil pollution and acid rain on growth and water use efficiency of a young model forest ecosystem. Plant Soil 297:171–183

Mohren GMJ, van de Veen JR (1995) Forest growth in relation to site conditions. Application of the model FORGRO to the Solling spruce site. Ecol Model 83:73–183

Neilsen WA (1981) Effect of simulated browsing on survival and growth of Pinus radiata (D. Don) seedlings. Aust Forest Res 11:47–53

Niklas KJ, Enquist BJ (2002) Canonical rules for plant biomass partitioning and annual allocation. Am J Bot 89:812–819

Nikolova P, St R, Andersen C, Mainiero R, Blaschke H, Matyssek R, Häberle K-H (2009) Effects of extreme drought in 2003 on soil respiration in a mixed forest. Eur J Forest Res 128:87–98

Noguchi K, Konopka B, Satomura T, Kaneko S, Takahashi M (2007) Biomass and production of fine roots in Japanese forests. J Forest Res 12:83–95

Ogaya R, Penuelas J (2007) Tree growth, mortality, and above-ground biomass accumulation in a holm oak forest under a five-year experimental field drought. Plant Ecol 189:291–299

Peterken CF, Mountford EP (1996) Effects of drought on beech in Lady Park Wood, an unmanaged mixed deciduous woodland. Forestry 69:125–136

Pichler P, Oberhuber W (2007) Radial growth response of coniferous forest trees in an inner Alpine environment to heat-wave in 2003. Forest Ecol Manage 242:688–699
Polomski J, Kuhn N (1998) Wurzelsysteme. Birmensdorf, Eidgenössische Forschungsanstalt für Wald, Schnee und Landschaft. Paul Haupt, Bern, 290 p
Pretzsch H (1997) Analysis an modeling of spatial stand structure. Methodological considerations based on mixed beech-larch stands in Lower Saxony. Forest Ecol Manage 95:237–253
Pretzsch H (2009) Forest dynamics, growth and yield. Springer, Berlin, 664 p
Pretzsch H, Grote R, Reineking B, Rötzer T, Seifert S (2008) Models for forest ecosystem management: a European perspective. Ann Bot 101:1065–1087
Rausher MD (1996) Genetic analysis of coevolution between plants and their natural enemies. Trends Genet 12:212–217
Rhoades DF (1979) Evolution of plant chemical defense against herbivores. In: Rosenthal GA, Janzen DH (eds) Herbivores: their interaction with secondary plant metabolites. Academic, Orlando, FL, pp 3–54
Rötzer T, Grote R, Pretzsch H (2004) The timing of bud burst and its effects on tree growth. Int J Biometeorol 48:109–118
Rötzer T, Grote R, Pretzsch H (2005) Effects of environmental changes on the vitality of forest stands. Eur J Forest Res 124:349–362
Rötzer T, Seifert T, Pretzsch H (2009) Modelling above and below ground carbon dynamics in a mixed beech and spruce stand influenced by climate. Eur J Forest Res 128:171–182
Rötzer T, Leuchner M, Nunn AJ (2010) Simulating stand climate, phenology, and photosynthesis of a forest stand with a process based growth model. Int J Biometeorol 54(4):449–464
Running SW, Coughlan JC (1988) A general model of forest ecosystem processes for regional applications 1. Hydrological balance, canopy gas exchange, and primary production processes. Ecol Model 42:125–154
Seifert T (2007) A simulation model to determine the extent of decay in stems of Norway spruce caused by annosum root rot. Forest Ecol Manage 248:95–106
Seifert T, Müller-Starck G (2009) Impacts of fructification on biomass production and correlated genetic effects in Norway spruce (*Picea abies* L. [Karst.]). Eur J Forest Res 128(2):155–169
Simms EL (1992) Costs of plant resistance to herbivory. In: Fritz RS, Simms EL (eds) Plant resistance to herbivores and pathogens: ecology, evolution and genetics. University of Chicago Press, Chicago, IL, pp 392–425
Stamp N (2003) Out of the *quagmire* of plant *defense* hypotheses. Q Rev Biol 78(1):23–55
Tyree MT, Alexander JD (1993) Plant water relations and the effects of elevated CO_2: a review and suggestions for future research. Vegetatio 104(105):47–62
Vivin P, Guehl JM (1997) Changes in carbon uptake and allocation patterns in Quercus robur seedlings in response to elevated CO_2 and water stress: an evaluation with ^{13}C labelling. Ann Forest Sci 54:597–610
Vogt KA, Vogt DJ, Palmitto PA, Boon P, O'Hara J, Asbjornsen H (1996) Review of root dynamics in forest ecosystems grouped by climate, climatic forest type and species. Plant Soil 187:159–219

Chapter 19
Predictability of Plant Resource Allocation: New Theory Needed?

R. Matyssek, S. Gayler, W. zu Castell, W. Oßwald, D. Ernst, H. Pretzsch, H. Schnyder, and J.-C. Munch

R. Matyssek (✉)
Chair of Ecophysiology of Plants, Technische Universität München, Hans-Carl-von-Carlowitz-Platz 2, 85354 Freising, Germany
e-mail: matyssek@wzw.tum.de

S. Gayler
Institute of Soil Ecology, Helmholtz Zentrum München, Ingolstädter Landstr. 1, 85764 Neuherberg, Germany

Water & Earth System Science Competence Cluster, University of Tübingen, Keplerstr. 17, 72074 Tübingen, Germany

W. zu Castell
Research Unit Scientific Computing, German Research Center for Environmental Health, Helmholtz Zentrum München, Ingolstädter Landstraße 1, 85764 Neuherberg, Germany

W. Oßwald
Phytopathology of Woody Plants, Technische Universität München, Hans-Carl-von-Carlowitz-Platz 2, 85354 Freising, Germany

D. Ernst
Institute of Biochemical Plant Biology, Helmholtz Zentrum München, Ingolstädter Landstr. 1, 85764 Neuherberg, Germany

H. Pretzsch
Chair of Forest Growth and Yield Science, Technische Universität München, Hans-Carl-von-Carlowitz-Platz 2, 85354 Freising, Germany

H. Schnyder
Lehrstuhl für Grünlandlehre, Technische Universität München, Alte Akademie 12, 85350 Freising, Germany

J.-C. Munch
Institute of Soil Ecology, Helmholtz Zentrum München, Ingolstädter Landstr. 1, 85764 Neuherberg, Germany

R. Matyssek et al. (eds.), *Growth and Defence in Plants*, Ecological Studies 220,
DOI 10.1007/978-3-642-30645-7_19,

19.1 The Challenge: Both to the Plant, and to the Plant Scientist

The research field featured in this book pursues the challenge to clarify the potential dilemma in plants of coping with the need for growth as an expression and prerequisite of competitiveness (e.g. Chaps. 10–13), without neglecting stress defence (cf. Chaps. 2–4), however, for preventing decline (see Preface and Chap. 1). Clarification was based in this book on the disclosure of mechanisms which enable plants to regulate energy and substrate fluxes between the physiological demands, and which, as a consequence, determine the resource partitioning amongst plants in stands under the prevalent ecological scenario. Obviously, the issue is about mechanisms in resource allocation across diverse spatio-temporal scales. Such mechanisms representing cause/effect-based interrelationships differ in scale-dependent ways, regarding spatio-temporal resolution and functional specificities, i.e. within and between the levels of cells, organs, whole-plants and stands, between ontogenetic stages and, in ecological terms, between growth (or site) conditions (see Chap. 1; Sandermann and Matyssek 2004; Baldocchi 1993). The scale-dependent appearance and functionality of mechanisms in resource allocation related to growth and defence demands is one central message of the contributions to this book. Links were demonstrated between molecular and whole-plant processes and their stand-level integration, while underlining the need for considering plant, parasite and mycorrhizospheric interactions as a functional unity that drives the plant's resource allocation (Matyssek et al. 2005). Given the associated functional complexity and, thus, encountered variability in plant performance, to what extent

- Is predictability in plant responsiveness to stress manifested?
- Do trade-offs actually exist, or become apparent, in the plant's regulation of resource allocation between growth and defence?
- Do conflicting findings on plant resource allocation impede theory development?
- Does new evidence require novel theoretical concepts?

These are the questions underlying this chapter, exemplifying the explanatory capacity of the *growth–differentiation-balance theory* (GDB), as introduced in Chap. 1 (also see Table 19.1 for explanation of frequently used expressions and abbreviations). This theory claims a trade-off in parallel to increasing resource availability (primarily nitrogen, or nutrients in general) and primary productivity between growth and defence-related metabolism, in particular, emphasizing defence against parasitic stress. GDB as propagated by Loomis (1953) and Lorio (1988), and extended by Herms and Mattson (1992) and Matyssek et al. (2002, 2005), was valuated in Chap. 1 to have reached evidential comprehensiveness to an extent that may foster mechanistic understanding and predictability of one crucial process underlying plant performance and persistence, namely, the regulation of resource allocation. Can it be generalized that plants under stress regulate their resource allocation in conflicting ways between the demands for growth and competitiveness *versus* those of defence?

Table 19.1 Expressions and abbreviations frequently used in this book volume and their explanation

Expression, abbreviation	Explanation (in relation to the rationale of this book volume)	Chapter with details
CBSC	Carbon-based secondary compound	1 and 17
Defence,		1 and 3
constitutive	Defence mechanisms which are pre-formed in healthy plants;	
induced	defence mechanisms activated by the plant upon pathogen attack	
GPP	Gross primary productivity	Preface and 1
GDB	*Growth–differentiation-balance hypothesis*, viewed in this book volume as immature theory	Preface and 1
Opportunity costs	Foregone opportunities by the plant upon resource investment in one out of several metabolic alternatives, reaching beyond the equivalent effect of resource partitioning *per se*	1
PLATHO	Acronym of numeric model, standing for "*PLAnts as Tree and Herb Objects*"	15 and 17
PR proteins	*Pathogenesis-related* proteins	3
SVM	Support vector machine: classification algorithm belonging to the class of supervised machine learning algorithms. SVM reflect the theoretical results from statistical learning theory	16
Trade-off	Two alternatives which exclude each other in terms of an inverse relationship	Preface and 1

19.2 "Opportunities" of Plants in Resource Allocation

The conflicting regulation suggested by GDB between the growth and defence-related resource allocation provided the inspiration of this book. However, do plants strictly follow such a kind of trade-off? Recent findings reported in this book emphasize the extent of such a trade-off to be circumstantial, depending both on plant-internal and external determinants. Distinctness apparently is an issue of the considered metabolites or metabolite class, by which the trade-off is presumed to be represented (cf. Koricheva et al. 1998, Chap. 1). Phenylpropanoids are an example, constituting a starting point of defence metabolism, while being linked closely to growth metabolism through their precursor phenylalanine. Hence, apple trees, which rely on phenylpropanoids in defence, conspicuously reflected the resource trade-off between growth and defence (Rühmann et al. 2002). Trade-off distinctness also is an issue of the hierarchical process level in the plant's metabolic organization (Koricheva et al. 1998), becoming apparent in the internal carbon flux as a whole rather than at the highly resolved underlying level of functionally particular metabolites (which, in addition, may not be readily detectable and individually quantifiable due to methodological constraints). Such an impression appears to be in accordance with GDB *sensu stricto*, in postulating the trade-off to occur between growth and *constitutive* defence at the whole-plant level (Herms and Mattson 1992), at which the common currency of both demands is carbon and associated nutrient elements, in particular, nitrogen. At this level, constitutive defence is a plant function that both competes internally with and is part of the growth-related metabolism, as this variant of defence is based both on structural resistance and preformed biochemical means (Oßwald 1995; Elstner et al. 1996). Some defence compounds may act as growth regulators and bias trade-offs (Taylor and Grotewold 2005). Remarkably, at the level of branches, defence did not vary in relation to the branch growth rate, as demonstrated in apple trees (cf. Chap. 3).

This brings us to viewing the other extreme, when presumed trade-offs do not become apparent, i.e. seemingly or actually do not exist. In fact, polyphenols as defence metabolites do not appear to incite high costs, so that trade-offs may stay minor. The latter may also be true at the whole-plant level, as the defence capacity can vary between the plant organs in relation to their risk of injury—perhaps representing "economic efficiency" in plant defence. Trade-off was not observed in beech (*Fagus sylvatica*) as infected by the root pathogen *Phytophthora citricola* (cf. Chap. 3). Here, beech trees which had survived the infection were concluded to possess a high degree of constitutive resistance. The latter appeared to have developed in parallel to (or even as an intrinsic part of) growth already *prior to* the infection, with the costs of growth and constitutive defence, in this case, being served by GPP sufficiently high to prevent mutual limitation. As mechanical robustness *via* cell wall biosynthesis (potentially including lignification) serves both mechanical requirements (i.e. "growth") and constitutive defence (cf. Lerdau and Gershenzon 1997; Sibly and Vincent 1997), it is conducive that costs of growth and defence are not distinguishable—and hence, a trade-off neither is plausible nor presentable. Or in other terms, any resource investment in plant structure as some means of constitutive defence intrinsically fosters plant competitiveness, and hence, favours growth through enabling the plant for space occupation and related resource exploitation (cf. Chaps. 10–12).

In addition, vigour in growth can be a defence strategy *per se*, as long as increments of stress-targeted plant tissues or organs are able to (over-)compensate for the loss of respective biomass upon stress impact (cf. Maurer and Matyssek 1997). On such grounds, the postulated trade-off is not existent, as in general, the plant's efforts in defending its above or belowground biomass may vanish, if adverse stress impacts stay minor relative to the abundance and intactness of the entire targeted biomass, e.g. the foliage or fine-root system as a whole, respectively (Zangerl and Bazzaz 1992).

Referring again to the beech trees mentioned above, which had survived pathogen impact in the absence of the claimed trade-off (cf. Chap. 3), they even displayed increase in photosynthetic performance (Fleischmann et al. 2010), which is interpreted as a means for warranting the defence costs required for survival. Such costs were spent for enhanced fine-root growth, compensating for the injury inflicted by the root pathogen. Apparently, a C sink was induced by the demand of defence that exerted stimulation on photosynthesis. The sink-driven stimulation de-escalates the claimed trade-off through raising GPP in favour of defence. Such a response reflects costliness in defence in beech, even in the absence of a trade-off with the growth-related metabolism, although the costliness was expressed in a way that differed from that in apple trees (see above). These did follow the trade-off scheme (Rühmann et al. 2002). Hence, at high defence costs both in beech and apple trees, GDB was validated in contrary ways. A response consistent with that in beech was found in potato (Ros et al. 2004), where high N availability promoted the growth metabolism to an extent that related genes were not repressed, even though defence genes were activated upon pathogen infection.

Complementary to the response mechanisms addressed above, comprehensive resistance can be achieved by the plant through *stress-induced defence*. This latter variant tends to be specific against particular stressors and may be less costly—at a first glance, at least—being activated only on demand, as opposed to the unspecific prevalence of constitutive defence. Capability of induced defence, therefore, might be a selective advantage of plants (Walters et al. 2005). The low-cost premise, moreover, is based on the typically micro-scale restriction of induced defence to the site of stress impact, in addition, to the "baseline" of constitutive defence, although the number of induced genes can be high (Ros et al. 2004). The premise appears to be supported by the observation that the amount of resources disposable between growth and defence-related metabolism may actually be low (Häberle et al. 2009, see Chap. 11). Micro-scale trade-offs in resource allocation between growth and defence perhaps occur under induced defence at the cell or tissue level, however, they hardly become manifested in the entire plant. Again, the claimed trade-off appears to be a matter of scale, and given the scope of induced defence, to lie beyond the whole-plant perspective of GDB. Hence, a restriction of GDB is its focus on constitutive defence (Herms and Mattson 1992; Matyssek et al. 2005).

The low-cost premise of induced defence only holds, however, if all related costs actually have been recognized and are accessible to quantification. In terms of a "full-cost analysis" (Lerdau and Gershenzon 1997), hidden or indirect costs potentially exist that are related to storage and transport of defence compounds or their precursors, warranting the readiness of induced defence on instantaneous demand. If so, such costs are hard to define and assess. However, the local restriction of induced defence often encountered in plants at least signalizes that the immediate costs may not dominate whole-plant resource allocation. More importantly, the account shows that plants have many options for coping concurrently with growth and defence in resource allocation, by de-escalating or even circumventing conflicts as reflected by trade-offs. Such options—or "opportunities"—apparently are not only associated with "opportunity costs", as incurring from the foregone opportunity upon following one alternative of a trade-off, but also provide means of escaping trade-offs between growth and defence. However, the multitude of options makes it very hard to predict plant behaviour under prevalent stress scenarios.

19.3 Enhancing Predictability

The predictability of resource allocation in operating between growth and defence is apparently restricted, given the range of regulatory "opportunities" plants possess. One may ask for means, therefore, that can enhance the predictability of plant resource allocation. Basically, two perspectives appear to be viable: (1) empirical molecular research, i.e. at the ultimate highly resolved scale that controls whole-plant performance, perhaps revealing the initiation of consistent response patterns to stress, mechanistically bridging the levels of gene expression, protein synthesis and metabolic activity (cf. Sandermann and Matyssek 2004); and (2) mathematical modelling. The latter option may be conducive, if plant response is not readily accessible to validation of GDB through empirical analysis. Such latter reasons are

- difficulties in ascertaining resource availabilities plant-internally in their relevance for defence. Challenging, in particular, is the plant's operation under resource limitation, given the parabolic relationship in such a case between defence and resource availability (cf. Chap. 1, Fig. 1.1),
- restrictions in controlled experimentation on resource availability because of complex resource interdependences, and in warranting coherence in theory evaluation across diverse ecological scenarios and spatio-temporal scales; and
- uncertainties about relevant defence pools (i.e. constitutive vs. induced, whole-plant vs. organ level; see above).

Such shortcomings may result in premature or contrasting hypothesis evaluations. The question arises, therefore, if modelling can serve as a complementary approach, which may set the empirical shortcomings into perspective. Respective capacities will be elucidated after highlighting, in the following, capacities of empirical molecular research.

19.3.1 Molecular Analysis

Capacities of empirical molecular research in enhancing the predictability of plant responsiveness are exemplified by means of O_3 effects (cf. Chap. 2). Starting point in integration is microarray analysis, which reflected coordinated regulation of all shikimate pathway genes under O_3 stress, and in the case of two enzymes (3-deoxy-D-*arabino*-heptulosonate-7-phosphate synthase 3 and 3-dehydroquinate dehydratase/shikimate dehydrogenase), transcript and protein levels were consistently increased. In addition, upon gene expression of salicylic/gentisic acid conjugates, metabolic end products were up-regulated. Similarly, a consistent chain reaction of altered gene and metabolite expression for ethylene biosynthesis and changed physiological and structural leaf differentiation were demonstrated. Down-regulated under O_3 stress were transcript levels related to mesophyll cell structure and photosynthesis (Calvin cycle), extending in the latter case to reduced protein levels, based on proteome analysis. Although direct transcript-protein overlap was not detected, overall down-regulation of primary metabolism upon O_3 impact was apparent. Two overlaps emerged (functional category disease/defence and transcription), however, in roots of European beech infected with *Phytophthora citricola*.

Transcriptome analysis of O_3-treated beech yielded gene grouping similar to that in herbaceous plants (cf. Chaps. 2 and 16). Gene expression was more strongly affected by O_3 impact than endophyte infestation, although pathogenic effects distinctly raised transcript levels. The latter responded in leaves to ozone and to pathogenic infection in roots, becoming assignable to similar functional categories. In addition, in beech, apple trees and potato plants, genes encoding PR proteins were identified. In conclusion, most genes of the defence category were up-regulated in the different plant species upon O_3 or pathogenic impact, corroborating the view on ozone as an "abiotic model pathogen" (cf. Matyssek et al. 2005). Transcriptional responses were more distinctly reflected in juvenile than mature trees, and gene expression typically mirrored leaf type (i.e. sun vs. shade-adapted) more distinctly than O_3 impact. Empirical molecular research does have capacities for unveiling consistent plant response patterns, fostering predictability in resource allocation between growth and defence.

19.3.2 Mechanistic Modelling

An approach complementary to empirical research is modelling, in particular, if employed as mechanistic numerical simulation models and based, in view of GDB, on the presumed parabolic relationship between resource availability and allocation to defence (cf. Fig. 1.1 in Chap. 1, Chap. 15). Given the plants' regulatory "opportunities", models must mirror the dynamics in resource allocation along the source–sink gradients of growth and defence-related metabolism, as determined by phenological and ontogenetic influences, and most importantly, by the internal availabilities of carbohydrates and nutrients (namely, nitrogen with respect to GDB). The internal availabilities need to respond to the resources outside, ensuring variation in interaction, as affected by fluctuating uptake capacities and variable factorial impacts. Competition with neighbouring plants needs to be considered as a crucial determinant (cf. Chap. 12), and baseline assumptions, derived from established knowledge, are to be integrated on CBSC physiology and biochemistry (cf. Chap. 1). Such requirements are comprehended in the novel PLATHO model, as introduced in Chap. 17. The advancement of PLATHO relative to the conceptual framework underlying GDB is the process-based, quantitative assessment of the plant's operation along internal resource gradients and under the influence of ecologically relevant site conditions. Hence, PLATHO represents a quantitative and dynamic extension of GDB.

Given the functional comprehensiveness and mechanistic character of PLATHO, parameterized and validated on a broad experimental data basis (see Chap. 17), the model performance under diverse simulation scenarios is summarized in Table 19.2 as contributing to the evaluation of GDB and a related working hypothesis. The latter is conceived for reasons of comparison more "liberal" in allowing carbon, nutrients and water as driving resources but ignores the parabolic relation to defence. This kind of relation, however, is intrinsic to GDB, while focusing on nutrients (namely nitrogen, cf. Koricheva 2002; Herms and Mattson 1992; Chap. 1, Fig. 1.1) as drivers. In summary:

- Starting with plant growth that suffers from N limitation, a scenario that leads to enhanced plant-internal N availability (as mediated through fertilization or changed competition between plants) will favour growth while diminishing the

Table 19.2 The plant's resource allocation between growth and defence, as reflected by the modelling approach of PLATHO

Plant *limited* by	Scenario	Effects on growth and CBSCs	Evaluation of working hypothesis (see text) and GDB
N availability	+N	Plant growth increased CBSC concentration decreased	Both confirmed
	+C	No or small stimulating effect on plant growth (decreasing N concentration) CBSC concentration increased	Hypothesis rejected, but GDB confirmed
C availability	+N	Small stimulating effect on plant growth (enhancement of N-concentration) CBSC concentration decreased	Both confirmed
	+C	Plant growth increased Stimulatory effect on CBSC concentration	Hypothesis rejected, but GDB confirmed at low resource availability

pool of CBSCs. This outcome confirms both the working hypothesis and GDB. A scenario, however, that augments the internal C instead of N availability in the plant, will not (or just negligibly) affect growth while enhancing the pool of CBSCs. The latter result rejects the working hypothesis introduced above, but still supports GDB.

– Conversely, taking C-limited growth as a starting point (as, e.g., under shading or incipient foliage development), a scenario that raises the N availability in the plant will moderately stimulate growth, but cause the CBSC pool to decline. This outcome is in agreement with both the working hypothesis and GDB. However, if instead the C limitation is released at low N availability, then growth and the CBSC pool will increase. Although the latter result rejects the hypothesis, it does not conflict with GDB, if the initial C supply was too low for N to induce growth, and if the enhanced C availability is higher than can be "consumed" by N for growth. This latter case is mediated at low N availability through the parabolic relation to defence (cf. Chap. 1).

Using information derived from modelling, one needs to caution that models, including PLATHO, represent integrated lines of hypotheses themselves, i.e. these form the basis of model functioning, so that in principle, hypothesis evaluation by model employment is not possible. However, if mechanistic models are able to explain a broad range of experimental findings during extensive validation, as was the case of PLATHO, agreement with the empirical evidence can be taken as an indirect confirmation of the reliability of underlying presumptions. The outcome that, contrasting with GDB, the GDB-derived working hypothesis was to be rejected in some cases indicates the differential view on carbon as a driving resource and the mathematical function describing defence. Such aspects are noteworthy, as validating GDB in the past has also suffered from inadequately accounting for underlying definitions (cf. Chap. 1). Procedures followed by PLATHO did comply, however, with fundamental requirements of science theory, namely ensuring processable evaluation of explicitly defined experimental scenarios, and as a consequence, providing new evidence about initial presumptions.

Experimental results yielded five response patterns after increasing C availability of plants, characterized by either stimulations or no response in both growth and defence, or stimulation of just one of the two plant functions—or, conforming to the trade-off presumption, decline in defence at increasing growth (cf. Chap. 17). Simulations manifested, in agreement with GDB, that low N in relation to C availability, even if the latter increased, was not reflected to promote growth, while allocation to defence could even decline (reflecting the parabolic relation between defence and nutrition). This means as a consequence, however, and still in compliance with GDB, that increasing N availability can drive both growth and defence. A general trend of favoured CBSC allocation upon increasing C availability was paralleled, in other cases, with unchanged or even declining CBSC levels. Such an increase in CBSCs was remarkably related to high N vs. C availability ratios, at a first glance conflicting with GDB. However, allocation to CBSC was reflected to become saturated at high C supply, with the saturation being reached the later the higher N availability was, and with the surplus C being allocated then to reserve storage rather than to defence. Such plant behaviour shows higher regulatory complexity than presumed by GDB. Additional cases showed stimulation in CBSC allocation, however, in the absence of growth response. Such cases are assumed to reflect growth to be determined by causes other than N availability. The five response patterns were similar in beech and spruce, although in the latter species the dependence on resource availabilities was hardly assessable, given lesser extents of C and N fluctuations relative to each other than observed in beech.

As an essential aspect of PLATHO, stand structure, and hence competition, proved to be crucial modifiers of the factorial impacts that drive the plant's resource allocation between growth and defence. A strong effect was reflected by simulations on C availability and CBSC allocation, depending on whether growth under CO_2 and O_3 regimes occurred in beech and spruce at mixed or pure stand conditions (complying with empirical observations by Kozovits et al. 2005 that "*competition dominates CO_2 and O_3 effects*"). Competition effects were mediated through changed light regimes (cf. Chaps. 8, 11–13), favouring in mixed stands the N uptake in spruce at the expense of that in beech (cf. Kozovits et al. 2005). The strong impact of competition on resource availability in beech is mirrored in the simulations (Fig. 19.1a) by a narrower range of data points both along the N and the C axis in pure than mixed stands.

Even more importantly, changes in competition turned out in the simulations to affect CBSC pools in contrasting ways, depending on species, which provides clues, why GDB evaluation may become contradictory. Modelling as exemplified by PLATHO can resolve such conflicts on mechanistic grounds, showing that hypothesis acceptance or rejection, and support of theories like GDB, can be decided by the prevalent ecological settings. The explanatory basis is provided by the variable three-dimensional relationship between plant-internal C and N availabilities and CBSC allocation. The three-dimensionality also explains

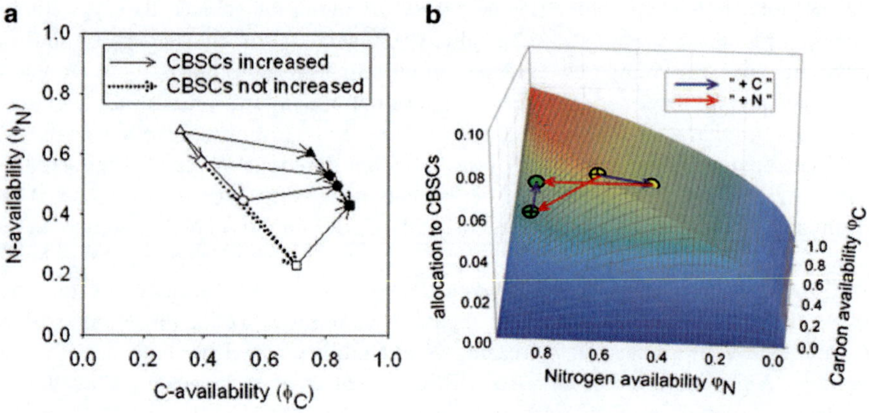

Fig. 19.1 (**a**) Positioning of experimental scenarios with juvenile beech (*Fagus sylvatica*) regarding plant-internally relative carbon (ϕ_C) and nitrogen (ϕ_N) availabilities. *Open symbols*: growth in mixture with spruce (*Picea abies*), *closed symbols*: pure beech stand; (*open circle*, *closed circle*): control, (*open triangle*, *closed triangle*): +O_3, (*open square*, *closed square*): +CO_2, (*open diamond*, *closed diamond*): +O_3/+CO_2. Vectors link simulation outcomes of two treatments, merged in the graph into one data point each. Direction of vectors point towards enhancement of CH_2O availability; *drawn vectors*: increased allocation into carbon-based secondary compound formation (CBSC), *dashed vectors*: CBSCs not increased. (**b**) Visualization of function in PLATHO describing the allocation of CH_2O into the CBSC pool as depending on plant-internal relative C (φ_C) and N (φ_N) availabilities. The *symbols* represent the predicted allocation to the pool of CBSCs in a fictive scenario, where relative carbon (φ_C) and nitrogen (φ_N) availabilities are averaged over the period of leaf development. The simulated scenario refers to an experiment with juvenile beech, in which nitrogen supply as well as atmospheric carbon dioxide concentration were varied: *Yellow symbols*: low N supply, *green symbols*: high N supply, *open symbols*: 700 μl l^{-1} CO_2, *crossed symbols*: 350 μl l^{-1} CO_2. *Red arrows* indicate the effect of enhanced nitrogen supply, *blue arrows* the effect of elevated carbon dioxide on plant-internal resource availabilities and allocation to CBSCs

inconsistent plant response to abiotic factors like, e.g., CO_2 supply (Fig. 19.1b). Ranges of different system behaviour become apparent along C and N availabilities, conforming along the N axis to the parabolic response function of defence, while the C axis communicates the saturation range of CBSC allocation, as already addressed above. The linear slope of CBSC allocation with C availability is consistent with GDB, although the latter does not claim a range of CBSC saturation. The complexity of plant resource allocation within the factorial three-dimensionality during seasonal courses was demonstrated in Chap. 17. One example is given in Fig. 19.1b, in that CBSC allocation may be increased or decreased at high CO_2 supply, depending on the level of concurrent N availability. The capacity of opposing plant behaviour in CBSC allocation becomes apparent, the more so as operation ranges of plants may change within the functional three-dimensionality during ontogenetic and phenological time courses (cf. Chap. 17).

Such kind of modelling exemplifies that judgement about hypotheses may be premature, and gain in evidence be missed, unless the mechanistic basis of plant response is unveiled thoroughly. Modelling further ensures implications of ecologically relevant settings to be accounted for in hypothesis evaluation, and in this way, warrants biologically meaningful conclusions on plant functionality. The modelling-based analysis extends beyond conventional GDB examinations, in that resource availabilities and dynamics become quantifiable under defined ecological scenarios, accounting for variable plant-external and internal influences, i.e. seasonal as well as ontogenetic and metabolic ones. The challenge and need of such an approach was underlined by Koricheva et al. (1998) and Stamp (2003).

The other novel process-based, numeric model demonstrated in this book is BALANCE (cf. Chap. 18). This model may be viewed as an extension of the scope of PLATHO to the stand level, in particular pursuing a temporal perspective related to the prolonged lifespans of forest trees. Simulations addressed the "opportunity costs" of trees, associated with the growth-related competitiveness in mixed forests, while investing resources in pathogen defence. Here, the costs of foregone opportunities, in this case at the expense of growth and competition, were valued given the need for staying defensive *sensu* GDB trade-off—rather than opportunities of "escaping" such a trade-off (see above). Opportunity costs were expressible through interest calculation, showing that the annual compound interest of "lost opportunities" after 14 years can range in beech up to about 4 % at defence allocation of 50 % of the C pool. This kind of outcome supports GDB. Competing, non-infected spruce again was the profiteer, similar to outcomes from PLATHO, raising biomass production by up to 140 %. The trade-off between growth and defence apparently also occurring at the stand level is species-dependent, i.e. driven by the species mixture. BALANCE assesses the buffering capacities of mixed forests regarding growth/defence trade-offs, exemplifying spruce to profit most from the defence costs of competing beech, in particular, if both species grow in a randomized single-tree arrangement.

Given again the strong impact of competition on resource allocation, BALANCE reflects a non-linear allometric relationship of stem diameter *versus* aboveground biomass at the stand level, with this ratio being determined through the proportions of tree species in mixture. The structural heterogeneity of stands is implied to increase at limiting resource supply. Under stress (e.g. as by drought), the growth performance of dominant tree individuals turns out to be species-dependent. BALANCE shows the tendency, however, that at increasing resource supply the size growth of dominant trees is over-proportionally favoured. Since, correspondingly fewer resources are left for the growth of subdominant trees, competition becomes exacerbated, turning its mode from symmetric towards asymmetric interaction (cf. Chaps. 12 and 14, Pretzsch and Dieler 2012). In general, competition, resource limitation and stress limitation affect tall trees more than their smaller neighbors, which is in line with GDB at the individual tree level. Beech in addition was indicated to pursue minimization of the aboveground resource demand when maximizing belowground resource uptake. Such an outcome conforms beyond circumstantial support of GDB even to another theory on plant growth, namely that on "*optimized plant allocation*" (cf. Chap. 18).

Resuming theory examination by numeric modelling, eventually brings us to two concluding questions:

1. Would it be a drawback, if hypothesis and theory evaluation stayed controversial? The question can be negated in view of science theory, if the scenarios leading to divergent evaluation are adequately analyzed in terms of unveiling the factorial determinants. In such cases, branching in system behaviour becomes comprehensible, and hence, the controversial outcome augments system understanding. It can be stated based on the evidence presented in this book that this kind of requirement is fulfilled.
2. Would it be a drawback then, if hypotheses and theories were predominantly confirmed, as shown, e.g. in Table 19.2, given the claim of science theory that rejection promotes evidence (Popper 1969)? In its absoluteness, this question may be negated as well, as long as confirmation allows consolidation of mechanistic system understanding and strengthening of validity across ecological scenarios. Such requirements were met, regarding the findings reported in this book and in view of the subsequent considerations.

In total, process-based, numerical modelling proves to be a powerful and complementary tool in linkage with empirical research, both having the capacity in jointly enhancing predictability of plant system performance as related to prevalent ecological scenarios.

19.3.3 Statistical Modelling

As numerical modelling linked with empirical research proves conducive in strengthening predictability, the means of coping with the challenge addressed in this book are not exhausted yet. One further approach is statistical modelling (cf. Chap. 16), in particular, if the focus is on the degree of generality or universality of empirical findings. Typically, these originate from manifold and contrasting observational scenarios in the absence of one over-arching research concept. This is the situation, by which current knowledge becomes available in plant research on resource allocation, and such grounds have been recognized as a major impediment in fostering respective theory development (Chap. 1).

One novel means of statistical modelling is based on the theory of "*unsupervised learning*", dating back to Vapnik (1995), as introduced in Chap. 16, which provides prediction for a yet unforeseen information input in determining the generalization error. The approach balances complexity *versus* accuracy, making use of machine learning algorithms, the so-called "*support vector machines*" (SVM), aimed at optimizing this kind of balancing in separating, i.e. classifying, different datasets by "hyperplanes". The ultimate outcome is the identification of such variables, which most distinctly respond to same driving factors under different scenarios. The misclassification error is improved in accuracy by repeated cross-validation of test and training data randomly chosen from the database. Variables with highest prediction accuracy are combined with any other variable and its prediction accuracy of the database (according to a Greedy Variant, see Chap. 16), iteratively increasing the pool of variables under analysis until a ranking list of accuracies related to variables is established. The list yields the optimum of prediction accuracy at a given set of combined variables. The crucial point is that the procedure does not start from a preset hypothesis, rather information is "learned" from the data, i.e., relevancy is disclosed for those data which are most indicative of the entire dataset.

Employing the SVM approach to patterns of gene expression within the data pool of diverse empirical investigations on the book subject (cf. Chaps. 2 and 16), the outcome of performed classification procedures was related to and confirmed by identified genes. As a result, hypotheses were derivable on the functional assignment of unidentified genes as based on similarities of expression patterns. The precision quality of such a kind of cluster assessment was higher as compared to conventional approaches with preceding functional classification, i.e., relating identified genes to regulatory metabolic pathways. Eventually, unknown genes become organized by their probabilities of cluster affiliation, now aiding the selection of demonstrative genes which appear to be compelling for further functional clarification. Viewing the outcome as immediately obtained from the SVM analysis across the diverse compared experimental scenarios provides the general conclusion that the number of differentially expressed genes is remarkably low, although gene response tends to be scenario-specific.

Another application of SVM analysis was demonstrated for beech in response to variable CO_2 supply under different growth scenarios (cf. Chaps. 3, 12, 16). The separating hyperplane was defined by the *C*/*N* ratio of leaves and the amount of cell wall-bound phenolic compounds in fine roots. This separation confirms GDB to the extent that resource limitation, as reflected by increasing *C*/*N*, is associated with an increase in phenolics. Maximum prediction accuracy was found to be related to six variables, namely, cell wall-bound and soluble phenolics in fine roots and leaves each, along with *C*/*N* and dry mass of leaves. Variables with largest effects on prediction accuracy were cell wall-bound phenolics of fine roots, and leaf *C*/*N* and dry mass, being those variables which represent most the resource-driven trade-off between growth and defence *sensu* GDB. The SVM analysis, therefore, conforms to the generally presumed concept that the plant's resource allocation is mainly determined by the regulation between the growth and defence-related metabolism.

19.4 Need for New Theory?

In view of the new evidence presented in this book, namely, the manifold "opportunities" plants apparently do have in balancing the resource demands of growth and defence, do arguments emerge for a new theory on plant resource allocation, i.e. replacing GDB? It might be tempting in arguing so, as cases exist, that do not disclose growth/defence trade-offs *sensu* GDB. Conversely, it cannot be denied either that other cases support GDB. Reasons for interpreting the ambiguity, i.e. for functionally "explaining" either outcome in a case-specific way, have been detailed in this book and earlier in this chapter. According to science theory, however, the encountered situation appears to justify the rejection of GDB (Popper 1969).

Nevertheless, a respective decision would seem to be premature, acknowledging that even a scope of investigations as introduced by this book may not be adequately comprehensive for thoroughly challenging a fundamental theory in plant science such as GDB, i.e. to the full breadth of conceivable ecological scenarios. Also, the fact that plants can apparently choose from many "opportunities", i.e. that complexity *sensu* plasticity in response is intrinsic to plant behaviour, means that resolving the growth-defence conflict through trade-offs in resource allocation is just one amongst several plant options. And it is one of these, for which GDB does hold. Attaining this kind of evidence, what GDB loses then is its claim for generality in elucidating the plant's balance in resource allocation between growth and defence. Or, in other terms, the applicability of GDB becomes restricted, being a matter of spatio-temporal scales, of specific mechanisms in the hierarchy of internal organization and of external specificities in multi-factorial ecological settings. Hence, theory validity is to be defined, in particular, for mechanistically linking adjacent spatio-temporal scales of biological organization.

Despite this restriction, even in cases where not applying, GDB can still give orientation for designing empirical research and modelling, the latter representing one tool for hypothesis validation and theory development (see above). Such kind of orientation gives guidance to understanding plant behaviour even beyond GDB, provided the attained evidence allows the identification of functional branching points by which plants leave GDB *sensu stricto* or its defined validity ranges. Such alternative branching pathways in plant behaviour, which apparently do exist, open the wide field of functional plasticity, which is an intrinsic organismic feature of perhaps the highest evolutionary value for plant persistence and successful stress adaptation. Examples of plasticity in plant response beyond GDB are, e.g. capacities of enhancing GPP in support of defence, in using growth vigour as a means of defence strategy, or in keeping defence locally restricted at the tissue or organ level without afflicting whole-plant metabolism. Since plant plasticity beyond GDB is that flexible, hardly a new theory can currently be posed to unify the diverse observed or even further "opportunities" on mechanistic grounds. The pre-requisite is to comprehend the cause–effect based interrelationships that underlie plasticity. This book has posed, however, a guiding perspective on the grounds of advanced theory building beyond GDB.

Within the above view, GDB has not become obsolete in giving orientation as long as two requirements will be met, whatever outcome be obtained, to permit mechanistic clarification, (1) elaboration of new substantial knowledge about relevant physiological and ecological processes, and (2) assessment of the ecological settings in each individual case study. To this end, revision of GDB is mandatory, however, to warrant a more advanced rationale than currently prevailing. Quite immediately and precisely, the extended rationale must aim at the disentangling of growth and constitutive defence, "full-cost" oriented clarification of induced defence within the whole-plant metabolism, mechanistically linking the process levels of metabolic control (i.e. the molecular responses within the genome) and metabolic activity (i.e. the biochemical and physiological response level), and the overall integration into the ecological interactions at the stand and ecosystem scales. The spectrum of biotic interactions and their mechanistic quantification is to be considered above and belowground in theory development, widening the conventional scope of host–parasite systems by the dimensions of competition and/or facilitation and the mycorrhizospheric interrelationships. As a result, the functional understanding of the varying degrees of trade-offs in plant resource allocation will be strengthened. Such a widening of the rationale appears to be conducive also towards a "holobiontic view" on the range of involved biotic interactions (i.e. beyond those between plants and micro-organisms) and approaches which expand "systems biology" to a comprehensive coverage of the relevant resource and information flows, integrating the molecular into the whole-plant system and ecosystem scale. Hence, "systems biology" must find its completion in an ecosystem biology.

Most importantly, the variability in growth/defence trade-off manifestation, ranging between distinctness to quantitative irrelevance or even absence, demands for attention on the highly dynamic and multi-functional regulatory capacity, as an expression of plasticity, in plant response. GDB-related research in the past has been fixed predominantly to steady-states in resource allocation under often monofactorial influences. What have been overlooked were transitions, non-linearities, multi-factorial interactions and hysteretic cause–effect relationships in plant performance as well as evident branching points, at which resource allocation commences to depart from the conventional scope of GDB. The "static" way of thinking about GDB impeded a mechanistic view beyond the claimed plant-internal dilemma between growth and defence-related metabolism, creating seeming conceptual conflicts in cases of unforeseen plant response. The required extended view must comprise growth and defence, therefore, as part of the plasticity intrinsic to resource allocation, and must unravel mechanisms that control plasticity. Upon reaching such an achievement, then the presumed "dilemma" claimed by GDB would be functionally recognizable as just one facet embedded into the overall continuum of the regulation range in resource allocation to growth and defence demands. Attaining such a stage might then allow new theory formulation. The perspectives have been shown, with revised GDB appearing meanwhile as a conducive conceptual tool. Progress towards universality in attained evidence and knowledge will be fostered by theoretical approaches as represented by the diverse kinds of modelling introduced in the book and accentuated earlier in this chapter. Such latter tools are prone to forward unification across empirical observations, as enabling for sublimating response patterns that overarch the range of case study scenarios. In such respect, the linkage between modelling and empirical research has the capacity of promoting new theory building through enhancing predictability in plant system behaviour, as demonstrated in this book.

For warranting the perspective to new theory building, the mechanistically founded, even though not unifying character of GDB is to be strengthened further as discussed above. The "flaw" of not being unifying yet continues to render GDB falsifiable, and the more GDB will become mechanistically founded, the more falsification will provide mechanistic clarification and gain in evidence. By this, the field of comprehensively understanding plasticity in plant responsiveness increasingly becomes accessible. It is the challenge in research to forward the functional variability of stress responses as an intrinsic feature of plant system biology towards process-based understanding. This book has created a basis for doing so, exemplifying conducive new evidence. Hence, mechanistic comprehension of plasticity becomes the key to plant system understanding.

References

Baldocchi DD (1993) Scaling water vapor and carbon dioxide exchange from leaves to a canopy: rules and tools. In: Ehleringer JR, Field CB (eds) Scaling physiological processes, leaf to globe. Academic, San Diego, pp 77–114

Elstner EF, Oßwald WF, Schneider I (1996) Phytopathologie: Allgemeine und biochemische Grundlagen. Spektrum Akademischer, Heidelberg

Fleischmann F, Raidl S, Oßwald W (2010) Changes in susceptibility of beech (*Fagus sylvatica*) seedlings towards *Phytophthora citricola* under the influence of elevated atmospheric CO_2 and nitrogen fertilization. Environ Pollut 158:1051–1060

Häberle KH, Nunn A, Reiter I, Werner H, Heller W, Bahnweg G, Gayler S, Lütz C, Matyssek R (2009) Variation of defence-related metabolites in the foliage of adult beech and spruce: a conceptual approach to approximating traded-off carbon. Eur J Forest Res 128:99–108

Herms DA, Mattson WJ (1992) The dilemma of plants: to grow or defend. Q Rev Biol 67:283–335

Koricheva J (2002) Meta-analysis of sources of variation in fitness costs of plant antiherbivore defenses. Ecology 83:176–190

Koricheva J, Larsson S, Haukioja E, Keinänen M (1998) Regulation od woody plant secondary metabolism by resource availability: hypothesis testing by means of meta-analysis. Oikos 83:212–226

Kozovits AR, Matyssek R, Blaschke H, Göttlein A, Grams TEE (2005) Competition increasingly dominates the responsiveness of juvenile beech and spruce to elevated CO_2 and/or O_3 concentrations throughout two subsequent growing seasons. Glob Chance Biol 11:1387–1401

Lerdau M, Gershenzon J (1997) Allocation theory and chemical defense. In: Bazzaz FA, Grace J (eds) Plant resource allocation. Academic, San Diego, pp 265–277

Loomis WE (1953) Growth and differentiation – an introduction and summary. In: Loomis WE (ed) Growth and differentiation in plants. Iowa State College Press, Ames, pp 1–17

Lorio PL Jr (1988) Growth and differentiation balance relationships in pines affect their resistance to bark beetles (Coleoptera: Scoloytidae). In: Mattson WJ, Levieux J, Bernard-Dagan C (eds) Mechanisms of woody plant defenses against insects: search for pattern. Springer, New York, pp 73–92

Matyssek R, Schnyder H, Elstner E-F, Munch J-C, Pretzsch H, Sandermann H (2002) Growth and parasite defence in plants: the balance between resource sequestration and retention. Plant Biol 4:133–136

Matyssek R, Agerer R, Ernst D, Munch J-C, Osswald W, Pretzsch H, Priesack E, Schnyder H, Treutter D (2005) The plant's capacity in regulating resource demand. Plant Biol 7:560–580

Maurer S, Matyssek R (1997) Nutrition and the ozone sensitivity of birch (*Betula pendula*), II. Carbon balance, water-use efficiency and nutritional status of the whole plant. Trees 12:11–20

Oßwald WF (1995) Die Wirt-Parasit-Beziehungen – Bakterien und Pilze als Parasiten. In: Hock B, Elstner EF (eds) Schadwirkungen auf Pflanzen, vol 3. Spektrum Akademischer, Heidelberg, pp 315–369

Popper KR (1969) Logik der Forschung. J.C.B. Mohr (Paul Siebeck), Tübingen, p 441

Pretzsch H, Dieler J (2012) Evidence of variant intra- and interspecific scaling of tree crown structure and relevance for allometric theory. Oecologia. doi:10.1007/s00442-011-2240-5

Ros B, Thümmler F, Wenzel G (2004) Analysis of differentially expressed genes in a susceptible and moderately resistant potato cultivar upon *Phytophthora infestans* infection. Mol Plant Pathol 5:191–201

Rühmann S, Leser C, Bannert M, Treutter D (2002) Relationship between growth, secondary metabolism, and resistance of apple. Plant Biol 4:137–143
Sandermann H, Matyssek R (2004) Scaling up from molecular to ecological processes. In: Sandermann H (ed) Molecular ecotoxicology of plants, vol 170, Ecological studies. Springer, Berlin, pp 207–226
Sibly RM, Vincent JFV (1997) Optimality approaches to resource allocation in woody tissue. In: Bazzaz FA, Grace J (eds) Plant resource allocation. Academic, San Diego, pp 143–160
Stamp N (2003) Theory of plant defensive level: example of process and pitfalls in development of ecological theory. Oikos 102:672–678
Taylor LP, Grotewold E (2005) Flavonoids as developmental regulators. Curr Opin Plant Biol 8:317–323
Vapnik VN (1995) Statistical learning theory. Wiley, New York
Walters D, Walsh D, Newton A, Lyon G (2005) Induced resistance for plant disease control: maximizing the efficacy of resistance elicitors. Phytopathology 95:1368–1373
Zangerl AB, Bazzaz FA (1992) Theory and pattern in plant defense allocation. In: Fritz RS, Simms EL (eds) Plant resistance to herbicides and pathogens. University of Chicago Press, Chicago, pp 363–391

Part IV
Outlook

Chapter 20
Conclusions and Perspectives

R. Matyssek

20.1 Gain in Knowledge

Facet richness and complexity in evidence on the subject of this book volume, i.e. the plants' manoeuvring between growth (warranting competitiveness in resource acquisition) and stress defence (enabling for sustained resource use), challenge the functional integration, in particular, if the perspective extends across multiple spatio-temporal scales. Such a challenge may be met by extending process scaling beyond conventional scopes, which have typically focused on cause–effect related transitions between cells and organs, on the latter as components of the whole-plant system, and on the plant as interlinked into the ecosystem's resource and energy fluxes. One step in widening such scopes is the integration of further scaling dimensions as defined by stages in plant ontogeny and extents of factorial complexity in growth conditions (i.e. controlled settings vs. multi-factorial interactions in the field: Matyssek et al. 2005; cf. Chap. 1). Treatises of this book have shown, however, that coverage of even further scaling dimensions is required in approaching generality in mechanistic understanding (cf. Chaps. 16 and 19): These are the transitions between empirically acquired evidence and theory-based simulations by modelling, and between the specificities of individual case studies and their degree of universality within the context of available knowledge (Fig. 20.1).

These two latter scaling dimensions integrate new empirical evidence into the currently known, functional continuum spanned by species systems, ontogenetic stages and growth conditions (see Prologue and Preface). Simulations by cause–effect based modelling provide transparency to such scenarios, which are experimentally inaccessible due to their spatio-temporal complexity, and may

R. Matyssek (✉)
Chair of Ecophysiology of Plants, Technische Universität München,
Hans-Carl-von-Carlowitz-Platz 2, 85354 Freising, Germany
e-mail: matyssek@wzw.tum.de

R. Matyssek et al. (eds.), *Growth and Defence in Plants*, Ecological Studies 220,
DOI 10.1007/978-3-642-30645-7_20, © Springer-Verlag Berlin Heidelberg 2012

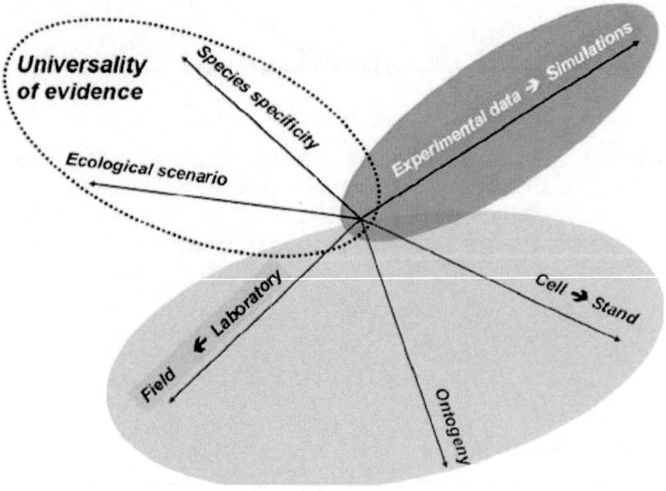

Fig. 20.1 Multi-dimensionality of evidence (according to W. zu Castell, pers. comm.)

resolve seeming conflicts between the evidence from discrete individual case studies (cf. Chaps. 15, 17–19). Novel statistical approaches can sublimate functional patterns from plant performance, examining their capacity for unifying and reconciling the diversity in response as encountered under the spectrum of ecological scenarios and, by this, directing empirical research to such mechanisms, which appear to be most decisive for comprehending biological universality (Chaps. 16 and 19). One prerequisite for reaching this kind of comprehension is analysing the spectrum of ecological scenarios under long-term perspectives. Such a requirement may pose, though, particular methodological and logistic challenges for research. Nevertheless, the integration of theory, experimentation and modelling, as has been demonstrated in this book, warrants the "magic tripod" in research for gaining novel evidence and substantially augmenting knowledge (see Prologue).

Given the "multi-dimensionality" represented by Fig. 20.1, it is likely that the distinctness of trade-offs in resource allocation between growth and defence processes is an issue of the spatio-temporal scale under consideration, and that such trade-offs, if pursued across scales, get subject to non-linearity and may lose their "sharpness", perhaps becoming non-predictable, as being blurred by varying factorial impacts. These can superimpose "third-party trade-offs", enhancing the complexity in plant response. Mechanisms change their manifestation in dependence on the position, i.e. the specific "knot" or "edge", from which an analysis is commenced within cause–effect networks that expand across spatio-temporal scales (see Fig. 3 in Prologue). This is the case for plants and their intrinsic mechanisms as components of ecosystems. Complexity augments "opportunities", however, in the regulation of resource allocation with associated structural and functional costs and benefits (cf. Chap. 1), which gives rise to what is termed "plasticity" in the plants' resource husbandry.

In view of the above outlined multi-dimensionality, this book has underlined response plasticity as a fundamental biological principle, which for becoming accessible to mechanistic comprehension requires an extension of the concept of "systems biology": Respective research must integrate all such cause–effect relationships and information flows amongst spatio-temporal scales which are relevant for plant functionality and persistence under site conditions. This means, "systems biology" must reach beyond the scale of cellular analysis under controlled environments.

Plasticity does not only become apparent, however, as an aspect of genetical constitution and arising capacities of physiological acclimation, but also in response to biotic interactions (cf. Part II of this book), i.e. between plants (competition/facilitation) or within holobiontic systems (parasitism/mutualism between plants and micro-organisms or, *sensu lato*, insects; cf. Zilber-Rosenberg and Rosenberg 2008). Negligence of such interactions bears high risk of misjudgement about plant performance and priorities in resource allocation between growth and stress defence, as has been exemplified by the book contributions. Such a conclusion mandates analyses of plant resource allocation without ignoring the shaping influence of biotic interactions, which is a realistic view, as plant evolution is driven by multi-factorial site conditions. Regarding the extended scope of "systems biology", attention needs to be directed to multi-organismic genotype/species networks and their resource fluxes and information signalling, integrating science theory, experimentation, bio-statistical concepts and modelling approaches. On such grounds, clarification of natural response variability is consolidated both in mechanistic terms and in ecologically relevant ways, which has been demonstrated by this book. This rationale commands, as a directive for research, resource allocation in plants to be emphasized as an inherent constituent of biotic interactions along the spatio-temporal scales.

20.2 Spin-offs to Practice

Spin-offs from the "empirical leg" of the "magic tripod of research" (see Prologue) on the book topic are methodological advances and gains in evidence in the fields of molecular biology in combination with such in plant biochemistry and physiology (cf. Part II of this book). Such advances have immediate relevance for plant breeding and engineering towards yield enhancement, increased pathogen resistance and resource-efficient plantation and stand management in agriculture and forestry, respectively. Not less important are spin-offs from the more "theoretical leg" of modelling (cf. Part III), both regarding advances in statistical analysis and in the development of novel mechanistic modelling approaches. Regarding the latter, new models PLATHO (Chap. 17) and BALANCE (Chap. 18), and in addition, compartmentation models on the dynamics of carbon pools (Chap. 7), have become ideal decision support tools in the fields of agriculture and forestry, in evaluating resource allocation at the plant and stand level. They assist judgement about effects of nutritional status, fertilization and CO_2 availability on plant competitiveness and the allocation into carbon-based secondary metabolites, such information being crucial for valuating plant resistance against diseases and tolerance of global change conditions. Risk assessment of forest productivity is aided regarding species compositions of mixed forest systems and stand-level trade-offs between yield formation and stress defence. Simulations give orientation to the long-term consequences of agricultural and silvicultural practices. For agriculture, evaluation of approaches is aided for mastering the challenge to increase productivity in face of the undamped over-population development of mankind. Decision making in forestry, in particular, profits from the outcome of mechanistic modelling: It is the long-lived character and factorial complexity of forests which prevents developing decision strategies on the basis of short-term experimentation.

20.3 Theory Development

Which are the lessons to be derived for the "*growth–differentiation balance theory*" (GDB)? It became apparent from the book contributions that the core of the theory cannot be rejected, regarding trade-offs in plant resource allocation between growth and stress defence (cf. Chap. 19). Nevertheless, it can be posed that GDB must be founded on a rationale, which is more advanced in mechanistic terms and ecologically differentiated to higher extents than has been the case to date. Three directives are postulated for the further development of GDB:

- The variable manifestation of trade-offs in plant resource allocation, ranging between distinct biological and statistical apparentness and concealment by compensatory effects or quantitative irrelevance, demands for an evaluation that copes with the highly dynamic and multi-functional regulatory capacity, i.e. plasticity, in plant response.
- Ranges of theory validity are to be defined in view of ecological scenarios and associated processes that are relevant for mechanistically linking adjacent spatio-temporal scales of biological organization.
- The spectrum of biotic interactions and their mechanistic quantification is to be considered above- and belowground in theory development, extending the conventional scope of host–parasite systems, and by this, the functional understanding of the varying degrees of trade-offs in plant resource allocation. Widening of the rationale appears to be conducive towards a "holobiontic view" on the range of involved biotic interactions (i.e. beyond those restricted to plants and micro-organisms) and approaches which expand "system biology" to a comprehensive coverage of the relevant resource and information flows.

Advanced approaches of this kind warrant ecologically, and hence, biologically relevant gain in evidence on plant performance and conclusions for practical implications, with the perspective of extending GDB towards mechanistically advanced new theory building.

References

Matyssek R, Agerer R, Ernst D, Munch J-C, Osswald W, Pretzsch H, Priesack E, Schnyder H, Treutter D (2005) The plant's capacity in regulating resource demand. Plant Biol 7:560–580

Zilber-Rosenberg I, Rosenberg E (2008) Role of microorganisms in the evolution of animals and plants: the hologenome theory of evolution. FEMS Microbiol Rev 32:723–735

Index

R. Matyssek et al. (eds.), *Growth and Defence in Plants*, Ecological Studies 220,
DOI 10.1007/978-3-642-30645-7, © Springer-Verlag Berlin Heidelberg 2012

Ecological Studies

Further volumes can be found at springer.com.

Volume 203
The Continental-Scale Greenhouse Gas Balance of Europe (2008)
H. Dolman, R. Valentini, and A. Freibauer (Eds.)

Volume 204
Biological Invasions in Marine Ecosystems: Ecological, Management, and Geographic Perspectives (2009)
G. Rilov and J.A. Crooks (Eds.)

Volume 205
Coral Bleaching: Patterns, Processes, Causes and Consequences (2009)
M.J.H. van Oppen and J.M. Lough (Eds.)

Volume 206
Marine Hard Bottom Communities: Patterns, Dynamics, Diversity, and Change (2009)
M. Wahl (Ed.)

Volume 207
Old-Growth Forests: Function, Fate and Value (2009)
C. Wirth, G. Gleixner, and M. Heimann (Eds.)

Volume 208
Functioning and Management of European Beech Ecosystems (2009)
R. Brumme and P.K. Khanna (Eds.)

Volume 209
Permafrost Ecosystems: Siberian Larch Forests (2010)
A. Osawa, O.A. Zyryanova, Y. Matsuura, T. Kajimoto, and R.W. Wein (Eds.)

R. Matyssek et al. (eds.), *Growth and Defence in Plants*, Ecological Studies 220,
DOI 10.1007/978-3-642-30645-7, © Springer-Verlag Berlin Heidelberg 2012

Volume 210
Amazonian Floodplain Forests: Ecophysiology, Biodiversity and Sustainable Management (2010)
W.J. Junk, M.T.F. Piedade, F. Wittmann, J. Schöngart, and P. Parolin (Eds.)

Volume 211
Mangrove Dynamics and Management in North Brazil (2010)
U. Saint-Paul and H. Schneider (Eds.)

Volume 212
Forest Management and the Water Cycle: An Ecosystem-Based Approach (2011)
M. Bredemeier, S. Cohen, D.L. Godbold, E. Lode, V. Pichler, and P. Schleppi (Eds.)

Volume 213
The Landscape Ecology of Fire (2011)
D. McKenzie, C.S. Miller, and D.A. Donald (Eds.)

Volume 214
Human Population: Its Influences on Biological Diversity (2011)
R.P. Cincotta and L.J. Gorenflo (Eds.)

Volume 215
Plant Desiccation Tolerance (2011)
U. Lüttge, E. Beck, and D. Bartels (Eds.)

Volume 216
Forest Hydrology and Biogeochemistry: Synthesis of Past Research and Future Directions (2011)
D.F. Levia, D. Carlyle-Moses, and T. Tanaka (Eds.)

Volume 217
Peatlands of the Western Guayana Highlands, Venezuela: Properties and Paleogeographic Significance of Peats (2011)
J.A. Zinck and O. Huber (Eds.)

Volume 218
Aralkum – A Man-Made Desert: The Desiccated Floor of the Aral Sea (Central Asia) (2012)
S.-W. Breckle, W. Wucherer, L.A. Dimeyeva, and N.P. Ogar (Eds.)

Volume 219
Seaweed Biology: Novel Insights into Ecophysiology, Ecology and Utilization (2012)
C. Wiencke and K. Bischof (Eds.)

Volume 220
Growth and Defence in Plants: Resource Allocation at Multiple Scales (2012)
R. Matyssek, H. Schnyder, W. Oßwald, D. Ernst, J.C. Munch, and H. Pretzsch (Eds.)